# Phosphorus-31 NMR
*Principles and Applications*

# Phosphorus-31 NMR

## Principles and Applications

*Edited by*

*David G. Gorenstein*
Department of Chemistry
University of Illinois at Chicago
Chicago, Illinois

*1984*

ACADEMIC PRESS, INC.
(HARCOURT BRACE JOVANOVICH, PUBLISHERS)
Orlando   San Diego   San Francisco   New York   London
Toronto   Montreal   Sydney   Tokyo   São Paulo

COPYRIGHT © 1984, BY ACADEMIC PRESS, INC.
ALL RIGHTS RESERVED.
NO PART OF THIS PUBLICATION MAY BE REPRODUCED OR
TRANSMITTED IN ANY FORM OR BY ANY MEANS, ELECTRONIC
OR MECHANICAL, INCLUDING PHOTOCOPY, RECORDING, OR ANY
INFORMATION STORAGE AND RETRIEVAL SYSTEM, WITHOUT
PERMISSION IN WRITING FROM THE PUBLISHER.


ACADEMIC PRESS, INC.
Orlando, Florida 32887


*United Kingdom Edition published by*
ACADEMIC PRESS, INC. (LONDON) LTD.
24/28 Oval Road, London NW1 7DX


Library of Congress Cataloging in Publication Data

Main entry under title:

Phosphorus-31 NMR, principles and applications.

    Includes index.
    1. Nuclear magnetic resonance.  2. Phosphorus--
Isotopes.  3. Biological chemistry--Technique.
I. Gorenstein, David G.
QH324.9.N8P48  1983    574.19'285    83-7154
ISBN 0-12-291750-2

PRINTED IN THE UNITED STATES OF AMERICA

84 85 86 87    9 8 7 6 5 4 3 2 1

# Contents

Contributors　xi
Preface　xiii

Introduction
*David G. Gorenstein*

    Text　1
    References　3

## PART 1. Phosphorus-31 Chemical Shifts and Spin–Spin Coupling Constants: Principles and Empirical Observations

### 1. Phosphorus-31 Chemical Shifts: Principles and Empirical Observations
*David G. Gorenstein*

    I. Introduction and Basic Principles　7
    II. Applications of $^{31}P$ Chemical-Shift Theory and Empirical Observations　9
    References　33

### 2. Phosphorus-31 Spin–Spin Coupling Constants: Principles and Applications
*David G. Gorenstein*

    I. Introduction　37
    II. Directly Bonded Coupling Constants: $^{1}J_{PX}$　38
    III. Two-Bond Coupling Constants: $^{2}J_{PX}$　42
    IV. Three-Bond Coupling Constants: $^{3}J_{PX}$　43
    References　51

## PART 2. Theory and Applications of Phosphorus-31 NMR to Biochemistry

3. Phosphorus-31 NMR of Enzyme Complexes
   *B. D. Nageswara Rao*
   - I. Introduction   57
   - II. Theoretical and Experimental Considerations   59
   - III. Experimental Studies of Specific Enzyme Systems   69
   - IV. Computer Calculation of NMR Line Shapes under the Influence of Exchange   96
   - V. Conclusion   100
   - References   102

4. $^{31}$P-NMR Studies of Phosphoproteins
   *Hans J. Vogel*
   - I. Introduction   105
   - II. Classification of Phosphoproteins   106
   - III. Theoretical and Experimental Considerations   113
   - IV. NMR Studies   123
   - V. Conclusions   144
   - References   149

5. Paramagnetic Probes of Enzyme Complexes with Phosphorus-Containing Compounds
   *Joseph J. Villafranca*
   - I. Introduction   155
   - II. General Considerations   156
   - III. Properties of Cr(III)– and Co(III)–Nucleotide Complexes   157
   - IV. Studies with Co(III)–Nucleotide Complexes   160
   - V. Studies with Cr(III)–Nucleotide Complexes   166
   - VI. Studies with Cr(III)– and Co(III)–Pyrophosphate Complexes   169
   - References   173

6. Use of Chiral Thiophosphates and the Stereochemistry of Enzymatic Phosphoryl Transfer
   *Ming-Daw Tsai*
   - I. Stereochemical Problems Studied with Phosphorothioates   175
   - II. $^{18}$O Isotope Shifts in $^{31}$P NMR   178
   - III. $^{17}$O Quadrupolar Effects in $^{31}$P NMR   181
   - IV. Prochiral Centers: $^{31}$P NMR   185
   - V. Pro-Prochiral Centers: $^{31}$P($^{18}$O) NMR or $^{31}$P($^{17}$O) NMR   187
   - VI. Pro-Pro-Prochiral Centers: $^{31}$P($^{18}$O) NMR and $^{31}$P($^{17}$O) NMR   190
   - References   196

7. Use of Chiral [$^{16}$O,$^{17}$O,$^{18}$O]Phosphate Esters to Determine the Stereochemical Course of Enzymatic Phosphoryl Transfer Reactions
*John A. Gerlt*

    I. Introduction    199
    II. Syntheses of Oxygen Chiral Phosphate Esters    201
    III. Configurational Analyses of Oxygen Chiral Phosphate Esters    210
    IV. Selected Examples    220
    V. Summary    230
    References    230

8. DNA and RNA Conformations
*Chi-Wan Chen and Jack S. Cohen*

    I. Introduction    233
    II. Helix–Coil Transitions    234
    III. Sequence Dependence of Double-Stranded DNA Conformations    239
    IV. Conformational Transitions of Double-Stranded DNA    253
    V. Dynamic Behavior of RNA and DNA    255
    VI. Nucleosomal DNA    256
    VII. Biological and Genetic Significance    259
    References    260

9. High-Resolution $^{31}$P-NMR Spectroscopy of Transfer Ribonucleic Acids
*David G. Gorenstein*

    I. Introduction    265
    II. Spectral Comparison of Different Acceptor tRNAs    268
    III. $Mg^{2+}$ Dependence of $^{31}$P Spectra of tRNA$^{Phe}$    276
    IV. $Mn^{2+}$ Effects on $^{31}$P Spectra of tRNA$^{Phe}$    280
    V. Assignment of $^{31}$P Signals    281
    VI. Conformational Transitions    287
    VII. Spermine Effects    291
    VIII. $^{31}$P NMR of Yeast tRNA$^{Phe}$ · *E. coli* tRNA$^{Glu}$ Complex: Spectral Changes    292
    IX. Conclusions    294
    References    295

10. Phosphorus-31 NMR of Drug–Nucleic Acid Complexes
*David G. Gorenstein and Evelyn M. Goldfield*

    I. Introduction    299
    II. Drug–Double-Helical Nucleic Acid Complexes: Intercalation    300
    III. Electrostatic Mode of Interaction    310
    IV. tRNA·Ethidium $^{31}$P Spectra    311
    References    315

11. Phosphorus Relaxation Methods: Conformation and Dynamics of Nucleic Acids and Phosphoproteins
   *Phillip A. Hart*

   I. Introduction     317
   II. General Considerations     319
   III. The Phosphorus–Proton Nuclear Overhauser Effect     322
   IV. Phosphorus Relaxation Times     328
   V. Summary     344
   References     345

12. Relaxation Behavior of Nucleic Acids: Dynamics and Structure
    *Thomas L. James*

    I. Introduction     349
    II. Theory     350
    III. Structure and Possible Molecular Motions in Nucleic Acids     368
    IV. Considerations for Nucleic Acid Dynamics and Structure     373
    V. Relaxation Studies of Nucleic Acids     377
    References     398

13. Solid-State Phosphorus-31 NMR: Theory and Application to Nucleic Acids
    *Heisaburo Shindo*

    I. Introduction     401
    II. Basic Concepts of Solid-State $^{31}$P NMR     402
    III. Studies of Oriented DNA Fibers     408
    IV. Studies of DNA Complexes     418
    V. Concluding Remarks     421
    References     421

14. Phosphorus-31 NMR of Phospholipids in Micelles
    *Edward A. Dennis and Andreas Plückthun*

    I. Introduction and Perspective     423
    II. Spectral Characteristics of Phospholipids in Mixed Micelles with Detergents     425
    III. $T_1$, Nuclear Overhauser Effect, and Quantitative Analysis     431
    IV. Solubilization of Phospholipids by Detergents     433
    V. Critical Micelle Concentration Determinations and Micellization of Monomeric Phospholipids by Detergents     435
    VI. Lysophospholipids: Acyl and Phosphoryl Migration     438
    VII. Phospholipases: Specificity and Kinetics     442
    References     444

## 15. Phosphorus-31 NMR of Phospholipids in Membranes
*Ian C. P. Smith and Irena H. Ekiel*

    I. Properties of Membranes    447
   II. $^{31}$P-NMR Spectra of Ordered Systems    449
  III. Obtaining $^{31}$P-NMR Spectra    457
  IV. Applications to Biomembranes    461
    V. Conclusion    472
        References    472

## PART 3. Application to Biology and Medicine: Future Directions

## 16. Two-Dimensional Phosphorus-31 NMR
*William C. Hutton*

    I. Introduction    479
   II. General Principles of Two-Dimensional NMR: The Homonuclear, *J*-Correlated Experiment    484
  III. Applications of Two-Dimensional Spectroscopy to $^{31}$P NMR    492
  IV. Concluding Remarks    509
        References    510

## 17. Identification of Diseased States by Phosphorus-31 NMR
*Michael Bárány and Thomas Glonek*

     I. Introduction    512
    II. Skeletal Muscle    513
   III. Heart    530
   IV. Kidney    533
    V. Brain    534
   VI. Eye    537
  VII. Mammalian Fluids    540
VIII. Perspectives    543
       References    544

## PART 4. Selected Data

## 18. Appendixes: Selected Compilation of $^{31}$P-NMR Data
*David G. Gorenstein and Dinesh O. Shah*

    Appendix   I.  Doubly Connected Phosphorus (P$\leq$)    550
    Appendix  II.  Triply Connected Phosphorus (P$\leq$)    551
    Appendix III.  Quadruply Connected Phosphorus ($\geq$P$\leq$)    563

Appendix IV.  Quintuply Connected Phosphorus ($>\!\!\underset{|}{P}\!\!<$)    586

Appendix V.  Sextuply Connected Phosphorus ($\geqq\!P\!\leqq$)    588

References    589

Index    593

# Contributors

Numbers in parentheses indicate the pages on which the authors' contributions begin.

Michael Bárány (511), Department of Biological Chemistry, Health Sciences Center, College of Medicine, University of Illinois at Chicago, Chicago, Illinois 60612

Chi-Wan Chen (233), Developmental Pharmacology Branch, National Institute of Child Health and Human Development, National Institutes of Health, Bethesda, Maryland 20205

Jack S. Cohen (233), Developmental Pharmacology Branch, National Institute of Child Health and Human Development, National Institutes of Health, Bethesda, Maryland 20205

Edward A. Dennis (423), Department of Chemistry, University of California at San Diego, La Jolla, California 92093

Irena H. Ekiel (447), Division of Biological Sciences, National Research Council of Canada, Ottawa, Ontario K1A 0R6, Canada

John A. Gerlt (199), Department of Chemistry, Yale University, New Haven, Connecticut 06511

Thomas Glonek (511), Nuclear Magnetic Resonance Laboratory, Chicago College of Osteopathic Medicine, Chicago, Illinois 60615

Evelyn M. Goldfield (299), Department of Chemistry, University of Illinois at Chicago, Chicago, Illinois 60680

David G. Gorenstein (1, 7, 37, 265, 299, 549), Department of Chemistry, University of Illinois at Chicago, Chicago, Illinois 60680

Phillip A. Hart (317), School of Pharmacy, University of Wisconsin, Madison, Wisconsin 53706

William C. Hutton[1] (479), Department of Chemistry, University of Virginia, Charlottesville, Virginia 22901

---

[1]Present address: Research Division, Monsanto Agricultural Products Co., St. Louis, Missouri 63167.

*Thomas L. James* (349), Department of Pharmaceutical Chemistry, School of Pharmacy, University of California, San Francisco, California 94143

*Andreas Plückthun* (423), Department of Chemistry, University of California at San Diego, La Jolla, California 92093

*B. D. Nageswara Rao* (57), Department of Physics, Indiana University–Purdue University at Indianapolis, Indianapolis, Indiana 46223

*Dinesh O. Shah* (549), Department of Chemistry, University of Illinois at Chicago, Chicago, Illinois 60680

*Heisaburo Shindo* (401), Tokyo College of Pharmacy, Tokyo 192-03, Japan

*Ian C. P. Smith* (447), Division of Biological Sciences, National Research Council of Canada, Ottawa, Ontario K1A 0R6, Canada

*Ming-Daw Tsai* (175), Department of Chemistry, The Ohio State University, Columbus, Ohio 43210

*Joseph J. Villafranca* (155), Department of Chemistry, The Pennsylvania State University, University Park, Pennsylvania 16802

*Hans J. Vogel* (105), Department of Physical Chemistry 2, University of Lund, 5220 07 Lund, Sweden

# Preface

With the recent development of high-field, multinuclear, Fourier-transform NMR spectrometers, phosphorus-31 NMR has become a widely applied spectroscopic probe of the structure and dynamics of phosphorus-containing compounds. The field has expanded greatly, particularly in its biochemical and medical applications, since Van Wazer and co-workers published the only other monograph[1] entirely devoted to $^{31}$P-NMR spectroscopy. Although other reviews have recently been published on various aspects of $^{31}$P-NMR spectroscopy, this treatise represents the first effort to bring together most of the current theory and applications in this disparate field. We trust it will serve as a reference and resource for the NMR spectroscopist as well as the researcher in fields of application of $^{31}$P-NMR spectroscopy, such as organophosphorus chemistry, biochemistry, and medicine. Perhaps this book will stimulate even further application and development in $^{31}$P NMR and interest scientists not familiar with these techniques.

In a collaborative effort with many research laboratories, we develop the basic principles of $^{31}$P chemical shifts, coupling constants, and relaxation parameters and then apply these spectroscopic probes to increasingly more complex molecular and biological systems. The chapters have been written by experts in various research areas of phosphorus NMR. The literature has been exhaustively surveyed from the earliest beginnings of phosphorus NMR in the late 1950s to the end of 1982. However, because of the tremendous expansion of the literature in this field, the presentation is selective, and much of the discussion is based on the more recent literature.

In Part 1 the basic principles and empirical observations on $^{31}$P chemical shifts and coupling constants are introduced. In Part 2, additional theory and applications of $^{31}$P NMR provide unique information on the structure, interactions, and dynamics of enzyme complexes, nucleic acids, and phospholipids. B. D. Nageswara Rao surveys the broad literature on $^{31}$P NMR of enzyme complexes, and Hans J. Vogel describes studies on phosphoproteins. Chapters on the use of

[1]Crutchfield, M. M., Dungan, C. H., Letcher, L. H., Mark, V., and Van Wazer, J. R. (1967). *Top. Phosphor. Chem.* **5**, 1–457.

paramagnetic probes of enzyme complexes by Joseph J. Villafranca, on the use of chiral thiophosphates and $^{17}$O quadrupolar effects in enzymatic phosphoryl transfer by Ming-Daw Tsai, and on the use of chiral [$^{16}$O,$^{17}$O,$^{18}$O]phosphates to determine the stereochemical course of enzymatic phosphoryl transfer reactions by John A. Gerlt complete this section on $^{31}$P applications to enzyme systems.

Chi-Wan Chen and Jack S. Cohen introduce the application of $^{31}$P NMR to DNA and RNA conformations. High-resolution $^{31}$P NMR of transfer ribonucleic acids and $^{31}$P studies on drug–nucleic acid complexes (prepared by the editor with Evelyn M. Goldfield) are then presented. Phillip A. Hart introduces phosphorus relaxation methods, which are shown to provide important information on the conformation and dynamics of nucleic acids and phosphoproteins. Thomas L. James delves further into relaxation behavior of solution-state nucleic acids, and Heisaburo Shindo completes this section by describing solid-state $^{31}$P NMR of nucleic acids.

The chapters on $^{31}$P NMR of phospholipids in micelles (Edward A. Dennis and Andreas Plückthun) and membranes (Ian C. P. Smith and Irena H. Ekiel) provide details of the structure and dynamics of molecules in these systems.

William C. Hutton describes the development of two-dimensional $^{31}$P NMR in chemical, biological, and future medical applications. Finally, Michael Bárány and Thomas Glonek's contribution on identification of diseased states by $^{31}$P NMR clearly emphasizes the important new medical applications of NMR. Whole cells, organs, and even intact human subjects are now "routinely" studied by $^{31}$P NMR.

The volume is completed by appendixes presenting a selective compilation of $^{31}$P-NMR chemical shifts and coupling constants from the literature (prepared by the editor with Dinesh O. Shah).

A treatise such as this, with contributions coming from many different laboratories, ultimately owes acknowledgment to many collaborators. I wish to express my appreciation to the students and co-workers in our own laboratory for their many contributions and to the staff of Academic Press for their assistance in the preparation of this volume.

# Introduction

*David G. Gorenstein*
Department of Chemistry
University of Illinois at Chicago
Chicago, Illinois

The first nuclear magnetic resonance (NMR) signal was detected in 1945 by Bloch, Hansen, and Packard (1946) at Stanford and by Purcell, Torrey, and Pound (1946) at Harvard. The first $^{31}$P-NMR signals of phosphorus compounds were reported by Dickenson (1951) and Gutowsky and co-workers (1951). Indeed, the pioneering studies of Gutowsky on phosphorus compounds containing strongly coupled magnetic nuclei such as $PF_3$ and $HPF_4$ contributed greatly to the development of the theory of spin–spin coupling constants as well as the theory of the chemical shift.

The development of commercial multinuclear NMR spectrometers by 1955, particularly due to the effort of Varian Associates, led to the recognition that $^{31}$P NMR could serve as an important analytical tool for structural elucidation. By 1956, chemical shifts on several hundred phosphorus-containing compounds had already been reported (Muller and Goldenson, 1956; Van Wazer *et al.*, 1956), and some success had been achieved in correlating structure with $^{31}$P chemical shifts (Parks, 1957; Callis *et al.*, 1957). Early spectrometers (pre-1963) generally required neat samples in large nonrotating tubes (8–12 mm OD). These spectrometers recorded only a single scan of the spectrum, and dilute samples gave very weak signals. In 1963 the introduction of signal averaging through a computer of average transients (CAT) and the availability in the middle 1960s of more sensitive, higher field electromagnets (2.3 tesla fields, equivalent to 100 MHz resonance frequency for protons) led to further rapid growth in the number of reported $^{31}$P chemical shifts and coupling constants. By 1962, Jones and Katritzky (1962) had compiled all of the known $^{31}$P data, consisting of 59 references. In 1966, Van Wazer and co-workers (Crutchfield *et al.*, 1967) published the first monograph entirely devoted to $^{31}$P-NMR spectroscopy. Van Wazer presented a rather complete treatise on the theory of $^{31}$P chemical shifts and tabulated $^{31}$P chemical shifts and proton–phosphorus coupling constants for ~3000 compounds. This and two reviews by Mavel (1966, 1973) covered the field from the beginnings of phosphorus NMR in the late 1950s to December 1969. Mavel (1966) reported on ~450 publications through 1965 and ~1500 papers during the years 1966–1969 alone.

With the introduction by 1970 of signal-averaging, Fourier-transform (FT), and high-field superconducting-magnet NMR spectrometers about this time, $^{31}$P-NMR spectroscopy expanded from the study of small organic and inorganic compounds to biological phosphorus compounds as well. Mildred Cohn's original continuous-wave $^{31}$P-NMR spectrum of adenosine triphosphate (Fig. 1a) observed in 1958 set the stage for many important later biological applications (Cohn and Hughes, 1960). The latest, routine, multinuclear FT-NMR spectrometers (80–500 MHz, proton frequency) have reduced if not eliminated the one serious limitation to the widespread utilization of phosphorus NMR, which is the low sensitivity of the phosphorus nucleus (6.6% at constant field compared to $^1$H NMR). Today, routinely, millimolar (or lower) concentrations of phosphorus nuclei in as little as 0.3 ml of solution are conveniently monitored. The $^{31}$P nucleus has other convenient NMR properties suitable for FT NMR: spin $\frac{1}{2}$ (which avoids problems associated with quadrapolar nuclei), 100% natural abundance, moderate relaxation times (providing relatively rapid signal averaging and sharp lines), and a wide range of chemical shifts (>600 ppm).

Biological and medical applications of $^{31}$P-NMR spectroscopy grew dramatically during the 1970s, and numerous reviews on biological $^{31}$P NMR covered this burgeoning field (Burt *et al.*, 1977; Gadian *et al.*, 1979; Hollis, 1979; Ugurbil *et al.*, 1979; Cohn and Rao, 1979; O'Neill and Richards, 1980; Jardetzky and Roberts, 1981; Gorenstein, 1978, 1981; Gorenstein and Goldfield, 1982). Separate reviews on $^{31}$P-NMR spectroscopy in metal complexes of phosphorus ligands (Meek and Mazanec, 1981; Pregosin and Kunz, 1979) have also appeared. I have surveyed nonbiological aspects of $^{31}$P NMR covering the years since the last Mavel and Van Wazer reviews (1969), to June 1982 (Gorenstein, 1983).

Why then yet another review—indeed a whole book—devoted to the field? Phosphorus-NMR applications in chemistry, biology, and medicine have now grown so widespread and cover so many different aspects of the NMR experiment that it seemed important to draw together in one complete treatise much of this widely disparate field. This book can in fact cover only a fraction of the current theory and applications of phosphorus-NMR spectroscopy.

NMR spectroscopic information includes the resonant line positions (chemical shifts $\delta$), the spin–spin coupling constants $J$, the nuclear Overhauser effect (NOE), line intensities, spin–lattice ($T_1$) and spin–spin ($T_2$) relaxation times, the rotating-frame spin–lattice time in an off-resonance rf

# Introduction

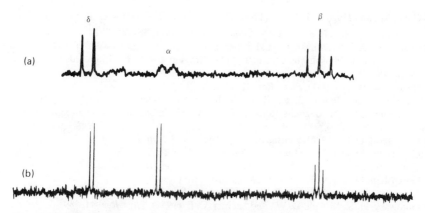

**Fig. 1.** High-resolution spectrum of ATP. (a) 1959: Single-scan, 24.3-MHz, 5-mm tube, continuous-wave operation of a Varian 4302 dual-purpose NMR spectrometer. [ATP] = 500 m$M$. (b) 1976: 10 scans, 72.9 MHz, 20-mm tube, in the Fourier-transform mode of operation, 20-sec repetition rate, and proton decoupling on a Bruker WH 180 NMR spectrometer. [ATP] = 1 m$M$. From Cohn (1979).

field ($T_{1\rho}^{\text{off}}$), and off-resonance intensity ratio $R$. In a collaborative effort from many research laboratories, we present in this volume basic theory on each of these important NMR spectroscopic probes of the phosphorus nucleus, discuss various empirical observations and general correlations, and show through numerous applications the important structural and dynamic information that $^{31}$P-NMR spectroscopy provides.

It is now impossible to cover all of the available literature in a single monograph. The *Specialist Periodical Reports* of the Chemical Society (United Kingdom) (S. Trippett, ed.) listed in the physical methods section of that series during the period 1969–1980 over 900 *selected* publications related to $^{31}$P-NMR spectroscopy. A computer literature search based on *Chemical Abstracts* from January 1979 to June 1982 revealed over 800 citations to phosphorus-NMR spectroscopy alone. Clearly, the field is well established and still growing dramatically!

## References

Block, F., Hansen, W. W., and Packard, M. E. (1946) *Phys. Rev.* **69**, 127.
Burt, C. T., Glonek, T., and Barany, M. (1977). *Science,* **195** 145–149.
Callis, C. F., Van Wazer, J. R., Shoolery, J. N., and Anderson, W. A. (1957). *J. Am. Chem. Soc.* **79**, 2719.
Cohn, M. (1979). *In* "NMR and Biochemistry" (S. J. Opella and P. Lu, eds.), pp. 7–27, Marcel Dekker, New York.

Cohn, M., and Hughes, T. R., Jr. (1960). *J. Biol. Chem.* **237**, 3250-3253.
Cohn, M., and Rao, B. D. N. (1979). *Bull. Magn. Reson.* **1**, 38-60.
Crutchfield, M. M., Dungan, C. H., Letcher, L. H., Mark, V., and Van Wazer, J. R. (1967). *In* "Topics in Phosphorus Chemistry" (M. Grayson and E. F. Griffin, eds.), Vol. 5, pp. 1-487. Wiley (Interscience), New York.
Dickenson, W. C. (1951). *Phys. Rev.* **81**, 717.
Gadian, D. G., Radda, G. K., Richards, R. E., and Seeley, P. J. (1979). *In* "Biological Applications of Magnetic Resonance" (R. G. Shulman, ed.), pp. 463-535. Academic Press, New York.
Gorenstein, D. G. (1978). *Jerusalem Symp. Quantum Chem. Biochem.* **11**, 1-15.
Gorenstein, D. G. (1981). *Annu. Rev. Biophys. Bioeng.* **10**, 355-386.
Gorenstein, D. G., and Goldfield, E. M. (1982). *Mol. Cell. Biochem.* **46**, 97-120.
Gorenstein, D. G. (1983). *Prog. NMR Spectrosc.* (in press).
Gutowsky, H. S., and McCall, D. W. (1951). *Phys. Rev.* **82**, 748.
Gutowsky, H. S., McCall, D. W., and Slichter, C. P. (1951). *Phys. Rev.* **84**, 589.
Hollis, D. P. (1979). *Bull. Magn. Reson.* **1**, 27-37.
Jardetzky, O., and Roberts, G. C. K. (1981). "NMR in Molecular Biology." Academic Press, New York.
Jones, R. A. Y., and Katritzky, A. R. (1962). *Angew. Chem., Int. Ed. Engl.* **1**, 32.
Mavel, G. (1966). *Progr. NMR Spectrosc.* **1**, 251-373. Pergamon Press, Oxford.
Mavel, G. (1973). *In* "Annual Reports on NMR Spectroscopy" (E. F. Mooney, Ed.), Vol. 5b, pp. 1-350. Academic Press, New York.
Meek, D. W., and Mazanec, T. J. (1981). *Accts. Chem. Res.* **14**, 266-274.
Muller, N., and Goldenson, J. (1956) *J. Am. Chem. Soc.* **78**, 3557.
O'Neill, I. K., and Richards, C. P. (1980). *In* "Annual Reports on NMR Spectroscopy" (W. A. Webb, ed.), Vol. 10a, pp. 133-236. Academic Press, New York.
Parks, J. R. (1957). *J. Am. Chem. Soc.* **79**, 757.
Purcell, E. M., Torrey, H. C., and Pound, R. V. (1946). *Phys. Rev.* **69**, 37.
Pregosin, P. S., and Kunz, R. W. (1979). "$^{31}$P and $^{13}$C NMR of Transition Metal Phosphine Complexes." Springer-Verlag, Berlin.
Ugurbil, K., Shulman, R. G., and Brown, T. R. (1979). *In* "Biological Applications of Magnetic Resonance" (R. G. Shulman, ed.), pp. 537-589. Academic Press, New York.
Van Wazer, J. R., Callis, C. F., Shoolery, J. N., and Jones, R. C. (1956). *J. Am. Chem. Soc.* **78**, 5715.

# PART 1

*Phosphorus-31 Chemical Shifts and Spin–Spin Coupling Constants: Principles and Empirical Observations*

CHAPTER 1

# Phosphorus-31 Chemical Shifts: Principles and Empirical Observations

*David G. Gorenstein*
Department of Chemistry
University of Illinois at Chicago
Chicago, Illinois

| | |
|---|---:|
| I. Introduction and Basic Principles | 7 |
| II. Applications of $^{31}$P Chemical-Shift Theory and Empirical Observations | 9 |
|    A. $^{31}$P Chemical Shifts in Phosphoryl Compounds | 10 |
|    B. Bond-Angle Effects on $^{31}$P Chemical Shifts in Nontetracoordinated Phosphorus Compounds | 14 |
|    C. Stereoelectronic Effects on $^{31}$P Chemical Shifts | 19 |
|    D. Polar and Resonance Effects on $^{31}$P Chemical Shifts | 23 |
|    E. Extrinsic Effects on $^{31}$P Chemical Shifts | 27 |
|    F. Additional Effects and Conclusions | 31 |
|    References | 33 |

## I. Introduction and Basic Principles

Nuclear magnetic resonance (NMR) spectroscopy is now so widely applied throughout chemistry that no attempt is made here to discuss the general details of NMR theory. Numerous monographs on NMR spectroscopy may be consulted for further details (Pople *et al.*, 1959; James, 1975; Jardetzky and Roberts, 1981). However, some basic theory for each of the $^{31}$P-NMR spectroscopic probes is introduced in the appropriate chapter. Here we present an introduction to $^{31}$P chemical-shift theory. Note that throughout this volume an attempt has been made to reference all $^{31}$P shifts to 85% H$_3$PO$_4$ and always to follow the IUPAC convention [*Pure Appl. Chem.* **45**, 217 (1976)] so that *positive values are to high frequency* (low field):

$$\delta = \frac{v - v_{\text{ref}}}{v_{\text{ref}}} 10^6 \qquad (1)$$

where $v$ is the frequency of the signal of interest and $v_{\text{ref}}$ the frequency of the

85% $H_3PO_4$ reference standard. We urge the reader to interpret reported $^{31}P$ chemical shifts cautiously because the early literature (pre-1970s) and even many later papers use the *opposite* sign convention.

The interaction of the electron cloud surrounding the phosphorus nucleus with an external applied magnetic field $H_o$ gives rise to a local magnetic field. This induced field shields the nucleus, with the shielding proportional to the field $H_o$ so that the effective field $H_{eff}$ felt by the nucleus is given by

$$H_{eff} = H_o (1 - \sigma) \qquad (2)$$

where $\sigma$ is the shielding constant. The difference in chemical shift $\delta$ between two lines is simply the difference in shielding constants $\sigma$ of the nuclei giving rise to the two lines, and from Eq. (1), increased shielding corresponds to more negative chemical shifts.

For a nonspherical charge distribution about the phosphorus nucleus, the shielding constant largely results from a local diamagnetic shielding contribution $\sigma_d$ and a local paramagnetic shielding contribution $\sigma_p$ (Pople *et al.*, 1959):

$$\sigma = \sigma_d + \sigma_p \qquad (3)$$

The $\sigma_d$ can be calculated from the Lamb (1941) equation

$$\sigma_d = \frac{e^2}{3mc^2} \sum_\mu P_{\mu\mu} \langle \phi_\mu | r^{-1} | \phi_\mu \rangle \qquad (4)$$

with the usual notation where the $P_{\mu\mu}$ are the electron populations in the $\phi_\mu$ orbitals on phosphorus and $\langle \phi_\mu | r^{-1} | \phi_\mu \rangle$ the quantum mechanical expectation values of the inverse distance $r^{-1}$ of the electron from the nucleus.

Because the charge distribution in a phosphorus molecule will be far from spherically symmetrical, the major contribution to the $^{31}P$ chemical shift comes from the paramagnetic term (Prado *et al.*, 1979). This is a deshielding contribution originating from the small amount of mixing of paramagnetic excited states into the ground-state wave function. Unfortunately, this is the most difficult term to calculate theoretically, and most $^{31}P$ chemical-shift calculations have relied on the Karplus–Das average excitation energy approximation

$$\sigma_p = -\frac{e^2\hbar^2}{2m^2c^2} \langle r^{-3} \rangle_{3p} (\Delta E_{av})^{-1} \zeta_p \qquad (5)$$

In this equation, $\langle r^{-3} \rangle_{3p}$ is the mean inverse cube radius for the phosphorus $3p$ orbitals, $\Delta E_{av}$ the average electronic excitation energy between ground and excited states, $\hbar$ Planck's constant divided by $2\pi$, and $\zeta_p$ is given by

$$\zeta_p = P_{xx} + P_{yy} + P_{zz} - \tfrac{1}{2}(P_{zz}P_{yy} + P_{xx}P_{yy} + P_{xx}P_{yy})$$
$$+ \tfrac{1}{2}(P_{yz}P_{zy} + P_{xy}P_{yx} + P_{xz}P_{zx}) \qquad (6)$$

with the P-matrix elements taken from the charge–bond-order matrix (Letcher and Van Wazer, 1967). The chemical shift will thus be affected by changes in bond overlap and hybridization (via the $\zeta_p$ term or orbital "unbalancing"), changes in atomic charges (via mainly the $\langle r^{-3} \rangle$ term), as well as changes in $\Delta E_{av}$. The often observed correlation of chemical shifts and charge densities probably derives from the expansion or contraction of the $p$ orbitals with charge variations (hence altering $\langle r^{-3} \rangle$). Although this theory generally accounts for major trends in heavy-atom chemical shifts, small variations frequently are only poorly reproduced. In fact, Gorenstein and Kar (1975) noted that $^{31}$P chemical shifts in phosphate esters are just as well correlated with calculated phosphorus electron densities as the full paramagnetic term [Eq. (5)].

Asymmetry in the charge distribution implies that the $^{31}$P chemical shifts (or shielding constants) vary as a function of the orientation of the molecule relative to the external magnetic field. This gives rise to a chemical-shift anisotropy that can be defined by three principal components $\sigma_{11}$, $\sigma_{22}$, and $\sigma_{33}$ of the shielding tensor (Pople et al., 1959). For molecules that are axially symmetrical, with $\sigma_{11}$ along the principal axis of symmetry, $\sigma_{11} = \sigma_\parallel$ (parallel component), and $\sigma_{22} = \sigma_{33} = \sigma_\perp$ (perpendicular component). These anisotrophic chemical shifts are only observed in oriented samples (and are important in large molecular assemblies such as discussed in later chapters; see Shindo, Chapter 13), whereas for small molecules in solution, rapid tumbling averages the shift. The average chemical shielding $\sigma_{av}$ is given by the trace of the shielding tensor or

$$\sigma_{av} = \tfrac{1}{3}(\sigma_{11} + \sigma_{22} + \sigma_{33}) \tag{7}$$

and the anisotropy $\Delta\sigma$ is given by

$$\Delta\sigma = \sigma_{11} - \tfrac{1}{2}(\sigma_{22} + \sigma_{33})$$

or, for axial symmetry,

$$\Delta\sigma = \sigma_\parallel - \sigma_\perp \tag{8}$$

## II. Applications of $^{31}$P Chemical-Shift Theory and Empirical Observations

Several authors have attempted to develop a unified theoretical foundation for $^{31}$P chemical shifts of phosphorus compounds (Gutowsky and McCall, 1954; Letcher and Van Wazer, 1966, 1967; Müller et al., 1955; Purdela, 1971; see also Rajzmann and Simon, 1975; Wolff and Radeglia, 1980; Weller et al., 1981; Radeglia and Grimmer, 1982). In one of the more

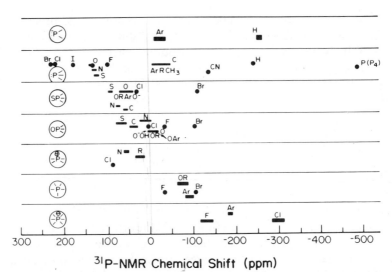

**Fig. 1.** Range of ³¹P chemical shifts for substituted phosphorus compounds. Group indicated is equivalently substituted about phosphorus.

successful theoretical approaches, Letcher and Van Wazer (1966, 1967), using approximate quantum-mechanical calculations, demonstrated that three factors appear to dominate $^{31}$P chemical-shift differences $\Delta\delta$, as shown by

$$\Delta\delta = -C\Delta\chi_x + k\Delta n_\pi + A\Delta\theta \qquad (9)$$

where $\Delta\chi$ is the difference in electronegativity in the P—X bond, $\Delta n_\pi$ the change in the $\pi$-electron overlap, $\Delta\theta$ the change in the $\sigma$-bond angle, and $C$, $k$, and $A$ are constants.

In this chapter, we attempt to use theories such as Letcher and Van Wazer's that relate $^{31}$P shift changes to structural and electronic parameters in discussing the wide variation of $^{31}$P shifts (Fig. 1).

## A. $^{31}$P Chemical Shifts in Phosphoryl Compounds

For phosphoryl compounds, Letcher and Van Wazer (1967) concluded that changes in the $\sigma$-bond angles make a negligible contribution ($|A| < 1$) to the $^{31}$P chemical shift, with electronegativity effects apparently predominating. Purdela (1971) has suggested a correlation (although poor) between X—P—X bond angles and chemical shifts for a wide variety of phosphoryl compounds. Contrary to the theory of Letcher and Van Wazer, Kumamoto et al. (1956) and Blackburn et al. (1971) have argued on the basis of cyclic phosphate ester shifts that phosphorus bond angles must play some role in

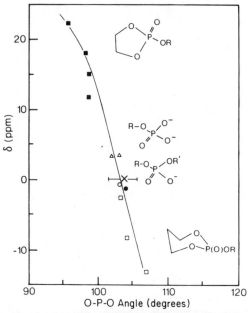

**Fig. 2.** Phosphorus-31 chemical shift of phosphate esters versus O—P—O bond angle. ■, Five-membered cyclic esters; △, monoester dianions; x, monoester monoanions; ○, acyclic diester monoanions; ●, acyclic diester free acids; □, six-membered cyclic esters. From Gorenstein (1975). Copyright 1975 American Chemical Society.

$^{31}$P chemical shifts. A change in $d_\pi$-$p_\pi$ bonding resulting from bond-angle changes was suggested as an explanation for these shifts. Blackburn et al. (1971) compiled all of the known cyclic ester chemical shifts and concluded that these ring shifts must arise from a complex stereoelectronic effect not explicable by present theory.

Our laboratory (Gorenstein, 1975) extended these observations and proposed an empirical correlation between $^{31}$P chemical shifts and O—P—O bond angles in phosphates. (Note that success here depends on the fact that we are dealing with only a limited structural variation: the number and chemical type of R groups attached to a tetrahedron of oxygen atoms surrounding the phosphorus nucleus.) As shown in Fig. 2, for a wide variety of different alkyl phosphates (mono-, di-, and triesters, cyclic and acyclic, neutral, monoanionic, and dianionic esters), a decrease in the smallest O—P—O bond angle (obtained from X-ray data) in the molecule results in a deshielding (downfield shift) of the phosphorus nucleus. The correlation possibly provides an explanation for the unusual downfield shift observed on ionizing an acyclic monoanion. The ~3° reduction in the O—P—O bond angle of the dianionic phosphate is consistent with the 4-ppm down-

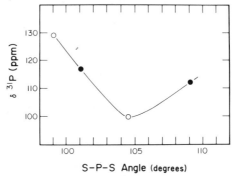

Fig. 3. Phosphorus-31 chemical shifts of 2-thioxo-2-*tert*-butyl-1,3,2-dithiaphospha compounds versus S—P—S bond angle. From Martin and Robert (1981). Copyright 1981 John Wiley & Sons, Ltd.

field shift. Charge alone does not appear responsible for the deshielding, because the acyclic monoanion and free acid have similar chemical shifts (and, significantly, similar O—P—O bond angles).

Martin and Robert (1981) have shown that a similar correlation of S—P—S bond angles and $^{31}$P chemical shifts in 2-thioxo-2-*tert*-butyl-1,3,2-dithiaphospha compounds likely exists (Fig. 3). The nonmonotonic variation of $\delta$ and S—P—S bond angle observed here is quite similar to Gorenstein's (1975) original empirical correlation. However, the point for $PO_4^{3-}$ included in the original plot has been eliminated from Fig. 2. If this point were included ($\delta = +5$ ppm, O—P—O angle = 109.5°), then a nonmonotonic variation would also be observed.

Gorenstein and Kar (1975) have attempted to calculate the $^{31}$P chemical shifts for a model phosphate diester in various geometries to confirm theoretically the bond-angle correlation. Using CNDO/2 SCF molecular orbital calculations (for details see Gorenstein and Kar, 1975; Gorenstein, 1983), a correlation was drawn between calculated phosphorus electron densities and $^{31}$P chemical shifts, and indeed deshielding of the phosphorus atom with decreasing O—P—O bond angles was found. A slightly better correlation was achieved between observed and calculated $^{31}$P chemical shifts using both the *p*-orbital unbalancing term $\zeta_p$ and the charge density on phosphorus (Gorenstein, 1983). No improvement was achieved by including the *d*-orbital unbalancing term in the analysis. The results of the analysis fit the normal expectation that increased electron density on phosphorus leads to increased shielding, and increased *p*-orbital unbalancing leads to increased deshielding. The calculations suggest that the reason the charge on the phosphate (neutral, mono-, di-, or trianion) has little effect on the $^{31}$P chemical shift is that the $\zeta_p$ and charge density both change in the same ratio, hence their effect on the paramagnetic term is canceled.

## TABLE I

Phosphorus-31 Chemical Shifts of Aliphatic Phosphate Triesters: $(RO)_3P=O$

| R | $\delta^{31}P$ [a] |
|---|---|
| $CH_3$ | +2.1 |
| $C_2H_5$ | −1.0 |
| $n\text{-}C_3H_7$ | −0.7 |
| $i\text{-}C_3H_7$ | −3.3 |
| $n\text{-}C_4H_9$ | −1.0 |
| $t\text{-}C_4H_9$ | −13.3 |

[a] See original citations in Crutchfield et al. (1967).

This explanation for the $^{31}P$ chemical-shift variation based on bond-angle differences could also explain the results shown in Table I. As the alkyl group in the series of trialkyl phosphates gets bulkier, hence as the ester (R)OPO(R) bond angle gets larger, the phosphorus atom is more shielded.

Bond-angle changes and hence distortion from tetrahedral symmetry in tetracoordinated phosphorus compounds should affect the chemical-shift anisotropy as well. Dutasta et al. (1981) experimentally have verified this bond-angle effect in a solid-state $^{31}P$-NMR study on a series of cyclic thioxophosphonates (1–4). As shown in Table II, the shielding tensors are

## TABLE II

Principal Values $\sigma_{11}$, $\sigma_{22}$, $\sigma_{33}$ of the $^{31}P$ Chemical-Shift Tensor for Cyclic Thioxophosphonates[a,b]

| Structure | $\sigma_{av}$ | $\eta$ | $\sigma_{11}$ | $\sigma_{22}$ | $\sigma_{33}$ | $\Delta\sigma$ | $\alpha$(degrees) |
|---|---|---|---|---|---|---|---|
| 1 | 114 | 95 | 244 | 117 | −19 | 263 | 98.0 |
| 2 | 92 | 55 | 202 | 124 | −50 | 252 | 103.5 |
| 3 | 104 | 42 | 198 | 143 | −29 | 227 | 105.1 |
| 4 | 92 | 29 | 173 | 136 | −33 | 206 | 108.2 |

[a] From Dutasta et al. (1981).
[b] $\sigma_{ii}$ values are in parts per million with positive values downfield from $H_3PO_4$ (85%). $\sigma_{av}$ is the isotropic solid-state chemical shift; $\eta$ (%) = $(\sigma_{22} - \sigma_{11})/(\sigma_{33} - \sigma_{iso})$ is the asymmetry parameter; $\Delta\sigma = \sigma_{11} - \sigma_{33}$ represents the anisotropy.

very sensitive to geometrical changes, and in fact, a linear correlation appears to exist between the asymmetry parameter $\eta$ and the intracyclic bond angle $\alpha$. The anisotropy is also correlated to the bond angle whereas the average shielding tensor shows a much poorer correlation.

## B. Bond-Angle Effects on $^{31}$P Chemical Shifts in Nontetracoordinated Phosphorus Compounds

There is conflicting evidence supporting a direct bond-angle effect on the $^{31}$P chemical shifts of nontetracoordinated phosphorus compounds. As shown in Fig. 4A in data compiled by Dennis (1968), all of the five-membered ring phosphoryl compounds are deshielded relative to six-membered ring or acyclic structural analogs. This presumably reflects the bond-angle dependence of $^{31}$P shifts demonstrated in Figs. 2 and 3. In fact, in *all* classes of phosphorus compounds, the $^{31}$P shift of the five-membered ring structure is downfield of the six-membered ring analog, as shown in Fig. 4B. The five-membered ring effect is quite large (13–26 ppm) (Blackburn *et al.*, 1971; Dennis, 1968; Allen and Tebby, 1970) with the exception of the phosphites, where the ring effect is only modest, 3–13 ppm.

In acyclic trivalent phosphorus compounds, however, the bond-angle effect on $^{31}$P shifts does not follow the same pattern observed in tetracovalent phosphorus compounds; in fact, it is often opposite to the five-membered ring effect. Thus, from the data in Table III, Mann (1972) concluded

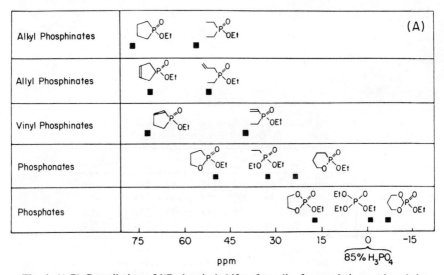

**Fig. 4.** (A,B) Compilation of $^{31}$P chemical shifts of acyclic, five- and six-membered-ring

# 1. ³¹P Chemical Shifts

that in phosphines an increase in steric bulk of the alkyl group, which should increase the R—P—R' bond angle, results in a *deshielding* of the phosphorus atom (opposite to the trend in phosphoryl compounds). Quin and Breen (1973) have pointed out correctly that bond-angle changes alone cannot be the only explanation for this trend. Thus bond-angle effects are inconsistent with the downfield shift of tri-*n*-propylphosphine (−33 ppm) compared to

phosphorus compounds. Partially derived from (A) Dennis (1968), and (B) Gorenstein (1983).

**TABLE III**

Phosphorus-31 Chemical Shifts[a] of Phosphines

| Compound | $\delta^{31}$P (ppm)[b] | <RPR (degrees) |
|---|---|---|
| PH$_3$ | −240[c] | 93.8[d] |
| PMe$_3$ | −62 | 98.9[d] |
| PEt$_3$ | −20.1 | — |
| P($n$Pr)$_3$ | −33.0[e] | — |
| P($i$Bu)$_3$ | −40[e] | — |
| P(CH$_3$)$_2$($t$Bu) | −28.7[e] | — |
| P(CH$_3$)($t$Bu)$_2$ | +12.0[e] | — |
| P($i$Pr)$_3$ | +20.0[d] | — |
| P($t$Bu)$_3$ | +61.9[e] | 105.7[d] |

[a] In parts per million; positive shifts are downfield from 85% H$_3$PO$_4$.
[b] From Mann (1972) unless otherwise noted.
[c] From Crutchfield et al. (1967).
[d] From Tolman (1977).
[e] From Quin and Breen (1973).

triisobutylphosphine (−45.5 ppm). Quin and Breen noted that the bond-angle argument of Mann (1972) would predict the opposite result. Mann's bond-angle hypothesis for these shifts is also inconsistent with the $^{31}$P shifts of symmetric phosphonium ions, where tetrahedral symmetry requires that the bond angles all be 109.5°: Me$_4$P$^+$I$^-$, +25.3 ppm; Et$_4$P$^+$I$^-$, +40 ppm; and $n$Bu$_4$P$^+$Br$^-$, +33.9 ppm (McFarlane and McFarlane, 1978).

Quin and Breen (1973) have proposed that these substituent effects on phosphine and phosphonium ion chemical shifts are likely similar to the well-known $\alpha$, $\beta$, and $\gamma$ substituent effects in $^{13}$C chemical shifts. In $^{13}$C NMR, substituting a hydrogen on an $\alpha$ carbon by a carbon atom ($\alpha$ effect) leads to deshielding, whereas substituting a hydrogen on a $\beta$ carbon along a chain by a carbon atom ($\beta$ effect) leads to shielding, and substituting a hydrogen on a $\gamma$ carbon further along the chain by a carbon atom ($\gamma$ effect) also leads to shielding. Quin and Breen (1973) proposed similar $\alpha$, $\beta$, and $\gamma$ substituent effects for aliphatic phosphorus compounds, where the $^{31}$P chemical shift can be calculated from

$$\delta^{31}\text{P} = \delta_{\text{parent}} + m\beta + n\gamma \tag{10}$$

with $m$ the number of $\beta$ carbons and $n$ the number of $\gamma$ carbons. Some of Quin and Breen's $^{31}$P chemical-shift parameters are given in Table IV. The origin of these $\alpha$, $\beta$, and $\gamma$ constants is still being debated, and it is most likely that several factors are operative in each of the effects, as discussed shortly for the $\gamma$ effect.

**TABLE IV**

Phosphorus-31 Chemical-Shift Parameters[a]

| Compound class | $\delta^{31}P$ of parent[b] (R = CH$_3$) | $\beta$ Constant[b,c] | $\gamma$ Constant[b,c] |
|---|---|---|---|
| RPH$_2$ | −163.5 | +35.5 | −7 |
| R$_2$PH | −99 | +22 | −10 |
| R$_3$P | −62 | +13.5 | −4 |
| R$_3$PH$^+$ | −3.2 | +8.6 | −3 |
| R$_4$P$^+$ | +25.3 | +3.7 | −1.5 |
| R$_3$PO | +36.2 | +4 | −1 |

[a] From Quin and Breen (1973).
[b] In parts per million; positive shifts are downfield from 85% H$_3$PO$_4$.
[c] Per carbon; negative value shielding.

Obviously, care in analyzing substituent effects is required to separate opposing trends such as the $\beta$ and $\gamma$ constants. It is probably safer to compare $^{31}$P chemical-shift differences between five- and six-membered rings in an attempt to separate only bond-angle from other effects (although a six-membered ring also has two $\gamma$ effects relative to a five-membered ring, which could also explain the upfield shift in the six-membered ring).

The $^{31}$P chemical shifts of acyclic phosphites depend little on the steric bulk of the alkyl group (Table V). On the other hand, the bicyclic phosphite **5** with one five-membered ring ($\delta = +105$ ppm) is shifted downfield by 13.5

ppm from the bicyclic phosphite **6** with only six-membered rings ($\delta = +91.5$ ppm). Both, however, are *shielded* relative to the acyclic phosphites (Table V). Verkade (1974, 1976; Bertrand et al., 1970) has argued that the $^{31}$P shifts of the constrained bicyclic esters are due in part to a "hinge effect" that decreases the P—O $\pi$ bonding. However, it is difficult then to explain the "near normal" $^{31}$P chemical shift of the adamantane phosphite (**7**) (+137.7 ppm) (Verkade, 1974). Clearly, several factors must be operative here and are discussed in the next section.

Other empirical correlations suggest a dependence of $^{31}$P shifts to bond angles; although in the case of phosphine oxides (Hays and Peterson, 1972), phosphine sulfides (Maier, 1972), and cyclophosphines (Smith and Mills, 1976; Mills, 1977) as with phosphines (Mann, 1972; Tolman, 1977) a

### TABLE V

Phosphorus-31 Chemical Shifts of Acyclic Phosphites[a]

| Compound | $\delta^{31}P$[b] |
|---|---|
| P(OMe)$_3$ | +139.7 |
| P(OEt)$_3$ | +137.6 |
| P(O-$i$Pr)$_3$ | +137.5 |
| P(O-$t$Bu)$_3$ | +138.1 |

[a] From Tolman (1977).
[b] In parts per million; positive shifts are downfield from 85% H$_3$PO$_3$.

monotonic upfield shift with decreasing bond angles is observed. As discussed before for phosphines, these changes may arise from $\alpha$, $\beta$, and $\gamma$ substituent effects.

Richman and Flay (1981) and Richman et al. (1981) have also shown that the $^{31}$P chemical shifts of a series of tetravalent cyclic phosphonium chloride salts and fluorophosphoranes **8** and **9** vary with ring size, as shown in Table VI.

### TABLE VI

Phosphorus-31 Chemical Shifts of **8** and **9** as a Function of Bridging Methylene Groups

| Compound[a] | Peripheral ring size | $\delta^{31}P$ (ppm)[b] | $\delta^{31}P$ (ppm)[c] |
|---|---|---|---|
| 2,2,2,2 | 12 | — | −14.2 |
| 3,2,2,2 | 13 | +65.6 | −32.2 |
| 3,3,2,2 | 14 | +43.2 | −42.8 |
| 3,2,3,2 | 14 | +31.4 | −46.3 |
| 3,3,3,2 | 15 | +25.4 | [d] |
| 3,3,3,3 | 16 | +15.0 | [d] |

[a] Numbers represent bridging methylenes ($n,n',n'',n'''$).
[b] Chlorophosphonium ion **8**.
[c] Fluorophosphorane **9**.
[d] Ionic fluoride.

# 1. ³¹P Chemical Shifts

The (3,3,3,3) phosphonium chloride structure with four six-membered rings can achieve tetrahedral geometry about phosphorus. As each of the six-membered 1,3,2-diazaphosphorinane rings is reduced to a five-membered ring, the $^{31}$P chemical shift of the macrocycle ionic chloride moves downfield by ~10–30 ppm.

The macrocyclic fluorophosphoranes show a similar multiple five-membered ring effect on the $^{31}$P chemical shifts. As additional five-membered rings with smaller N—P—N bond angles are introduced into the pentacovalent structure, a downfield shift of 10–20 ppm per ring is observed. Other pentacovalent oxyphosphoranes behave similarly (see Gorenstein and Shah, Chapter 18).

## C. Stereoelectronic Effects on $^{31}$P Chemical Shifts

The semiempirical calculations of Gorenstein and Kar (1975) suggest that $^{31}$P chemical shifts are also dependent on P—O ester torsional angles $\omega,\omega'$. [The two torsional angles $\omega$ and $\omega'$ are defined by the R—O—P—O(R') dihedral angles (Sundaralingam, 1969; see Fig. 5).] These chemical-shift calculations indicated that a phosphate diester in a gauche, gauche (g,g)[1] conformation should have a $^{31}$P chemical shift substantially upfield (by at least several ppm) from a phosphate diester in more extended conformations, such as gauche, trans (g,t) or trans, trans (t,t) conformations. Chemical shifts of other conformations have been calculated and have been used to generate a $^{31}$P chemical-shift torsional-angle contour map (Fig. 6), which we use in later chapters of the book.

Prado et al. (1979) have also described a torsional-angle sensitivity to the

**Fig. 5.** Definition of phosphate diester torsional angles $\omega,\omega'$: −g,−g (left); −g,t (right).

---

[1] We should mention that for purposes of conveniently describing the P—O ester torsional angle dependence on chemical shifts, we generally make no distinction between R—O—P—O(R') torsional angles +60° (+g) or −60° (−g). In addition, $\omega,\omega'$ torsional angles (see Fig. 5) of g,t (60°, 180°); −g,t; t,g; and t,−g will often be grouped together as g,t. Similarly, g,g includes conformers −g,−g; g,−g; and −g,g, although the latter two conformers do have $^{31}$P chemical shifts that are different from g,g (Gorenstein and Kar, 1975).

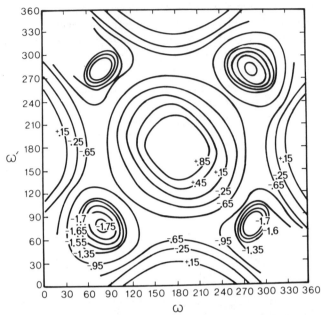

**Fig. 6.** Phosphorus-31 chemical-shift torsional-angle contour map for a phosphate diester. "Isoshift" contours are in parts per million from 85% $H_3PO_4$. From Gorenstein and Luxon (1979). Copyright 1979 American Chemical Society.

$^{31}$P chemical shifts with generally more accurate *ab initio*, gauge-invariant-type, molecular orbital, chemical-shift calculations. They found that the $^{31}$P chemical shift of a dimethyl phosphate in a g,g conformation is 3–6 ppm upfield from a phosphate in a g,t conformation.

Actually, the bond-angle effect and torsional-angle effect are not unrelated phenomena, because Gorenstein *et al.* (1976b, 1977), Gorenstein and Kar (1977), and Perahia and Pullman (1976) established a coupling of phosphate ester bond angles with these torsional angles. Molecular orbital calculations and comparisons with X-ray structures show that rotation about the P—O ester bond produces about an 11° bond-angle distortion. The coupling of these two geometric parameters is most significant for $^{31}$P-NMR studies on phosphate esters because it means that a perturbation in the $^{31}$P chemical shift will generally describe a single structural change. Thus ambiguity over the origin of a given $^{31}$P shift perturbation is often eliminated. This coupling fits nicely with the proposed downfield shift in trans versus gauche conformations, because the bond angle is reduced in the trans conformation; we have termed this a stereoelectronic $^{31}$P effect. Unfortunately, present NMR theory is inadequate to identify whether bond- or torsional-angle (or the combination, stereoelectronic) effects are most significant or which term in Eq. (5) is responsible for these shifts, although calculations by C. Giessner-Prettre (personal communication) have since suggested that the bond-angle-distortion effect is more important.

Interestingly, the C—O(H—C—O—P) torsional angle also appears to influence $^{31}$P shifts, although not as much as the P—O ester torsional angle (C. Giessner-Prettre, private communication).

Experimentally, the stereoelectronic effect on $^{31}$P shifts can be confirmed by using six-membered-ring model systems in which the torsional angles are rigidly defined by some molecular constraint, such as the two diastereomeric phosphate triesters **10** and the phosphoramidates **11** (Gorenstein, 1978,

1981; Gorenstein and Rowell, 1979, 1980). Generally those diastereomeric phosphates with an axial ester group have $^{31}$P chemical shifts as much as 6 ppm upfield from these isomeric phosphates with an equatorial ester group (Mosbo and Verkade, 1972; Bentrude and Tan, 1973; Maryanoff et al., 1979; Day et al., 1983; Gorenstein, 1978, 1983; also Gorenstein and Shah, Chapter 18 for other related structures and references.) In **10b** the equatorial ester group is locked into a trans conformation relative to the endocyclic P—O ester bond. Again, trans esters are downfield of gauche esters such as **10a**.

This $^{31}$P shielding for the gauche conformation may be another manifestation of the $\gamma$ effect often noted in $^{13}$C chemical shifts, (Grant and Cheney, 1967; Stothers, 1972; Gorenstein, 1977; also described as above by Quin for phosphines). Thus the $^{13}$C chemical shifts of the $\gamma$-methylene carbon and the methyl carbon in axially substituted (gauche conformation) methylcyclohexane are shifted 2–5 ppm upfield relative to the signals in equatorially substituted methylcyclohexane. Although Grant and Cheney (1967) have proposed that these shifts originate from steric interaction, Gorenstein (1977) has suggested the alternative stereoelectronic explanation for a generalized $\gamma$ effect in $^{13}$C and $^{31}$P (and $^{19}$F and other heavy atom) chemical shifts, based on the coupling of bond-angle distortion with torsional angles. However, Schneider and Weigand (1977) have argued in favor of the Grant and Cheney steric origin for the $^{13}$C $\gamma$ effect.

Surprisingly, diastereomeric phosphoramidates **11a** and **11b** show only small $^{31}$P chemical-shift differences. The gauche ester **11a**, with an axial ester group, is in fact 0.06 ppm downfield of the trans ester, **11b**, with an equatorial ester group. Resolution of this discrepancy from the proposed stereoelectronic $^{31}$P theory was achieved by conformational analysis of the two diastereomers. Apparently, the anomeric preference (Romers et al., 1969; Gorenstein and Rowell, 1979, 1980) in molecules such as **10** and **11** for placing an electronegative substituent (—OAr) in the axial rather than the equatorial position is so strong that equatorial phosphoramidate **11b** prefers a twist-boat conformation (**11c**). An X-ray structure and NMR conformational analysis of other six-membered-ring esters have supported this conclusion (Bajwa et al., 1979; Hutchins et al., 1979; Gorenstein and Rowell, 1979). In both chair **11a** and boat **11c** conformations, the ester groups are axial and in a gauche conformation and therefore, according to the stereoelectronic $^{31}$P theory, should have similar $^{31}$P chemical shifts. In fact, this anomeric preference in **10b** is strong enough to populate the twist-boat conformation (50–100%) preferentially for different aryl cyclic triesters, and this explains why this pair (as others in the literature) sometimes has only modest (1–3 ppm) chemical-shift differences.

Holmes, Gorenstein, and co-workers (Day et al., 1983) have confirmed

# 1. $^{31}$P Chemical Shifts

experimentally in an X-ray study on epimeric six-membered-ring phosphate triesters **10** that the O—P—O bond angles are coupled to torsional angles. Thus in axial esters (gauche conformation) such as **10a**, the sum of the three (R)O—P—O(R′) ester angles is on average 5.3° larger than the sum of the three ester angles in equatorial esters (trans conformation) such as **10b**. This decrease in ester O—P—O bond angles $b$ must, of course, be compensated by an increase in R(O)P=O bond angles $a$ as represented here:

$$\begin{array}{c} O \\ \parallel a \\ RO\text{---}P \\ RO\ \ b\ \ OR \end{array}$$

In the phosphorinane esters, the bond angles are significantly distorted from tetrahedral values, and this distortion is greatest for the equatorial esters. In these esters, the ester bond angles $b$ contract to an average of 102.9°, whereas the phosphoryl ester angles increase to an average of 115.4°. This stereoelectronic distortion from tetrahedral symmetry will lead to phosphorus hybridization differences in the axial and equatorial esters and changes in the paramagnetic term [Eq. (5)] that are quite likely responsible for the 4- to 6-ppm stereoelectronic difference in epimeric ester $^{31}$P chemical shifts.

This stereoelectronic interpretation probably provides an explanation for the $^{31}$P shift difference between axially and equatorially substituted phosphites **12a** and **12b**, where axial epimers such as **12a** are 2.7–4 ppm upfield of equatorially substituted cyclic phosphites (Haemers et al., 1973).

**12a**

+ 129.3 ppm

**12b**

+ 132.0 ppm

## D. Polar and Resonance Effects on $^{31}$P Chemical Shifts

As discussed earlier, the bond-angle (including stereoelectronic $^{31}$P) effect is only one of at least three factors that influence $^{31}$P chemical shifts. Electronegativity effects and π-electron overlap contribute significantly (if not dominantly) to $^{31}$P shifts in a number of classes of phosphorus compounds. Whereas theoretical calculations have largely supported these conclusions, the equations relating $^{31}$P shift changes to structural and substituent changes are not generally applicable. Also, because $^{31}$P shifts are

**TABLE VII**

Examples of Additive Constants $\sigma_P$

| R | $\sigma_P$ | R | $\sigma_P$ |
|---|---|---|---|
| $CH_3$ | 0 | $C_6H_5CH_2$ | +17 |
| $C_2H_5$ | +14 | $H_2C=CH-CH_2$ | + 9 |
| $i\text{-}C_3H_7$ | +27 | $NCCH_2CH_2$ | +13 |
| $C_6H_5$ | +18 | CN | −24.5 |
| | | $(C_2H_5)_2N$ | − 1 |

influenced by at least these three factors, empirical correlations can only be applied to classes of compounds that are similar in structure. Within this limitation, a number of empirical observations and correlations, however, have been established and have proved useful in predicting $^{31}$P chemical-shift trends.

Grim and co-workers (1967; Grim and McFarlane, 1965) have established correlations for various P(III) and P(IV) compounds. Thus for tertiary phosphines,

$$\delta = -62 + \sum_1^3 \sigma_P$$

for secondary phosphines,

$$\delta = -99 + 1.5 \sum_1^2 \sigma_P$$

and for primary phosphines,

$$\delta = -163.5 + 2.5\sigma_P$$

where some of the additive constants $\sigma_P$ are given in Table VII.

In such composite additive constants, electronegativity, bond angle (including $\alpha$-, $\beta$-, and $\gamma$-effect parameters), and even $\pi$-electron overlap effects are going to be inseparably intermixed. The $\sigma_P$ parameters do not even correlate significantly with Taft or Hammett constants.

Although early attempts to correlate $^{31}$P shifts with various Taft or Hammett substituent constants met with failure, such empirical correlations were later more successful when structural variation in the class of compounds was limited. Thus the $^{31}$P chemical shifts of substituted phenylphosphonic dichlorides correlate reasonably well with either Hammett $\sigma$ or Taft $\sigma^*$ values (Grabiak *et al.*, 1980). Grabiak *et al.* also observed an upfield shift (~ 1.0 ppm) for the *o*-tolylphosphonic dichloride relative to the meta and para isomers. An even larger shielding (~ 5.2 ppm) was observed for the *o*-methoxy substituent and was interpreted in terms of a steric $\gamma$ effect. Interestingly, *shielding* of the phosphorus signal was observed for meta and para electron-withdrawing groups. Normally, meta- and para-substituted electron-withdrawing groups would be expected to increase the positive charge on the atom and lead to its deshielding. (Letcher and Van Wazer, 1966, 1967; Gorenstein and Kar, 1975). An explanation was offered suggesting that electron-withdrawing groups increase the P=O bond order by stabilizing the major resonance structure **I** (Grabiak *et al.*, 1980).

Supposedly, an increase in the $d_\pi - p_\pi$ bond order was responsible for an increase in the *d*-orbital occupation on phosphorus and a shift of the $^{31}$P signal to higher field (Grabiak *et al.*, 1980; Tarasevich *et al.*, 1971; see also Ionin, 1968). This common interpretation of the Letcher and Van Wazer theory, however, is much too simplistic because electron withdrawal from phosphorus can lead to shielding *or* deshielding. In fact, Maier (1974) and Koslov and Gaidamaka (1972) noted that the Letcher and Van Wazer theory requires that increased $\pi$ bonding to the phosphorus should result in greater deshielding. Confusion in some of the literature apparently arises from the lack of separation of the *p*- and *d*-orbital effects in paramagnetic term equations such as Eq. (5). Letcher and Van Wazer assume that *s* and *p* orbitals are involved in the $\sigma$-bonding framework whereas *d* orbitals on phosphorus are involved in $d_\pi - p_\pi$ bonding. Equation (5) represents the paramagnetic contribution owing to *p*-orbital occupation only, and a similar term can be written for the *d* orbitals as well (Jameson and Gutowsky, 1964). The chemical shift is then a sum of the *p*-orbital term ($\delta_p$) and the *d*-orbital term ($\delta_d$):

$$\delta = \delta_0 + \delta_p + \delta_d$$

where $\delta_0$ is a constant determined by the reference standard, and $\delta_p$ and $\delta_d$ have forms given by the paramagnetic shielding terms $\sigma_p$ of Eq. (5). For example, for molecules of structural type MPZ$_3$ (M=O, S, or lone pair), the following equation has been derived (Letcher and Van Wazer, 1967):

$$\delta = -11{,}828.5 + 7940 k_1 \zeta_p + 8.87 k_2 \zeta_d$$

where $\zeta_p$ and $\zeta_d$ are the *p*- and *d*-orbital unbalancing terms, respectively, and $k_1 = 0.972$ and $k_2 = 2.80$ for phosphines (M = lone pair) and $k_1 = k_2 = 1.000$ for M = S or O. The *d*-orbital term $\zeta_d$ is proportional to the number of electrons in the $d_\pi - p_\pi$ bond ($n_\pi$) with $\zeta_d = 16.8\, n_\pi$. The *p*-orbital term

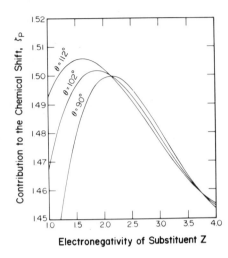

Fig. 7. $p$-Orbital unbalancing term contribution $\zeta_p$ to the theoretical chemical shift for molecules of the form $MP_3$ as a function of the electronegativity of the Z substituent and the Z—P—Z bond angle $\theta$. Derived from Letcher and Van Wazer (1967).

involved in $\sigma$ bonding is a function of the electronegativity of the Z substituent and the Z—P—Z bond angle $\theta$, as shown in Fig. 7, according to the Letcher and Van Wazer theory. The $p$-orbital term reaches a maximum for the substituent electronegativity near the electronegativity of phosphorus. Hence for M and Z that are more electronegative than phosphorus, an increase in electronegativity will *decrease* the $p$-orbital unbalancing term $\zeta_p$ and lead to greater shielding. However, for substituents less electronegative than phosphorus, an increase in electronegativity should *increase the p-orbital unbalancing term and lead to greater deshielding.* Grabiak et al. (1980) and others presumably should argue that electronegativity effects influence and observed $^{31}P$ shift changes through the $p$-orbital unbalancing term rather than the claimed $\pi$-bonding effect. Confusing all of this even more is the realization that $\pi$ bonds can be of the $p_\pi$—$p_\pi$ type, and that actual $d$-orbital occupancy in phosphorus compounds is probably much less than early molecular orbital calculations suggested (Perahia et al., 1975).

Similar Hammett or Taft correlations have been shown for para-substituted triphenylphosphines (although poor), substituted phenylphosphonic difluorides (Szafraniec, 1974; slope $\rho = -5.96$), substituted phenylphosphonic acids (Mitsch et al., 1970), aryl-substituted diethyl phenyl phosphonates (Mitsch et al., 1970), substituted phenylthiophosphonic dichlorides (where resonance and polar contributions are separated; Maier, 1974), and others (Anttimanninen, 1981). In all cases electron-withdrawing substituents lead to increased shielding of phosphorus. If the Letcher and Van Wazer theory is correct (and criticism exists, cf. Kozlov and Gaidamaka, 1972; Ionin, 1968; Murray et al., 1971; Schmidpeter and Schumann, 1970), then an explanation based solely on increased $\pi$-bond order (increased $n_\pi$) and increased shielding with increasing electronegativity must be incorrect.

Substituent effects will thus be a complex interplay of electronegativity, bond angle, and occupation changes of the phosphorus $d$ and $p$ orbitals. [For a more complete discussion of these effects, the reader is encouraged to consult Letcher and Van Wazer (1967).] No simple pattern for $^{31}P$ shift changes with substituent electronegativity or structure change is found.

## E. Extrinsic Effects on $^{31}P$ Chemical Shifts

Environmental effects on $^{31}P$ chemical shifts are generally smaller than the intrinsic effects discussed in previous sections. Lerner and Kearns (1980) have shown that $^{31}P$ shifts of phosphate esters are modestly sensitive to solvent effects (Fig. 8), and Costello et al. (1976) have noted similar sensitivity of $^{31}P$ shifts of orthophosphate, diethyl phosphate, and monoethyl orthophosphate to salt (Fig. 9).

Gorenstein et al. (1976a,b; see also Gorenstein, 1978, 1981) have shown that $^{31}P$ shifts are also sensitive to temperature (Fig. 10), although they have analyzed this effect in terms of the stereoelectronic $^{31}P$ shift effect. Thus the $^{31}P$ shift of five-membered cyclic phosphate 2′,3′-cyclic cytidine monophosphate (cCMP) shows little change with temperature relative to the $D_2O$

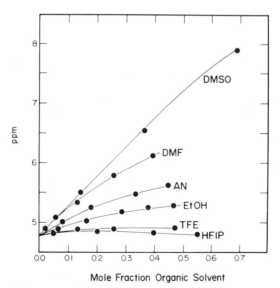

**Fig. 8.** Effect of solvent on $^{31}P$ chemical shift of 3′,5′-cAMP from Lerner and Kearns (1980). Copyright 1980 American Chemical Society.

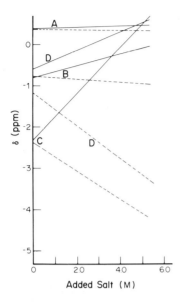

Fig. 9. Effect of salt on $^{31}$P chemical shift of (A) triethyl phosphate, (B) diethyl phosphate, (C) monoethyl phosphate, and (D) orthophosphate. —, Represents added tetramethylammonium chloride; ---, represents added sodium chloride. From Costello et al. (1976). Copyright 1976 American Chemical Society.

lock. The cyclic ester, of course, is relatively rigidly constrained. Apparently, the intrinsic downfield shift at higher temperature owing to increased population of higher vibronic levels of the D$_2$O lock signal and the $^{31}$P signal (see Jameson and Jameson, 1973) of this conformationally restricted phosphate diester are similar. The smaller but real *additional* downfield shift for the simple acyclic diesters, dimethyl and diethyl phosphate, under D$_2$O-lock conditions likely arises from a change in the average conformation of these molecules as a function of temperature. The lowest energy structure is g,g, and thus at low temperature more of the molecules are in the g,g conformation than in either g,t or t,t conformations.

As discussed in further detail by Chen and Cohen (Chapter 8), Gorenstein and Goldfield (Chapter 10), the temperature dependence of the $^{31}$P signals of nucleic acids (Fig. 11) provides further support for the stereoelectronic origin of these $^{31}$P shift effects. Briefly, the $^{31}$P chemical shifts of the double (or triple) helix $^{31}$P signal of a 1:1 complementary base-pairing mixture of polyadenylic acid and oligouridylic acid (Gorenstein, 1978, 1981, 1983; Gorenstein *et al.*, 1981, 1982) or of the numerous $^{31}$P signals of yeast phenylalanine tRNA (see Gorenstein and Goldfield, Chapter 10) undergo a major transition at the melting temperature (~50°C) of the nucleic acids (see also Patel and Canuel, 1979; Shindo *et al.*, 1980). Below this melting transition, the poly(A)·oligo(U) is largely in the double-helix state, with the phosphates largely locked into the helical,−g,−g phosphate ester conformation. At a temperature above the cooperative melting transition, the

# 1. ³¹P Chemical Shifts

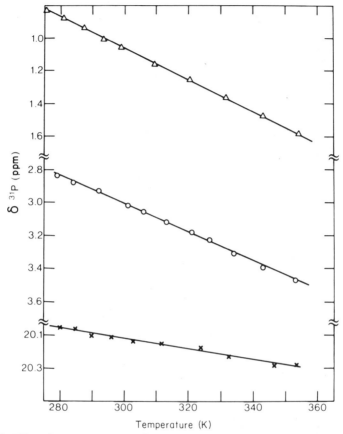

**Fig. 10.** Effect of temperature on the ³¹P chemical shift of dimethyl phosphate (O), diethyl phosphate (△), and 2′,3′-cyclic cytidine monophosphate (x). Partially derived from Gorenstein *et al.* (1976a).

double helix is disrupted and the single-strand nucleic acids are in a random-coil conformation with a mix of various phosphate ester conformations. The ~0.5-ppm downfield shift between 40–60°C with increasing temperature (Fig. 11, dashed curve) is attributed to the decrease in the population of the −g,−g conformation (Gorenstein, 1978, 1981, 1983; Chen and Cohen, Chapter 8; Gorenstein, Chapter 9; Gorenstein and Goldfield, Chapter 10).

Below the melting transition, the tRNA is also largely in its native conformation, greatly constraining the individual phosphate ester conformations and bond angles. Peak M (Fig. 11) of the tRNA also shifts downfield by a total of 0.8–0.9 ppm with increasing temperature. This signal, assigned

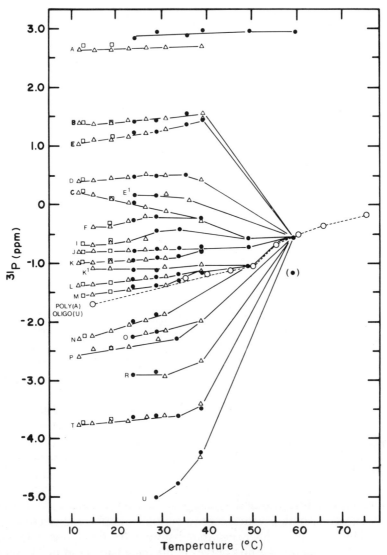

**Fig. 11.** Temperature dependence of the $^{31}$P chemical shift of the double-helix signal of the 1:1 mixture of oligouridylic acid [oligo(U)] and polyadenylic acid [poly(A)] (O, ---) and $^{31}$P signals of yeast phenylalanine-tRNA (—). For poly(A)·oligo(U) sample: pH 7, 20% D$_2$O, 0.2 $M$ NaCl, 10 m$M$ cacodylate, 1 m$M$ EDTA. 24 mg/ml total nucleotide. For tRNA: 0.1 $M$ NaCl, 10 m$M$ cacodylate, 10 m$M$ MgCl$_2$, 1 m$M$ EDTA, 10–20% D$_2$O, pH 7. From Gorenstein, *et al.* (1982). Copyright 1982 American Chemical Society.

to the double-helical phosphates in the stems of the native tRNA, behaves very much like the poly(A)·oligo(U) signal (and for similar reasons; see Gorenstein and Goldfield, Chapter 10). The $^{31}$P chemical shifts of most of the other tRNA signals do not shift very much below the melting transition. Again, this supports the stereoelectronic $^{31}$P shift hypothesis because the tRNA largely retains its native conformation at lower temperature and greatly constrains the phosphate ester conformational freedom. Magnesium ion further stabilizes the tRNA conformation, and as shown in Figs. 1–6 in Gorenstein and Goldfield, (Chapter 10), the $^{31}$P chemical shift for most of the signals are remarkably insensitive to temperature below the melting transition.

Gordon and Quin (1976) have noted that trivalent and tetravalent organophosphorus compounds (e.g., phosphines, phosphites, phosphonates) may show upfield or downfield shifts with increasing temperature. These shift variations may also relate to P—O and P—C torsional-angle changes.

## F. Additional Effects and Conclusions

Under most solvent and salt conditions, the intrinsic (and particularly stereoelectronic) effects appear to dominate $^{31}$P chemical shifts of phosphate esters. As discussed briefly before, this idea is utilized in later chapters to probe the structure of double-helical nucleic acids (Chen and Cohen, Chapter 8), drug–nucleic acid complexes (Gorenstein and Goldfield, Chapter 10), and the transfer ribonucleic acids (Gorenstein, Chapter 9). Other possible factors that could affect $^{31}$P shifts in phosphate esters have been found generally to be relatively unimportant. Thus ring-current effects associated with the bases are expected to have only small (<0.5 ppm) perturbations on the $^{31}$P signals. This diamagnetic contribution ot the $^{31}$P chemical shift influences $^{1}$H and heavy-atom chemical shifts to the same extent and is strongly distance dependent. The phosphorus nucleus is shielded by a tetrahedron of oxygens, and therefore nucleic acid bases can never approach close enough to cause any marked shielding or deshielding.

Small differences in ring-current effects and hydrogen-bonding interactions could be responsible for the 0.4- to 0.6-ppm dispersion in $^{31}$P signals observed in model single- and double-stranded nucleic acids as discussed by Chen and Cohen (Chapter 8) and Gorenstein (Chapter 9). Alternatively, differences in averaged P—O conformational states could also be at least partially responsible for these differences.

Most surprisingly, association of hydrogen-bonding donors often appears to have little effect on the $^{31}$P chemical shift other than that explained by a shift in the p$K$. Secondary ionization of a phosphate monoester does

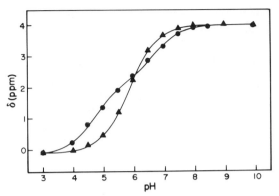

Fig. 12. Phosphorus-31 chemical shift versus pH for 3'-CMP in 1 mM EDTA/H$_2$O. ▲, Represents the shift of the inhibitor free in solution; ●, represents the chemical shift of the bovine pancreatic ribonuclease A complex (1:1), pH 7, 1 mM EDTA, 36.4 MHz. From Gorenstein et al. (1976c). Copyright 1976 American Chemical Society.

produce ~4-ppm downfield shift of the $^{31}$P signal (Cohn and Hughes, 1960; Jones and Katritzky, 1960; Crutchfield et al., 1967; Blumenstein and Raftery, 1972; Sarma and Mynott, 1972; Moon and Richards, 1973; Gorenstein and Wyrwicz, 1973; Haar et al., 1973), but as noted earlier (Gorenstein, 1975), this is possibly attributable to an O—P—O bond-angle effect. Thus the pH dependence of the $^{31}$P signal of pyrimidine nucleotides, both free in solution and when bound to bovine pancreatic ribonuclease A (RNase A), demonstrates this point (Fig. 12) (Harr et al., 1973; Gorenstein and Wyrwicz, 1973; Gorenstein et al., 1976c). The $^{31}$P chemical shift of the free solution, cytidine 3'-monophosphate (3'-CMP), follows a simple titration curve, and the ionization constant derived from the $^{31}$P shift variation agrees with potentiometric titration values.

The $^{31}$P chemical-shift titration curve for the 3'-CMP·RNase A complex, however, cannot be analyzed in terms of a single ionization process. The two inflections in Fig. 12 suggest two ionizations with $pK_1 = 4.7$ and $pK_2 = 6.7$. The 5'-CMP, 2'-CMP, and 3'-UMP complexes of RNase A also exhibit similar biphasic $^{31}$P titration curves (Haar et al., 1973; Gorenstein et al., 1976c).

These results suggest that the nucleotides bind around neutral pH in the dianionic ionization state. Thus the 3'-CMP·RNase A complex is shifted upfield less than 0.3 ppm from the free 3'-CMP between pH 6.5 and 7.5, whereas monoprotonation of the free dianion results in a 4-ppm upfield shift. Furthermore, the addition of the first proton to the nucleotide complex ($pK_2 = 6.0-6.7$) must occur mainly on some site other than the dianionic phosphate because the $^{31}$P signal is shifted upfield by only 1-2 ppm. The addition of a second proton ($pK_1 = 4.0-5.7$) to the complex shifts the $^{31}P$ signal further upfield so that at the lowest pHs, the phosphate finally appears to be in the monoanionic ionization state.

On the basis of X-ray (Richards and Wyckoff, 1971) and $^1H$-NMR (Meadows *et al.*, 1969; Markley, 1975) studies, it is known that the nucleotides are located in a highly basic active site with protonated groups histidine-119, histidine-12, and, probably, lysine-41, quite close to the phosphate. This suggests that $pK_1$ is associated with ionization of the monoanionic inhibitor and $pK_2$ with ionization of a protonated histidine residue that hydrogen bonds to the phosphate.

This highly positive active site, which is capable of perturbing the $pK$ of the phosphate from 6 to 4.7 (or to 4.0 for 2′-CMP), must have one or more hydrogen bonds to the phosphate over the entire pH region. Yet at the pH extrema, little if any perturbation of the $^{31}P$ chemical shift is found. Apparently, the $^{31}P$ chemical shift of the phosphate esters is mostly affected only by the protonation state and *not* by the highly positive local environment of the enzyme.

Additional studies have supported this suggestion, although in some instances binding of an ionizable phosphate to a protein does produce $^{31}P$ shifts that cannot be rationalized in terms only of a shift in the $pK$ of the phosphate (Gorenstein, 1981). Perhaps the most dramatic demonstration of enzyme-associated $^{31}P$ perturbation is the observation that the phosphate covalently bound to alkaline phosphatase (Bock and Sheard, 1975; Hull *et al.*, 1976; Chlebowski *et al.*, 1976) is shifted 6-8 ppm downfield from inorganic phosphate in solution. However, Chlebowski *et al.* (1976) have suggested that this originates from bond-angle strain in the enzyme complex (see also Vogel, Chapter 4).

Evans and Kaplan (1979) have shown that $^{31}P$ chemical shifts in a series of 5′-nucleotides and related phosphate monoesters are only modestly sensitive to intramolecular hydrogen bonding between the base RNH and dianionic phosphate. The 0.4-ppm shielding resulting from this hydrogen-bonding interaction correlates nicely with the population of the g,g conformation about the C-4′—C-5′ bond.

# References

Allen, D. W., and Tebby, J. C. (1970). *J. Chem. Soc. B* pp. 1527-1529.
Anttimanninen, P. (1981). *Acta Chem. Scand., Ser. B* **B35**, 13-18.
Bajwa, G. S., Bentrude, W. G., Pantaleo, N. S., Newton, M. G., and Hargis, J. H. (1979). *J. Am. Chem. Soc.* **101**, 1602-1604.
Bentrude, W. G., and Tan, H. W. (1973). *J. Am. Chem. Soc.* **95**, 4666-4675.

Bertrand, R. D., Verkade, J. G., White, D. W., Gagnaire, G., Robert, J. B., and Verrier, J. (1970). *J. Magn. Reson.* **3**, 494–502.
Blackburn, G. M., Cohen, J. S., and Weatherall, I. (1971). *Tetrahedron* **27**, 2903–2912.
Blumenstein, M., and Raftery, M. A. (1972). *Biochemistry* **11**, 1643–1654.
Bock, J. L., and Sheard, B. (1975). *Biochem. Biophys. Res. Commun.* **66**, 24–30.
Chlebowski, J. F., Armitage, I. M., Tusa, P. P., and Coleman, J. E. (1976). *J. Biol. Chem.* **254**, 1207–1216.
Cohn, M., and Hughes, T. R., Jr. (1960). *J. Biol. Chem.* **237**, 3250–3253.
Costello, A. J. R., Glonek, T., and Van Wazer, J. R. (1976). *J. Inorg. Chem. Soc.* **15**, 972–974.
Crutchfield, M. M., Dungan, C. H., Letcher, L. H., Mark, V., and Van Wazer, J. R. (1967). *Top. Phosphorus Chem.* **5**, 1–487.
Day, R. O., Gorenstein, D. G., and Holmes, R. R. (1983). *Inorg. Chem.* **22**, 2192–2195.
Dennis, E. D. (1968). Ph.D. Thesis, Harvard University, Cambridge, Massachusetts.
Dutasta, J. P., Robert, J. B., and Wiesenfeld, L. (1981). *In* "Phosphorus Chemistry" (L. D. Quin and J. G. Verkade, eds.), pp. 581–584. Am. Chem. Soc., Washington, D.C.
Evans, F. E., and Kaplan, N. O. (1979). *FEBS Lett.* **105**, 11–14.
Gordon, M. D., and Quin, L. D. (1976). *J. Magn, Reson.* **22**, 149–153.
Gorenstein, D. G. (1975). *J. Am. Chem. Soc.* **97**, 898–900.
Gorenstein, D. G. (1977). *J. Am. Chem. Soc.* **99**, 2254–2258.
Gorenstein, D. G. (1978). *Jerusalem Symp. Quantum Chem. Biochem.* **11**, 1–15.
Gorenstein, D. G. (1981). *Annu. Rev. Biophys. Bioeng.* **10**, 355–386.
Gorenstein, D. G. (1983). *Prog. NMR Spectrosc.* (in press).
Gorenstein, D. G., and Kar, D. (1975). *Biochem. Biophys. Res. Commun.* **65**, 1073–1080.
Gorenstein, D. G., and Kar, D. (1977). *J. Am. Chem. Soc.* **99**, 672–677.
Gorenstein, D. G., and Luxon, B. A. (1979) *Biochemistry* **18**, 3796–3804.
Gorenstein, D. G., and Rowell, R. (1979). *J. Am. Chem. Soc.* **101**, 4925–4928.
Gorenstein, D. G., and Rowell, R. (1980). *J. Am. Chem. Soc.* **102**, 5077–5081.
Gorenstein, D. G., and Wyrwicz, A. M. (1973). *Biochem. Biophys. Res. Commun.* **54**, 976–982.
Gorenstein, D. G., Findlay, J. B., Momii, R. K., Luxon, B. A., and Kar, D. (1976a). *Biochemistry.* **15**, 3796–3803.
Gorenstein, D. G., Kar, D., Luxon, B. A., and Momii, R. K. (1976b). *J. Am. Chem. Soc.* **98**, 1668–1673.
Gorenstein, D. G., Wyrwicz, A. M., and Bode, J. (1976c). *J. Am. Chem. Soc.* **98**, 2308–2314.
Gorenstein, D. G., Luxon, B. A., and Findlay, J. B. (1977). *Biochim. Biophys. Acta* **475**, 184–190.
Gorenstein, D. G., Goldfield, E. M., Chen. R., Kovar, K., and Luxon, B. A. (1981). *Biochemistry* **20**, 2141–2150.
Gorenstein, D. G., Luxon, B. A., Goldfield, E. M., Lai, K., and Vegeais, D. (1982). *Biochemistry* **21**, 580–589.
Grabiak, R. C., Miles, J. A., and Schwenzer, G. M. (1980). *Phosphorus Sulfur* **9**, 197–202.
Grant, D. M., and Cheney, B. V. (1967), *J. Am. Chem. Soc.* **89**, 5315–5318.
Grim, S. O., and McFarlane, W. (1965). *Nature (London)* **208**, 995–996.
Grim, S. O., McFarlane, W., and Davidoff, E. F. (1967), *J. Org. Chem.* **32**, 781–784.
Gutowsky, H. S., and McCall, D. W. (1954). *J. Chem. Phys.* **22**, 162–164.
Haar, W., Thompson, J. C., Mauer, W., and Ruterjans, H. (1973). *Eur. J. Biochem.* **40**, 259–266.
Haemers, M. Ottinger, R., Zimmerman, D., and Reisse, J. (1973). *Tetrahedron* **29**, 3539–3545.
Hays, H. R., and Peterson, D. J. (1972). *In* "Organic Phosphorus Compounds" (G. M. Kosolapoff and L. Maier, eds.), Vol. 3, pp. 341–500. Wiley (Interscience), New York.

Hull, W. E., Halford, S. E., Gutfreund, H., and Sykes, B. D. (1976). *Biochemistry* **15**, 1547–1561.
Hutchins, R. O., Maryanoff, B. E., Castillo, M. J., Hargrave, K. D., and McPhail, A. T. (1979). *J. Am. Chem. Soc.* **101**, 1600–1602.
Ionin, B. I. (1968). *Zh. Obshch. Khim.* **38**, 1659–1662.
James, T. L. (1975). "Nuclear Magnetic Resonance in Biochemistry: Principles and Applications." Academic Press, New York.
Jameson, A. K., and Jameson, C. J. (1973). *J. Am. Chem. Soc.* **95**, 8559–8561.
Jameson, C. J., and Gutowsky, H. S. (1964). *J. Chem. Phys.* **40**, 1714–1724.
Jardetzky, O., and Roberts, G. C. K. (1981) "NMR in Molecular Biology." Academic Press, New York.
Jones, R. A. Y., and Katritzky, A. R. (1960). *J. Inorg. Nucl. Chem.* **15**, 193–197.
Kozlov, E. S., and Gaidamaka, S. N. (1972). *Theor. Exp. Chem.* **8**, 351.
Kumamoto, J., Cox, J. R., Jr., and Westheimer, F. H. (1956). *J. Am. Chem. Soc.* **78**, 4858–4860.
Lamb, W. E. (1941). *Phys. Rev.* **60**, 817–819.
Lerner, D. B., and Kearns, D. R. (1980). *J. Am. Chem. Soc.* **102**, 7611–7612.
Letcher, J. H., and Van Wazer, J. R. (1966). *J. Chem. Phys.* **44**, 815–829.
Letcher, J. H., and Van Wazer, J. R. (1967). *Top. Phosphorus Chem.* **5**, 75–266.
McFarlane, H. C. E., and McFarlane, W. (1978). *J. Chem. Soc., Chem. Commun.* pp. 531–532.
Maier, L. (1972). *In* "Organic Phosphorus Compounds" (G. M. Kosolapoff and L. Maier, eds.) Vol. 4, pp. 1–73. Wiley (Interscience), New York.
Maier, L. (1974). *Phosphorus* **4**, 41–47.
Mann, B. E. (1972). *J. Chem. Soc., Perkin Trans.* 2 pp. 30–34.
Markley, J. H. (1975). *Acc. Chem. Res.* **8**, 70–80.
Martin, J., and Robert, J. B. (1981). *Org. Magn. Reson.* **15**, 87–93.
Maryanoff, B. E., Hutchins, R. O., and Maryanoff, C. A. (1979) *Top. Stereochem.* **11**, 187–326.
Meadows, D. H., Roberts, G. C. K., and Jardetzky, O. (1969). *J. Mol. Biol.* **45**, 491–511.
Mills, J. L. (1977). *In* "Homoatomic Rings, Chains and Macromolecules of Main-Group Elements" (A. L. Theinfold, ed.). Am. Elsevier, New York.
Mitsch, C. C., Freedman, L. D., and Moreland, C. G. (1970). *J. Magn. Reson.* **3**, 446–455.
Moon, R. B., and Richards, J. H. (1973). *J. Biol. Chem.* **248**, 7276–7278.
Mosbo, J. A., and Verkade, J. G. (1972). *J. Am. Chem. Soc.* **94**, 8224–8225.
Müller, N., Lauterbur, P. C., and Goldenson, J. (1955). *J. Am. Chem. Soc.* **78**, 3557–3561.
Murray, M., Schmutzler, K. Grundemann, E., and Teichmann, H. (1971). *J. Chem. Soc. B* pp. 1714–1719.
Patel, D. J., and Canuel, L. L. (1979). *Eur. J. Biochem.* **96**, 267–276.
Perahia, D., and Pullman, B. (1976). *Biochim. Biophys. Acta* **435**, 282–289.
Perahia, D., Pullman, A., and Berthod, H. (1975). *Theor. Chim. Acta* **40**, 47–60.
Pople, J. A., Schneider, W. G., and Bernstein, H. J. (1959). "High Resolution Nuclear Magnetic Resonance." McGraw-Hill, New York.
Prado, F. R., Geissner-Prettre, C., Pullman, B., and Daudey, J.-P. (1979). *J. Am. Chem. Soc.* **101**, 1737–1742.
Purdela, D. (1971). *J. Magn. Reson.* **5**, 23–26.
Quin, L. D., and Breen, J. J. (1973). *Org. Magn. Reson.* **5**, 17–19.
Radeglia, R., and Grimmer, A. R. (1982). *Z. Phys. Chem. (Leipzig)* **263**, 204–207.
Rajzmann, M., and Simon, J.-C. (1975). *Org. Magn. Reson.* **7**, 334–338.
Richards, R. F. M., and Wyckoff, H. W. (1971). *In* "The Enzymes" (P. D. Boyer, ed.), 3rd ed., Vol. 4, Chapter 24. Academic Press, New York.

Richman, J. E., and Flay, R. B. (1981). *J. Am. Chem. Soc.* **103**, 5265.
Richman, J. E., Gupta, O. D., and Flay, R. B. (1981). *J. Am. Chem. Soc.* **103**, 1291–1292.
Romers, C., Altona, C., Buys, H. R., and Havinga, E. (1969). *Top. Stereochem.* **4**, 39.
Sarma, R. H., and Mynott, R. J. (1972). *Org. Magn. Reson.* **4**, 577–584.
Schmidpeter, A., and Schumann, K. (1970). *Z. Naturforsch., B: Anorg. Chem., Org. Chem., Biochem., Biophys., Biol.* **25B**, 1364–1370.
Schneider, H.-J., and Weigand, E. F. (1977). *J. Am. Chem. Soc.* **99**, 8362–8363.
Shindo, H., McGhee, J. D., and Cohen, J. S. (1980). *Biopolymers* **19**, 523–537.
Smith, L. R., and Mills, J. L. (1976). *J. Am. Chem. Soc.* **98**, 3852–3857.
Stothers, J. B. (1972). "Carbon-13 NMR Spectroscopy," Chapters 3–5. Academic Press, New York.
Sundaralingam, M. (1969). *Biopolymers* **7**, 821–860.
Szafraniec, L. L. (1974). *Org. Magn. Reson.* **6**, 565–567.
Tarasevich, A. S., and Egorov, Yu. P. (1971). *Theor. Exp. Chem.* **7**, 676–679.
Tolman, C. A. (1977). *Chem. Rev.* **77**, 313–348.
Verkade, J. G. (1974). *Bioinorg. Chem.* **3**, 165–182.
Verkade, J. G. (1976). *Phosphorus Sulfur* **2**, 251–281.
Weller, T., Deininger, D., and Lochman, R. (1981). *Z. Chem.* **21**, 105–106.
Wolff, R., and Radeglia, V. R. (1980). *Z. Phys. Chem. (Leipzig)* **261**, 726–744.

CHAPTER 2

# Phosphorus-31 Spin–Spin Coupling Constants: Principles and Applications

*David G. Gorenstein*
Department of Chemistry
University of Illinois at Chicago
Chicago, Illinois

| | |
|---|---:|
| I. Introduction | 37 |
| II. Directly Bonded Coupling Constants: $^1J_{PX}$ | 38 |
|     A. Theory | 38 |
|     B. Applications | 41 |
| III. Two-Bond Coupling Constants: $^2J_{PX}$ | 42 |
| IV. Three-Bond Coupling Constants: $^3J_{PX}$ | 43 |
|     A. General | 43 |
|     B. Conformational Analysis of Phosphate Esters | 47 |
|     C. Applications to Nucleic Acid Structure | 50 |
|     References | 51 |

## I. Introduction

Scalar spin–spin coupling constants provide important information on the nature of bonding and structure in molecules. Mavel (1966, 1973) has extensively reviewed the theoretical and empirical studies dealing with spin–spin coupling constants in phorphorus compounds. Whereas in Van Wazer's review (Crutchfield *et al.,* 1967) only phosphorus–phosphorus (PP) and phosphorus–hydrogen (PH) coupling constants were reported, in Mavel's later review other heteroatom– ($^{13}C$, $^{19}F$, etc.) phosphorus constants were discussed as well. Note that included in Gorenstein and Shah (Chapter 18) are one-, two-, and three-bond and some long-range phosphorus–hydrogen, phosphorus–phosphorus, and other phosphorus–heteroatom coupling constants. An additional compilation of phosphorus coupling constants may be found in Gorenstein (1983). In this chapter, we review the more recent literature on one-, two-, and three-bond coupling constants to phosphorus and discuss several specific applications to conformational analysis of organophosphorus compounds and nucleic acids.

## II. Directly Bonded Phosphorus Coupling Constants: $^1J_{PX}$

### A. Theory

Theoretical calculations (Cowley and White, 1969; Jameson and Gutowsky, 1969; Gray and Albright, 1977; Safiullin et al., 1975; Albrand et al., 1976a,b; McFarlane and McFarlane, 1978) on the variation of the magnitude of one-bond P–X coupling constants have generally been successful and rationalized in terms of a dominant Fermi-contact term. Even the sign reversal in heteroatom directly bonded phosphorus coupling constants has been rationalized reasonably well.

Because the Fermi-contact spin–spin coupling mechanism involves the electron density at the nucleus (hence the $s$-orbital electron density), an increase in the $s$ character of the P—X bond is generally associated with an increase in the coupling constant:

$$J_{P-X} = \frac{A a_P^2 a_X^2}{1 + S_{P-X}^2} + B \tag{1}$$

where $A$ and $B$ are constants, $a_P^2$ and $a_X^2$ the percent $s$ character on phosphorus and atom X, respectively, and $S_{PX}$ the overlap integral for the P—X bond (Gray and Albright, 1977; Gorenstein, 1977). The percent $s$ character is determined by the hybridization of atoms P and X, and as expected $sp^3$-hybridized atoms often have $^1J_{PX}$ larger than $p^3$ hybridized atoms. Thus $^1J_{PH}$ for phosphonium cations of structure $PH_n R_{4-n}^+$ with $sp^3$ hybridization are $\sim 500$ Hz, whereas $^1J_{PH}$ for phosphines $PH_n R_{3-n}$ with phosphorus hybridization of approximately $p^3$ are smaller, $\sim 200$ Hz. Furthermore, as the electronegativity of atom X increases, the percent $s$ character of the P—X bond increases, and the coupling constant becomes more positive. In many cases, however, these simple concepts fail to rationalize experimental one-bond P–X coupling constants (Table I). It has been shown recently that other spin–spin coupling mechanisms (orbital and spin–dipolar terms) can also contribute significantly to the coupling constant, particularly in $^1J_{PC}$ of trivalent phosphorus compounds (Beer and Grinter, 1978). This would lead to a breakdown in the relationship between percent $s$ and $^1J_{PX}$.

It is not surprising then that in cyclic phosphines (Centofani, 1976) $^1J_{PC}$ does not follow ring size monotonically, as shown in Table II. Note, however, that $^1J_{PC}$ to C-1 of the phenyl ring does become less negative as the ring size increases.

Hybridization (percent $s$) also does not appear to be the major factor in the variation of the $^1J_{PH}$ coupling constants for a series of protonated cyclic phosphite triesters (Vande Griend et al., 1977). The variation of $^1J_{PH}$ does, however, appear to follow changes in the charge on phosphorus. Other

## TABLE I

One-Bond Phosphorus Spin–Spin Coupling Constants $^1J_{PX}$

| Structural class (or structure) | $^1J$ (Hz)[a] | Structural class (or structure) | $^1J$ (Hz)[a] |
|---|---|---|---|
| P(II) | | P(IV) (continued) | |
| $PH_2^-$ | 139 | >P(O)H | 460–1030 |
| P(III) | | =P(M)F | 1000–1400 |
| >PH | 180–225 | (M = O, S) | |
| $PH_3$ | +182 | $P(O)F_3$ | 1055–1080 |
| >P—C | 0–45 | $P(S)F_3$ | 1170–1184 |
| $P(CH_3)_3$ | −13.5 to −14.0 | $F_2P(O)OP(O)F_2$ | −1063 |
| >PF | 820–1450 | >P(S)H | 490–650 |
| $P_2F_4$ | −1194 | P(V) | |
| >P—P< | 100–400 | >PH | 700–1000 |
| $H_2PPH_2$ | −108 | $PF_4H$ | 1075 |
| P(IV) | | >P—F | 530–1100 |
| >PH (+) | 490–600 | $PF_5$ | 938 |
| $PH_4$ (+) | 548 | $P(CH_3)_2F_2$ | 541 |
| =PC | 50–305 | P(VI) | |
| $P(CH_3)_4$ (+) | +56 | $PF_6^-$ | 706 |
| $P(S)(CH_3)_3$ | +56.1 | | |
| $P(O)(OC_2H_5)_2C\equiv CCH_3$ | 304 | | |

[a] Largely derived from Mavel (1973). For structural classes, only absolute value for $J$ is given.

examples may be given to show that $^1J_{PX}$ coupling constants are not always directly related to bond-angle (and percent $s$ character) changes (see Martin and Robert, 1981), although Mann (1972) has made such a claim for tertiary phosphines.

Albright (1976) has carried out extensive theoretical calculations on the Fermi-contact contribution to $^1J_{PC}$ coupling constants using the finite perturbation method (Pople et al., 1968) and CNDO/2 MO wave functions. Good calculated agreement is found with the experimentally observed increase in $^1J_{PC}$ on going from ethyl- to vinyl- to propynylphosphonium cations (Table III). These results suggest that at least in these compounds the Fermi-contact contribution does dominate $^1J_{PC}$. The intuitive belief that $^1J_{PC}$ does increase with percent $s$ in the P—C bonding orbital in going from alkyl to alkenyl to alkynyl ($sp^3 \to sp^2 \to sp$) appears correct. In fact, for tetravalent phosphorus, a very good correlation is found between the phos-

## TABLE II

$^1J_{PC}$ Coupling Constants for Phosphines of Varying Ring Size

| Compound | $^1J_{PC}$ (ring) (Hz) | $^1J_{PC}$ (phenyl) (Hz) |
|---|---|---|
| (triangle)P—C$_6$H$_5$ | −39.7 | −38.7 |
| (square)P—C$_6$H$_5$ | 0.6 | −35.4 |
| (pentagon)P—C$_6$H$_5$ | −14.0 | −25.0 |
| (hexagon)P—C$_6$H$_5$ | −14.8 | −19.1 |
| (heptagon)P—C$_6$H$_5$ | −15.0 | −15.6 |
| $n$-Bu$_2$P—C$_6$H$_5$ | −13.4 | −17.7 |

phorus 3s–carbon 2s bond orders and $^1J_{PC}$ (Albright, 1976). Calculations on trivalent phosphorus compounds are *not* successful, however (see also Lequan *et al.*, 1975) and suggest that the Fermi-contact contribution only dominates tetravalent phosphorus compounds.

For the methyl phosphonium ion, calculated s bond orders and $^1J_{PC}$ coupling constants were found to vary with both P—C bond length and H—P—C bond angle (Fig. 1). A ~10-Hz decrease in $^1J_{PC}$ per 5° bond-angle decrease is found. These results also help explain Gray *et al.*'s (1976) observation that $^1J_{PC}$ coupling constants for cyclic phosphonium ions vary little with ring size. Apparently, bond-length and bond-angle effects on $^1J_{PC}$ oppose each other as the ring size increases.

Stec (1974), Stec *et al.* (1976), Stec and Zielinski (1980), and Gorenstein (1977) have noted that conformationally dependent, directly bonded coupling constants may also be attributed to a stereoelectronic effect as discussed earlier for $^{31}$P chemical shifts (Chapter 1). Thus, in the phosphorinane ring system (**1**) (Gray and Cremer, 1972; Stec, 1974; Gorenstein, 1977; Buchanan and Morin, 1977) and phosphonates (**2**), the $|^1J_{PX}|$ coupling

**1**      **2**

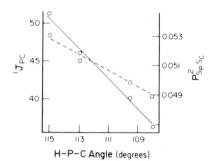

Fig. 1. Calculated $^1J_{PC}$ and $P^2_{S_PS_C}$ versus the H—P—C bond angle. —, $^1J_{PC}$; ---, $P^2_{S_PS_C}$. From Albright (1976). Copyright 1976 John Wiley & Sons, Ltd.

constants are 5–10 Hz smaller in the axially (gauche) substituted rings than in equatorically (trans) substituted rings. A similar difference is found in the pseudoaxially and equatorically substituted four-membered-ring trivalent phosphetan system **3** (Gray and Cremer, 1972). Inexplicably, the $^1J_{PC}$ values

**3**

for the phosphetan oxide **4** (Buchanan and Bowen, 1977) do not obey this simple rule.

$^1J_{CP} = 40.9$ Hz (CH$_3$)     $^1J_{CP} = 36.9$ Hz (CH$_3$)

**4**

Bond-angle changes coupled to the torsional-angle differences should change the hybridization and should change the percent $s$ character of the P—C bond [as shown by Albright's (1976) calculations]. Torsional-angle dependence to $^1J_{PP}$ (Albrand et al., 1976a,b), $^1J_{PN}$ (Gray and Albright, 1977; Gray and Cremer, 1972), $^1J_{PO}$ (Gray and Albright, 1977), $^1J_{PF}$ (Stec, 1974), and $^1J_{PSe}$ (Stec and Zielinski, 1980) have also been noted. In $^1J_{PP}$ adjacent lone-pair orientation rather than hybridization changes have been suggested as an explanation for the directly bonded coupling constant (Albrand et al., 1976a,b) dependence on conformation.

## B. Applications

One-bond P–H coupling constants appear always to be positive and vary from about +120 to +1180 Hz (Mavel, 1973). Other heteroatom one-bond P–X coupling constants vary over a similar wide range and can be either

**TABLE III**

Calculated and Observed $^1J_{PC}$ for Phosphonium Ions[a]

| Compound | $^1J_{PC}$ (Hz) | |
|---|---|---|
| | Experimental[b] | Calculated[c] |
| $R_3\overset{+}{P}CH_2CH_3$ | 41.9 | 51.6 |
| $R_3\overset{+}{P}HC=CH_2$ | 72.3 | 80.3 |
| $R_3\overset{+}{P}C\equiv CCH_3$ | 176.1 | 191.7 |

[a] From Albright (1976).
[b] R = Phenyl.
[c] R = H, calculated with $d$ orbitals on phosphorus.

positive or negative. Typical examples are given in Table I, and more examples may be found in Gorenstein and Shah (Chapter 18) and in Mavel (1973).

## III. Two-Bond Coupling Constants: $^2J_{PX}$

As shown in Table IV, both positive and negative $^2J_{P-X}$ coupling constants have been reported. The stereospecificity of the two-bond coupling constants has proved to be most useful. Thus in the *cis*- and *trans*-phosphorinanes (**5, 6**), the $^2J_{PC}$ constants are 0 and 5.1 Hz in the cis and trans isomers,

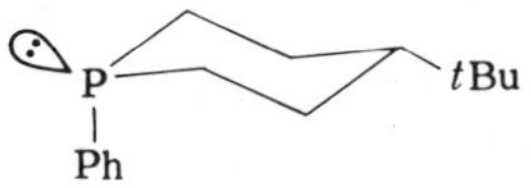

**5** 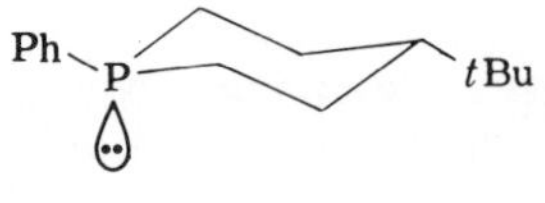  **6**

respectively (MacDonnell *et al.*, 1978). The smaller coupling constant is attributed to the trans orientation of the lone pair to the C-3 or C-5 atoms. Similar results for other phosphorinane ring systems have been noted (Featherman and Quin, 1973; Breen *et al.*, 1975; Venkataramu *et al.*, 1978; Gray and Cremer, 1972).

The $^2J_{PCH}$ and $^2J_{PCF}$ constants also are stereospecific and a Karplus-like dihedral dependence to the two-bond coupling constant (H or F)—C—P—X (X = lone pair or heteroatom) has been described (Mavel, 1973; McFarlane and Nash, 1969)

**TABLE IV**

Two-Bond Phosphorus Spin–Spin Coupling Constants $^2J_{PX}$

| Structural class (or structure) | $^2J$ (Hz)[a] | Structural class (or structure) | $^2J$ (Hz)[a] |
|---|---|---|---|
| P(III) | | P(IV) (continued) | |
| ⟩PCH | 0–18 | =⁺PCH | 12–18 |
| P(CH$_3$)$_3$ | +2.7 | ⁺P(CH$_3$)$_4$ | −14.6 |
| P(Cl)$_2$CH$_3$ | 17.5 | =⁺PCC | 0–40 |
| ⟩PCF | 40–149 | ⁺P(C$_2$H$_5$)$_4$ | −4.3 |
| P(CF$_3$)$_3$ | 85.5 | ⟩POP | 0–23 |
| ⟩PNC | | F$_2$(O)POP(O)F$_2$ | 0–4 |
| P[N(CH$_3$)$_2$]$_3$ | 17 | ⟩POC | ~6 |
| ⟩PNH | 13–28 | P(O)(OCH$_3$)$_3$ | −5.8 |
| ⟩PCC | 12–20 | P(S)OCH$_3$)$_3$ | −5.6 |
| P(C$_2$H$_5$)$_3$ | +14.1 | =PSC | |
| P(C$_6$H$_5$)$_3$ | 19.6 | P(S)(SCH$_3$)$_3$ | <10 |
| ⟩POC | 10–12 | P(V) | |
| P(OCH$_3$)$_3$ | +10.0 | ⟩PCH | 10–18 |
| ⟩PXP | 70–90 | P(CH$_3$)$_3$F$_2$ | 17.2 |
| (X = S, C) | | ⟩PCF | 124–193 |
| P(IV) | | P(CF$_3$)Cl$_4$ | 154 |
| ⟩P(O)CH | 7–30 | P(VI) | |
| P(O)(CH$_3$)$_3$ | −12.8, −13.4 | =PCF | 130–160 |
| ⟩P(S)CH | 11–15 | P(CF$_3$)F$_5^-$ | 145 |
| P(S)(CH$_3$)$_3$ | −13.0 | | |
| ⟩P(O)CF | 100–130 | | |
| P(O)(CF$_3$)$_3$ | 113.4 | | |

[a] Largely derived from Mavel (1973). For structural classes, only absolute value for $J$ is given.

## IV. Three-Bond Coupling Constants: $^3J_{PX}$

### A. General

Representative values for $^3J_{PX}$ coupling constants through intervening C, N, O, or other heteroatoms are presented in Table V. As might be expected from the dihedral-angle dependence of vicinal $^3J_{HCCH}$ coupling constants,

**TABLE V**

Three-Bond Phosphorus Spin–Spin Coupling Constants $^3J_{PX}$

| Structural class (or structure) | $^3J$ (Hz)[a] | Structural class (or structure) | $^3J$ (Hz)[a] |
|---|---|---|---|
| P(III) | | P(IV) (continued) | |
| ⟩POCH | 0–15 | ⟩P(O)OCH | 0–13 |
| P(OCH$_3$)$_3$ | 10.8–11.8 | P(O)(OCH$_3$)$_3$ | 10.2–11.4 |
| ⟩PCCH | 10–16 | ⟩P(M)CCH | 14–25 |
| P(C$_2$H$_5$)$_3$ | +13.7 | (M = O, S) | |
| ⟩PCCC | | P(O)(C$_2$H$_5$)$_3$ | +18 |
|  |  | P(S)($t$–C$_4$H$_9$)$_3$ | +14 |
| P(C$_4$H$_9$)$_3$ | 12.5 | ⟩P(S)OCH | 10–14 |
| ⟩PSCH | 2–20 | P(S)(OCH$_3$)$_3$ | 13.4 |
| P(SCH$_3$)$_3$ | 9.8 | ⟩POCC | |
| ⟩PNCH | 3–14 | P(O)(OC$_2$H$_5$)$_3$ | +6.8 |
| P[N(CH$_3$)$_2$]$_3$ | 8.8–9.0 | P(S)(OC$_2$H$_5$)$_3$ | +6.4 |
| P(IV) | | P(V) | |
| ⟩P$^+$OCH | 7–11 | ⟩PCCH | 20–27 |
| P$^+$(OCH$_3$)$_4$ | 11.2 | P(C$_2$H$_5$)F$_3$H | 26.4 |
| ⟩P$^+$CCH | 15–22 | ⟩PSCH | 13–25 |
| P$^+$(C$_2$H$_5$)$_4$ | +18.1 | P(SCH$_3$)F$_4$ | 25.2 |
| ⟩PSCH | 16–20 | ⟩POCH | 12–17 |
| P$^+$(SCH$_3$)(NH$_2$)$_3$ | 16.2 | P(OCH$_3$)$_5$ | 12 |
| P(O)(SCH$_3$)$_3$ | 15.1 | ⟩PNCH | 2–15 |
| ⟩PNCH | 9–17 |  |  |
| P$^+$[N(CH$_3$)$_2$]$_3$Cl | 13.0 | P[N(CH$_3$)$_2$]$_3$F$_2$ | 10.6 |
| P(O)[N(CH$_3$)$_2$]$_3$ | 9.4–9.5 |  |  |
| P(S)[N(CH$_3$)$_2$]$_3$ | 11.0–11.3 |  |  |

[a] Largely derived from Mavel (1973). For structural classes, only absolute value for $J$ is given.

similar Karplus-like curves have been established for vicinal $^3J_{POCH}$ coupling (Figs. 2 and 3; Mavel, 1973; Kainosho and Nakamura, 1969; Kung et al., 1977), $^3J_{PCCH}$ (Fig. 4; Bothner-by and Cox, 1969; Benezra, 1973), and $^3J_{PCCC}$ (Figs. 5 and 6) (Quin et al., 1980; Duncan and Gallagher, 1981). The curves may be fitted to the general Karplus equation

$$J(\phi) = A \cos^2 \phi + B \cos \phi + C \qquad (2)$$

2. $^{31}$P Spin–Spin Coupling Constants

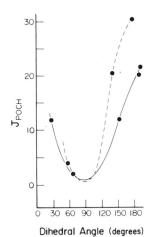

**Fig. 2.** $^3J_{\text{POCH}}$ versus phosphate ester POCH dihedral angles; values derived from cyclic 2′,5′-arabinosylcytidine (---) or various unstrained phosphates (—). From Kung et al. (1977). Copyright 1977 American Chemical Society.

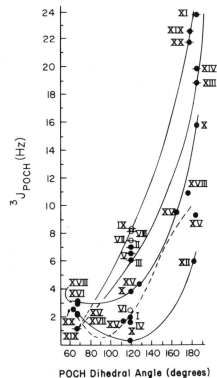

**Fig. 3.** $^3J_{\text{POCH}}$ versus POCH dihedral angles in phosphites (○, ●) and phosphates and thiophosphates (⊖, ⦁). See White and Verkade (1970) for structures. Dashed curve from Kainosho and Nakamura (1969).

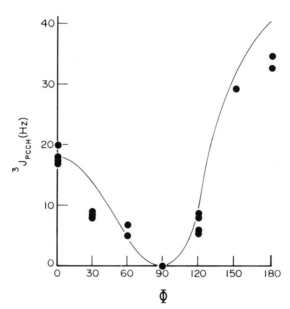

**Fig. 4.** $^3J_{PCCH}$ versus PCCH dihedral angles in phosphonates. From Benezra (1973). Copyright 1973 American Chemical Society.

where $\phi$ is the dihedral angle and $A$, $B$, and $C$ are constants for the particular molecular framework. Caution is recommended when attempting to apply these Karplus equations and curves to classes of phosphorus compounds that have not been used in establishing these relationships (Guimaraes *et al.*, 1978; White and Verkade, 1970; Buchanan and Bowen, 1977; Buchanan and Benezra, 1976) because separate correlations and values for the constants $A$, $B$, and $C$ in Eq. (2) likely exist for each structural class (see Figs. 2

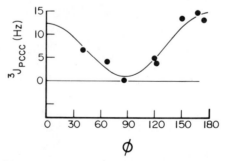

**Fig. 5.** $^3J_{PCCC}$ versus PCCC dihedral angle in RP(S)(CH$_3$)$_2$ compounds. From Quin *et al.* (1980). Copyright 1980 American Chemical Society.

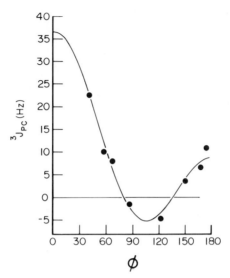

**Fig. 6.** $^3J_{PC}$ versus PCCC dihedral angle in $RP(CH_3)_2$ compounds From Quin et al. (1980). Copyright 1980 American Chemical Society.

and 3). A dihedral-angle dependence of vicinal PSCH coupling constants also likely exists since there is a marked stereospecificity to $^3J_{PSCH}$ as observed in the 1,3,2-dithiaphosphorinane (7). Axial $^3J_{PSCH_a}$ (dihedral angle

~60°) is 0.5 Hz, whereas equatorial $^3J_{POCH_e}$ (dihedral angle ~180°) is 23.5 Hz (Martin et al., 1976).

In all cases, a minimum in the vicinal coupling constant versus dihedral-angle plot is found around 90°. Semiempirical MO (Giessner-Prettre and Pullman, 1974; Albright, 1976) and *ab initio* MO (Giessner-Prettre and Pullman, 1978) calculations have approximately reproduced the torsional-angle dependence to the PC and PH three-bond coupling constants.

## B. Conformational Analysis of Phosphate Esters

As an example of the utility of $^{31}P$ coupling constants, their application to the conformational analysis of epimeric pairs of 2-aryloxy-2-oxy-1,3,2-dioxaphosphorinanes **8a,b–11a,b** and 1,3,2-oxazaphosphorinanes **12a,b** is described (Gorenstein and Rowell, 1979; Gorenstein et al., 1980):

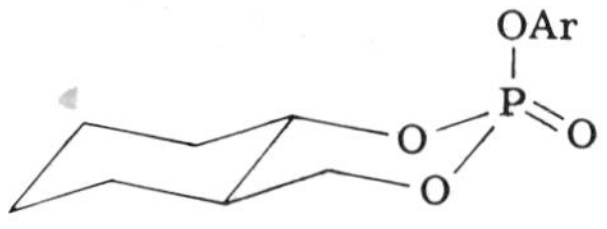

**8a-11a**

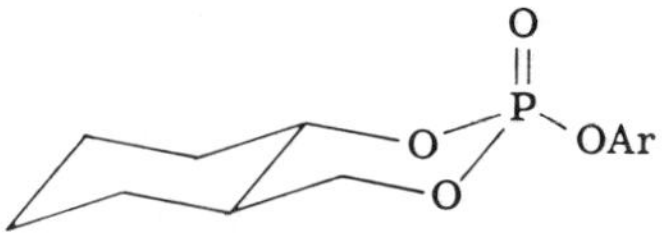

**8b-11b**

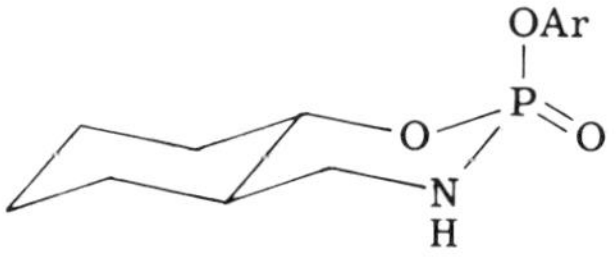

**12a**

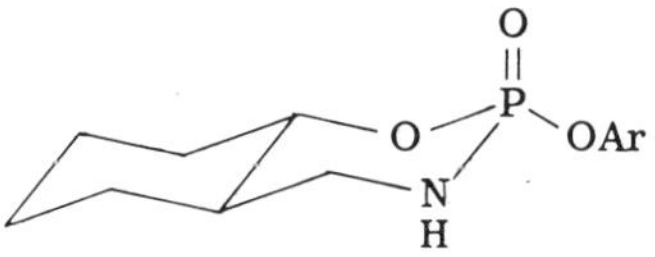

**12b**

Coupled $^{31}$P- and $^1$H-NMR spectra were simulated using a Laocoön-type spectral-simulation program; representative $^{31}$P and $^1$H coupling constants derived from this analysis are shown in Table VI. The three-bond coupling constants (particularly $^3J_{2P} \sim 24.5$ Hz) for all aryl esters **8a–11a** (**8**, Ar = *p*-methoxyphenyl; **9**, Ar = phenyl; **10**, Ar = *p*-nitrophenyl; **11**, Ar = 2,4-dinitrophenyl; **12**, Ar = *p*-nitrophenyl) are consistent with an axial-ester chair

TABLE VI

Selected $^1$H- and $^{31}$P-NMR Spectral Parameters for **8a,b–12a,b**

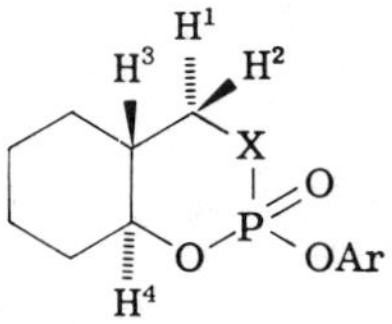

| Compound | $\delta^{31}$P Chemical shift (ppm) | Coupling constants (Hz) | | | | | |
|---|---|---|---|---|---|---|---|
| | | $J_{13}$ | $J_{23}$ | $J_{1P}$ | $J_{2P}$ | $J_{3P}$ | $J_{4P}$ |
| 8a | −12.56 | 11.1 | 4.6 | 0 | 24.4 | 0.0 | 1.0 |
| 9a | −13.03 | 11.1 | 4.6 | 0 | 24.4 | 0.0 | 1.0 |
| 10a | −13.89 | 11.4 | 4.4 | 1.0 | 24.6 | 0.0 | ~0 |
| 11a | −14.93 | 11.0 | 4.6 | 0.5 | 24.9 | 0.0 | ~0 |
| 12a | −2.59 | 10.7 | 4.4 | 2.0 | 27.9 | 0.0 | ~0 |
| 8b | −10.20 | 11.0 | 5.0 | 5.5 | 18.5 | 0.0 | 2.0 |
| 9b | −10.65 | 11.0 | 5.0 | 5.5 | 18.5 | 0.0 | 2.0 |
| 10b | −11.76 | 12.5 | 5.0 | 5.5 | 18.0 | 0.0 | 2.0 |
| 11b | −14.24 | 10.5 | 5.2 | 10.9 | 11.4 | 0.0 | ~0 |
| 12b | −2.84 | 11.1 | 5.5 | 13.6 | 8.8 | 0.0 | 11.5 |

conformation with a trans dihedral angle for $H^2$—C—O—P. The $^3J_{1P}$ and $^3J_{4P}$ coupling constants ($\sim 0$) indicate a gauche conformation of $H^1$ and $H^4$ relative to phosphorus, confirming the chair conformation assigned to these axial esters (Gorenstein *et al.*, 1980).

In contrast to the doublets found in the $^{31}$P proton-coupled spectra of **8a–12a** (due to the single large $^3J_{2P}$), the $^{31}$P spectra of epimers **8b–11b** appear as triplets (although the intensity of the peaks are not in the 1:2:1 ratio). As can be seen from Table VI for **11b**, $J_{1P}$ (10.9) and $J_{2P}$ (11.4) are quite similar and consistent with a twist-boat conformation **13** (dihedral

(X = O, NH)

**13**

angle $\sim 120°$ between phosphorus and $H^1$ and $H^2$). A slightly different twist form was suggested for **12b**, whereas compounds **8b–10b**, which have intermediate $^3J_{1P}$ and $^3J_{2P}$ coupling constants, are mixtures of chair and twist-boat conformations. The percent twist form was calculated from

$$P_{TB} + P_C = 1 \tag{3}$$

and

$$J_{obsd} = J^{2P}_{TB}(P_{TB}) + J^{2P}_C(P_C) \tag{4}$$

where $J_{obsd}$ is the observed coupling constant for $H^2$ and phosphorus in Table VI; $J^{2P}_C$ and $J^{2P}_{TB}$ are the $H^2$–phosphorus coupling constants for pure chair and twist-boat, respectively; and $P_{TB}$ and $P_C$ are fractions of twist-boat and chair conformation, respectively. The $J^{2P}_C$ value for the axial ester (assumed 100% chair) was assumed to be the same for the chair equatorial ester. The value of 11.4 Hz is given to $J^{2P}_{TB}$ on $J_{2P}$ (11.4 Hz) for **11b** and $J_{4P}$ (11.5 Hz) for **12b**, which are both 100% twist-boat. The population of the twist-boat conformer for **8b–10b** was estimated to be between 50 and 80% by this analysis (Gorenstein *et al.*, 1980).

The $^{13}$C-NMR spectra were also consistent with this conformational analysis (Table VII). In flipping from a chair to a boat, the C-5—C-4—O—P torsional angle changes from trans to gauche. In **8a–11a**, the carbon–phosphorus coupling constant $^3J_{5P}$ is 8.8–9.8 Hz (trans), whereas in **11b**, $^3J_{5P}$ is 6.1 Hz (gauche). In **8b–11b**, $^3J_{5P}$ is intermediate (7.8–8.5 Hz), consistent with the rapid equilibration between chair and boat conformations.

Three-bond coupling to C-3 (via C-3—C-4—O—P or C-3—

**TABLE VII**

Selected $^{13}C-^{31}P$ Coupling Constants

|  | $\delta^{13}C-^{31}P$ Coupling constants (Hz) | | | | | | | | | |
|---|---|---|---|---|---|---|---|---|---|---|
| Carbon | 8a | 9a | 10a | 11a | 12a | 8b | 9b | 10b | 11b | 12b |
| 6 | 2.44 | 2.44 | 2.65 | 2.44 | 2.44 | 2.44 | 2.45 | 1.95 | 2.45 | 2.92 |
| 5 | 9.76 | 9.76 | 8.84 | 9.77 | 8.55 | 8.54 | 8.54 | 7.82 | 6.10 | 6.84 |
| 3 | 4.98 | 6.11 | 6.20 | 6.10 | 6.10 | 7.32 | 7.32 | 8.78 | 12.21 | 12.70 |
| 7 | 7.33 | 7.32 | 7.96 | 7.32 | 2.44 | 6.10 | 6.10 | 6.84 | 7.32 | 0.0 |
| 4 | 7.32 | 7.32 | 7.07 | 7.33 | 7.33 | 4.88 | 4.89 | 5.86 | 7.33 | 7.82 |

C-7—O—P) is small (5–6.2 Hz) in **8a–11a** because in the chair conformation C-3—C-4—O—P and C-3—C—O—P torsional angles are gauche (note the similarity between $^3J_{5P}$ in boat **11b** and $^3J_{3P}$ in chair **8a–11a**: ~6 Hz). In one of the twist-boat forms, C-3—C-4—O—P or C-3—C-7—O—P torsional angles are more trans-like, and in **11b**, $^3J_{3P}$ is very large (12.2 Hz). In **8b–11b**, $^3J_{3P}$ is 7.3–8.8 Hz, indicating both trans and gauche coupling patterns. Note also the long-range four-bond $^{13}C-^{31}P$ coupling constant to C-6 (Table VII).

## C. Applications to Nucleic Acid Structure

The Karplus-like relationship between HCOP and CCOP dihedral angles and $^3J_{HP}$ and $^3J_{CP}$ three-bond coupling constants, respectively, has been used to determine the conformation about the ribose–phosphate backbone of nucleic acids in solution. Torsional angles about both the C-3'—O-3' and C-5'—O-5' bonds in 3',5'-phosphodiester linkages have been determined from the coupled $^1H$- and $^{31}P$-NMR spectra (see Tsuboi et al., 1969; Jardetzky and Roberts, 1981; Cheng and Sarma, 1977; Hruska et al., 1970; Altona et al., 1976; Neumann et al., 1979; Fang et al., 1971; Kondo and Danyluk, 1976; Schleich et al., 1976; Cozzone and Jardetzky, 1976). Conformations for mono-, di-, and trinucleotides and polynucleotides have been reported.

Within the limitations just described for the general application of the Karplus relationship, the best values for the nucleotide coupling constants appear to be $^3J_{HP}$ (trans) = 23 Hz, $^3J_{HP}$ (gauche) = 2.1 Hz (Lee and Sarma, 1976), $^3J_{CP}$ (trans) = 10 Hz, and $^3J_{CP}$ (gauche) = 2.5 Hz (Schleich et al., 1976). Because of the conformational flexibility about the C—O bonds and the large number of conformers that may be significantly populated, analysis of a unique structure(s) from the coupling constant data is potentially problematical. Generally, only a few conformations are assumed to be important, and the observed coupling constants are analyzed to yield conformational populations. Thus the $J_{3'P}$ coupling constants for adenylyl-3',5'-adenosine (ApA) is $\sim 8.0$ Hz (Tsuboi et al., 1969; Lee et al., 1974). Assuming that *only* the trans and gauche conformations are allowed, this result suggests that the gauche conformation is preferred. However, it is generally not possible to distinguish between the $-g$ or $+g$ conformations.

Coupling constants to both 5' protons are analyzed in order to determine conformations about the C-5'—O bond. Again, with small H-5'-P and H-5"-P coupling constants (3–7 Hz), the gauche conformation appears to be preferred. The observed similarity in the two C-5' proton–phosphorus coupling constants suggests that the favored conformation has the phosphorus trans to the C-4' atom ($-g$ to one of the C-5' protons and $+g$ to the other):

```
            P
            |
    H5'   / \   H5"
         ( C5' )
            |
           C4'
```

The fractional population $P$ of this conformation may be estimated from the coupling constant sum $\Sigma = J_{5'P} + J_{5''P}$ and (Wood et al., 1974; Jardetzky and Roberts, 1981)

$$P = \frac{J_t + J_g - \Sigma}{J_t - J_g}$$

For ApA with $J_{5'P}$ and $J_{5''P}$ both 4.3 Hz, it would appear that this gauche conformer population is $\sim 80\%$ (Lee et al., 1974). For the deoxy-ApA, the gauche conformer population is even higher (94%).

Applications such as these to the elucidation of the conformation of oligonucleotides generally have been quite successful and serve as some of the most important structural probes of the solution conformation of these molecules.

## References

Albrand, J.-P., Robert, J.-B., and Goldwhite, H. (1976a). *Tetrahedron Lett.* **12**, 949–952.
Albrand, J.-P., Faucher, H., Cagnaire, D., and Robert, J.-B. (1976b). *Chem. Phys. Lett.* **38**, 521–523.

Albright, T. A. (1976). *Org. Magn. Reson.* **8**, 489–499.
Altona, C., Van Boom, J. H., and Haasnoot, C. A. G. (1976). *Eur. J. Biochem.* **71**, 557–562.
Beer, M. D., and Grinter, R. (1978). *J. Magn. Reson.* **31**, 187–193.
Benezra, C. (1973). *J. Am. Chem. Soc.* **95**, 6890–6894.
Bothner-by, A. A., and Cox, R. H. (1969). *J. Phys. Chem.* **73**, 1830–1834.
Breen, J. J., Lee, S. O., and Quin, L. D. (1975). *J. Org. Chem.* **40**, 2245–2248.
Buchanan, G. W., and Benezra, C. (1976). *Can. J. Chem.* **54**, 231.
Buchanan, G. W., and Bowen, J. H. (1977). *Can. J. Chem.* **55**, 604.
Buchanan, G. W., and Morin, F. G. (1977). *Can. J. Chem.* **55**, 2885–2892.
Centofanti, L. F. (1976). *Inorg. Chem.* **12**, 1131–1133.
Cheng, D. M., and Sarma, R. H. (1977). *J. Am. Chem. Soc.* **99**, 7333–7348.
Cowley, A. H., and White, W. D. (1969). *J. Am. Chem. Soc.* **91**, 1913–1917, 1917–1922.
Cozzone, P., and Jardetzky, O. (1976). *Biochemistry* **15**, 4853–4859.
Crutchfield, M. M., Dungan, C. H., Letcher, L. H., Mark, V., and Van Wazer, J. R. (1967). *Top. Phosphorus Chem.* **5**, 1–487.
Duncan, M., and Gallagher, M. J. (1981). *Org. Magn. Reson.* **15**, 37–42.
Fang, K. N., Kondo, N. S., Miller, P. S., and Ts'o, P. O. P. (1971). *J. Am. Chem. Soc.* **93**, 6647–6656.
Featherman, S. I., and Quin, L. D. (1973). *Tetrahedron Lett.* pp. 1955–1958.
Giessner-Prettre, C., and Pullman, B. (1974). *J. Theor. Biol.* **48**, 425–443.
Giessner-Prettre, C., and Pullman, B. (1978). *J. Theor. Biol.* **72**, 751–756.
Gorenstein, D. G. (1977). *J. Am. Chem. Soc.* **99**, 2254–2258.
Gorenstein, D. G. (1983). *Prog. NMR Spectrosc.* (in press).
Gorenstein, D. G., and Rowell, R. (1979). *J. Am. Chem. Soc.* **101**, 4925–4928.
Gorenstein, D. G., Rowell, R., and Findlay, J. (1980). *J. Am. Chem. Soc.* **102**, 5077–5081.
Gray, G. A., and Albright, T. A. (1977). *J. Am. Chem. Soc.* **99**, 3243–3250.
Gray, G. A., and Cremer, S. E. (1972). *J. Org. Chem.* **37**, 3470–3475.
Gray, G. A., Cremer, S. E., and Marsi, K. L. (1976). *J. Am. Chem. Soc.* **98**, 2109–2118.
Guimaraes, A. C., Robert, J.-B., Taieb, C., and Tabony, J. (1978). *Org. Magn. Reson.* **11**, 411–417.
Hruska, F. E., Grey, A. A., and Smith, I. C. P. (1970). *J. Am. Chem. Soc.* **92**, 214–215.
Jameson, C. J., and Gutowsky, H. S. (1969). *J. Chem. Phys.* **51**, 2790–2803.
Jardetzky, O., and Roberts, G. C. K. (1981). "NMR in Molecular Biology," Chapter 6. Academic Press, New York.
Kainosho, M., and Nakamura, A. (1969). *Tetrahedron* **25**, 4071–4081.
Kondo, N. S., and Danyluk, S. S. (1976). *Biochemistry* **15**, 756–768.
Kung, W.-J., Marsh, R. E., and Kainosho, M. (1977). *J. Am. Chem. Soc.* **99**, 5471–5477.
Lee, C.-H., and Sarma, R. H. (1976). *J. Am. Chem. Soc.* **98**, 3541–3548.
Lee, C.-H., Evans, F. E., and Sarma, R. H. (1974). *FEBS Lett.* **51**, 73–79.
LeQuan, R. M., Pouet, M. J., and Simionnin, M. P. (1975). *Org. Magn. Reson.* **7**, 392–400.
MacDonnell, G. D., Berlin, K. D., Baker, J. R., Ealick, S. E., Van der Helm, D., and Marsi, K. L. (1978). *J. Am. Chem. Soc.* **100**, 4535–4540.
McFarlane, H. C. E., and McFarlane, W. (1978). *J. Chem. Soc., Chem. Commun.* pp. 531–532.
McFarlane, W., and Nash, J. A. (1969). *J. Chem. Soc., Chem. Commun.* pp. 127–128.
Mann, B. E. (1972). *J. Chem. Soc., Perkin Trans.* **2**, 30–34.
Martin, J., and Robert, J. B. (1981). *Org. Magn. Reson.* **15**, 87.
Martin, J., Robert, J. B., and Taieb, C. (1976). *J. Phys. Chem.* **80**, 2417–2421.
Mavel, G. (1966). *Prog. NMR Spectrosc.* **1**, 251–373.
Mavel, G. (1973). *Annu. Rep. NMR Spectrosc.* **5b**, 1–350.

Neumann, J.-M., Guschlbauer, W., and Tran-Dinh, S. (1979). *Eur. J. Biochem.* **100,** 141–148.
Pople, J. A., McIver, J. W., Jr., and Ostlund, N. S. (1968). *J. Chem. Phys.* **49,** 2960–2964, 2965–2970.
Quin, L. D., Gallagher, M. J., Cunkle, G. T., and Chesnut, D. B. (1980). *J. Am. Chem. Soc.* **102,** 3136–3143.
Safiullin, R. K., Aminova, R. M., and Samitov, Yu. Yu. (1975). *Zh. Strukt. Khim.* **16,** 42.
Schleich, T., Cross, B. P., and Smith, I. C. P. (1976). *Nucleic Acid Res.* **3,** 355–370.
Stec, W. J. (1974). *Z. Naturforsch. B: Anorg. Chem., Org. Chem.* **29B,** 109.
Stec, W. J., and Zielinski, W. S. (1980). *Tetrahedron Lett.* **21,** 1361–1364.
Stec, W. J., Kinas, R., and Okruszek, A. (1976). *Z. Naturforsch., B: Anorg. Chem., Org. Chem.* **31B,** 393–395.
Tsuboi, M., Takahiski, S., Kyogoku, Y., Hayatsu, H., Ukita, T., and Kainosho, M. (1969). *Science* **166,** 1504.
Vande Griend, L. J., Verkade, J. G., Pennings, J. F. M., and Buck, H. M. (1977). *J. Am. Chem. Soc.* **99,** 2459–2463.
Venkataramu, S., Berlin, K. D., Ealick, S. E., Baker, J. R., Nichols, S., and Van der Helm, D. (1978). *Phosphorus Sulfur* **4,** 133.
White, D. W., and Verkade, J. G. (1970). *J. Magn. Reson.* **3,** 111–116.
Wood, D. J., Hruska, F. E., and Ogilvie, K. K. (1974). *Can. J. Chem.* **52,** 3353–3363.

# PART 2

*Theory and Application of Phosphorus-31 NMR to Biochemistry*

CHAPTER 3

# Phosphorus-31 NMR of Enzyme Complexes

### B. D. Nageswara Rao
Department of Physics
Indiana University–Purdue University at Indianapolis
Indianapolis, Indiana

| | |
|---|---:|
| I. Introduction | 57 |
| II. Theoretical and Experimental Considerations | 59 |
|    A. Theoretical | 59 |
|    B. Experimental | 67 |
| III. Experimental Studies of Specific Enzyme Systems | 69 |
|    A. Enzyme Complexes of Substrates, Coenzymes, and Inhibitors | 69 |
|    B. Enzyme-Bound Equilibrium Mixtures | 87 |
| IV. Computer Calculation of NMR Line Shapes under the Influence of Exchange | 96 |
|    A. Arginine Kinase and Creatine Kinase | 97 |
|    B. Variation of Line Shapes with Exchange Rates | 98 |
|    C. Adenylate Kinase | 99 |
| V. Conclusion | 100 |
|    References | 102 |

## I. Introduction

In this chapter studies in which the NMR spectra of $^{31}$P nuclei contained in substrate molecules noncovalently bound to enzymes are considered. A variety of enzymes in key biochemical pathways use compounds containing phosphate groups, such as nucleotides, as substrates or cofactors, and the phosphate groups are often the participant moieties that become rearranged or transferred in a number of enzymatic reactions. The interaction of the substrate with the enzyme can therefore be studied close to the point of enzymatic action by observing the $^{31}$P NMR of the substrate in the presence of the enzyme. The effects produced by the enzyme on the substrate are studied in a straightforward manner by observing the substrate in the presence of enzyme concentrations sufficiently in excess of that of the substrate, so that on the basis of known dissociation constants, the substrate will be found predominantly in its enzyme-bound complexes. It may not be possible to achieve this condition in some cases because of limitations on availability or solubility of the enzyme or on the lowest phosphate concentrations that can be detected in a reasonable time of signal averaging, or because of unfavorable dissociation constants of substrates from their enzyme-bound complexes or a combination of all these factors. However, in cases where such experiments are feasible, the information obtained by $^{31}$P NMR may well be incisive and significant in understanding the mechanism of enzymatic action. In addition to some natural properties that the $^{31}$P nucleus has such as 100% abundance, spin $\frac{1}{2}$, and appreciable sensitivity, for the purpose of observing it by NMR spectroscopy, $^{31}$P NMR has other features attractive for studying enzyme reactions: (1) the enzyme itself produces no background signals;[1] (2) there are usually a limited number of $^{31}$P-containing moieties in the substrates, and these give rise to NMR signals that can be readily assigned to specific groups in a straightforward manner; and (3) the NMR parameters of these resonances (chemical shifts, spin–spin coupling constants, and factors that determine the line shape) often exhibit a measurable sensitivity to pH and metal chelation. The sensitivity of $^{31}$P-NMR signals to metal chelation is especially useful because a large number of enzymes of interest, in the present context, require cations for activation.

The evolution of modern high-frequency, high-resolution Fourier-transform NMR technology provided considerable scope to exploit the advantages offered by $^{31}$P NMR. A large body of research literature has accumulated even in the specific area of studying enzyme systems. [For previous reviews on the subject, see Cohn and Nageswara Rao (1979) and Nageswara Rao (1983).] This chapter is limited to the studies of diamagnetic[2] (noncovalently bound) enzyme complexes in which the phosphate groups are not modified either by oxygen isotopic labeling or by replacement of oxygen atoms by sulfur.[3] The basic theoretical considerations relevant to the analysis and interpretation of the NMR spectra normally encountered in the studies of enzyme-bound complexes are summarized in Section II,A, followed by a discussion of the various factors that need to be appreciated for optimal design of the experiments in Section II,B. Section III presents experimental results on a number of examples of $^{31}$P-NMR investigations of enzyme systems chosen with a view to illustrate the theoretical and experi-

---

[1] Unless the enzyme has phosphate groups present in it such as in phosphoenzymes. In these cases, however, $^{31}$P NMR may still be utilized to study the enzyme by observing the spectra of the covalently attached phosphate itself (see Chapter 4 by Vogel).

[2] See Chapter 5 by Villafranca for studies of paramagnetic metal-ion effects.

[3] These aspects are covered in separate chapters by Tsai (Chapter 6) and Gerlt (Chapter 7).

mental aspects discussed in Section II and to provide some insight into the information that may be gathered from such studies. The line shapes observed in NMR spectra of enzyme-bound complexes, especially in enzyme-bound equilibrium mixtures in which the resonances from $^{31}$P nuclei in substrates and products are observed as the interconversion proceeds on the surface of the enzyme, are governed by chemical-exchange processes implicit in the reactions. An accurate and systematic analysis of such spectra requires computer simulation of the line shapes under the influence of chemical exchange. Such analyses have provided worthwhile additional kinetic information on the enzyme reactions. Examples of the computer-simulated line-shape analysis used for some enzymes are described in Section IV. The chapter concludes with a brief discussion in Section V of the prospects and limitations of the type of investigations considered thus far.

## II. Theoretical and Experimental Considerations

### A. Theoretical

#### 1. Changes in Spectral Parameters

The $^{31}$P-NMR spectral parameters (chemical shifts $\delta$ and spin–spin coupling constants $J$) are readily measureable from the spectra of enzyme-bound complexes. The values of $\delta$ of most phosphates used in enzymatic systems lie within a range of $\sim 4-10$ ppm on the low-field side and $\sim 30$ ppm on the high-field side of the signal from 85% $H_3PO_4$, which is often used as an external reference to express the values of $\delta$. Whereas the exact values of $\delta$ for $^{31}$P nuclei in different phosphate moieties depend on a variety of factors such as pH, metal chelation, and complexation with an enzyme, the signals of typical phosphates occur in identifiable regions. [See Gorenstein (Chapters 1 and 2) for some of the general principles regarding $\delta$ and $J$, and Gorenstein and Shah (Chapter 18) for a compilation of $^{31}$P-NMR data.]

The changes in the values of $\delta$ (and $J$) associated with a substrate in an enzyme–substrate complex, relative to those in free substrates, arise from the effect of environment or conformation or both (changes in $J$, if any observed, are more often likely to be due to conformational changes). The $^{31}$P chemical shifts of phosphate groups are also usually affected by factors such as pH and metal chelation (see, for example, Cohn and Hughes, 1960, 1962; Moon and Richards, 1973), and this dependence may undergo a change when the substrate is bound to the enzyme. Phosphorus-31 chemical-shift measurements of substrates free in solution and bound to enzymes in various complexes as a function of pH and metal-ion concentration may allow a differentiation of the effects of pH and metal chelation on and off the enzyme. It may sometimes be possible to separate the overall change in $\delta$ resulting from binding to the enzyme into environmental and conformational factors. The change in $\delta$ that is exclusively due to substrate conformation on the enzyme may then be related to the structure of the enzyme–substrate complex. However, such interpretations are not possible on the basis of the present theories of $\delta$ and $J$ in $^{31}$P NMR. Some empirical correlations between $\delta$ and structural parameters were made earlier (Gorenstein and Kar, 1975), and attempts have been made at performing *ab initio* quantum mechanical calculations of the $^{31}$P chemical-shift tensor of a phosphate group (Ribas Prado *et al.*, 1979). A reliable theory of $\delta$ (and/or $J$) applicable for many interesting biological molecules is sorely needed but appears somewhat remote (however, see also Gorenstein, Chapters 1 and 2).

Phosphorus-31 chemical-shift changes of phosphate moieties in substrates and inhibitors because of complexation with enzymes measured thus far are not usually very large. The values are usually less than $\sim 2$ ppm, occasionally they are as large as 4–5 ppm (Cohn, 1979), and in one case a change of $\sim 7$ ppm is observed (Nageswara Rao and Cohn, 1977b). In typical experiments in which such shifts are measured there are accompanying line-broadening effects arising from binding to a macromolecule or chemical exchange or both. If the line broadening is comparable to the chemical-shift change, the accuracy in the determination of chemical-shift changes will be impaired. Chemical-exchange effects might arise from a number of causes, for example exchange of substrate between bound and free forms or between different conformations of the enzyme-bound complexes or interconversion of substrates and products on the surface of the enzyme ($E \cdot S_1 \cdot S_2 \rightleftarrows E \cdot P_1 \cdot P_2$) catalyzing the reaction and so on. If the chemical exchange is present along with (or without) concomitant line broadening, an analysis of the line shape may usually be required to deduce the chemical shift. Although this may appear to be a complication, such an analysis has the utility of providing information on the exchange rates involved as well as the chemical shifts of the species in different states between which exchange takes place.

Spin–spin couplings involving $^{31}$P nuclei occur in a number of molecules used in studies of enzymatic systems. Phosphorus-31 nuclei in phosphates may be spin-coupled to $^1$H nuclei in adjacent groups (e.g., $\alpha$-P of ATP with the 5′ protons in the ribose), and $^{31}$P nuclei in adjacent phosphate groups in a phosphate chain (e.g., in ADP and ATP) may be coupled to each other. The presence of spin–spin multiplets in the spectrum of a substrate molecule offers the advantage of monitoring changes in the coupling constants, for example, on binding to an enzyme, although such cases have been rather

# 3. ³¹P NMR of Enzyme Complexes

rare in ³¹P-NMR studies thus far (Nageswara Rao *et al.*, 1978a; Ulrich and Markley, 1980; Groneborn *et al.*, 1981b; Brauer and Sykes, 1981b). On the other hand, the presence of the multiplet structure in resonances that participate in chemical-exchange processes makes the analysis of the line shapes considerably more complex than the cases where spin–spin coupling is absent. This complexity cannot be obviated because ignoring spin-coupling effects leads to erroneous results. Some theoretical details pertinent to the analysis of lineshapes in the presence of spin–spin coupling and under the influence of chemical exchange are discussed.

## 2. Line Shapes: Chemical Exchange Effects

³¹P-NMR linewidths of phosphates covalently bound to the enzymes are governed by relaxation processes (dipolar interactions, anisotropic chemical shifts, and paramagnetic metal-ion effects, if any) and the correlation times associated with these processes. As discussed in Section II,A,1, noncovalently bound substrates often exchange between different states, either on the enzyme or between the free and enzyme-bound states. The resultant lineshapes of ³¹P resonances of these substrate molecules depend on the spectral parameters ($\delta$ and $J$), the linewidths in the different states in the absence of exchange, and the exchange rates between these states for the chosen set of experimental conditions.

The qualitative and quantitative features of exchange effects for "simple lines" in the absence of spin–spin coupling have been known for a long time and are understood in a straightforward manner (see Kaplan and Fraenkel, 1980, and references cited therein). If a nucleus exchanges between two sites ($B \rightleftarrows B'$) with resonance frequencies $\omega_B$ and $\omega_{B'}$ and lifetimes $\tau_B$ and $\tau_{B'}$ (corresponding exchange rates are $\tau_B^{-1}$ and $\tau_{B'}^{-1}$), the effect of chemical exchange on the line shape may be summarized by considering three conditions:

1. If $|\omega_B - \omega_{B'}| \gg \tau_B^{-1} \tau_{B'}^{-1}$, called "slow exchange," the resonances at $\omega_B$ and $\omega_{B'}$ merely acquire additional widths given by $(\pi\tau_B)^{-1}$ and $(\pi\tau_{B'})^{-1}$ (in units of Hz) respectively.

2. If $|\omega_B - \omega_{B'}| \ll \tau_B^{-1} \tau_{B'}^{-1}$, called "fast exchange," the two resonances merge into a single resonance with a frequency and linewidth given by the average[4] of these quantities in the absence of exchange weighted by the fractional concentrations of the two species.

---

[4] It may be noted that within the general constraint of $|\omega_B - \omega_{B'}| \ll \tau_B^{-1} \tau_{B'}^{-1}$ averaging of the peak positions is obtained at exchange rates slower than those at which the linewidths are averaged. To be more precise, while the above condition describes averaging of peak positions, the averaging of the linewidths requires $(\omega_B - \omega_{B'})^2 \tau_B$ and $(\omega_B - \omega_{B'})^2 \tau_{B'}$ to be negligible compared to the linewidths of B and B' resonances, respectively, in the absence of exchange.

3. If $|\omega_B - \omega_{B'}| \simeq \tau_B^{-1}$ and $\tau_{B'}^{-1}$, called "intermediate exchange," both frequency shifts and linewidth changes occur giving rise to a complex line shape.

All of these features can be described quantitatively by means of a simple theory based on Bloch equations.

Line shapes simulated for various values of exchange rates using these equations for two single resonances occuring (in the absence of exchange) at $-19.0$ ppm and $-21.5$ ppm in an 81-MHz $^{31}$P-NMR spectrum, assuming equal concentrations of B and B' (therefore, $\tau_B = \tau_{B'}$) are shown in Fig. 1. (The chemical-shift values are all reported with negative values to high field, and in this example are chosen nearly to equal those for $\beta$-P(MgATP) and $\beta$-P(ATP), respectively, at pH 8.0 to provide a comarison with the case discussed next, where spin–spin coupling is present). The exchange rate ($\tau^{-1}$) is expressed as a fraction of the chemical shift $\Delta\omega$ in units of rad s$^{-1}$. The line shapes in Fig. 1 resemble those widely available in NMR literature and the qualitative features of slow, intermediate, and fast exchanges described earlier are displayed quite clearly in the spectra. It may also be noted that although a coalescence of the resonances with a peak at the average position results for an exchange rate comparable to the chemical-shift difference, the resonance exhibits significant residual broadening. At faster exchange rates this averaged resonance begins to narrow, and at very large exchange rates the resonance has a frequency and a linewidth given by the weighted averages of the respective values in the absence of exchange. The exchange rate cannot be measured from the position or shape of such an averaged resonance. A lower limit of the fast exchange rate may, however, be estimated by comparison to theoretical simulations.

If spin–spin coupling is present in the exchanging system, this description is quantitatively inadequate. The qualitative features discussed here are applicable for every component of a spin–spin multiplet that is undergoing exchange if (1) the spin system remains weakly coupled (chemical shift between any pair of nuclei is large compared with the coupling constant between them) and (2) the spin–spin interactions are not significantly modified by the exchange process. The weak coupling condition is required because the simple theory is incapable of taking strong coupling effects into account. The second requirement relates to a rather common feature in exchange processes involving spin–spin coupled nuclei present in itinerant chemical moieties such as those in (reversible) enzymatic reactions. The spin–spin couplings involving such nuclei are turned on and off as the moiety is transferred back and forth between reactants and products. Such an exchange would cause a "washing out" of those particular spin–spin interactions from the spectrum at sufficiently large exchange rates and,

## 3. ³¹P NMR of Enzyme Complexes

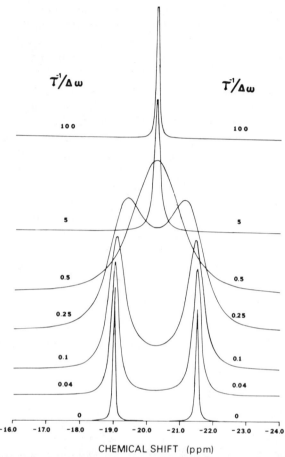

**Fig. 1.** Computer-simulated spectra for the two-site exchange (B ⇌ B') for simple resonances with no multiplicity from spin–spin coupling. The chemical shifts in the absence of exchange are chosen to be nearly equal to those for $\beta$-P of ATP ($-21.5$ ppm) and MgATP [or CaATP ($-19$ ppm)]. The exchange rate is expressed as a fraction of the chemical shift (in units of radians per second) at 81 MHz. The concentrations of the two species are taken to be equal, that is, [B] = [B']. Compare with a similar calculation in the presence of spin–spin coupling presented in Fig. 2. Shifts upfield are negative (K. V. Vasavada and B. D. Nageswara Rao, unpublished results).

consequently, produces some line-shape features of its own at lower exchange rates. The theoretical description of line shapes under the influence of exchange in the presence of spin–spin couplings is thus a rather complex problem. The Bloch-equation approach can be used only under the restricted set of conditions described above. In general, however, a complete theory such as the density matrix theory is required for computation of line shape (Kaplan and Fraenkel, 1980). Regardless of which theory is used, it will be necessary to use computer programs to simulate the theoretical spectra in view of the large dimension of the problem arising from the fact that many different NMR transitions that comprise the spin–spin multiplets are now participating in the exchange process. It is, therefore, preferable to use the density matrix theory with the help of a computer program for all cases because such an approach obviates the need for examining the question of applicability of the theory for each individual case and introduces minimal additional complexities in the computations.

Figure 2 shows the line shapes calculated for the same exchange rates as in Fig. 1 but with a 1:2:1 spin multiplet each at $-19.0$ and $-21.5$ ppm. The coupling constants are chosen to be 16.0 and 19.5 Hz for the two cases corresponding closely to the values for metal–ATP complexes with $Mg^{2+}$ (and $Ca^{2+}$) and ATP (Cohn and Hughes, 1960, 1962; Nageswara Rao and Cohn, 1977a), respectively. The $\beta$-P resonance is a 1:2:1 triplet because the coupling constants $J_{\alpha\beta}$ and $J_{\beta\gamma}$ are equal to each other both for ATP and MgATP (or CaATP) complexes. A comparison of the line shapes in Fig. 2 to those in Fig. 1 reveals the changes that arise from spin–spin coupling (in weakly coupled spectra). The spectra for values of $0.1 \lesssim (\Delta\omega\tau)^{-1} \lesssim 1.0$ are somewhat deceptive in that they show no evidence of spin multiplet structure and the line shape is very similar to that in Fig. 1. The exchange exemplified by the association and dissociation of the divalent diamagnetic metal ion does not destroy the spin–spin interaction between the $^{31}P$ nuclei in ATP [except for a reduction change in the coupling by $\sim 20\%$ in the MgATP (or CaATP) complex], and the line shapes at large exchange rates once again exhibit the multiplet structure. A comparison of the line-shape calculation with $^{31}P$-NMR experiments on MgATP and CaATP in exchange with free ATP allows a good estimate of the off and on rates of these metal ions from ATP (Vasavada et al., 1984).

This discussion describes qualitative features of line shapes produced by chemical exchange and the manner in which spin–spin interactions alter the lineshapes. In exchanges characteristic of enzymatic reactions, certain groups are transferred back and forth and the spin–spin interactions involving nuclei in the itinerant moieties become randomized. For example, the $^{31}P$-NMR studies of kinase reactions exemplify a case in which the exchange destroys some of the couplings and preserves the others. These line shapes are considered in Section IV.

## 3. Relaxation Times

Two mechanisms primarily contribute to spin–lattice and spin–spin relaxation ($T_1$ and $T_2$, respectively) of $^{31}P$ nuclei in diamagnetic enzyme

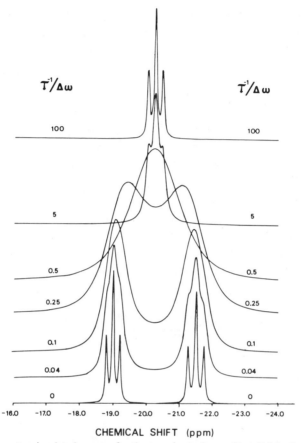

**Fig. 2.** Computer-simulated spectra for the two-site exchange (B ⇌ B′) in the presence of weak spin–spin coupling. The spin–spin coupling constants used are 16.0 and 19.5 Hz for metal–ATP complexes and ATP, respectively. All other parameters are identical to those in Fig. 1 (81 MHz, pH 8.0). Contrast with the spectra in Fig. 1 to see the effect of spin–spin coupling on the line shape. Shifts upfield are negative (K. V. Vasavada, B. D. Ray, and B. D. Nageswara Rao, to be published).

complexes that are of interest in this chapter and also described elsewhere in this volume (James, Chapter 12; Shindo, Chapter 13; and Smith, Chapter 15): dipole–dipole interactions, and $^{31}$P chemical-shift anisotropy. Dipolar interactions are with other magnetic nuclei in the complex including $^{31}$P. The general theory for relaxation mediated by these mechanisms is described in Abragam's (1961) book. The expressions for $T_1$ and $T_2$ of $^{31}$P nuclei arising from these mechanisms are summarized in the following equations assuming (1) isotropic rotational motion (single correlation time

$\tau_c$) and (2) slow tumbling (i.e., $\omega\tau_c \gg 1$ for all nuclei; $\omega$ is the Larmor frequency).

*a. Chemical Shift Anisotropy Mechanism.*

$$(T_1)^{-1}_{\text{CSA}} = \frac{2}{15}\frac{(\Delta\sigma)^2}{\tau_c}\left(1 + \frac{\eta^2}{3}\right) \qquad (1)$$

$$(T_2)^{-1}_{\text{CSA}} = \frac{4}{45}\omega_P^2(\Delta\sigma)^2\left(1 + \frac{\eta^2}{3}\right)\tau_c \qquad (2)$$

where $\Delta\sigma = \frac{3}{2}\delta_z$, $\eta = (\delta_x - \delta_y)/\delta_z$, where $\delta_x$, $\delta_y$, and $\delta_z$ are the principal values of the traceless part of the chemical-shift tensor (i.e., $\delta_x + \delta_y + \delta_z = 0$), with $\delta_z$ as the maximum value, and $\eta$ is known as the asymmetry parameter. Note that $(T_1)^{-1}$ is independent of frequency $\omega_P$, whereas $(T_2)^{-1}$ increases quadratically with $\omega_P$ in this limit.

*b. Heteronuclear Dipolar Interaction.* Intramolecular (or within a single complex) dipolar interaction of $^{31}$P with a nucleus of a different species such as $^1$H leads to a coupled equation for the magnetizations of the two spins. A $T_1$ can be defined if the second nucleus ($^1$H) relaxes quickly, independent of this interaction. In this condition and with $\omega\tau_c \gg 1$,

$$(T_1)^{-1}_{\text{HP-dipolar}} = \frac{3.53\gamma_P^2\gamma_H^2\hbar^2}{10r_{\text{HP}}^6\,\omega_P^2\,\tau_c} \qquad (3)$$

and

$$(T_2)^{-1}_{\text{HP-dipolar}} = \frac{\gamma_P^2\gamma_H^2\hbar^2}{5r_{\text{HP}}^6}\tau_c \qquad (4)$$

where $\gamma_i$ is the gyromagnetic ratio of nucleus i and $r_{ij}$ the distance between nuclei i and j.

*c. Homonuclear Dipolar Interactions.* This interaction is usually rather weak for phosphate groups. $T_1$ and $T_2$ can be defined for this mechanism for nuclei with overlapping resonances. For such nuclei (with $\omega\tau_c \gg 1$),

$$(T_1)^{-1}_{\text{PP-dipolar}} = \frac{3}{5}\frac{\gamma_P^4\hbar^2}{r_{\text{PP}}^6\,\omega_P^2\,\tau_c} \qquad (5)$$

and

$$(T_2)^{-1}_{\text{PP-dipolar}} = \frac{9}{20}\frac{\gamma_P^4\hbar^2}{r_{\text{PP}}^6}\tau_c \qquad (6)$$

Note that intermolecular dipolar interactions (i.e., between two enzyme complexes) are not included in Eqs. (3)–(6). Expressions similar to Eqs. (1)–(6) are also given by Brauer and Sykes (1981b).

Whenever contaminating paramagnetic ions are present in the solutions, they tend to bind phosphate groups and can cause significant relaxation rates comparable to the contribution of the mechanisms discussed before, even at minute levels of contamination. When enzyme-bound complexes are of interest, depending on the experimental conditions, the substrate molecule may be undergoing chemical exchange as discussed earlier. The effect of chemical exchange on line shapes (or implicitly $T_2$) are discussed at some length in Section II,A,2. Chemical-exchange effects also need to be considered in the interpretation of the measured values of $T_1$. It may not always be possible to define a unique $T_1$ or $T_2$ when chemical exchange is present; strictly speaking, multiple-site exchanges yield multiple values for $T_1$ and $T_2$, leading to nonexponential decays. Under some limiting conditions, however, it is possible to define $T_1$ and $T_2$ for the purpose of comparison with experimental results [see, for example, Leigh (1971) for the case of two-site exchange].

## B. Experimental

For most of the biological magnetic resonance studies, the design of the experiment requires optimization both from NMR and biological points of view; $^{31}$P-NMR experiments are no exception. Commercially available NMR instrumentation spans the operating frequency range 24–240 MHz. The sensitivity of the NMR signals as well as the chemical-shift dispersion increase with increasing operating frequency. However, if anisotropic chemical shift contributes significantly to the linewidths of the $^{31}$P-NMR signals (because this contribution increases quadratically with the applied magnetic field), the increase in sensitivity may be substantially offset by the increase in linewidth of the signal. In such cases an intermediate frequency of operation may be optimal for the experiments.

Typical $^{31}$P-NMR probes accept sample tubes of 10- to 12-mm o.d. requiring sample volumes of 1.5–2.5 ml. Smaller volumes of sample are possible by using microcells that contain bulbs that fill the region of the receiver coil in the probe. Probes that accept 20-mm o.d. sample tubes requiring normal sample volumes ~12 ml are available for special purposes. The choice of the appropriate sample volume is dictated by the available material, solubility, and concentration chosen for the experiment. Whereas it is preferable to use as high a concentration as possible from the consideration of NMR signal sensitivity, the concentration used may have to be kept low either on the basis of the available material or because of biochemical constraints on the sample. Sometimes lower concentrations are required for reasons of solubility limitations of materials or enzymes. In experiments on enzyme complexes, the enzyme concentrations exceed that of the substrate, and therefore the highest usable substrate concentration is limited by the solubility of the enzyme. Lowering the concentration has the effect of requiring longer times for data accumulation that in turn might vitiate the prevailing biochemical conditions at which the experiments must be performed. When sufficient quantities of material are available, these disadvantages of low concentrations may be partly compensated by using larger sample tubes (volumes). In $^{31}$P-NMR studies of enzyme complexes, typical low concentrations are seldom below 1 mM, typical accumulation times are seldom over 3–4 h, and typical sample-tube sizes tend to be 10–12 mm with sample volumes ranging from 1.0 to 2.5 ml (which may be reduced to ~0.5 ml using microcells). In favorable cases it may be possible to lower the concentrations to 0.5 m$M$. Biochemical experiments in which the $^{31}$P-containing substrates are at concentrations much inferior to those stated are thus outside the range of the experimental sophistication available at present.

The time required for a typical experiment will be determined, aside from the aforementioned factors of spectrometer frequency, sample volume, and concentration, by the slowest $^{31}$P relaxation times in the sample of interest. In Fourier-transform NMR, where the experimental data are acquired as a function of time following a radio-frequency pulse, it is desirable to have the $^{31}$P spins almost in thermal equilibrium with the surroundings prior to the pulse. The equilibrium is reached in a time that is a factor of 3 or more larger than the spin–lattice relaxation time. This waiting time between pulses can be optimized by using shorter pulse lengths (i.e., flip angles <90°). Compromising on the equilibrium condition usually results in intensity and lineshape distortion but leaves the peak positions unchanged in the spectrum. If rapid acquisition of data is important because of the nature of the sample, some spectral information on the fast-relaxing spins can be obtained by rapid pulsing, thus sacrificing the signals from the slow-relaxing spins.

When all the factors discussed here are well optimized, the reproducibility and reliability of the experiments depend on the high purity of enzymes and other chemicals used and on the control of the biochemical conditions. Experiments on enzyme-bound complexes place strict requirements on the purity of the enzyme because the sample contains a large amount of enzyme, and the substrates are in interaction with the enzyme for a good deal of time often required to perform the experiments. Thus any contaminating enzyme at levels that may ordinarily be considered minute produces noticeable and unanticipated side effects that interfere with the desired experiment. In experiments dealing with linewidths and relaxation data in phosphate compounds, it is of crucial importance to remove scrupulously paramagnetic impurities by using suitable chelating agents (such as Chelex-100) or extracting procedures. Paramagnetic cations often tend to bind in the vicinity of phosphate groups that have negatively charged oxygen atoms owing to deprotonation at conventional pH values, and a concentration of paramagnetic ions $10^{-4}$–$10^{-5}$ times that of the molecules with the phosphate groups, such as the adenine nucleotides, is sufficient to cause measureable changes in linewidths and relaxation times.

## III. Experimental Studies of Specific Enzyme Systems

Experimental results obtained by using $^{31}$P NMR of enzyme complexes of some specific enzyme systems are described in this section. These results are presented in two categories: (1) inert enzyme-bound complexes of substrates, inhibitors, cofactors, and their analogs, and (2) equilibrium mixtures in the enzyme-bound form (i.e., where the substrates and products are observed as they interconvert on the enzyme). The studies include a number of different enzymes; however, the list is not complete. The enzyme systems chosen for description are expected to illustrate the experimental and theoretical considerations presented in Section II and indicate the kind of information that might be extracted by $^{31}$P-NMR studies of enzyme complexes and the limitations and difficulties of the method.

### A. Enzyme Complexes of Substrates, Coenzymes, and Inhibitors

Enzyme complexes of substrates, coenzymes, and inhibitors are the simplest complexes to study for the purpose of gaining information on the manner in which the enzymes facilitate catalytic action. Usually one looks for changes either in $^{31}$P-NMR spectral parameters ($\delta$ and $J$) or in dynamical parameters (line shape and $T_1$) of the substrate (inhibitor or analogs) nuclei on binding the enzyme and on making further specified additions to the complexes or making modifications to the enzyme, and so on. Results of this type on several different enzyme systems are discussed.

### 1. Phosphoryl Transfer Enzymes (Kinases)

Kinases catalyze the reversible transfer of $\delta$-P of ATP to a second substrate S:

$$ATP + S \underset{}{\overset{Mg^{2+}}{\rightleftharpoons}} ADP + SP \qquad (7)$$

All kinases require $Mg^{2+}$ as an obligatory component of the reactions. Phosphorus-31 NMR is a convenient method to study the enzyme–substrate interactions for this class of enzymes because at least three of the four substrates contain a total of six phosphate groups (if S does not contain any phosphate groups) that yield signals readily identified in the NMR spectrum. Enzyme-bound complexes of a number of different kinases have been studied by $^{31}P$ NMR: arginine kinase (Nageswara Rao et al., 1976; Nageswara Rao and Cohn, 1977a), creatine kinase (Nageswara Rao and Cohn, 1981), 3-phosphoglycerate kinase (Nageswara Rao et al., 1978b), pyruvate kinase (Nageswara Rao et al., 1979), and adenylate kinase (Nageswara Rao and Cohn, 1977b; Nageswara Rao et al., 1978a). These papers also include equilibrium mixtures of enzyme-bound substrates and products that are considered in Section II,B. In considering the enzyme complexes of substrates and inhibitors, since several enzymes of this class are studied, the results on nucleotide binding to these enzymes are presented together, followed by results pertaining to specific complexes relevant to individual enzymes.

*a. Chemical Shifts and Coupling Constants of Nucleotides Bound to Kinases.* The chemical shifts of ATP and ADP in enzyme-bound complexes of different kinases are summarized in Table I. ATP chemical shifts are practically unaffected by binding to the enzymes with one exception, namely, $\beta$-P(MgATP) bound to adenylate kinase. $\beta$-P(MgADP) is the only nucleotide $^{31}P$ resonance that shifts on all the enzymes except pyruvate kinase. Although there is a downfield shift in most cases, an upfield shift occurs on 3-P-glycerate kinase. Thus the $^{31}P$ chemical shifts of nucleotide–kinase complexes do not reveal a systematic change that might signify gross

**TABLE I**

Phosphorus-31 Chemical Shifts of Free and Enzyme-Bound Nucleotides for Kinases at pH 7.0[a]

| Enzyme | MgADP | | MgATP | | |
|---|---|---|---|---|---|
| | $\alpha$-P | $\beta$-P | $\alpha$-P | $\beta$-P | $\gamma$-P |
| (None − Mg) | −10.8 | −7.5 | −11.0 | −21.8 | −7.3 |
| (None + Mg) | −9.9 | −5.9 | −10.8 | −19.2 | −5.6 |
| Arginine kinase | −11.0 | −3.3 | −11.0 | −19.4 | −5.6 |
| Creatine kinase | −11.0 | −3.8 | −10.9 | −19.0 | −5.4 |
| Adenylate kinase | −10.2 | −3.5,[b] −6.7 | −10.7 | −17.8 | −6.1 |
| Phosphoglycerate kinase | −11.0 | −7.5 | −11.0 | −19.4 | −6.0 |
| Pyruvate kinase | −10.0 | −5.7 | −10.9 | −19.2 | −5.5 |

[a] Parts per million from 85% $H_3PO_4$, upfield shifts negative.
[b] Second ADP (bound by Mg) visible under conditions of slow exchange.

similarities in environment or conformation (or both) of the phosphate chain in these complexes. This shift may possibly be related to the distinction between different enzymes as reflected, for example, in the specificity for the second substrate (i.e., the phosphoryl acceptor; Cohn, 1979; Nageswara Rao, 1979).

The $^{31}P-^{31}P$ spin–spin coupling constants in the enzyme-bound complexes of ATP and ADP are equal to those of the nucleotides free in solution within experimental error. An exception occurs for ATP bound to adenylate kinase for which the values are $J_{\alpha\beta} = 17.3$ Hz and $J_{\beta\gamma} = 21.0$ Hz, whereas $J_{\alpha\beta} = J_{\beta\gamma} = 19.5$ Hz for free ATP. In E·MgATP complex, however, these values become 15.1 and 14.7 Hz, respectively, very nearly equal to 15.5 Hz for both coupling constants in free ATP. The significance of these changes and of the fact that they occur for this enzyme alone is not clear.

*b. Arginine Kinase and Creatine Kinase [S = Arginine and Creatine, Respectively, in Eq. (7)].* It is interesting to compare the results of these two enzymes because of their similarity in catalyzing phosphoryl transfer to a guanidino nitrogen. Arginine kinase is sometimes referred to as the invertebrate analog of creatine kinase. Furthermore, both these enzymes are known to form transition-state analog complexes in the presence of the planar anion $NO_3^-$ occupying the position of the transferrable phosphoryl group (Milner-White and Watts, 1971; Buttlaire and Cohn, 1974); for example, E·MgADP·$NO_3^-$·arginine is considered to mimic the transition state of the arginine kinase reaction. Note, however, that arginine kinase is a monomer whereas creatine kinase is a dimer.

Evidence for possible homology between arginine kinase and creatine kinase is revealed by the $^{31}$P-NMR results on transition-state analog complexes (see Fig. 3). The $\beta$-P(MgADP) resonance from E·MgADP·$NO_3^-$·creatine complexes shows a significant upfield shift in the direction of the $\beta$-P(MgATP) resonances, and for this shift to occur, the presence of all the components of these complexes was mandatory. Exactly the same result was obtained for E·MgADP·$NO_3^-$·arginine in the studies with arginine kinase (Nageswara Rao and Cohn, 1977a). The shift of the $\beta$-P(MgADP) resonance in the direction of the $\beta$-P(MgATP) is strong evidence for the formation of a transition-state analog complex since the electronic shielding of the $\beta$-P(MgADP) resonances exhibits a change toward its value in the final product, namely, $\beta$-P(MgATP).

The $^{31}$P-NMR spectra of E·MgADP·creatine and the transition-state analog complex shows another feature, the significance of which is somewhat unclear. The $\beta$-P(MgADP) resonance in both these complexes shows two broad peaks of nearly equal intensity (see Fig. 3) at low temperatures (~4°C). These two peaks merge into a single peak at an average position at temperatures above 25°C, possibly owing to sufficiently rapid chemical

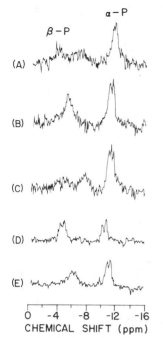

Fig. 3. $^{31}$P-NMR spectra (at 40.3 MHz) of enzyme-bound, dead-end, and transition-state analog complexes of creatine kinase containing MgADP. Shifts upfield are negative (A) E + MgADP + creatine; 4°C. (B) As in (A) (pH 8.0), but at 32°C. (C) As in (A), but plus nitrate (4°C). (D) As in (C), but plus EDTA (4°C). (E) As in (A), but plus formate (25°C) (Nageswara Rao and Cohn, 1981).

exchange (Fig. 3B). This observation was independently reported earlier by Milner-White and Rycroft (1977). Creatine kinase is a dimer, and possible dissimilarities between two MgADP binding sites on this enzyme were noted previously (McLaughlin, 1974; Price and Hunter, 1976). The $^{31}$P-NMR results suggest a distinction between the binding sites on the two subunits and possibly some intersubunit interaction (McLaughlin, 1974). However, that may not be the case because a similar line shape of the $\beta$-P(MgADP) resonance was noted in the experiments with arginine kinase, which is a monomer (Nageswara Rao and Cohn, 1977a). At any rate, these spectra reveal some conformational heterogeneity and immobilization in the vicinity of the active site, even in the absence of the planar anion required for the formation of the transition-state analog complex. It is interesting to note that the characteristic line shape described here occurs only in the presence of $Mg^{2+}$.

Although the nucleotide chemical shifts of the various binary, ternary, quaternary, and transition-state analog complexes of arginine kinase and creatine kinase show significant resemblance, the similarity ends in the pH dependence of the $\beta$-P resonance of the E·MgADP complex. For MgADP bound to arginine kinase, this resonance shows a $pK_a \sim 1.5$ units higher than free MgADP, whereas the chemical shift was independent of pH in the range

3. ³¹P NMR of Enzyme Complexes

6.0–9.0 for the complex with creatine kinase. The chemical shifts of the $^{31}$P resonances from MgADP$\beta$S[5] and MgATP bound to creatine kinase are also unchanged in the pH range 6.0–9.0. Thus there appears to be a broad similarity in the structure or conformation (or both) of the phosphate chain of the nucleotides (to the extent these features are reflected by the chemical shifts) in the complexes bound to arginine kinase and creatine kinase, whereas the environments of these moieties as indicated by their protonation behavior seem to be significantly different.

*c. Pyruvate Kinase [S = Pyruvate; SP = P-Enolpyruvate in Eq. (7)].* This glycolytic enzyme is unique among the kinases in that it needs a monovalent cation K$^+$ in addition to Mg$^{2+}$ for activation. Moreover, two Mg$^{2+}$ ions are bound per reaction complex, one to the protein and the other to the nucleotide. The $^{31}$P-NMR data of E·ATP complexes in the presence of Mg$^{2+}$ corroborate the requirement for these two Mg$^{2+}$ ions. Mg$^{2+}$ binds to the protein site tighter than to the nucleotide. Aside from this, the $^{31}$P chemical shifts of nucleotides are not altered much as a result of binding to pyruvate kinase. For this enzyme, binding influences the P-enolpyruvate chemical shift the most (Nageswara Rao *et al.*, 1979). The $^{31}$P chemical shifts of enzyme-bound substrate show a pattern in that the shifts of the substrate that binds strongest are altered most on binding to the enzyme. The signal from E·P-enolpyruvate is shifted ∼1.0 ppm downfield from the position of free P-enolpyruvate, in addition to being significantly broadened. Addition of Mg$^{2+}$ to this complex causes a further downfield shift of ∼2 ppm and considerably more line broadening. In the equilibrium mixture (i.e., when ADP is added to the complex), the signal is shifted a further 1.3 ppm downfield and is somewhat narrowed. Addition of K$^+$ has no discernible effect on the signal. These changes in the P-enolpyruvate resonance on the enzyme illustrate the complex nature of the structural changes involved in the interaction of pyruvate kinase with this tightly bound substrate. If a good theoretical basis is available for relating the changes in these NMR parameters to structure, considerably more information could be extracted from these data.

*d. 3-P-Glycerate Kinase [S = 3-P-Glycerate in Eq. (7)].* This is another glycolytic enzyme that has been studied (Nageswara Rao *et al.*, 1978b). Chemical-shift changes were observed for $^{31}$P signals from 3-P-glycerate and

---

[5] The pH dependence of the chemical shift of MgADP$\beta$S bound to arginine kinase has been determined (C. L. Lerman and M. Cohn, unpublished). The variation indicates that the p$K_a$ is below 6.5, and, furthermore, the change with pH is in the opposite sense to that of the MgADP complex. Consequently, the original sugestion (Nageswara Rao and Cohn, 1977a) that the p$K_a$ of the chemical shift of $\beta$-P of MgADP bound to arginine kinase was due to an ionization of amino acid residue at the active site is an oversimplification.

1,3-bis-P-glycerate in the enzyme-bound equilibrium mixture (see Section III,B). All three signals are shifted ~1.5 ppm downfield with respect to their positions in free solution. In the presence of sulfate ion, which is known to affect the kinetics of this enzyme, the shift of 3-P-glycerate in the equilibrium mixture changes to a value close to that in free solution. The $^{31}$P experiments also provide distinct evidence for ATP binding to this enzyme at two sites; to one of these sites — probably a noncatalytic site — $Mg^{2+}$ binds much more weakly than to the other.

*e. Adenylate Kinase [S = AMP in Eq. (7)].* This kinase, which yields two ADPs as products [see Eq. (7)], is ubiquitous and is essential for the production of adenine nucleotides beyond the monophosphate level in the cell. It is also unique among the kinases in that it catalyzes a phosphoryl transfer from ADP as donor to another ADP as the acceptor in the reverse reaction. The enzyme has only one $Mg^{2+}$ ion per reaction complex. A detailed analysis of $^{31}$P-NMR spectra of the enzyme-bound equilibrium mixtures of adenylate kinase (Nageswara Rao *et al.*, 1978a), described briefly in Section III,B, showed that (1) the donor and acceptor ADP resonances are distinguishable in the enzyme complexes and (2) the $Mg^{2+}$ ion plays an unmistakable role in differentiating the two ADPs in that the enzyme selectively binds metal nucleotides only at the acceptor ADP (or ATP) sites.

The distinction between the two nucleotide binding sites is strikingly demonstrated in the $^{31}$P-NMR spectra of adenylate kinase complexes of diadenosine pentaphosphate ($Ap_5A$), a symmetrical molecule and a potent inhibitor of the porcine adenylate kinase reaction. If the two nucleotide binding sites are very different, it suggests that in the enzyme–inhibitor complex $Ap_5A$ may not be symmetrical. This asymmetry is clearly depicted in the $^{31}$P-NMR spectrum of the $E \cdot Ap_5A$ complex compared to the spectrum of $Ap_5A$ or $Mg \cdot Ap_5A$ free in solution (see Fig. 4A–C). In the spectrum of the $E \cdot Ap_5A$ complex (Fig. 4C), the resonances of 1-P and 5-P (about −11 ppm) are resolved, and those of the three middle phosphate groups 2-P, 3-P, and 4-P are also distinct from each other. The resonance of 3-P is in the middle of the three upfield resonances. The resonances of 2-P and 4-P, on either side of the 3-P resonance, are broadened because 2-P and 4-P can interchange when $Ap_5A$ dissociates from $E \cdot Ap_5A$ and binds again. This exchange preserves the identity of 3-P. Addition of $Mg^{2+}$ to the sample of Fig. 4C highly accentuates the asymmetry of the enzyme-bound $Ap_5A$ in the $E \cdot MgAp_5A$ complex (Fig. 4D). It has been shown (Price *et al.*, 1973) that only one metal ion binds readily to an $Ap_5A$ molecule. All five phosphates in this complex resonate at clearly distinct positions. The 2-P and 4-P resonances are no longer significantly broadened compared to the 3-P reso-

# 3. ³¹P NMR of Enzyme Complexes

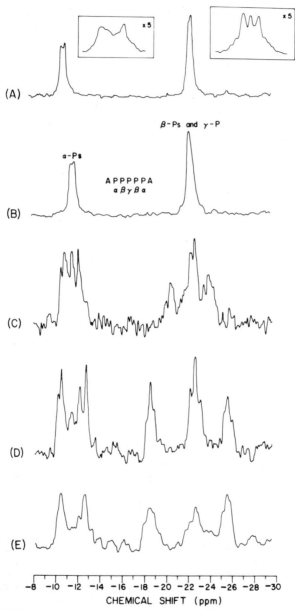

**Fig. 4.** ³¹P-NMR spectra (at 40.3 MHz) of Ap$_5$A and MgAp$_5$A free and bound to adenylate kinase (porcine; 4°C; pH 8.0). (A) Free Ap$_5$A. (B) Free MgAp$_5$A. (C) Ap$_5$A bound to adenylate kinase. (D) MgAp$_5$A bound to adenylate kinase. (E) As in (D), but with Mn²⁺ added ([Mn]:[Mg] = 1:50). Shifts upfield are negative. From Nageswara Rao and Cohn (1977b).

nance. Because, as mentioned earlier, the enzyme has a preferred site for metal-bound nucleotides, the exchange of 2-P and 4-P (and 1-P and 5-P) in the $E \cdot MgAp_5A$ complex cannot proceed with the same ease as before the addition of $Mg^{2+}$. This exchange is now slower because an additional step has been added; after $MgAp_5A$ dissociates from the enzyme, the metal ion may shift from one side of $Ap_5A$ to the other before $MgAp_5A$ reassociates with the enzyme again. Attempts to assign 2-P and 4-P unambiguously by the addition of $Mn^{2+}$ to the sample of Fig. 4D (see Fig. 4E) have not yielded definitive results. The chemical shift of ~7 ppm between the 2-P and 4-P resonances in Fig. 4D is one of the largest $^{31}P$ chemical shifts observed thus far for enzyme-bound substrates (noncovalent) and is comparable to the shifts observed in alkaline phosphatase (Bock and Sheard, 1975).

## 2. Methionyl-tRNA Synthetase

This enzyme catalyzes the aminoacylation of tRNA$^{Met}$ in a two-step reaction, the first step (activation reaction) of which involves nucleotidyl transfer:

$$E + \text{Methionine} + ATP \xrightleftharpoons{Mg^{2+}} E \cdot \text{Met-AMP} + PP_i$$
$$E \cdot \text{Met-AMP} + tRNA^{Met} \rightleftharpoons E + AMP + \text{MET-tRNA}^{Met}$$
(8)

$Mg^{2+}$ is an obligatory component for the first step of this reaction. $^{31}P$-NMR studies of enzyme complexes pertinent to the activation reaction allowed a comparison of the $^{31}P$-NMR parameters of diamagnetic enzyme-bound complexes of phosphoryl transfer and nucleotidyl transfer enzymes (Fayat et al., 1980). Most of the experiments were performed on a monomeric trypsin-modified enzyme that has a single ATP binding site. The modification does not cause any impairment of the activity of the activation reaction.

Binding to the enzyme causes relatively small chemical shifts (< 1 ppm) in $\beta$-P and $\gamma$-P resonances of ATP and MgATP; the largest change occurs in the chemical shift of the $\alpha$-P resonance (0.7–1.5 ppm). Enzyme-bound ATP and MgATP were also observed in the presence of methioninol, an inhibitor of the enzyme. The chemical-shift changes for $\alpha$-P are larger for both ATP (2 ppm) and MgATP (1.6 ppm) in the ternary complexes with methioninol, primarily because the binding constants of ATP and MgATP are increased in the presence of methioninol and the nucleotides are almost entirely in the enzyme-bound species in these complexes (Fayat et al., 1977). The addition of the inhibitor methioninol also causes increased changes in chemical shifts of the $\beta$-P and $\gamma$-P resonances of MgATP in the $E_M \cdot MgATP \cdot$ methioninol complex. However, these changes remain smaller than that of the $\alpha$-P.

Changes in the shift of γ-P may arise from a change of $pK_a$ of this phosphate at the active site of the enzyme.

Phosphorus-31 chemical shifts of dead-end complexes of various combinations of ligands bound to modified methionyl-tRNA synthetase (including $PP_i$, $MgPP_i$, and AMP), methioninol, methionine, and adenosine differ <1 ppm from those in free substrates. The largest change for the $PP_i$ chemical shift is found in the $E_M \cdot PP_i \cdot$ adenosine $\cdot$ methioninol complex and that for the AMP chemical shift in the $E_M \cdot$ AMP $\cdot$ methioninol complex. The resonance of $MgPP_i$ is not shifted with respect to free $MgPP_i$ in either $E_M \cdot MgPP_i \cdot$ adenosine, $E_M \cdot MgPP_i \cdot$ adenosine $\cdot$ methioninol, or $E_M \cdot MgPP_i \cdot$ adenosine $\cdot$ methionine complex.

Of all the $^{31}P$ resonances observed, the largest changes in chemical shift on binding to methionyl-tRNA synthetase were found for Met-AMP and the α-P of ATP, the former being shifted upfield ~2.5 ppm in the equilibrium mixture and the latter being shifted downfield ~1.5 ppm. A significant shift of the α-P of MgATP has not been observed in any of the MgATP complexes with the kinases. Large shifts of the α-P of nucleotides have been previously observed only in the case of binding of $MgAp_5A$ to adenylate kinase (Nageswara Rao and Cohn, 1977b) and of GDP and GTP to elongation factor Tu (G. E. Wilson, Jr. and M. Cohn, unpublished; see also Cohn, 1979). However, in these latter cases, the magnitude of the shift of the β-P resonances of these nucleotides on binding to protein are even greater than for the α-P, and such shifts exhibit a correlation with very tight nucleotide binding ($K_D = 10^{-7} - 10^{-9} M$) that may impose constraints on the geometry of the polyphosphate chain. In the case of the methionyl-tRNA synthetase, a similar explanation cannot be invoked for the large chemical shift for the α-P of ATP or MgATP on binding because the change in chemical shift on binding is large only for the α-P; furthermore, the value of $K_D$ for ATP or MgATP in the ternary enzyme complexes is only of the order of $10^{-4} M$. The tight binding ($K_D = 10^{-6} - 10^{-8} M$) of Met-AMP may be related to the large shift in its $^{31}P$ resonance induced on binding to the enzyme. A similar phenomenon occurs with P-enolpyruvate for which the $^{31}P$ chemical shift in the enzyme-bound complex ($K_D = 10^{-7} M$) in the equilibrium mixture of the pyruvate kinase reaction changes by ~4 ppm (Nageswara Rao et al., 1979).

### 3. Dihydrofolate Reductase

Dihydrofolate reductase catalyzes the reduction of dihydrofolate to tetrahydrafolate using NADPH as coenzyme. As part of their continuing and intensive investigation of this enzyme (from *Lactobacillus casei*) by various

NMR techniques [see, for example, the papers of Birdsall et al. (1980a,b), Hyde et al. (1980a,b), Groneborn et al. (1981a,b), and two reviews of Roberts et al. (1978) and Feeney et al. (1978), and other papers cited therein], the MRC group in England used $^{31}$P NMR to study the spectra of NADP$^+$ and NADPH and a number of their structural analogs in their binary and ternary complexes with the enzyme (the third component of the ternary complexes was either the substrate folate or one of two potent inhibitors of the enzyme methotrexate or trimethoprim).

$^{31}$P-NMR spectra of NADP$^+$ free in solution and bound to dihydrofolate reductase are shown in Fig. 5. The middle and bottom spectra in Fig. 5 are obtained with a coenzyme–enzyme concentration ratio of ~2:1. Free and bound resonances are observed simultaneously, indicating that the exchange between the two forms is slow (Hyde et al., 1980b). The 2'-P that resonates at −0.22 ppm shifts downfield by 2.9 ppm on binding to the enzyme. The two pyrophosphate resonances occur with chemical shifts +14.2 and +14.5 ppm and a coupling constant $J_{PP}$ = 20 Hz in free NADP$^+$; as a result they form a strongly coupled two-spin system (AB), and the outer lines of the AB quartet are low in intensity. Binding to the enzyme causes a substantial upfield shift of ~1.7 ppm for one of these resonances and a much smaller shift of ~0.3 ppm for the other. Addition of folate to the enzyme–NADP$^+$ complex leads to a sharpening of all the resonances and a further upfield shift of the pyrophosphate resonances. Because the presence of folate tightens the binding of NADP$^+$ to the enzyme, the additional linewidth of the signals in the binary complex is attributed to the effect of chemical exchange between bound and free forms, and the exchange rates are estimated from these linewidths.

The pyrophosphate $^{31}$P nuclei are spin-coupled through three bonds (P—O—C—H) to the respective 5'-CH$_2$ protons. This coupling is not resolved as a fine structure but causes a broadening of the $^{31}$P resonances owing, at least in part, to the short $T_2$ of the CH$_2$ protons. For the enzyme-bound NADPH, which exhibits a significant difference in linewidths of the two pyrophosphate signals (assuming a $T_2$ of 20–30 ms), these coupling constants were estimated. It is found that while the coupling constants of the high-field pyrophosphate signals are similar to those in free coenzyme (<5 Hz) the low-field pyrophosphate signals show a coupling constant ~13 Hz. The appreciable difference in these $^1$H–$^{31}$P coupling constants indicates significant difference in the conformation about C-5'—O bond in which the two C-4'—C-5'—O—P dihedral angles differ by ~50°. By comparing this results with crystallographic data on enzyme-bound NADPH (Feeney et al., 1975; Mathews, 1979), the two low-field pyrophosphate signals are assigned to adenosine 5'-phosphate and the high-field signal to the nicotinamide riboside 5'-phosphate.

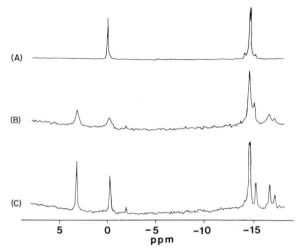

**Fig. 5.** $^{31}$P-NMR spectra (at 40.5 MHz) of NADP$^+$ free in solution (A), in the presence of dihydrofolate reductase (B), and in the ternary complex with folate (C). Samples for (B) and (C) spectra contained a twofold excess of NADP$^+$ over the enzyme so that separate resonances are observed for free and bound coenzymes. From Hyde et al., 1980b. Copyright 1980 American Chemical Society.

$^{31}$P-NMR spectra were obtained for a number of coenzyme analogs bound to the enzyme. The analogs used were nicotinamide hypoxanthine dinucleotide phosphate (NHDP$^+$), 3-acetylpyridine adenine dinucleotide (APADP$^+$), thionicotinamide adenine dinucleotide phosphate (TNADP$^+$), and the methyl β-riboside of 2′-phosphoadenosine-5′-diphosphoribose (PADPR-OMe). The 2′-P resonance shifted downfield to the same extent by $\sim 2.7-2.9$ ppm for all the analogs in both the binary and ternary complexes. The line shape of this particular resonance was analyzed on the basis of exchange to determine dissociation rate constants. Folate and methotrexate tend to increase substantially the binding of the coenzyme and its analogs to the enzyme and consequently enhance the dissociation rate constants. Such a change in binding is not accompanied by a change in the environment of the 2′-phosphates sensed by the $^{31}$P chemical shift in these compounds. The pyrophosphate resonances showed a greater degree of variation than the 2′-P resonances. Their chemical shifts vary over a range of $\sim 4$ ppm from one complex to another. The $^{31}$P chemical shifts in the binary and ternary complexes with the enzyme and other results of $^1$H-NMR experiments on these complexes suggest that the analogs APADP$^+$ and TNADP$^+$ bind the enzyme quite differently from the natural coenzyme.

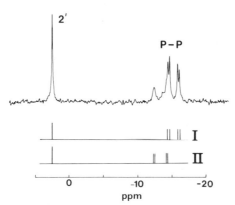

**Fig. 6.** $^{31}$P-NMR spectra (at 4.50 MHz) of E·trimethoprim·NADP$^+$ complex showing the mixture of two conformations of **I** and **II** (Gronenborn et al., 1981b). The theoretical spectra for **I** and **II** are shown by stick plots.

Addition of the inhibitor methotrexate produces changes in the pyrophosphate chemical shifts that are noticeably different from those produced by the substrate folate. Furthermore, interesting and striking differences are noticed between the ternary complexes of methotrexate and those of trimethoprim, another potent inhibitor of the enzyme (Gronenborn et al., 1981a,b). The spectrum of the ternary complex E·NADP$^+$·trimethoprim is shown in Fig. 6. Clearly this complex exists in two conformations: one has pyrophosphate chemical shifts nearly the same as those for E·NADP$^+$·methotrexate and the other has shifts nearly the same as those for E·TNADP$^+$·trimethoprim. The two conformations interconvert at a slow rate, estimated to be $\sim 0.5$ s$^{-1}$ with the change in protein conformation as the rate-limiting step. The two-bond $^{31}$P–$^{31}$P coupling constant in the first conformation is 21.0 Hz, nearly equal to its value in free coenzyme (20.8 Hz), whereas in the other conformation its value is reduced to 11 Hz. Structural explanation of this difference is not possible because of the lack of proper theory for these coupling constants. Chemical shifts and spin–spin coupling constants determined for enzyme-bound ternary complexes of trimethoprim with NADP$^+$ and its analogs and those for the ternary complex of methotrexate with NADP$^+$ are summarized in Table II. These data suggest that whereas the adenine end of NADP$^+$ binds in the same way in both conformations, the nicotinamide end does not. This difference is most likely to be associated with the effect of the enzyme on the pyrophosphate in producing a rotation of $\sim 50°$ about the C-5′—O bond at the nicotinamide end of the coenzyme. The experiments on dihydrofolate reductase complexes not only demonstrate that conformational selectivity occurs when specific ligands (or drugs) bind the receptor enzyme but also that there can be more than one preferred conformation for such binding.

## TABLE II

Phosphorus-31 Chemical Shifts[a] and Coupling Constants[b] of Ternary Complexes of Coenzyme (and Analogs) Bound to Dihydrofolate Reductase

| Complex | Conformation I | | | | Conformation II | | | |
|---|---|---|---|---|---|---|---|---|
| | $\delta\,2'$-P | $\delta$ PP | $\Sigma^3 J_P$ | $^2 J_{PP}$ | $\delta\,2'$-P | $\delta$ PP | $\Sigma^3 J_{PH}$ | $^2 J_{PP}$ |
| E·trimethoprim·NADP+ | 2.7 | −14.9<br>−16.4 | ND[c] | 21.0 | 2.7 | −12.9<br>−14.9 | ND[c] | 11 |
| E·trimethoprim·NHDP+ | 2.8 | −15.0<br>−16.4 | ND[c] | 21.5 | 2.8 | −12.7<br>−15.0 | ND[c] | 10 |
| E·methotrexate·NADP+ | 2.7 | −14.9<br>−16.5 | ∼13<br><5 | 21.0 | — | — | — | — |
| E·trimethoprim·TNADP+ | — | — | — | — | 2.7 | −12.8<br>−14.8 | <5<br><5 | 11 |
| E·trimethoprim·APADP+ | — | — | — | — | 2.7 | −12.7<br>−14.6 | ND[c] | 12 |

[a] In parts per million from inorganic phosphate (pH* 8.0), upfield shifts negative; $\delta\,2'$-P, chemical shift of 2'-phosphate resonance; $\delta$ PP, chemical shifts of pyrophosphate resonances.
[b] In Hertz; $\Sigma^3 J_{PH}$, sum of the $^{31}P-O-O-5'-^1H_A$ and $^{31}P-O-C-5'-^1H_B$ coupling constants estimated from linewidths with and without $^1$H decoupling and from spectral simulation; $^2 J_{PP}$, two-bond $^{31}P-O-^{31}P$ coupling constant measured directly from the $^1$H-decoupled spectra.
[c] ND, Not determined.

### 4. Myosin Subfragment-1 (S-1)

Shriver and Sykes (1981a,b) studied the intermediates of the ATPase reaction catalyzed by myosin subfragment-1 (S-1) by observing the enzyme-bound complexes of MgADP and MgAMP·PNP (which is a nonhydrolyzable MgATP analog) using $^{31}$P NMR. At 25°C and pH 7.0 in the presence of 0.1 $M$ KCl, the enzyme-bound $\beta$-P(MgADP) resonance shifts 3.7 ppm downfield with respect to free $\beta$-P(MgADP). The bound $\beta$-P(MgADP) resonance is insensitive to pH from 6.6 to 8.0 (note that the chemical shift of $\beta$-P of free MgADP changes in this region of pH with a p$K_a \approx 6.0$; Nageswara Rao and Cohn, 1977a). Significantly, at 0°C the enzyme-bound $\beta$-P(MgADP) splits into two resonances of equal intensity, one of them at the same position as above and the other shifted further downfield by 0.7 ppm (Fig. 7A). This splitting was studied as a function of temperature between 0 and 25°C (Fig. 7B). The results were analyzed using lineshape simulation on the basis of two different forms of enzyme-bound MgADP with a temperature-dependent equilibrium constant. The $\alpha$-P(MgADP) resonance showed negligible chemical shift (<0.5 ppm) in the enzyme-bound form in the temperature range studied. Enzyme-bound complexes of the nonhydrolyzable MgATP analog MgAMP·PNP also show two forms

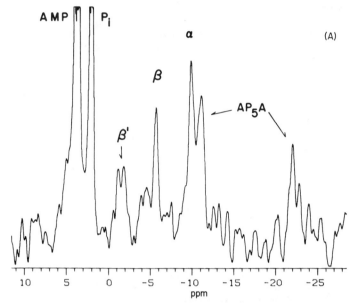

**Fig. 7.** (A) $^{31}$P-NMR spectra of myosin subfragment·ADP complex at 0°C. Assignments: free ADP, $\alpha$-P at $-9.83$ ppm and $\beta$-P at $-5.84$ ppm; bound ADP, $\alpha$-P at $-9.8$ ppm and $\beta$-P gives two signals at $-1.15$ and $-1.9$ ppm; S-1 ATPase coupled with contaminant adenylate kinase leads to the generation of AMP (3.5 ppm) and $P_i$ (2.0 ppm) signals. Ap$_5$A, added to inhibit adenylate kinase, gives signals at $-1$ and $-22$ ppm. (B) (Curve 1) Temperature dependence of the $\beta$-P resonance of ADP bound to myosin subfragment S-1 [see part (A)]. (Curve 2) Computer simulation of spectra shown in curve 1. The relative populations of the two forms derived from simulations were used to determine the ratio of the low-temperature (low-field) state to the high-temperature (high-field) state defined by $K$. From Shriver and Sykes (1981b). Copyright 1981 American Chemical Society.

with a temperature-dependent equilibrium. Shifts were observed in the bound complexes of MgAMP·PNP for all three phosphate resonances. The nucleotide (N) binding to myosin S-1 is thus represented by the scheme

$$M + N \rightleftharpoons M \cdot N \rightleftharpoons M\#N \rightleftharpoons M^*N$$

where M·N is the initial recognition complex in fast exchange with free nucleotide, and M#N and M*N are the two complexes observed by $^{31}$P NMR. The slow exchange between the resonances for the two enzyme-bound nucleotide conformations is consistent with the previous rate constants ($<20$ s$^{-1}$) measured by temperature-dependent fluorescence experiments on the system (Bechet *et al.*, 1979). The results of Shriver and Sykes (1981a) confirm other suggestions made earlier indicating two forms of myosin–nucleotide complexes. They have used this evidence to propose an ATPase scheme in which the presence of the conformations and the asso-

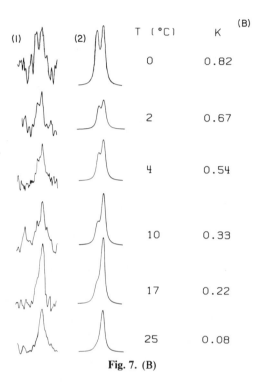

**Fig. 7. (B)**

ciated energetics and kinetics (Shriver and Sykes, 1981b) are related to the "power stroke" in a muscle fiber.

## 5. G-Actin

Actin, another important protein in muscle contraction, was studied by Brauer and Sykes (1981a,b, 1982). Monomeric G-actin has one high-affinity binding site for ATP and one high-affinity binding site for divalent cations. In order to avoid polymerization of G-actin at the high concentrations required by $^{31}$P NMR experiments, the protein was modified by nitration of tyrosine-69. This procedure does not significantly alter the spectrum of ATP bound to the protein (Fig. 8A). As seen in Fig. 8A, in the presence of excess ATP separate signals were observed for the bound and free ATP for the three phosphate groups of ATP indicating that slow-exchange conditions prevail. The chemical-shift differences were 1–2 ppm. From these values an upper limit to the exchange rate was estimated to be 480 s$^{-1}$ between free and bound forms at pH 7.8 and 4°C. The spectra were obtained at four different $^{31}$P-NMR frequencies: 36.4, 81.0, 109.3, and 162.0 MHz (Fig. 8B). The

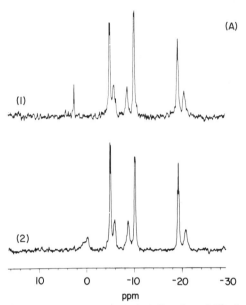

**Fig. 8.** (A) $^{31}$P-NMR spectra (at 109.3 MHz) of (1) G-actin and (2) nitrated G-actin at 4°C and pH 7.8. Both samples have free ATP in the molar ratio 0.5 mol free ATP to 1.0 mol protein. (B) $^{31}$P-NMR spectra of nitrated G-actin obtained at 162.0, 109.3, 81.0, and 36.4 MHz (top to bottom) at 4°C and pH 7.8. The spectra show (1) $\gamma$-P of free and protein-bound ATP, (2) $\alpha$-P of protein-bound and free ATP, and (3) $\beta$-P of free and protein-bound ATP. From Brauer and Sykes (1981b). Copyright 1981 American Chemical Society.

substantial increase in linewidths of bound ATP resonances with increasing $^{31}$P frequency indicates that $^{31}$P chemical-shift anisotropy is a mechanism of relaxation for these resonances (see Section II,A,3). Measurements of $T_1$ were made on these signals at different frequencies. Analysis of the results on the basis of chemical-shift anisotropy and dipole–dipole interaction mechanisms allows an estimate of the chemical-shift anisotropy factor $\Delta\sigma[1 + (\eta^2/3)]^{1/2}$ for the three signals to be ~250 ppm, larger than the 80–180 ppm determined for a number of model organophosphate compounds. Furthermore, from the $^{31}$P-NMR spectra of G-actin-bound ATP at 36.4 MHz, in which case the resonances are relatively narrow, it was observed that $J_{\alpha\beta} = 12.0 \pm 2$ Hz and $J_{\beta\gamma} = 20 \pm 2$ Hz, whereas for free ATP, $J_{\alpha\beta} = J_{\beta\gamma} = 19.6$ Hz. This change in coupling constant indicates a possible conformational change of the phosphate chain (Nageswara Rao *et al.*, 1978a).

## 6. *Ribonuclease $T_1$*

Ribonuclease (RNase) catalyzes the hydrolysis of 3′,5′-phosphodiester linkages of ribonucleic acids or nucleotide esters at the 5′-ester bond.

3. ³¹P NMR of Enzyme Complexes

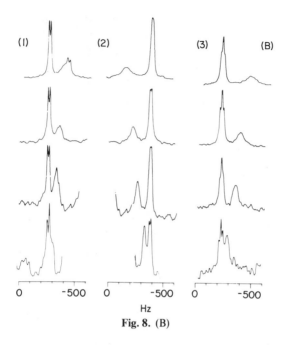

**Fig. 8.** (B)

Previous work on the enzyme–inhibitor complexes of RNase by ³¹P NMR led to some discrepancies in the results obtained by several different research groups (Cohn and Nageswara Rao, 1979; Jardetzky and Roberts, 1981). Nucleoside monophosphates were used as inhibitors, and attention was focused on determining the ionization state of the inhibitors in their enzyme-bound complexes. Arata et al. (1979) published a ¹H- and ³¹P-NMR study of RNase $T_1$ (from Takadiastase) that has three histidine residues at positions 27, 40, and 92, two of which, His-40 and His-92, were implicated in the active site. They used 3'-GMP (guanosine 3'-monophosphate), a strong competitive inhibitor, to probe the active site through ³¹P NMR and found significant perturbation of the pH variation of the chemical shift of the histidine C-2 proton peaks of His-40 and His-92 in the presence of the inhibitor. The pH dependence of the ³¹P chemical shift of 3'-GMP, which shows a p$K_a$ ~ 5.5, is altered when bound to the enzyme and displays a p$K_a$ of ~6.5. This suggests that at a pH of 5.5, at which the inhibitor binds strongest to the enzyme, the inhibitor may be in the monoanionic form, reinforcing previous evidence for stronger binding of the monoanionic than the dianionic form. Tritium incorporation at the C-2–H position takes place much faster for His-40 than for His-92, indicating that His-40 is more exposed to the solvent than His-92. On the basis of this and the ¹H- and ³¹P-NMR results, Arata et al. (1979) presented a model for the active site in which His-40, His-92, and Glu-58 are involved. His-92 is chosen for binding of the guanine base with high specificity, and His-40 and Glu-58 are responsible for the catalytic activity of the enzyme. These findings may be contrasted with the $^{31}$P- and $^1$H-NMR results on RNase A, for which the dianionic form of the inhibitor may be the more stable form (Gorenstein et al., 1976; see also Chapter 1), and two histidine residues His-12 and His-119 are implicated in the catalytic function (Jardetzky and Roberts, 1981).

## 7. Triosephosphate Isomerase

Triosephosphate isomerase is the glycolytic enzyme that interconverts D-glyceraldehyde 3-phosphate and 1,3-dihydroxyacetone phosphate. Campbell et al. (1979) studied $^{31}$P-NMR spectra of enzyme–substrate and enzyme–inhibitor complexes of this enzyme from chicken muscle. Complexes were observed with the substrate dihydroxyacetone phosphate and two inhibitors, glycerol 3-phosphate which is a substrate analog and 2-phosphoglycolate which is considered a potential transition-state analog. The $^{31}$P-NMR spectrum of a sample containing the enzyme and 2-phosphoglycolate showed separate resonances for the bound and free forms of the inhibitor. The chemical shift for the free species showed the normal variation with pH in the range 5.5–8.5, (p$K_a$ = 6.4), whereas that for the bound form did not change with pH in the same range. Since separate signals are observed, the condition for slow exchange is applicable and on the basis of a separation of 5 Hz between the bound and free signals at pH 8.5, an upper limit of 30 s$^{-1}$ was deduced for the dissociation rate constant. Furthermore, the increase in linewidth of the resonance of the free species owing to the exchange with the bound form, along with a measurement of the linewidth of the bound species allowed an estimate of 3–6 s$^{-1}$ for the dissociation rate constant. The absence of variation of the chemical shift of the bound species with the pH was used to conclude that the inhibitor is in the trianionic form and that a group on the protein, believed to be glutamate-165, is protonated in the process. Implicit in this deduction is the assumption that the variation of chemical shift with pH for a given ionic species is the same on and off the enzyme.

$^{31}$P-NMR spectra of a sample of rac-glycerol 3-phosphate in the presence of the enzyme showed two resonances. One of these belongs to the L-isomer which binds to the enzyme poorly and the other to the bound and free forms of D-isomer in fast exchange. The chemical shift of the signal from the D-isomer varies with pH, both monoanion and dianion bind the enzyme; the latter binds better. The substrate dihydroxyacetone phosphate behaves essentially like glycerol 3-phosphate. The keto form of the substrate binds an order of magnitude stronger than the hydrated form.

3. ³¹P NMR of Enzyme Complexes

## B. Enzyme-Bound Equilibrium Mixtures

Phosphorus-31 NMR was used to observe the substrates and products interconverting on the surface of the enzyme at equilibrium for the five phosphoryl transfer enzymes and one nucleotidyl-transfer enzyme discussed in Section III,A. In these experiments the concentrations of the enzyme (3–5 m$M$) and all the different substrates, whether they contain ³¹P or not (2–4 m$M$), are such that 80–90% of the reactants and products will be bound to the enzyme (for dissociation constants less than ~200 $\mu M$). The results obtained for the different enzymes are now summarized.

### 1. Arginine Kinase and Creatine Kinase

Figure 9A shows the ³¹P-NMR spectrum of an equilibrium mixture of the arginine kinase reaction established with catalytic quantities of enzyme (Nageswara Rao *et al.*, 1976). The signals of the six phosphate groups contained in the reactants and products are labeled. The $\gamma$-P(ATP) signal is a doublet due to spin–spin coupling with the ³¹P nucleus in $\beta$-P(ATP). The $\alpha$-P(ATP) is also a doublet due to coupling with $\beta$-P(ATP). The $\beta$-P(ATP) signal is a 1:2:1 triplet because of simultaneous spin–spin coupling with $\alpha$-P(ATP) and $\gamma$-P(ATP) with coupling constants that are equal. The $\beta$-P(ADP) and $\alpha$-P(MgADP) signals are also doublets owing to their mutual spin–spin coupling. The $\alpha$-P signals of ATP and ADP are both rather broad because of a small spin–spin coupling of the ³¹P nucleus in $\alpha$-P with the respective 5′ protons. From the areas under the different signals in Fig. 9A, the catalytic equilibrium constant

$$K_{eq} = \frac{[P_1][P_2]}{[S_1][S_2]} \tag{9}$$

where ADP and P-arginine are the products $P_1$ and $P_2$ and may be readily evaluated. At pH 7.2 and 12°C, $K_{eq} = 0.1$ for arginine kinase.

The spectrum in Fig. 9B is obtained by setting up an equilibrium mixture of the reactants and products of arginine kinase such that all the substrates are bound to the enzyme. It can be readily shown that most of the line broadening in Fig. 9B is caused by the participation of almost all the components of the system in the reaction most of the time. By adding EDTA to the sample of Fig. 9B to sequester the Mg²⁺ from the enzyme-bound complexes, the reaction was stopped. It is clear from Fig. 9C that binding to the enzyme in the absence of reaction does not cause excessive broadening of ³¹P signals. It may also be seen from Fig. 9C that the ³¹P chemical shifts of the different phosphate groups, with the exception of $\beta$-P(ADP), remain essentially unaltered by binding of the substrates to the enzyme. The

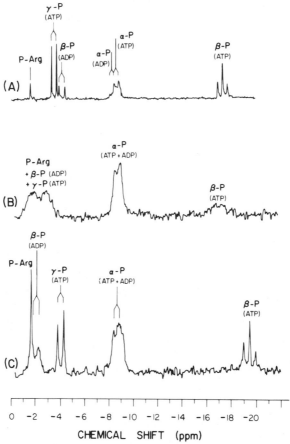

**Fig. 9.** ³¹P-NMR spectra (at 40.3 MHz) of the arginine kinase reaction (Nageswara Rao et al., 1976) at 12°C and pH 7.25. (A) Equilibrium mixture ([nucleotide]/[E] = 4000) of overall reaction, catalytic concentration of enzyme. (B) Equilibrium mixture ([nucleotide]/[E] = 0.97) of enzyme-bound substrates and products. (C) Same as (B) with EDTA added to chelate $Mg^{2+}$ and stop reaction, no chemical exchange. Shifts upfield are negative.

position of β-P(ATP) at a higher field than in Fig. 9B is due to the removal of $Mg^{2+}$; the same shift occurs for ATP of the enzyme. Furthermore, the equilibrium of the reaction of enzyme-bound substrates appears to favor P-arginine much more than is the case for catalytic levels of the enzyme (compare Fig. 9B with 9A).

The three limiting conditions of exchange defined in Section II,A,2 qualitatively explain the line shapes in Fig. 9B. The α-P(ATP) ⇌ α-P(ADP) exchange is in the fast-exchange limit, owing to the small chemical shift between the signals, and a single resonance is obtained for both these signals. The β-P(ATP) ⇌ β-P(ADP) exchange is in the slow-exchange limit because

## 3. ³¹P NMR of Enzyme Complexes

of the large chemical shift between the respective signals leading to a broadening of these resonances without a shift in frequency. The $\gamma$-P(ATP) $\rightleftarrows$ P(P-arginine) exchange falls in the intermediate-exchange condition accompanied by both a line broadening and a frequency shift.

From the linewidth of the $\beta$-P(ATP) signal in Fig. 9B relative to that in Fig. 9C, the lifetime of the E·MgATP·arginine complex in the reaction can be estimated. Furthermore, from the areas under the superposed $\alpha$-P signals of ATP and ADP and the isolated $\beta$-P(ATP) signals, the ratio of the concentrations [E·MgATP·arginine] and [E·MgADP·P-arginine] may be obtained. Thus the equilibrium constants and exchange rates associated exclusively with the interconversion step of the reaction $E \cdot S_1 \cdot S_2 \rightleftarrows E \cdot P_1 \cdot P_2$ are obtained from the experiment in a rather straightforward manner.

The ability to isolate and monitor exclusively the step of interconversion of the reactants and products on the surface of the enzyme is one of the attractive features of the ³¹P NMR of enzyme-bound substrates of these reactions. It must be noted, however, that the accuracy in the determination of the different parameters, equilibrium constants, and exchange rates is primarily governed by the feasibility of having the substrates present only in the active complexes, (e.g., in the case of Fig. 9B, all enzyme-bound complexes should be in either of the two forms E·MgATP·arginine or E·MgADP·P-arginine). Departure from this condition leads to the presence of a substantial fraction of enzyme-bound complexes that do not undergo the reaction. In such cases the interconversion rates determined from the spectra may only be taken as lower limits of the rates.

It should be noted that the $\beta$-P(MgATP) signal is a 1:2:1 triplet in the absence of exchange and as discussed in Section II,.A,2, the influence of chemical exchange on spin multiplets is rather complex and requires a detailed analysis for quantitative purposes. The results of such an analysis for the spectrum in Fig. 9B and other results obtained with some of the other enzymes are discussed in Section IV. For arginine kinase this analysis yields a value of 120 s⁻¹ for the reciprocal lifetime of E·MgATP·arginine complex (i.e., the rate of phosphoryl transfer).

A comparison of the interconversion rates with the overall rates of the reaction obtained under the same conditions, by enzymatic assay, allows one to determine if the interconversion step is the rate-limiting step for the reaction. The equilibrium constants of the enzyme-bound reactants and products are of physiological relevance in situations where the enzyme and substrate concentrations are comparable. For the specific case of arginine kinase, the equilibrium constant

$$K'_{eq} = \frac{[E \cdot P_1 P_2]}{[E \cdot S_1 \cdot S_2]} \simeq 1.2 \tag{10}$$

is an order of magnitude different from $K_{eq}$. The interconversion rates in either direction are an order of magnitude faster than the overall rate of the reaction, indicating that the interconversion is not the rate-determinating step of the arginine kinase reaction.

The results obtained for creatine kinase equilibrium mixture (Nageswara Rao and Cohn, 1981) are broadly similar to those for arginine kinase. At pH 8.0 and 4°C, $K_{eq} = 0.08$ whereas $K'_{eq} \simeq 1$. Computer simulations allow an estimate of 90 s$^{-1}$ for the interconversion rate on the enzyme. The interconversion step is not rate-limiting for the overall reaction. The accuracy in the determination of the equilibrium constants and rates of interconversion of enzyme-bound substrates for creatine kinase is somewhat impaired by two factors: (1) during the experiments there is an irreversible accumulation of $P_i$ owing to an inherent (or partly a contaminant) ATPase activity, and (2) the substrate and products do not bind creatine kinase tightly enough so that appreciable concentrations of free substrates are likely to be present in the sample.

## 2. Adenylate Kinase

The adenylate kinase reaction (Noda, 1973)

$$ATP + AMP \underset{}{\overset{Mg^{2+}}{\rightleftharpoons}} 2ADP$$

is particularly well suited for study by $^{31}P$ NMR because all four substrates contain $^{31}P$ nuclei. Of particular interest in this reaction was the attempt to establish the distinction, if any, between the acceptor and donor ADP molecules that are converted to ATP and AMP, respectively, in the reaction and to pinpoint the possible role of $Mg^{2+}$ in effecting such a distinction (Nageswara Rao et al., 1978a). The presence of two identical ADP molecules as substrates provides an additional chemical-exchange process that is unique to this enzyme. This exchange between the acceptor and donor ADP molecules (Fig. 10) is unrelated to the chemical exchanges occurring during the adenylate kinase reaction but should be considered in any detailed interpretation of the spectra.

The key experiment that provided the clue for the possible distinction between the donor and acceptor ADP is illustrated in Fig. 10. The $^{31}P$-NMR spectrum of an equilibrium mixture of enzyme-bound substrates of porcine adenylate kinase is obtained after $\sim 4$ h of NMR signal averaging (Fig. 10A). In contrast to the arginine kinase experiments, the spectrum in Fig. 10A is difficult to explain, particularly because no resonances are seen either in the region where AMP resonates ($\sim 4$ ppm) or in the region for $\beta$-P of MgATP (about $-19.0$ ppm). Note the presence of a small $P_i$ peak at $\sim 1.5$ ppm, indicating that some ATP hydrolysis occurred during the signal accumulation owing to the contaminant ATPase in the sample.

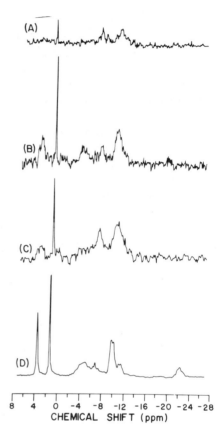

**Fig. 10.** $^{31}$P-NMR at (40.3 MHz) spectra of an equilibrium mixture (E + ATP + AMP + MgCl$_2$) of substrates and products fully bound to porcine adenylate kinase (Nageswara Rao *et al.*, 1978a) at 15°C and pH 7.0. (A) After ~4 h of accumulation. (B) As in (A), but after ~12 h; separate resonances for each ADP bound to enzyme. (C) As in (B), but with addition of exogenous ATP. (D) As in (C), but with EDTA added to stop reaction. Shifts upfield are negative.

Attempts to obtain a signal-to-noise ratio better than that in Fig. 10A by signal averaging for ~12 h on the same sample yielded the spectrum shown in Fig. 10B, which appears quite different from that in Fig. 10A. The P$_i$ resonance is considerably enhanced, there is an AMP resonance, and there are now two resonances in the region of the β-P resonance of ADP (−4 to −7 ppm) instead of one at −6.5 ppm, as in Fig. 10A. The appearance of the spectrum in Fig. 10B may be explained by noting that as the irreversible ATP hydrolysis proceeds concomitantly with the adenylate kinase reaction, the concentration of ATP is progressively depleted while the concentrations of AMP and P$_i$ increase. The resonance of β-P(MgADP) is in slow exchange with that of β-P (MgATP) in the adenylate kinase reaction, as in the arginine kinase reaction. As the ATP is progressively depleted, a greater fraction of MgADP would be found in abortive enzyme-bound complexes such as E·AMP·MgADP complexes. The MgADP resonance is then effectively narrowed and becomes observable.

This explanation implies that in Fig. 10A the resonance of MgADP is too broad to be observed because of chemical exchange in the presence of sufficient amounts of ATP in the sample. To verify this interpretation, a small quantity of ATP was introduced into the sample of Fig. 10B and the $^{31}$P-NMR spectrum recorded again. In the resulting spectrum (Fig. 10C) the MgADP resonance is significantly broadened. There should then be an appreciable concentration of ATP in the sample of Fig. 10C, although no resonance is observable in the vicinity of $-19$ ppm. Apparently this resonance is still too broad to be observable when turnover is occurring. However, addition of a sufficient quantity of EDTA to the sample of Fig. 10C to remove $Mg^{2+}$ from the reaction complexes, stops the reaction. The spectrum of the resulting sample shown in Fig. 10D does exhibit an unmistakable $\beta$-P(ATP) resonance. The contaminant ATPase, which is normally a nuisance in these experiments, has thus serendipitously provided the evidence for the distinction between the two nucleotide binding sites on this enzyme. Further demonstration of this differentiation in the $^{31}$P-NMR spectra of $E \cdot Ap_5A$ complexes was presented in Section III,A,1.

It may also be seen from Fig. 10D that the two $\beta$-P(ADP) signals of Fig. 10B merge into a single peak. This result, along with several other experiments, was used to show that a single resonance is obtained for the two $\beta$-P(ADP) groups because of fast exchange between the identical ADP molecules at the acceptor and donor sites on the enzyme. The rate of this exchange is severely reduced in the presence of $Mg^{2+}$, leading to two distinct resonances in Fig. 10B. The evidence strongly suggests that MgADP does not bind at the donor ADP site. Furthermore, independent experiments on ATP bound to the enzyme in the presence of varying concentrations of $Mg^{2+}$ show that whereas ATP or MgATP binds at the ATP site, only metal-free ATP binds at the AMP site. It has also been shown that the optimal $Mg^{2+}$ concentration for the reaction is $[Mg^{2+}] = [ATP] + \frac{1}{2}[ADP]$, that is, one metal ion per reaction complex. It is quite evident that in addition to facilitating catalysis of the reaction as in other kinases, $Mg^{2+}$ provides the distinction between acceptor and donor ADP molecules on the enzyme.

For adenylate kinase, $^{31}$P-NMR measurements yield $K_{eq} \approx 0.4$ and $K'_{eq} \approx 1.5$ at 4°C and pH 7.0. The rate of phosphoryl transfer was estimated to be 600–700 s$^{-1}$ (Nageswara Rao et al., 1978a), nearly an order of magnitude faster than the rate of overall reaction. Once again the interconversion rates are not rate-limiting for the reaction. Computer simulations of the line shapes of some of the spectra in Fig. 10 and others obtained on the spectrum yield not only a reliable phosphoryl transfer rate of 600 s$^{-1}$ on the enzyme but also provide evidence for and allow an estimate of the rate of exchange of the two ADP molecules between the acceptor and donor sites on the enzyme (see Section IV for a brief description of these results).

### 3. 3-P-Glycerate Kinase and Pyruvate Kinase

Both of these glycolytic enzymes are used to manufacture ATP in anaerobic glycolysis. The catalytic equilibrium for both cases overwhelmingly favors ATP by a factor of ~3000, that is, $K_{eq} = 3 \times 10^{-4}$. $^{31}$P-NMR spectra of catalytic and enzyme-bound equilibrium mixtures for both the enzymes are shown Fig. 11. The one-sidedness of the catalytic equilibrium is evidenced by the fact that for pyruvate kinase, the small signal for P-enolpyruvate in the spectrum (this signal would normally be dismissed as noise but for the fact that P-enolpyruvate is known to resonate at that position) is obtained by having a 15-fold excess of pyruvate over ATP; for 3-P-glycerate kinase no signal is visible for 1,3-bis-P-glycerate. However, in the enzyme-bound equilibrium mixture, substantial signals can be seen for all substrates containing $^{31}$P, indicating that $K'_{eq} \approx 1$ for both these enzymes. For the

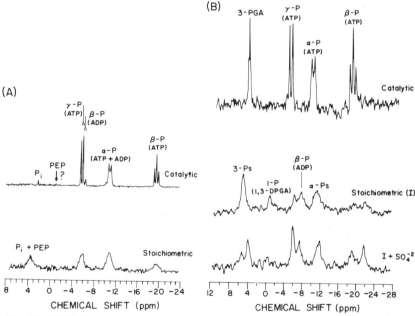

**Fig. 11.** $^{31}$P-NMR spectra of equilibrium mixtures of (A) rabbit muscle pyruvate kinase (Nageswara Rao et al., 1979) and (B) yeast 3-P-glycerate kinase (Nageswara Rao et al., 1978b) at catalytic and stoichiometric enzyme concentrations. Spectra for pyruvate kinase were obtained at 40.3 MHz, 15°C, and pH 8.0; for 3-P-glycerate kinase catalytic equilibrium mixture was recorded at 24 MHz and stoichiometric equilibrium mixtures at 109.3 MHz, 1°C, and pH 7.0. The appearance a small P-enolpyruvate signal in the catalytic equilibrium mixture of pyruvate kinase occurs in the presence of a 15-fold excess of pyruvate over ATP. Shifts upfield are negative.

3-P-glycerate kinase reaction, Fig. 11 also shows the effect of sulfate ion on the spectrum of the stoichiometric equilibrium mixture. That $K'_{eq}$ is nearly equal to 1 for arginine kinase, creatine kinase and adenylate kinase for which $K_{eq}$ indicates significant catalysis in either direction, or for 3-P-glycerate kinase and pyruvate kinase for which $K_{eq}$ overwhelmingly favors catalysis in the direction of ATP production is an interesting result, the implications of which are discussed at the end of this section.

## 4. Methionyl-tRNA Synthetase

A $^{31}$P spectrum of an equilibrium mixture of the activation reaction of this enzyme [Eq. (8)] is shown in Fig. 12. An ambiguity existed initially regarding the position of the enzyme-bound Met-AMP in this spectrum. The $^{31}$P resonance of chemically synthesized Met-AMP was found to be $-7.5$ ppm at pH 4.1. In order to arrive at the assignment shown in Fig. 12A,

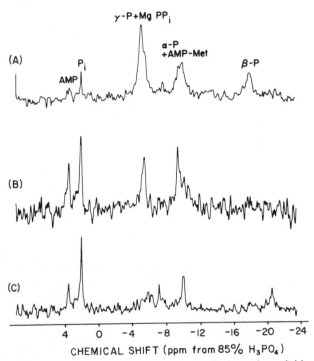

**Fig. 12.** $^{31}$P-NMR spectra (at 145.7 MHz) of (A) an equilibrium mixture initiated by mixing MgATP, methionine, and modified methionyl-tRNA synthetase ($K' =$ [E·AMP-Met·MgPP$_i$]/[E·Met·MgATP]) with [E] > [ATP]; at 20°C and pH 7.6; (B) after saturating the $\beta$-P(ATP) resonance by radio-frequency irradiation; and (C) as in (B), but with added excess EDTA. Shifts upfield are negative. From Fayat et al. (1980).

## 3. $^{31}$P NMR of Enzyme Complexes

a saturation transfer experiment was performed. A spectrum of the sample of Fig. 12A was recorded while the $\beta$-P(ATP) resonance was irradiated with a saturating radio-frequency field. When the reaction proceeds at a rate faster than the relaxation rates of $^{31}$P nuclei in the different complexes in equilibrium, the saturation of the $\beta$-P(ATP) resonance is transferred to $^{31}$P resonances of the chemical species into which this nucleus converts during reaction, that is, MgPP$_i$ in the forward direction and subsequently $\gamma$-P(ATP) on reversal. The result of the experiment is shown in Fig. 12B. The group of resonances in the region $-4$ to $-6$ ppm is now significantly reduced in intensity and the total intensity is the same as that of the group in the region $-9$ to $-11$ ppm. These results clearly indicate that the PP$_i$ and $\gamma$-P(ATP) resonances superpose and the Met-AMP resonance superposes that of $\alpha$-P(ATP). The spectrum in Fig. 12A also contains both P$_i$ and AMP resonances owing to contamination of inorganic pyrophoshphtase in the enzyme preparation. After the spectrum in Fig. 12B was recorded, EDTA was added to the sample to remove Mg$^{2+}$ from PP$_i$- and ATP-containing complexes, and a $^{31}$P-NMR spectrum of the resultant sample was recorded (Fig. 12C). The spectrum shows depletion of ATP because of the inorganic pyrophosphatase. However a distinct resonance is observed about $-10$ ppm for the residual E·Met-AMP. The results depicted by Fig. 12B,C confirm the assignment shown in Fig. 12A.

The equilibrium constant for the interconversion of the enzyme-bound substrates to enzyme-bound products for the methionyl-tRNA synthetase reaction has been found to have a value between 1 and 2 at pH 7.65 and 20°C. This value is in good agreement with that determined from stopped-flow experiments (Hyafil *et al.*, 1976), where equilibrium constants of 1.5 and 1.1 were found for the native and trypsin-modified enzyme, respectively. From the line width of the $\beta$-P resonance in the NMR spectrum, an approximate value of $360 \pm 60$ s$^{-1}$ is estimated as an upper limit for the rate of adenylylation reaction for the central complex of the modified synthetase. The validity of this value rests on the assumption that E·Met-AMP·MgPP$_i$ arises from E·MgATP·Met without the occurrence of any intermediate. Taking into account that the measured value is overestimated because the effect of spin–spin coupling on the line shape of the $\beta$-P resonance has been neglected (see Section IV), the rate from the NMR data agrees satisfactorily with the rate that was derived from the kinetic experiments ($259 \pm 9$ s$^{-1}$) (Hyafil *et al.*, 1976). Thus, the quantitative aspects of the $^{31}$P NMR of the enzyme-bound components of the methionyl-tRNA synthetase reaction at equilibrium support the earlier interpretation of the activation reaction scheme.

For all the kinase reactions studied thus far and the nucleotidyl transfer enzyme methionyl-tRNA synthetase, the equilibrium constants of enzyme-bound substrates and products is approximately unity. Similar results were obtained for yeast hexokinase (Wilkinson and Rose, 1979) and for muscle triosephosphate isomerase (Iyengar and Rose, 1982). These values of the equilibrium constants seem to be consistent with the arguments given by Albery and Knowles (1976a,b) in connection with their kinetic studies on triosephosphate isomerase, that for a perfect (or well-evolved) enzyme, this equilibrium constant should indeed approach unity (Knowles and Albery, 1977). A corollary to their argument is that interconversion step is not rate-limiting in the overall reaction, which is the case for all the enzymes discussed so far (see also Hassett *et al.*, 1982). If the arguments of Albery and Knowles are valid, the kinases and methionyl-tRNA synthetase discussed could all be considered to have attained a high degree of maturity in their evolution. Nevertheless, the validity of their suggestion seems to depend, among other things, on the assumption that evolutionary pressure on the enzymes in the cell manifests itself primarily through kinetics of the various steps of the respective enzymatic reactions. Furthermore, it appears difficult to determine the significance of the hypothesis unless an independent criterion could be established to assess evolutionary maturity.

## IV. Computer Calculation of NMR Line Shapes under the Influence of Exchange

In the analysis of $^{31}$P-NMR spectra of equilibrium mixtures of interconverting enzyme-bound reactants and products of the arginine kinase reaction (Nageswara Rao *et al.*, 1976), the increase in the linewidth of $\beta$-P(MgATP) resonance owing to the interconversion was used to estimate the reciprocal lifetime of the E·MgATP·arginine complex. This quantity is equal to the rate of phosphoryl transfer on the surface of the enzyme. This procedure, while qualitatively correct, implicitly ignores the effects of spin–spin coupling on the line shapes. A density-matrix theoretical framework for the calculation of the NMR line shapes of exchanging systems in the presence of spin–spin coupling has existed for some time (see Kaplan and Fraenkel, 1980, and references therein). A computational procedure for simulating the line shapes directly applicable to the $^{31}$P-NMR spectra of the enzyme-bound equilibrium mixtures of kinase reactions, in which the second substrate in the forward reaction does not contain a $^{31}$P nucleus, has been formulated (Vasavada *et al.*, 1980; Nageswara Rao, 1983). Such an exchange is of the general category

$$ABC \rightleftarrows A'B' + C'$$

where A, B, and C represent the $^{31}$P nuclei $\alpha$-P, $\beta$-P, and $\gamma$-P of ATP; A', B' represent those in ADP; and C′ represents that in the phosphorylated second substrate, respectively. The method involves point-by-point calculation of NMR intensities, covering the width of the entire spectrum, with the help of a computer. The parameters required for the calculation are six chemical shifts $\omega_A$, $\omega_B$, $\omega_C$, $\omega_{A'}$, $\omega_{B'}$, and $\omega_{C'}$, three nonvanishing, spin–spin coupling constants $J_{AB}$, $J_{BC}$, and $J_{A'B'}$, the linewidths of the different transitions in the absence of exchange, and the forward and reverse exchange rates $\tau_1^{-1}$ and $\tau_2^{-1}$, respectively. The ratio $\tau_1/\tau_2$ is available from the equilibrium constant determined from the areas under the different resonances (Nageswara Rao et al., 1976), so that only one exchange rate ($\tau_1^{-1}$) was to be chosen. Theoretical spectra may be obtained for various sets of parameters and compared with the experimental spectrum in order to pick the best set.

## A. Arginine Kinase and Creatine Kinase

Computer simulation of the $^{31}$P spectra of enzyme-bound equilibrium mixtures were made for the arginine kinase and creatine kinase reactions. It is difficult to determine accurately some of the chemical shifts in the reaction complex because of the complex line shapes arising from the interconversion of the reactants and products. Therefore, the chemical shifts, linewidths, and spin–spin coupling constants from the ternary enzyme complexes E·MgATP and E·MgADP were used as a starting point in the calculation. Figure 13 shows a comparison of the observed spectrum for the arginine kinase equilibrium mixture, along with the computed spectrum that agrees best with it. An accuracy of ~0.1–0.2 ppm in the chemical shifts

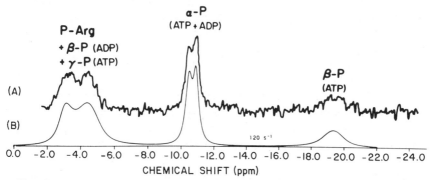

**Fig. 13.** Comparison of experimental (at 40.3 MHz) and computer-calculated $^{31}$P-NMR spectra for the equilibrium mixture of enzyme-bound substrates and products of arginine kinase (12°C; pH 7.25). (A) Experimental spectrum (Nageswara Rao et al., 1976). (B) Computer-simulated spectra with a phosphoryl transfer rate of 120 s$^{-1}$. Shifts upfield are negative. From Vasavada et al., (1980).

and ~5 Hz in linewidths and ~10 s$^{-1}$ in the exchange rates was readily obtained in the choice of the final set of parameters. It may be noted that the value of $\tau_1^{-1}$ chosen for arginine kinase on the basis of the present calculation is 120 s$^{-1}$, considerably smaller than the value 190 s$^{-1}$ estimated earlier (Nageswara Rao et al., 1976) on the basis of the linewidth change of the $\beta$-P resonance (located at about $-19$ ppm), ignoring spin–spin coupling effects. Furthermore, the chemical-shift values chosen for P-arginine and $\gamma$-P(MgATP) were $-2.9$ ppm and $-4.4$ ppm, significantly different from those in enzyme-bound complexes in the absence of the second substrate ($-3.6$ and $-5.6$ ppm, respectively). The chemical shifts chosen represent the values in the reaction complex and are therefore of some significance in understanding the effect of the enzyme on the substrates in facilitating catalysis. It may also be noted that the chemical shifts of P-arginine and $\gamma$-P(MgATP) in the reaction complex cannot be determined when exchange effects are as pronounced as in Fig. 13A because the individual resonance positions are no longer discernible in the spectrum. The parameters obtained from the computed spectra for creatine kinase, on the other hand, differ by <0.5 ppm from those used for a starting set.

## B. Variation of Line Shapes with Exchange Rates

Nucleoside triphosphate–nucleoside diphosphate exchanges occur in a variety of biochemical problems. The spin–spin coupling effects on the $^{31}$P-NMR line shapes in this system are not straightforward. In order to illustrate these effects for different exchange rates, the E·ATP·X $\rightleftarrows$ E·ADP·XP equilibrium mixture spectra of arginine kinase were simulated for a number of $\tau_1^{-1}$ values ranging from 10 to 600 s$^{-1}$ and are presented in Fig. 14. It may be noted, however, that these simulations are specific to an XP chemical shift corresponding to P-arginine, a $^{31}$P-NMR operating frequency of 40.3 MHz, and to the case in which the second substrate (X) does not contain a $^{31}$P nucleus and should be considered with these factors in mind.

The line shapes of $\beta$-P(ATP) resonance at about $-19$ ppm in the simulated spectra show evidence of spin–spin splitting up to $\tau^{-1} \simeq 40-60$ s$^{-1}$; for the values of $\tau^{-1} \simeq 80$ s$^{-1}$ and larger, the spin-coupling effects are smoothed out and the resonance resembles a Lorentzian. However, the linewidth change (with respect to $\tau^{-1} \simeq 0$) is implicitly contributed in part by the spin–spin coupling effects, and exchange rates estimated merely by considering linewidth changes are likely to be larger than the true value ($\tau^{-1}$). It is also interesting to note that the coalescence of the $\alpha$-P(ATP) and $\alpha$-P(ADP) resonances for large values of $\tau^{-1}$ does not destroy the doublet structure of this resonance, exhibiting the fact that the exchanges in the

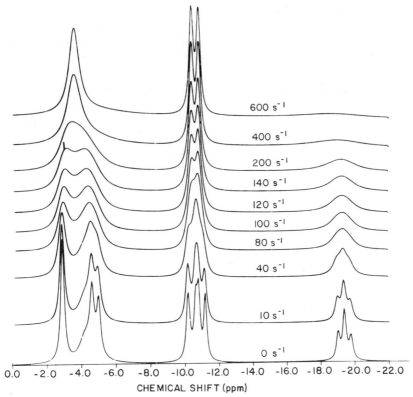

**Fig. 14.** Computer-simulated spectra for the ABC $\rightleftarrows$ A'B' + C' exchange obtained using the spectral parameters for arginine kinase and varying the phosphoryl transfer rate in the range 10–600 s$^{-1}$. The bottom spectrum is obtained with the exchange rate set equal to zero (i.e., no exchange). Shifts upfield are negative. From Vasavada et al. (1980).

kinase reaction do not randomize the spin–spin coupling between $\alpha$-P and $\beta$-P (except for the small difference (~2.5 Hz) in $J_{\alpha\beta}$ between ATP and ADP). Finally, it may be noted that for values of $\tau^{-1}$ approaching 400 s$^{-1}$ or higher, the $\beta$-P resonance is broadened to such an extent that it may not be detectable. The spectrum is somewhat deceptive in that it appears that there is no ATP in the sample (see, for example, the $^{31}$P-NMR spectra of enzyme-bound equilibrium mixtures of adenylate kinase in Fig. 10; Nageswara Rao et al., 1978a).

## C. Adenylate Kinase

Computer calculation of chemical-exchange effects were used (Vasavada et al., 1984) for the analysis of the $^{31}$P line shapes of enzyme-bound equilibrium mixtures and of other complexes relevant to the differentiation of the nucleotide-binding sites in adenylate kinase (see Section III,B,2). As discussed before, the enzyme catalyzes a reaction in which one ADP molecule reacts with another, and $Mg^{2+}$ is associated with the acceptor ADP. Furthermore, the interconversion rates on the enzyme are too rapid [and the consequent line broadening is large enough to make the $\beta$-P(MgATP) signal unobservable] to be estimated on the basis of linewidths, even by ignoring the spin–spin coupling. In addition, there is evidence that the donor and acceptor ADP molecules exchange their positions because they are indistinguishable off the enzyme, and the line shapes in the spectra suggest that this exchange is slowed down appreciably in the presence of $Mg^{2+}$. This reduction in exchange rate is in turn interpreted as a consequence of the selectivity of the donor ADP site to $Mg^{2+}$-free nucleotides.

The computer analysis not only substantiates the qualitative interpretations made earlier (Nageswara Rao *et al.*, 1978a) but also allows reliable estimates of the exchange parameters. It is estimated that the rate of phosphoryl transfer from ATP is $\sim 600$ s$^{-1}$ (the reverse rate is obtained by noting that $K'_{eq} = 1.6$); furthermore, it is found that in order to fit experimental spectra it is necessary to include an exchange rate of 100 s$^{-1}$ between acceptor and donor ADP molecules (in the equilibrium mixture, i.e., in the presence of $Mg^{2+}$) between their binding sites on the enzyme (through the medium of the solvent). The computer analysis also shows that the exchange rate between the two ADP molecules is enhanced to $\sim 1500$ s$^{-1}$ in the absence of $Mg^{2+}$, thus corroborating the interpretation made earlier. As mentioned, the slowing down of this exchange rate is related to the selectivity of the donor ADP site to $Mg^{2+}$-free ADP, in contrast to the case of ATP that requires $Mg^{2+}$ chelation to be a phosphoryl donor. It is interesting to note further that the presence of $Mg^{2+}$ brings the exchange rate to a value much lower (from a value significantly larger) than the interconversion rate, so that the catalytic step is not seriously affected by this exchange.

The computer simulation of line shapes for a thorough analysis of the spectra has thus proved to be a useful method of extracting quantitative information about several parameters, spectral as well as kinetic, associated with reaction complexes on the enzyme that may otherwise be difficult to obtain from the spectra.

## V. Conclusion

Ever since the $^{31}$P-NMR spectrum of ATP was first recorded (see Fig. 1, Gorenstein, Introductory chapter) and was observed to have measurable sensitivity to changes in pH and metal chelation, the potential for using this

# 3. $^{31}$P NMR of Enzyme Complexes

method for fruitful biochemical and biological research has existed. The technological advances in NMR methodology since then allowed this potential to fructify. The depth and breadth of research presented in this book on a variety of biochemical problems bear ample testimony to this. In the specific area of enzymology, $^{31}$P NMR is used quite effectively to obtain significant information, some of which was heretofore unavailable or difficult to obtain by other methods, on enzymes that use $^{31}$P-containing substrates, inhibitors, coenzymes, and so on.

A few significant barriers remain, however. It should be evident from a perusal of the research presented in this chapter that the experiments yield data on changes in $^{31}$P chemical shifts and coupling constants in a variety of cases and under a number of biochemically interesting conditions. However, whereas these changes may be qualitatively interpreted, and that in itself provides useful information, quantitative interpretation of such data in terms of molecular structure is not yet possible because no reliable theory exists for the same. The development of a theory has difficulties arising from (1) a formalistic point of view and (2) the dimension and volume of computations involved. The availability of larger and faster computers with time might perhaps obviate the second difficulty and provide incentive to overcome the first.

The present NMR methodology requires $^{31}$P concentrations of $\sim 1$ m$M$ for experimental study in a reasonable amount of time. The concentrations may sometimes be lowered to $\sim 0.5$ m$M$ but not much lower. In a variety of problems, the $^{31}$P-containing molecules exist at much lower concentrations and these problems are outside the scope of presently available NMR technology. Working at higher magnetic fields is unlikely to overcome this limitation because line broadening as a result of chemical-shift anisotropy may neutralize the gain in sensitivity in a number of cases.

An improvement in the theory of $^{31}$P chemical shifts and spin–spin coupling constants undoubtedly will make the information from the experiments incisive. A major breakthrough in increasing the sensitivity to enable the use of concentrations much lower than at present will bring a number of problems within the range of experimental investigation. Particularly for the $^{31}$P-NMR studies of enzyme complexes, which is the subject of this chapter, it is clear that any advance made in overcoming either of the limitations mentioned, will be invaluable.

## Acknowledgment

Thanks are due to Ms. Margo Page for typing the manuscript patiently and efficiently. The financial support received from the Research Corporation through the award of a Cottrell Research Grant and from the National Science Foundation grant no. NSF PCM 80-22075 are gratefully acknowledged. The years spent in the research group of Professor Mildred Cohn at the University of Pennsylvania, when I was introduced to this area of research, are remembered with pleasure.

## References

Abragam, A. (1961). "Principles of Nuclear Magnetism," Chapter 8. Oxford Univ. Press, London and New York.
Albery, W. J., and Knowles, J. R. (1976a). *Biochemistry* **15,** 5627–5631.
Albery, W. J., and Knowles, J. R. (1976b). *Biochemistry* **15,** 5631–5640.
Arata, Y., Kimura, S., Matsuo, H., and Narita, K. (1979). *Biochemistry* **18,** 18–24.
Bechet, J. J., Breda, C., Guinand, S., Hill, M., and d'Albis, A. (1979). *Biochemistry* **18,** 4080–4089.
Birdsall, B., Burgen, A. S. V., and Roberts, G. C. K. (1980a). *Biochemistry* **19,** 3723–3731.
Birdsall, B., Burgen, A. S. V., and Roberts, G. C. K. (1980b). *Biochemistry* **19,** 3732–3737.
Bock, J., and Sheard, B. (1975). *Biochem. Biophys. Res. Commun.* **66,** 24–30.
Brauer, M., and Sykes, B. D. (1981a). *Biochemistry* **20,** 2060–2064.
Brauer, M., and Sykes, B. D. (1981b). *Biochemistry* **20,** 6767–6775.
Brauer, M., and Sykes, B. D. (1982). *Biochemistry* **21,** 5934–5939.
Buttlaire, D. H., and Cohn, M. (1974). *J. Biol. Chem.* **249,** 5741–5748.
Campbell, I. D., Jones, R. B., Kiener, P. A., and Waley, S. G. (1979). *Biochem. J.* **179,** 607–621.
Cohn, M. (1979). *In* "NMR and Biochemistry" (S. J. Opella and P. Lu, eds.), pp. 3–27. Dekker, New York.
Cohn, M., and Hughes, T. R. (1960). *J. Biol. Chem.* **235,** 3250–3253.
Cohn, M., and Hughes, T. R. (1962). *J. Biol. Chem.* **237,** 176–181.
Cohn, M., and Nageswara Rao, B. D. (1979). *Bull. Magn. Reson.* **1,** 38–60.
Fayat, G., Fromant, M., and Blanquet, S. (1977). *Biochemistry* **16,** 2570–2579.
Fayat, G., Blanquet, S., Nageswara Rao, B. D., and Cohn, M. (1980). *J. Biol. Chem.* **255,** 8164–8169.
Feeney, J., Birdsall, B., Roberts, G. C. K., and Burgen, A. S. V. (1975). *Nature (London)* **257,** 564–566.
Feeney, J., Birdsall, B., Roberts, G. C. K., and Burgen, A. S. V. (1978). *In* "NMR in Biology" (R. A. Dwek, I. D. Campbell, R. E. Richards, and R. J. P. Williams, eds.), pp. 111–123. Academic Press, New York.
Gorenstein, D. G., and Kar, D. (1975). *Biochem. Biophys. Res. Commun.* **65,** 1073–1080.
Gorenstein, D. G., Wyrwicz, A. M., and Bode, J. (1976). *J. Am. Chem. Soc.* **98,** 2308–2314.
Gronenborn, A., Birdsall, B., Hyde, E. I., Roberts, G. C. K., Feeney, J., and Burgen, A. S. V. (1981a). *Nature (London)* **290,** 273–274.
Gronenborn, A., Birdsall, B., Hyde, E. I., Roberts, G. C. K., Feeney, J., and Burgen, A. S. V. (1981b). *Mol. Pharmacol.* **20,** 145–153.
Hassett, A., Blatter, W., and Knowles, J. R. (1982). *Biochemistry* **21,** 6335–6340.
Hyafil, F., Jacques, Y., Fayat, G., Fromant, M., Dessen, P., and Blanquet, S. (1976). *Biochemistry* **15,** 3678–3685.
Hyde, E. I., Birdsall, B., Roberts, G. C. K., Feeney, J., and Burgen, A. S. V. (1980a). *Biochemistry* **19,** 3738–3746.
Hyde, E. I., Birdsall, B., Roberts, G. C. K., Feeney, J., and Burgen, A. S. V. (1980b). *Biochemistry* **19,** 3746–3754.

Iyengar, R., and Rose, I. A. (1982). *Biochemistry* **20**, 1223–1229.
Jardetzky, O., and Roberts, G. C. K. (1981). "NMR in Molecular Biology," Chapter 9. Academic Press, New York.
Kaplan, J. I., and Fraenkel, G. (1980). "NMR of Chemically Exchanging Systems." Academic Press, New York.
Knowles, J. R., and Albery, W. J. (1977). *Acc. Chem. Res.* **10**, 105–111.
Leigh, J. S., Jr. (1971). *J. Mag. Res.* **4**, 308–311.
McLaughlin, A. C. (1974). *J. Biol. Chem.* **249**, 1445–1452.
Mathews, D. A. (1979). *Biochemistry* **18**, 1602–1610.
Milner-White, E. J., and Rycroft, D. S. (1977). *Biochem. J.* **167**, 827–829.
Milner-White, E. J., and Watts, D. C. (1971). *Biochem. J.* **122**, 727–740.
Moon, R. B., and Richards, J. H. (1973). *J. Biol. Chem.* **248**, 7276–7278.
Nageswara Rao, B. D. (1979). *In* "NMR and Biochemistry" (S. J. Opella and P. Lu, eds.), pp. 371–387. Dekker, New York.
Nageswara Rao, B. D. (1983). *In* "Biological Magnetic Resonance" (L. Berliner and J. Reuben, eds.), pp. 75–128. Plenum, New York.
Nageswara Rao, B. D., and Cohn, M. (1977a). *J. Biol. Chem.* **252**, 3344–3350.
Nageswara Rao, B. D., and Cohn, M. (1977b). *Proc. Natl. Acad. Sci. U. S. A.* **74**, 5355–5357.
Nageswara Rao, B. D., and Cohn, M. (1981). *J. Biol. Chem.* **256**, 1716–1721.
Nageswara Rao, B. D., Buttlaire, D. H., and Cohn, M. (1976). *J. Biol. Chem.* **251**, 6981–6986.
Nageswara Rao, B. D., Cohn, M., and Noda, L. (1978a). *J. Biol. Chem.* **253**, 1149–1158.
Nageswara Rao, B. D., Cohn, M., and Scopes, R. K. (1978b). *J. Biol. Chem.* **253**, 8056–8060.
Nageswara Rao, B. D., Kayne, F. J., and Cohn, M. (1979). *J. Biol. Chem.* **254**, 2689–2696.
Noda, L. (1973). *In* "The Enzymes" (P. D. Boyer, ed.), 3rd ed., Vol. 8, pp. 279–305. Academic Press, New York.
Price, N. C., and Hunter, M. G. (1976). *Biochim. Biophys. Acta* **445**, 364–376.
Price, N. C., Reed, G. H., and Cohn, M. (1973). *Biochemistry* **12**, 3322–3327.
Ribas Prado, F., Giessner-Prettre, C., Pullman, B., and Daudey, J.-P. (1979). *J. Am. Chem. Soc.* **101**, 1737–42.
Roberts, G. C. K., Feeney, J., Birdsall, B. Kimber, B., Griffiths, D. V., King, R. W., and Burgen, A. S. V. (1978). *In* "NMR in Biology" (R. A. Dwek, I. D. Campbell, R. E. Richards, and R. J. P. Williams, eds.), pp. 95–110. Academic Press, New York.
Shriver, J. W., and Sykes, B. D. (1981a). *Biochemistry* **20**, 2004–2012.
Shriver, J. W., and Sykes, B. D. (1981b). *Biochemistry* **20**, 6357–6362.
Ulrich, E., and Markley, J. (1980). *Int. Conf. Magn. Reson. Biol. Syst., 9th,* Abstract No. 100.
Vasavada, K. V., Kaplan, J. I., and Nageswara Rao, B. D. (1980). *J. Magn. Reson.* **41**, 467–482.
Vasavada, K. V., Ray, B. D., and Nageswara Rao, B. D. (1984). To be published.
Wilkinson, K. D., and Rose, I. A. (1979). *J. Biol. Chem.* **2545**, 12567–12572.

CHAPTER 4

# $^{31}$P-NMR Studies of Phosphoproteins

*Hans J. Vogel*

Department of Physical Chemistry 2
University of Lund
Lund, Sweden

| | |
|---|---|
| I. Introduction | 105 |
| II. Classification of Phosphoproteins | 106 |
|    A. Enzymatic Catalytic Intermediates | 107 |
|    B. Covalently Attached Coenzymes | 109 |
|    C. Regulatory Phosphorylations | 111 |
|    D. Metal-Ion Binding Proteins | 112 |
| III. Theoretical and Experimental Considerations | 113 |
|    A. Chemical Shift | 113 |
|    B. Coupling Constants | 116 |
|    C. Relaxation Behavior | 116 |
|    D. pH Titrations | 118 |
|    E. Practical Considerations: The Use of Analogs | 121 |
| IV. NMR Studies | 123 |
|    A. Phosphoamino Acids as Covalent Intermediates | 123 |
|    B. Role of the Phosphoryl Moiety of Prosthetic Groups | 131 |
|    C. Characterization of Regulatory Phosphorylation Sites | 138 |
|    D. Phosphoamino Acids Involved in Metal-Ion Binding | 140 |
| V. Conclusions | 144 |
|    A. Active Sites of Enzymes | 144 |
|    B. Regulatory and Metal-Ion Binding Sites | 145 |
|    C. Prospects | 148 |
| References | 149 |

## I. Introduction

Although the existence of phosphoproteins had already been demonstrated at the turn of the century (ovalbumin; Osborne and Campbell, 1900), it is only recently that phosphorus-31 NMR measurements have been utilized for the characterization of covalently attached phosphate groups in proteins. The first exploratory papers using the continuous-wave NMR method were reported in the late 1960s and addressed the study of the phosphate groups in the multiphosphorylated milk protein casein (Ho and Kurland, 1966; Ho et al., 1969). It was not until after the widespread introduction of Fourier-transform NMR (Farrar and Becker, 1971) that such studies became more common. In 1975 three papers reported on the phosphoserine residue of alkaline phosphatase (Bock and Sheard, 1975) and the pyridoxal phosphate prosthetic groups of glycogen phosphorylase (Busby et al., 1975) and aspartate aminotransferase (Martinez-Carrion, 1975). The observation of the first phosphohistidine was published 2 yr later (Gassner et al., 1977). Since then numerous experiments have been reported directly studying the covalently linked phosphoryl moieties of various enzymes and proteins; these are the topic of this chapter. At the same time, binding of phosphorylated ligands was studied by $^{31}$P NMR (Nageswara Rao, Chapter 3 in this volume; Cohn and Nageswara Rao, 1979; Sykes, 1983). A more recent development is the utilization of phosphates enriched with the oxygen isotopes $^{18}$O or $^{17}$O. Such studies can provide insight into the stereochemistry as well as the rates of phosphoryl transfer reactions and hence can provide indirect evidence about the role of phosphoamino acids (Tsai, Chapter 6; Gerlt, Chapter 7; Cohn, 1982). Before discussing the published $^{31}$P-NMR accounts, I first review some biochemical aspects of phosphoproteins and second, some theoretical and experimental considerations concerning $^{31}$P-NMR studies of such species.

## II. Classification of Phosphoproteins

One characteristic common to all natural phosphate groups is that the phosphorus atom is in the +5 oxidation state and hence is always surrounded by four substituents. In most cases the four neighbors are oxygen atoms as seen, for example, in the amino acid phosphoserine or in the coenzyme pyridoxal phosphate. However, other phosphoamino acids having a phosphoryl group ($PO_3^{2-}$) attached to a nitrogen atom (as observed in phosphohistidine) are also very common. Thus phosphoproteins could be classified according to the nature of the chemical linkage or, more specifically, by grouping them according to the observed phosphoamino acids which they contain. Evolutionary pressure has, however, not selected unique phosphorylated amino acids for specific biochemical functions. For example, both phosphoserine and phosphohistidine have been implicated for regulatory or catalytic functions (see below). Thus, in the next section I discuss the phosphoproteins according to their biochemical functions.

## A. Enzymatic Catalytic Intermediates

A large proportion of enzymatic phosphoryl-transferring reactions are thought to proceed via a covalent enzyme–phosphoryl intermediate (Knowles, 1980). Phosphorylation of a specific amino acid residue is usually brought about by the binding and reaction of ATP. On discovery of a new phosphorylated enzyme, tentative conclusions can be drawn about the nature of the phosphorylated amino acid residue from a study of the stability of the phosphoenzyme at extremes of pH (Bridger, 1973; Knowles, 1980). Phosphomonoesters ($RCH_2OPO_3^{2-}$) are generally acid-stable and base-labile. In contrast, phosphoramidates ($RCH_2NHPO_3^{2-}$) are acid-labile and stable to base. Acyl phosphates ($RCOOPO_3^{2-}$) are readily discerned by their lability at both pH extremes and their sensitivity toward hydroxylamine. Clearly, isolation of the phosphoamino acid will facilitate further identification, but possible migration of the phosphoryl group during the hydrolysis of the protein may give rise to anomalous results (Taborsky, 1974). Another possible route in order to identify the nature of the phosphorylated amino acid is via $^{31}P$ NMR, because different linkages will lead to characteristic differences in the chemical shifts observed in the spectra, as is discussed.

Once the nature of the phosphoamino acid is established, its catalytic competence has to be demonstrated. This can be done by kinetic experiments demonstrating the existence of double-displacement (or ping-pong) kinetics. However, if sequential kinetics is observed, it will have to be shown that the rates of the partial reactions leading to phosphorylation or dephosphorylation are comparable to the overall catalytic rate (Bridger *et al.*, 1968; Wimmer and Rose, 1978). Not in all cases where a phosphoamino acid was identified could this condition be fulfilled, as is demonstrated clearly by penetrating studies on hexokinase (Wimmer and Rose, 1978).

Two new techniques have been introduced that are especially useful when no stable intermediate can be isolated. Thiophosphates or phosphates with isotopically labeled oxygen atoms are chiral and can thus be used to probe the stereochemistry of the phosphoryl transfer event (Eckstein, 1979; Knowles, 1980). One in-line displacement leads to an inversion of configuration on the phosphorus atom. If one covalent intermediate is involved, retention of configuration should result from the double-displacement reaction. Using this technique, the existence of the acetate kinase acyl phosphate intermediate has been challenged (Knowles, 1980), whereas others (Spector, 1980) have interpreted the same results as being indicative of a triple-displacement mechanism.

In the other technique, known as "positional isotope exchange," scrambling of the $\beta,\gamma$-bridging ($^{18}O$) oxygen with the $\beta$-nonbridging ($^{16}O$) oxygen is followed by $^{31}$P NMR (Rose, 1979). Only when transient formation of phosphorylated intermediates takes place does such scrambling occur (Lowe and Sproat, 1978; Rose, 1979; Cohn, 1982).

Several enzymes carrying phosphoamino acids are known where most of the afore-mentioned criteria are fulfilled. Phosphoserine, phosphoaspartate, and $N^3$- and $N^1$- phosphohistidine residues can all serve as obligatory intermediates in catalysis. However, phosphohistidine is most often found, illustrating the fact that at neutral pH the imidazole ring is the most effective nucleophilic amino acid for attacking the electrophilic phosphorus atom (Walsh, 1979). Examples of enzymes with $N^3$-phosphohistidine intermediates are succinyl-CoA synthetase (Hultquist *et al.*, 1966), phosphoglycerate mutase (Rose, 1980), and acid phosphatase (Ostrowski, 1978). It is of interest that in general no amino acid homology is observed for phosphohistidine containing active-site peptides (Wang *et al.*, 1972; Goss *et al.*, 1980). In addition, the only two known phosphoserine-containing active-site peptides of alkaline phosphatase and phosphoglucomutase are not homologous (Vogel and Bridger, 1982c). In striking contrast to this is the fact that the phosphoaspartate containing active-site peptides of the $(Na^+,K^+)$-, the $Ca^{2+}$-, and the $(H^+,K^+)$-ATPase are virtually indistinguishable (Walderhaug *et al.*, 1982).

Another interesting property of phosphohistidine residues should be pointed out. The phosphoryl group is found more often on the N-3 than on the N-1 atom. This seems surprising in view of the demonstrations by $^{13}$C NMR (Reynolds *et al.*, 1973; Walters and Allerhand, 1980) and $^{15}$N NMR (Blomberg *et al.*, 1977; Roberts *et al.*, 1982) that the tautomeric form of histidine carrying the lone electron pair on the N-1 is strongly favored. This is considered to be the form in which it is normally present (Walters and Allerhand, 1980). The top of Fig. 1 illustrates this point. For N-3 to be able to act as the nucleophile, the unusual form with the lone electron pair carried on N-3 will have to be stabilized. This presumably happens, as is indicated in Fig. 1, with a salt linkage stabilizing the hydrogen on the N-1 position. Such linkages are known in serine proteases and may provide control over the $pK_a$ and orientation of the catalytic histidine residue in the active sites of enzymes (Kraut, 1977; Steitz and Shulman, 1982).

For certain enzymes it is possible to chemically modify a nucleophilic active-site amino acid with phosphate-containing ligands. A well-documented example of this is the inactivation of serine proteases by diisopropyl fluorophosphate (DIFP; Kraut, 1977; Steitz and Shulman, 1982). Although such residues are not really part of a section of covalent intermediates, the phosphorus atom provides a convenient and sensitive NMR probe in order to monitor changes in the active site.

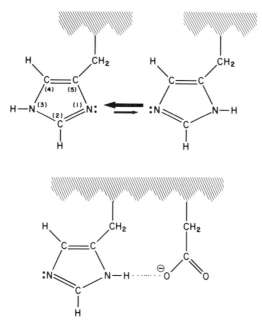

**Fig. 1.** The top of this figure shows that the tautomeric form of histidine most often found is that carrying the hydrogen on N-3. Hence N-1 would normally be expected to act as a nucleophile. Because N-3 usually reacts in this manner, it is likely that the hydrogen atom bound to N-1 is stabilized by an adjacent acidic side chain. The nomenclature indicated here is used throughout this chapter. From Vogel and Bridger (1983b).

## B. Covalently Attached Coenzymes

Very different prosthetic groups that are involved in enzymatic catalysis and that are usually covalently linked or very tightly bound to the enzyme can be found. However, all of those carrying phosphorus are part of the class of the B vitamins ($B_1$, $B_2$, $B_6$, and $B_{12}$). For example, vitamin $B_1$ (thiamin) is the precursor for synthesis of the prosthetic group thiamin pyrophosphate. It functions as a coenzyme, usually in conjunction with $Mg^{2+}$, which is thought to complex to the pyrophosphate group (Chauvet-Monges *et al.*, 1981) in enzyme-catalyzed reactions of aldehyde removal or transfer.

Vitamin $B_2$ (riboflavin) is a yellow prosthetic group that is usually found as riboflavin 5′-phosphate as the active coenzyme, also known as flavin mononucleotide (FMN; see Scheme 1). It can also be further esterified on the phosphate group with an AMP nucleotide to form the flavin adenine dinucleotide (FAD). These prosthetic groups can undergo reversible reduc-

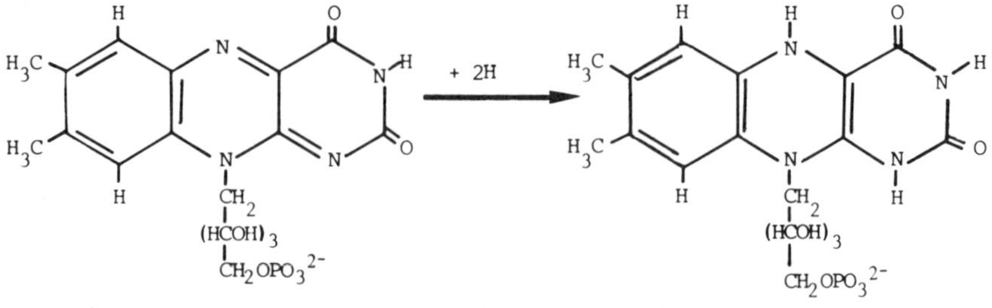

**Scheme 1.** (Left) FMN oxidized. (Right) FMN reduced.

tions in the modified isoalloxazine ring as a part of a flavoprotein or flavoenzyme to yield $FMNH_2$ or $FADH_2$ (Scheme 1). Such proteins function in electron transport or in oxidative degradative processes. Most often the moiety is tightly bound, but in some cases it is known to be covalently linked to a histidine group (Kearney, 1960; Weiner and Dickie, 1979).

A very important and versatile prosthetic group is pyridoxal phosphate, which is a derivative of vitamin $B_6$. It has been observed in the active site of

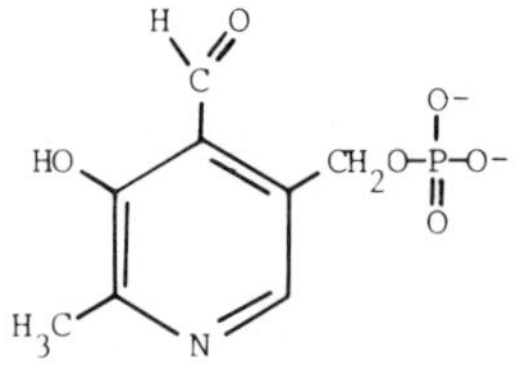

many different enzymes and has been implicated for reactions such as the decarboxylation and dehydration of amino acids but most often for transamination reactions.

Enzymes catalyzing transfer of the α-amino group of an amino acid to the α-carbon of an α-keto acid (often α-ketoglutarate) are known as transaminases or aminotransferases. Important intermediates in such reactions are a series of Schiff bases that can be "trapped" by reduction with sodium borohydride (Fischer *et al.*, 1958). These enzymes provide typical examples of double-displacement (ping-pong) mechanisms, whereby the pyridoxamine phosphate form is a discrete intermediate, and reaction with an α-keto acid is necessary to regenerate the pyridoxal form (Scheme 2). Although the

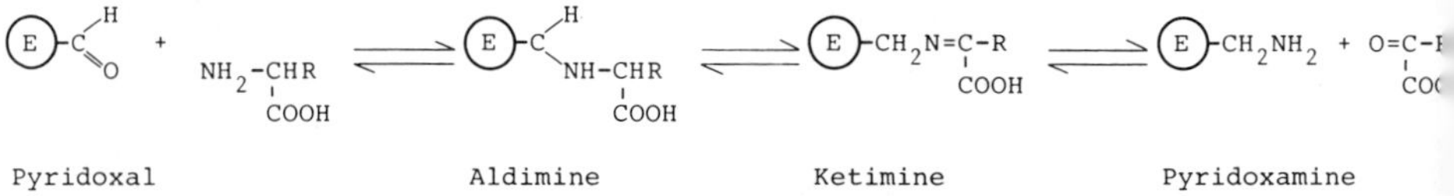

**Scheme 2**

phosphate group does not appear to be directly involved in catalysis, the pyridoxal form (dephospho form) of the prosthetic group generally does not support efficient catalysis (Mattingly *et al.*, 1982).

Very different from the transaminase mechanism and certainly more controversial is the role of the pyridoxal phosphate group in the active site of glycogen phosphorylase (Helmreich and Klein, 1980; Fletterick and Madsen, 1980). This is discussed in detail in Section IV because phosphorus-31 NMR measurements have provided some new insights (Withers *et al.*, 1981b).

A final prosthetic group that can be studied is the phosphate moiety of vitamin $B_{12}$ (5-deoxy adenosyl cobalamin). This group is instrumental for the shifting of hydrogen atoms as part of the mechanism of several enzymes.

## C. Regulatory Phosphorylations

A wide variety of enzymes and other proteins can become reversibly phosphorylated on specific serine (occasionally threonine) residues. These residues are remote from the active site of enzymes, and their modification usually results in marked changes in the activity of the enzyme (Krebs and Beavo, 1979; Cohen, 1980). Glycogen phosphorylase, for example, becomes strongly activated on modification of its serine-14 residue (Fletterick and Madsen, 1980), whereas phosphorylation of glycogen synthase, the enzyme catalyzing the opposite reaction of phosphorylase, results in a decrease in activity (Cohen, 1980). The activities of the protein kinases and phosphatases responsible for the covalent modifications are under hormonal control. Clearly, activation of the protein kinases leads to inhibition of all major anabolic processes and an activation of catabolic processes. Likewise, activation of the protein phosphatases results in the opposite response. Although such global effects on metabolism are now quite well understood, little is known about the mechanism by which these phosphorylations bring about the changes in activity. X-Ray crystallographic studies have indicated that phosphoserine-14 of phosphorylase *b* is complexed in a salt linkage with Arg-69 as well as Arg-43 of the other subunit and thus interacts with an α-helix running through the enzyme to the AMP activator site (Fletterick and Madsen, 1980). Analysis of the protein-surface topography indicated the presence of a ring of positive charges around the serine phosphate residue. These are presumably involved in the protein–protein interaction between phosphatase and its substrate phosphorylase (Fletterick *et al.*, 1979). Subsequent binding of the inhibitor glucose to the active site may expel the serine phosphate from the surface of the protein so that it becomes available to the phosphatase (Detwiler et al., 1977). Although these results imply that large domains on the surface of proteins do interact, studies with small peptides have shown that these can serve as reasonable substrates for both the protein kinase (Feramisco et al., 1980) and the protein phosphatase (Titanji et al., 1980), albeit at lower $K_m$ (but similar $V_{max}$). Such studies, combined with sequence data (Krebs and Beavo, 1979), have pinpointed the high degree of amino acid sequence homology for such regulatory sites.

In addition to phosphorylation of key regulatory enzymes of intermediary metabolism, some of the proteins involved in muscle contraction undergo a similar phosphorylation (Perry, 1979), although the role of such modification is not well understood at this moment. Examples of this are tropomyosin (Mak et al., 1978), the myosin light chains, and multiple sites within the troponin system. Functions for phosphorylation of such proteins, such as modulation of the contractile event (Perry, 1979) or regulation of proteolytic degradation rates (Toyo-Oka, 1982), have been proposed.

Other forms of regulatory phosphorylations have been documented as well. For example, DNA transcription and replication may be regulated by phosphorylation of histones on serine, histidine, or, lysine residues (Isenberg, 1979; Chen et al., 1974). Proteins with phosphotyrosine amino acids have often been found in the viral transformation processes (Hayman, 1981; Martensen, 1982). Also, 5'-adenylyl-O-tyrosine has been isolated as the known regulator in the adenylation of glutamine synthetase (Shapiro and Stadtman, 1968).

### D. Metal-Ion Binding Proteins

Phosphoamino acids that are part of proteins known to bind metal ions are posttranslational modifications introduced by specific protein kinases (Meggio et al., 1981; Vogel and Bridger, 1982c). The bovine milk protein casein and the hen egg-white protein ovalbumin, as well as possibly the human saliva acidic proline-rich proteins share sequence homology of their phosphorylated sites. Dephosphorylation of such sites by enzymatic phosphatase treatment usually reduces the affinity of such proteins for metal-ion binding (Bennick et al., 1981). Hence it is likely that dianionic phosphoryl moieties are directly involved in the complexation of metal ions. This seems particularly important for the two polyelectrolyte proteins that contain large amounts of phosphoserine residues, phosvitin purified from egg yolk (Taborsky, 1974), and the phosphoprotein purified from dentine (Linde et al., 1980).

## III. Theoretical and Experimental Considerations

Phosphorus-31 NMR has the advantage that the phosphorus nucleus has a spin $= \frac{1}{2}$ and is 100% abundant, thus alleviating the need for expensive enrichment procedures. This, combined with the reasonable sensitivity of the $^{31}$P nucleus for NMR experiments, has contributed to the popularity of $^{31}$P-NMR studies. Only a few phosphorus-containing compounds are present normally, so assignments usually can be made readily. However, the long spin–lattice relaxation times ($T_1$) and rather broad lines ($1/T_2$) force the experimentalist to use large amounts of proteins. As well, long accumulation times are often necessary to obtain spectra with a reasonable signal-to-noise ratio.

### A. Chemical Shift

The chemical shifts measured in the $^{31}$P-NMR spectra are related to shielding by the electron cloud around the phosphorus nucleus. Thus factors that influence the distribution of these electrons are expected to lead to changes in chemical shift. As pointed out by Van Wazer and Letcher (1967), changes in charge should not necessarily lead to changes in chemical shifts. The shift is more determined by other properties of the phosphorus atom: the bond geometry, the electronegativity of the substituents, and the relative amounts of $\pi$ bonding. Because change of charge can be accompanied by change in any of these three parameters, differences in chemical shifts are usually observed. Moreover, these three parameters are interdependent and their interrelationships are unfortunately not delineated. As seen throughout this chapter, the interpretation of such shifts, for those reasons, remains largely empirical. The importance of the effect of the bond geometry on the chemical shifts was demonstrated experimentally (Gorenstein, 1975, 1981; Gorenstein, Chapter 1). An other factor that may contribute to the chemical shifts of protein-bound residues is, for instance, ring-current shifts caused by binding of the phosphoryl moiety in the vicinity of an aromatic residue. These effects, however, can never be very large, because the phosphorus atom is usually surrounded by four oxygen atoms and thus will never get very close to an aromatic ring.

It has been suggested that the presence of salt linkages causes downfield shifts (Schnackerz *et al.*, 1979). Hydrogen bonding may also cause small changes in chemical shift (Evans and Kaplan, 1979). Phosphate binding sites in proteins are not only provided for by arginine residues (Riordan *et al.*, 1977; Vogel and Bridger, 1981) and metal ions (Mildvan, 1979) but are

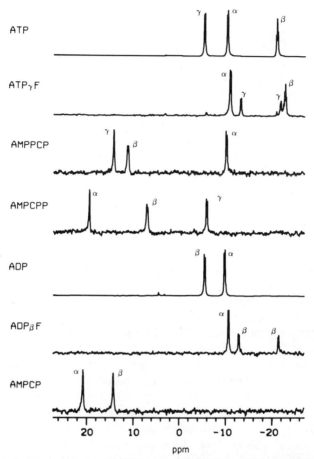

**Fig. 2.** $^{31}$P-NMR spectra of ATP, ADP, and their methylene and fluoro analogs. The assignment of the resonance is indicated. Note that only the substituted resonance shifts. Methylene substitution causes a 20- to 30-ppm downfield shift. Fluorine substitution causes an 11-ppm upfield shift as well as a splitting of the substituted resonance, caused by direct phosphorus–fluorine spin-coupling constant of 935 Hz. From Vogel and Bridger (1982a). Copyright 1982 American Chemical Society.

often found at the N-termini of $\alpha$-helices. The demonstration that this is caused by an interaction between the $\alpha$-helix dipole and the charged phosphate (Hol *et al.*, 1978) raises the question as to what effect such active-site dipole fields may have on the chemical shifts measured in $^{31}$P-NMR spectra.

The important influence that the electronegativity of the P—X bond has on the chemical-shift position is demonstrated nicely by studies on a series of substituted ATP analogs (Vogel and Bridger, 1982a, and references

**TABLE I**

Chemical Shifts and pH Titration Data for Representative Model Compounds

| Compound | Chemical shift (ppm)[a] | Titratable[b] | p$K_a$ | Reference[c] |
|---|---|---|---|---|
| **Phosphomonoesters** | | | | |
| Phosphoserine | 4.6 | + | 5.8 | 1 |
| Phosphothreonine | 4.0 | + | 5.9 | 1 |
| Pyridoxal phosphate | 3.7 | + | 6.2 | 2,3 |
| Pyridoxamine phosphate | 3.7 | + | 5.7 | 2,3 |
| Flavin mononucleotide | 4.7 | + | ~6.0 | 4 |
| **Phosphodiesters** | | | | |
| RNA, DNA, phospholipids | 0 to −1.5 | − | | 5 |
| **Diphosphodiesters** | | | | |
| Flavin adenine dinucleotide | −10.8/−11.3 | − | | 6 |
| **Phosphoramidates** | | | | |
| $N^3$-Phosphohistidine | −4.5 | − | | 7 |
| $N^1$-Phosphohistidine | −5.5 | − | | 7 |
| Phosphoarginine | −3.0 | + | 4.3 | 8 |
| Phosphocreatine | −2.5 | + | 4.2 | 8 |
| **Acyl phosphates** | | | | |
| Acetyl phosphate | −1.5 | + | 4.8 | 8 |
| Carbamoyl phosphate | −1.1 | + | 4.9 | 8 |

[a] All chemical shifts are reported with respect to an external 85% $H_3PO_4$ standard; upfield shifts are given a negative sign.

[b] Titrability: +, indicates that changes are observed in the chemical shift on changes in pH; for phosphomonoesters this change is 4 ppm, for phosphoramidates 2.5 ppm, and for acyl phosphates 5.1 ppm (Vogel and Bridger, 1983a); —, no changes observed.

[c] References: 1, Ho et al. (1969); 2, Martinez-Carrion (1975); 3, Feldmann and Helmreich (1976); 4, Moonen and Müller (1982); 5, Armstrong et al. (1981); 6, James et al. (1981); 7, Gassner et al. (1977); 8, Vogel and Bridger (1983a).

therein). This is illustrated in Fig. 2, where spectra for nonbridging fluorine-substituted, bridging-methylene-substituted, and natural nucleotides are compared. For such nucleotides, the value of the chemical shift decreased, going from S < O < F for nonbridging substitutions. Likewise for bridging substitutions, the order found was C < N < O. This result is exactly opposite to that predicted on the basis of a simple deshielding model, wherein the most electronegative atom is considered to be most effective in withdrawing electrons from the phosphorus. This anomaly has been noted earlier (Van Wazer and Letcher, 1967; Gorenstein, Chapter 1) and underlines once more the difficulties in interpreting $^{31}$P-NMR chemical shifts in the study of phosphoproteins because, for example, metal-ion binding should cause slight changes in the electronegativity of the atoms surrounding the phosphorus atom. Thus the values presented in Table I for the chemical shifts observed for model compounds can only be viewed as first indications when studying native phosphoproteins. Results obtained with denatured proteins usually agree better.

## B. Coupling Constants

It is not uncommon that resonances of protein-bound moieties are so broad that no coupling with adjacent protons can be observed. Existence of proton–phosphorus spin–spin ($J_{POCH}$) coupling can, in some cases, be deduced from the line narrowing observed when a proton-decoupling field is applied. For some mobile phosphoserine groups it has been possible to observe directly the $J_{POCH}$ (Chlebowski et al., 1976; Vogel and Bridger, 1982c). Detection is facilitated by using a low-field spectrometer, whereby the contributions of the chemical-shift anisotropy mechanism to the relaxation are minimal (see below). Information about the size of the $J_{POCH}$ is very useful because a Karplus-type relation exists, which allows one to deduct conformational information (Hall and Malcolm, 1972; Blackburn et al., 1973; Evans and Sarma, 1974; Gorenstein, Chapter 2). Three different rotamers are possible, as indicated in Scheme 3. Structure **2** has a $J_{POCH}$ of 3 Hz, and **1** and **3** each have a coupling constant of 25 Hz. For a phosphoserine standard, a value of 6.5 Hz is measured, indicating that **2** dominates but that rotation is allowed (Blackburn et al., 1973).

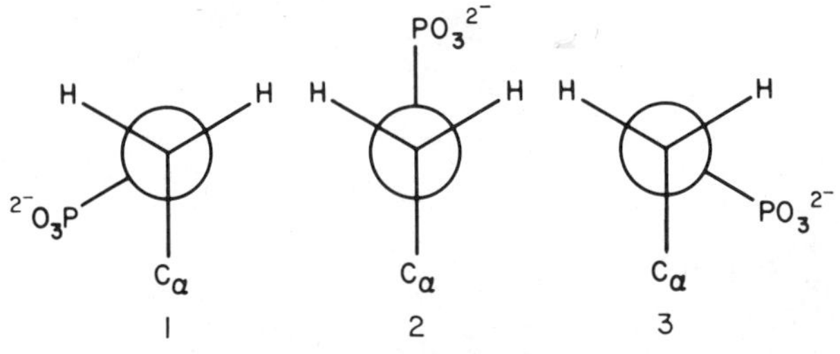

**Scheme 3**

Threonine phosphate gives rise to a doublet (Ho et al., 1969), and no coupling has been observed or reported for other phosphoamino acids. It should finally be stated that the coupling between the phosphorus and the proton also allows for the study of the proton with the use of two-dimensional NMR techniques (Bolton, 1982; see Hutton, Chapter 16).

## C. Relaxation Behavior

Although it had been known for some time that the chemical-shift anisotropy (CSA) mechanism could contribute significantly to the relaxation rates of phosphorus in phospholipid vesicles (Berden *et al.*, 1974) and transfer RNA (Guéron and Shulman, 1975; Gorenstein and Luxon, 1979), its contribution to the relaxation of phosphorus nuclei bound to phosphoproteins often has been underestimated, and many researchers have assumed that the relaxation processes are dominated by interactions with surrounding protons (Schnackerz *et al.*, 1979; Fossel *et al.*, 1981; Fujitaki *et al.*, 1981). Experiments have since shown that at higher magnetic field strength the effects of the CSA mechanism become considerable (Vogel *et al.*, 1982) and cause a decrease in experimental resolution and a reduction in signal-to-noise ratio, thus disproving claims that higher field strength should automatically give rise to higher sensitivity (Fossel *et al.*, 1981). Similar effects have been reported for DNA (Shindo, 1980), bacteriophages (Opella *et al.*, 1981) and nucleotides bound to proteins (Brauer and Sykes, 1981). Relaxation in small molecules seems to be controlled by other mechanisms, and there is still some debate as to what the major mechanisms are in this case (McCain and Markley, 1980; Nanda *et al.*, 1980; Bendel and James, 1982).

It should be realized, however, that chemical-shift anisotropy is not the sole mechanism that can cause relaxation for phosphoryl moieties or phosphoproteins. Proton–phosphorus dipole–dipole relaxation may play a role, especially at lower magnetic field strength, and it is necessary to unravel the two contributions in order to be able to deduce motional characteristics from relaxation measurements. Under conditions of nonextreme narrowing ($\omega\tau_c \gg 1$), where $\omega$ is the resonance frequency and $\tau_c$ the rotational correlation time, a condition that will generally hold for most phosphoproteins with immobilized phosphate residues, the following equations apply (Vogel *et al.*, 1982). [For a more complete discussion, see Hull and Sykes (1975); Shindo (1980) and Chapter 13; Hart, Chapter 11; James, Chapter 12.]

Chemical-shift anisotropy:

$$\Delta v = \frac{4}{45\pi} \omega_P^2 (\Delta\sigma)^2 \left(1 + \frac{\eta^2}{3}\right) \tau_c \qquad (1)$$

Proton–phosphorus dipole–dipole interaction:

$$\Delta v = \frac{1}{5\pi} \left(\frac{\gamma_P^2 \gamma_H^2 \hbar^2}{r_{PH}^6}\right) \tau_c \qquad (2)$$

where $\Delta v$ is the linewidth, $\Delta\sigma(1 + \eta^2/3)^{1/2}$ the "anisotropy term," $\gamma_P$ and $\gamma_H$ are the gyromagnetic ratio for different nuclei, and $r_{PH}$ is the distance between phosphorus and proton nuclei. The latter can be determined from X-ray studies of comparable model compounds (phosphoserine, Sundralingam and Putkey, 1970; phosphohistidine, Beard and Lenhert, 1968). The anisotropy term can be calculated from powder spectra obtained for compa-

**TABLE II**

Values of the Chemical-Shift Anisotropy Term Determined for Model Compounds[a]

| Class | Compound | $\Delta\sigma\left(1+\frac{\eta^2}{3}\right)$ (ppm) |
|---|---|---|
| Phosphomonoesters | Phosphoserine | 90–110 |
|  | Adenosine monophosphate (AMP) | 110–140 |
| Phosphodiesters | Polynucleotides | 170 |
|  | 3′,5′-cAMP | 190–220 |
| Phosphoramidates | Phosphoimidazole | 185 |
|  | Phosphocreatine | 180 |
| Acyl phosphates | Acetyl phosphate | 170 |

[a] Data obtained from Brauer and Sykes (1981), Vogel *et al.* (1982), and references therein. Some of these values may be overestimated by about 15% (Vogel *et al.*, 1982).

rable model compounds. Table II gives a small overview of the values that have been found. With this information, Eq. (1) can be used to calculate $\tau_c$. Conversely, if $\tau_c$ is known from other spectroscopic measurements (Yguerabide *et al.*, 1970), the anisotropy term can be calculated and compared to those of the model compounds. Equation (2) can be used to estimate a reasonable distance to the nearest proton(s) (Vogel *et al.*, 1982). A practical example of this is shown in Fig. 3. Four spectra obtained for one sample of succinyl-CoA synthetase (which has one phosphohistidine residue) are shown. A plot of the observed linewidth $\Delta\nu$ against the frequency squared results in a straight line as predicted by Eq. (1). This clearly demonstrates the dominance of chemical-shift anisotropy at high-field strength. The field-independent relaxation determined from the y intercept (Fig. 3b) gives the contribution from the proton–phosphorus dipole–dipole mechanism (Vogel *et al.*, 1982). This result shows that it is usually not unreasonable to assume that at low field the latter is the dominant mechanism. Similar analysis of $T_1$ data is also feasible (Shindo, 1980; Brauer and Sykes, 1981). Very similar results have been observed for the mobile phosphoryl moieties of ovalbumin (Vogel and Bridger, 1982c). A problem one may encounter with the indicated analysis is that exchange is also dependent on $\omega_p^2$, as observed for the phosphoserine residue of glycogen phosphorylase *a* (Vogel *et al.*, 1982).

## D. pH Titrations

A useful way of characterizing phosphoryl groups is via pH titration. Differently substituted moieties give rise to characteristic p$K_a$ values. Also,

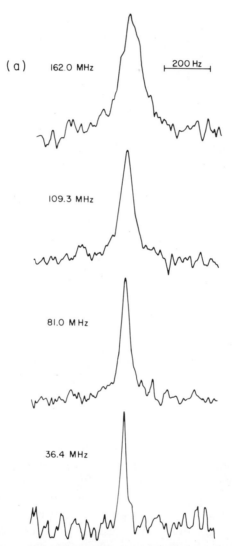

**Fig. 3.** (a) $^{31}$P-NMR spectra obtained for a sample of *E. coli* succinyl-CoA synthetase at four different magnetic field strengths. (b, next page) Plot of the linewidth observed for the spectra in (a) versus the resonance frequency squared. Note that at lowest field strength, the major contribution to the linewidth is caused by a field-independent mechanism. In contrast, at high-field strength 85% of the linewidth originates from the field-dependent mechanism. From Vogel *et al.* (1982). Copyright 1982 American Chemical Society.

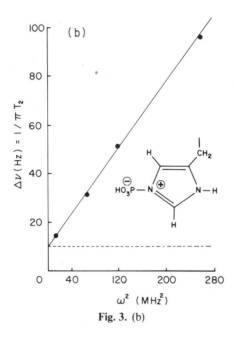

Fig. 3. (b)

the change in chemical shift that is observed over the titration can be a diagnostic parameter (Vogel and Bridger, 1983a). The results obtained have been summarized in Table I. Note that for the abundant phosphohistidine moieties no titration effect is observed (Gassner et al., 1977). Hence, absence of pH-dependent shifts for a phosphohistidine-containing protein does not indicate inaccessibility of the phosphoryl group. Although observation of a titration curve for a phosphoprotein resembling that of the free amino acid is a major indication of the exposed nature of the group, changes may be induced by titrating neighboring groups (Porubcan et al., 1979). Also, the possibility of pH-dependent protein conformational changes leading to a different chemical shift cannot always be excluded. Absence of a pH titration effect can be interpreted as inaccessibility or it may signal the presence of a salt linkage such that the protons compete ineffectively. Observation of a $pK_a$ lower than that observed for a model compound is also a good indication of the presence of a salt linkage.

An additional way of analyzing pH titration data is by calculating a Hill coefficient $n$ as originally proposed by Markley (1975) for the titration of histidine side chains. A low value for $n$ ($n < 1$) may be taken as an indication of the presence of proximal groups titrating in the same pH range, whereas values with $n > 1$ have been interpreted as signaling the presence of proximal positive charges (Schiselfeld et al., 1982; Vogel and Bridger, 1983a).

## E. Practical Considerations: The Use of Analogs

To obtain spectra in a reasonable time, it is often advisable to use high protein concentrations (>0.25 m$M$ in a 10-mm tube). Such concentrated solutions often may give rise to undesirable aggregation of the protein. Such problems will be less severe when using a wide-bore magnet and larger sample size with lower protein concentration. Some authors have also used chemically modified proteins in order to abolish aggregation (Hoerl et al., 1979; Brauer and Sykes, 1981). In cases where the available amount of protein is a limiting factor, horizontal probe designs with higher sensitivity may be used (Hoult, 1978). Proton decoupling is advisable in those cases where it provides line narrowing; however, the nuclear Overhauser enhancement for such studies is always extremely small (Brauer and Sykes, 1981; Vogel and Bridger, 1982c; Ray et al., 1977).

Active turnover by an enzyme can complicate the NMR studies, thus many researchers have utilized nonhydrolyzable analogs to avoid such complications. For example, imido-ATP analogs have been used (Tran-Dinh and Roux, 1977; Shriver and Sykes, 1981), whereas others have used methylene analogs that have the extra advantage that their lines are shifted downfield by ~20 ppm and thus do not overlap with other resonances in the spectra (Vogel and Bridger, 1982a,b). This is one advantage when utilizing $^{31}$P NMR, and many substituted analogs are often commercially available or their synthesis has been described. Most of these analogs have altered chemical shifts and thus can be used in complex mixtures so that all resonances can be studied without overlap. Table III presents an overview of several available analogs, with a mention of their NMR properties, that can be utilized to advantage. For example, the higher p$K_a$ of methylene phosphonate analogs makes them very suitable for pH titration studies in the range where most proteins are stable (Schnackerz and Feldmann, 1980; Vogel and Bridger, 1982a; Schiselfeld et al., 1982). Another interesting case is the deployment of cyclic phosphate analogs (Withers et al., 1981a). Because of the effects of bond strain, these analogs have altered chemical shifts (Gorenstein, 1975, 1981) and thus can be used to avoid overlap of resonances. It is unlikely that the isotope shift of $^{18}$O will be of much use because it is so small (0.02 ppm) that it will not be observable with broad resonances. It has been demonstrated that $^{17}$O-substituted phosphates can be used advantageously in the study of phosphoproteins. The coupling (scalar relaxation of the second kind) between the $^{31}$P and $^{17}$O nuclei collapses for protein-bound residues, leading to the unexpected observation that protein-bound residues are observable, whereas the resonance for small ligands is too broad to be observable (Tsai, Chapter 6; Markley et al., 1982).

**TABLE III**

Some Properties of Substituted Phosphate Compounds

| Compounds | Change in chemical shift (ppm)[a] | Change in $pK_a$[b] | Remarks |
|---|---|---|---|
| Thiophosphates | −40 | −1.5 | Available as AMP, ADP, and ATP analogs with the nonbridging thio substitution on the α, β, or γ position of the triphosphate of ATP. Thiophosphates can function normally as substrates in enzyme-catalyzed reactions (Eckstein, 1979; Jaffe and Cohn, 1978). |
| Fluorophosphates | +11 | No[c] | The terminal titratable hydroxyl group of AMP, ADP, and ATP has been replaced by a fluorine atom. These analogs have one negative charge and therefore resemble the protonated phosphoryl moiety at all pH values. They have a low affinity for metal ions (Yount, 1975; Vogel and Bridger, 1982a). |
| Imidophosphates | −10 | +0.9 | Nonhydrolyzable analogs of ATP (or GTP) with β,γ bridging linkage substituted (Tran Dinh and Roux, 1977). |
| Methylene phosphates | −25 | +1.5 | Nonhydrolyzable analogs of ADP and ATP (or other nucleotides) with bridging substitutions between either the α and β or the β and γ phosphorus atoms. They have a high affinity for metal ions (Schiselfeld et al., 1982; Vogel and Bridger 1982a). Pyridoxal phosphate analogs have been used as well (Hoerl et al., 1979; Schnackerz and Feldmann, 1980). |
| Cyclic phosphates | ? | No[c] | The extent of chemical shift depends on the O—P—O bond angle (Gorenstein, 1975, 1981). Note that such analogs are monoanionic and thus do not fully resemble natural compounds. |

[a] Negative values indicate downfield shifts.
[b] Negative values indicate a decrease in $pK_a$.
[c] No, Means that no pH titration occurs between pH 3 and 10.

## IV. NMR Studies

### A. Phosphoamino Acids as Covalent Intermediates

*1. Alkaline Phosphatase*

A wide variety of papers dealing with $^{31}$P-NMR studies of the dimeric enzyme (MW 94,000) purified from *Escherichia coli* have been reported, and two comprehensive reviews are available (Coleman and Chlebowski, 1979; Coleman *et al.*, 1979). The enzyme catalyzes the nonspecific hydrolysis of phosphomonoesters, and it is widely accepted that a covalent phosphoserine intermediate (E–P) and a noncovalent E · P complex play an important role in catalysis (Reid and Wilson, 1971). The enzyme is normally purified as a zinc(II) metalloenzyme, but the metal ion can be removed and replaced with cadmium, cobalt, copper, or manganese, giving rise to enzymes with different properties. Such species have been used to advantage for various spectroscopic studies including $^{31}$P-NMR experiments.

On titration of the Zn–enzyme with phosphate at pH 5.5, three resonances appear in the spectrum (Fig. 4a). These have been assigned to the covalent E–P intermediate (8.4 ppm), the noncovalent E · P complex (4.2 ppm), and the free phosphate (most upfield resonance, which shifts as a function of pH). At higher pH values (pH > 6.5), the noncovalent E · P complex mainly is present; at values of pH < 5.0, the E–P form is most stable and no E · P can be observed in the spectra (Bock and Sheard, 1975; Chlebowski *et al.*, 1976; Hull *et al.*, 1976). These measurements correlate well with earlier biochemical studies of the pH stability of the intermediates. It should be noted that the chemical shift of E–P lies outside the range normally observed for standards (Table I). This effect has been attributed to the presence of bond strain that would be introduced by the enzyme (Bock and Sheard, 1975; Chlebowski *et al.*, 1976). In studies of the cadmium-substituted enzyme, the covalent E–P form appeared at 8.4 ppm, identical to the Zn enzyme, but it turned out to be stable over a wide range of higher pH values (Bock and Sheard, 1975; Chlebowski *et al.*, 1976; Hull *et al.*, 1976; Otvos *et al.*, 1979a). Enzyme-bound phosphate cannot be observed for the cobalt- and manganese-substituted enzymes at an enzyme-to-phosphate ratio of 1:1. When more phosphate is added, a resonance attributable to E · P appears, suggesting that the first phosphate is bound very closely to the rigidly held metal ion (Bock and Sheard, 1975; Chlebowski *et al.*, 1976; Hull *et al.*, 1976; Weiner *et al.*, 1979). In fact, in subsequent studies with $^{113}$Cd-substituted enzyme (a nucleus with spin $\frac{1}{2}$), the E · P, but not E–P, appears as a doublet owing to the $^{113}$Cd–$^{31}$P spin coupling (Otvos *et al.*, 1979a).

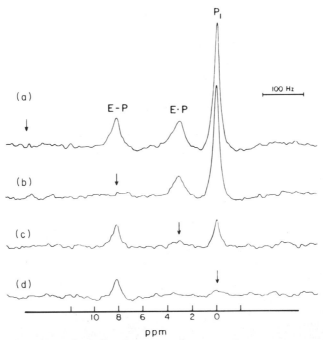

**Fig. 4.** $^{31}$P-NMR spectra of a 2.4 m$M$ solution of metal-saturated alkaline phosphatase with 9.6 m$M$ inorganic phosphate at pH 5.5. The assignments of the resonances are indicated in the figure. The arrows indicate the positions where a saturating pulse was applied: (a) Control. (b) Saturation of E–P leads to a reduction of 24% in E · P and 12% in $P_i$. (c) Saturation of E · P leads to 36% reduction in E–P and 81% in $P_i$. (d) Saturation of $P_i$ results in a decrease of E–P of 43% and abolishes E · P. From Otvos et al. (1979a).

Careful analysis of the linewidth of the E · $P_i$ complex has indicated that the dissociation of noncovalent phosphate is the rate-limiting step in the turnover of the enzyme (Hull et al., 1976). A study of the phosphorylated apoenzyme revealed that the linewidth observed was much narrower than that for the metal-containing enzyme. This suggested considerable flexibility for the residue (Chlebowski et al., 1976) in contrast to the holoenzyme, where the observed linewidth has been interpreted as that of a rigidly held moiety (Vogel et al., 1982). pH titrations showed that the residue in the apoenzyme cannot be protonated until the enzyme dissociates and unfolds at low pH (Chlebowski et al., 1976). The phosphorus nucleus in the apophosphoryl enzyme is coupled to the methylene protons of the serine with a coupling constant of 13 Hz, indicating that rotomers 1 or 3 are dominant and not much rotomer 2 is present, in contrast to the model compound (see **3,** Scheme 3) (Chlebowski et al., 1976).

In addition to the two firmly bound zinc ions that are normally present and thought to be the "catalytic" metal ions, the enzyme can be further activated and stabilized by the binding of two additional "structural" Zn ions and two Mg ions. Thus further studies have addressed the question of metal-ion content and phosphate-binding stoichiometry (Chlebowski et al., 1977; Bock and Kowalsky, 1978; Otvos et al., 1979b; Gettins and Coleman, 1983). It was found that the enzyme containing two metal ions showed absolute negative cooperativity because only 1 mol of E–P or E·P was formed. However, when further metal ion was added, 2 mol of E–P or E·P were formed (Otvos et al., 1979b). Subsequent $^{113}$Cd-NMR studies provided evidence for metal migration between subunits and suggested that only a subunit carrying two metal ions could give rise to active catalysis (Otvos and Armitage, 1980). Thus the strong negative cooperativity previously taken as evidence for an alternating-site (flip-flop) functioning of the enzyme (Lazdunski, 1972) may only be in operation when the enzyme is metal-deficient.

Information about the kinetics of the reaction has been obtained by using double-resonance saturation transfer measurements (Forsén and Hoffman, 1963a,b, 1964). The results of the experiment are shown in Fig. 4 (Otvos et al., 1979a). Saturation of E–P reduced the intensities of E·P and P$_i$ by 24 and 12%, respectively (Fig. 4, see legend). From such effects and a knowledge of the spin–lattice relaxation times $T_1$ of these resonances, kinetic information can be extracted:

$$E + P_i \underset{k_{-1}}{\overset{k_1}{\rightleftharpoons}} E \cdot P \underset{k_{-2}}{\overset{k_2}{\rightleftharpoons}} E{-}P + H_2O \tag{3}$$

It was found that $k_{-1}$ is $> 10^1$ s$^{-1}$, whereas $k_2$ and $k_{-2}$ are 0.19 and 0.23 s$^{-1}$, respectively (Otvos et al., 1979a). These values are in excellent agreement with those obtained by other methods.

### 2. Phosphoglucomutase

Another enzyme utilizing a covalent intermediate is phosphoglucomutase, which catalyzes the transfer of the phosphoryl group between the 1 and 6 hydroxyl group of glucose phosphate. The chemical shift measured for a 1.5 m$M$ solution of phosphorylated enzyme is 3.8 ppm, the same as measured for a dianionic phosphoserine residue. It does not shift with variations of pH, suggesting that inaccessibility or a strong salt linkage prevents protonation (Ray et al., 1977). No changes occur on addition of activating diamagnetic Mg$^{2+}$ ions. Addition of paramagnetic nickel ions results in a large broadening that could be interpreted as the metal ion binding within 6 Å of the phosphorus atom (Ray et al., 1977). In fact, $^{113}$Cd–$^{31}$P spin coupling has recently been observed, indicating direct coordination (Markley *et al.*, 1982). Proton-decoupling provided for some line narrowing, demonstrating coupling between methylene protons and the phosphorus nucleus; the nuclear Overhauser effect appeared to be zero. From measurements of $T_1$ and $T_2$, assuming proton–phosphorus dipole–dipole interactions as the dominant relaxation mechanism at this frequency (40.5 MHz), it was calculated that the phosphate group is immobilized on the surface of the protein (Ray *et al.*, 1977). This notion has found further support by the line narrowing observed for the $^{17}O$ phosphorylated enzyme as compared to small molecules (Markley *et al.*, 1982).

## 3. ATPases

Although membrane-embedded ATPases are rather large proteins (MW 250,000), the observation by $^{31}P$ NMR of the catalytic phosphoaspartate intermediate has been reported for detergent-solubilized protein (Fossel *et al.*, 1981). The resonance can be observed for low protein concentrations in an unusually short time, probably owing to a remarkably short $T_1 = 0.2$ s (normal $T_1$ values are usually a few seconds; Withers *et al.*, 1983; H. J. Vogel, unpublished observations). The resonance appears at $-17$ ppm, which is $\sim 15$ ppm unfield from the region where acyl phosphate resonances are normally observed. The authors presented three lines of evidence in support of this assignment. First, the resonance appears both after addition of the substrate ATP or the product $P_i$. Second, the resonance disappears on addition of $K^+$ (which earlier had been reported to destabilize the intermediate) and hydroxylamine (an agent commonly used to destabilize acyl phosphate linkages). Third, a similar chemical shift was observed for a model compound that was unfortunately not purified and poorly characterized. Furthermore, it is remarkable that no resonances are observed for the phospholipids that are known to copurify with these enzymes (Goldman and Albers, 1973).

## 4. PEP-Dependent Phosphotransferase System

Many bacteria utilize a sugar-uptake system that makes use of phosphoenolpyruvate (PEP) as the energy source and the phosphoryl source for the concommittant uptake and phosphorylation of sugars from the growth medium. Several enzymes and proteins are involved in this process (for review, see Robillard, 1982). All of these use phosphohistidine residues for transfer of the phosphoryl moiety between PEP and the phosphorylated sugar. The smallest one of these is a heat-stable protein (HPr) (MW 8000) that has been studied by phosphorus NMR (Gassner *et al.*, 1977; Dooijewaard *et al.*, 1979) and has also been a favorite subject for $^1$H-NMR studies (Maurer *et al.*, 1977; Gassner *et al.*, 1977; Dooijewaard *et al.*, 1979;

Schmidt-Aderjan et al., 1979; Rösch et al., 1981; Kalbitzer et al., 1982). For native phospho-HPr purified from *Staphylococcus aureus* (Gassner et al., 1977) and *E. coli* (Dooijewaard et al., 1979), a resonance is observed at 4.1 ppm. This shift does not coincide with that observed for chemically synthesized $N^1$- or $N^3$-phosphohistidine (Gassner et al., 1977). Only on alkaline denaturation did the resonance shift to 5.6 ppm, a position characteristic for an imidazole moiety phosphorylated at the N-1 position. Denaturation is irreversible, the resonance does not shift back when lowering the pH (Gassner et al., 1977; Dooijewaard et al., 1979). Chemically phosphorylated HPr does not support sugar transport and phosphorylation. $^{31}$P-NMR experiments showed that addition of phosphoramidate results in the formation of $N^3$-phosphohistidine HPr rather than the active $N^1$-phosphohistidine HPr (Gassner et al., 1977). The linewidth of the resonance is indicative of an immobilized residue (Vogel et al., 1982). Phosphorylation of HPr gives rise to a large rise in the $pK_a$ of the active-site histidine. This is thought to facilitate transfer of the phosphoryl group between all enzymes in the system (Dooijewaard et al., 1979). Proton transfer has been found to be an essential step in the phosphorylation of HPr (Hoving et al., 1981). Work on the $F_{III}$ phosphoryl carrier protein utilized $^1$H-NMR studies of a tryptic fragment in order to assign the nature of the $N^3$-phosphohistidine moiety of this protein (Kalbitzer et al., 1981).

## 5. Succinyl-CoA Synthetase

The enzyme from *E. coli* is a tetramer with an $(\alpha\beta)_2$ structure and an overall molecular weight of 140,000. The enzyme has two active sites that are arranged at the interface of an $\alpha$- and $\beta$-subunit (Bridger, 1981). Only one of these is phosphorylated at any one time at the N-3 position of one specific histidine residue located in the $\alpha$-subunit (Hultquist et al., 1966; Bridger, 1974). The catalytic mechanism is thought to proceed via this phosphorylated histidine residue and possibly also through a noncovalent succinyl phosphate intermediate. Frequency-dependent phosphorus NMR experiments (see Fig. 3a,b) indicated that the phosphoryl moiety on the histidine is rigidly held and immobilized on the protein; it is also thought to be in a monoanionic form (Vogel et al., 1982). The resonance is observed at −4.8 ppm (Fig. 5A), a position expected for an $N^3$-HisP resonance, suggesting that no large bond-angle distortions play a role as was observed, for example, for the active-site moiety of alkaline phosphatase (Bock and Sheard, 1975; Hull et al., 1976; Chlebowski et al., 1976). The resonance shifts upfield with addition of $Mg^{2+}$; titration of this effect indicated a dissociation constant of 4 m$M$, which is in good agreement with the amount known to be necessary for optimal activity (Gibson et al., 1967). Experi-

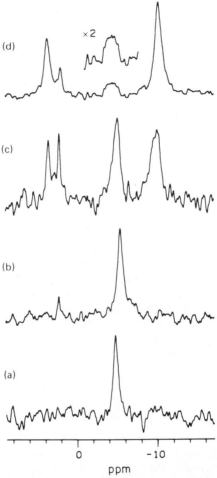

**Fig. 5.** $^{31}$P-NMR spectra of 0.2 m$M$ solution of *E. coli* succinyl-CoA synthetase at pH 7.2. (a) No additions. (b) Addition of 2.5 m$M$ CaCl$_2$. (c) Addition of 0.12 m$M$ coenzyme A. (d) Addition of 0.60 m$M$ coenzyme A. The assignments of the resonances are as follows: E–P phosphohistidine, between 4.1 and 5.3 ppm; coenzyme A 3′-phosphomonoester, 3 ppm; and pyrophosphodiester, 10 ppm; inorganic phosphate, 2.5 ppm. From Vogel and Bridger (1982b).

ments with paramagnetic Mn$^{2+}$ suggested that the metal ion binds within 10 Å of the phosphorus atom and could be coordinated to one of its oxygen atoms. Addition of the substrate coenzyme A to E–P gives rise to a broadening of the observed phosphohistidine resonance, as well as a downfield shift (Fig. 5C). This effect appears only complete on saturation of both binding sites with coenzyme A (Fig. 5D). Subsequent addition of a competi-

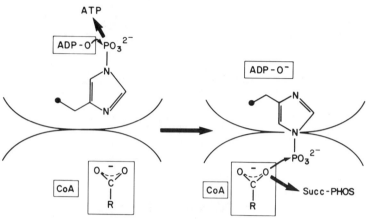

**Fig. 6.** Scheme of the active site of *E. coli* succinyl-CoA synthetase at the point of contact between the two subunit types. Binding of coenzyme A induces mobility for the phosphohistidine so that it can exchange between a conformation facing the ADP binding site and one facing the subunit site. When both coenzyme A and succinate are present, only the latter conformation is present. In the absence of coezyme A, only the former is found. From Vogel and Bridger (1982b).

tive substrate inhibitor, 2,2-difluorosuccinate, causes narrowing to the original linewidth and a small upfield shift (Vogel and Bridger, 1982b). These results can be interpreted by the scheme depicted in Fig. 6. Binding of coenzyme A is thought to permit two exchanging conformations for the phosphohistidine. One may allow for dephosphorylation by ADP and the other for transfer to the succinate residue to form succinyl phosphate. The phosphohistidine appears to be frozen into the latter conformation when both coenzyme A and 2,2-difluorosuccinate are present. In the absence of substrate, the other conformation appears to be favored.

Further studies showed that a resonance for the noncovalent succinyl phosphate resonance could only be observed when E–P, succinate, and $Mg^{2+}$, as well as ATP are present. When ATP was replaced by the nonhydrolyzable $\beta,\gamma$-methylene analog, chosen because its spectrum did not overlap with that of the phosphohistidine, no hydrolysis of the phosphohistidine and no formation of succinyl phosphate were detectable (Vogel and Bridger, 1982b). These results suggested that phosphorylation of one active site triggers the transfer of the phosphoryl group from histidine to succinate on the other (phosphorylated) site and provides evidence for alternate functioning of the two active sites. A model for this has been proposed (Vogel and Bridger, 1982b) and further evidence in addition to that obtained by $^{31}$P-NMR studies have been reviewed in detail elsewhere (Vogel and Bridger, 1983b).

## 6. Active-Site Labeling

Treatment of serine proteins with diisopropyl fluorophosphate (DIFP) results in the modification of the active-site serine residue, giving rise to an inactivated species (Kraut, 1977; Markley, 1979; Steitz and Shulman, 1982). Apart from this residue, one aspartate side chain and the imidazole group of a histidine play important roles in the hydrolysis of peptide linkages. The modified DIP-enzyme is an interesting species because a phosphorus atom is placed as a reporter group in the active site. Both Reeck *et al.* (1977) and Gorenstein and Findlay (1976) reported $^{31}$P-NMR studies of DIP-chymotrypsin. It was noted that the chemical shift observed for the modified zymogen occurs 2 ppm downfield from the native enzyme, suggesting that the active-site structure in native enzyme and in zymogen are dissimilar (Reeck *et al.*, 1977). These studies were extended to a variety of serine proteases and combined with pH titration studies (Porubcan *et al.*, 1979). A summary of those results is presented in Fig. 7. It becomes readily apparent that in all cases the zymogen resonance appears upfield from the native enzyme. The only exception may be $\alpha$-lytic protease, but this enzyme is known to have a somewhat different activation mechanism compared to the mammalian enzymes (James *et al.*, 1978). Although a phosphorus atom in a DIP linkage is not expected to change its chemical shift as a function of pH, the results of Fig. 7 show that pH titrations are observed for all

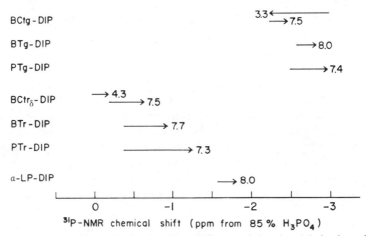

**Fig. 7.** Summary of $^{31}$P-NMR titration data of diisopropylphosphoryl derivatives of serine proteinases. The arrows represent the extent of titration shifts and the numbers indicate p$K_a$ values. Abbreviations: DIP, diisopropylphosphoryl; BCtg, bovine chymotrypsinogen A; BTg, bovine trypsinogen; PTg, porcine trypsinogen; BCtr$_\alpha$, bovine chymotrypsin A$_\alpha$; BTr, bovine trypsin; PTr, porcine trypsin; $\alpha$-LP, $\alpha$-lytic proteinase. From Porcubcan *et al.* (1979). Copyright 1979 American Chemical Society.

phosphorus moieties (note that the size of the change in chemical shift 0.1–0.8 ppm is small compared to those given in Table I). Identical $pK_a$ values were obtained for $^1$H-NMR titrations of the active-site histidine residue; hence the $^{31}$P-NMR reflects the titration of the active-site histidine residue rather than titration of the phosphorus moiety (Porcuban et al., 1979).

D. G. Gorenstein (personal communication) has utilized $^{31}$P-NMR spectroscopy to determine the stereochemistry of the phosphorylation reaction of a cyclic triester with α-chymotrypsin. As shown in Scheme 4, the axial

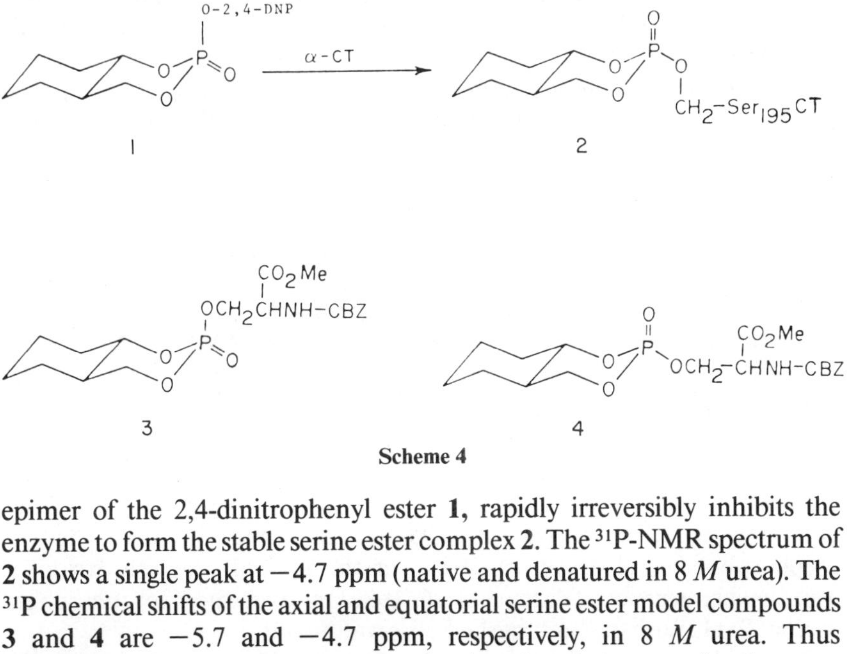

**Scheme 4**

epimer of the 2,4-dinitrophenyl ester **1**, rapidly irreversibly inhibits the enzyme to form the stable serine ester complex **2**. The $^{31}$P-NMR spectrum of **2** shows a single peak at −4.7 ppm (native and denatured in 8 M urea). The $^{31}$P chemical shifts of the axial and equatorial serine ester model compounds **3** and **4** are −5.7 and −4.7 ppm, respectively, in 8 M urea. Thus $^{31}$P NMR indicates that the enzymatic phosphorylation reaction has proceeded with *inversion* of configuration.

## B. Role of the Phosphoryl Moiety of Prosthetic Groups

### 1. Glycogen Phosphorylase

Breakdown of glycogen to glucose 1-phosphate requires the presence of phosphate as a cosubstrate and is catalyzed by glycogen phosphorylase, a pyridoxal phosphate-containing enzyme (MW 200,000). The enzyme is exceptional because in contrast to all other enzymes carrying this coenzyme, reduction with sodium borohydride does not result in reduction of a Schiff-base linkage (Fischer *et al.*, 1958). Reconstitution experiments with numerous substituted and modified pyridoxal phosphate analogs have implicated a role for the phosphorus moiety in active catalysis (for reviews, see Helmreich and Klein, 1980; Fletterick and Madsen, 1980). Most importantly, high-resolution X-ray studies showed that the phosphate group of the coenzyme points toward the substrate (Fletterick and Madsen, 1980), whereas others have shown that the inactive pyridoxal form (dephosphoform) of the reconstituted enzyme can be activated by phosphate analogs (Parrish *et al.*, 1977). This knowledge, combined with the fact that the enzyme is readily available in large quantities, provides an explanation for the fact that at least ten papers have dealt with $^{31}$P-NMR studies of this enzyme. The earliest studies by Busby *et al.* (1975) and Feldmann and Helmreich (1976) demonstrated that the pyridoxal phosphate could be observed and that one of the problems of study was that the resonances observed for this moiety appeared in the same region of the spectrum as those of the product phosphate, the substrate glucose phosphate, the activator AMP, and the regulatory, covalently phosphorlyated Ser-14 moiety. Hence a wide arsenal of analogs and substituted compounds have been used in order to avoid such overlap. With glycogen phosphorylase, phosphate can be replaced by arsenate (Feldmann and Helmreich, 1976; Feldmann and Hull, 1977), AMPS (thio-AMP) can be substituted for AMP (Feldmann and Hull, 1977; Withers *et al.*, 1979, 1981a), whereas thiophosphorylation of the phosphoserine moiety confers activity on the enzyme similar to phosphorylation (Gratecos and Fisher, 1974). Glucose 1-phosphate has been replaced by its 1,2-cyclic phosphate analog that has a different chemical shift because of the effects of bond strain (Withers *et al.*, 1981a). Moreover the pyridoxal phosphate can be replaced by analogs like the pyridoxal 5'-deoxymethylenephosphonate (Hoerl *et al.*, 1979), or the pyridoxal or the pyridoxal pyrophosphate groups (Withers *et al.*, 1983), or even the fluorophosphate analog (S. G. Withers, unpublished observations).

The early work of Busby *et al.* (1975) identified two forms for the enzyme-bound pyridoxal phosphate that could be interconverted by addition of substrate or activator or by phosphorylation. Withers *et al.* (1979, 1981a) also have identified two slowly exchanging and interconvertable resonances for the pyridoxal phosphate; three resonances are sometimes visible in the spectra (Feldmann and Hull, 1977; Hoerl *et al.*, 1979). Although the authors interpreted these as three different environments for the pyridoxal phosphate moiety, it cannot be excluded that resonance II is a minor contaminant of $P_i$ or free AMP, which would fit with its chemical-shift position and behavior on pH titration. The other two resonances appear at 3.8 ppm (form III) and 0.5 ppm (form I), and these also happen to be the chemical shifts observed for the dianionic and monoanionic residues as determined for model compounds (Martinez-Carrion, 1975; Feldmann and Helmreich, 1976). Therefore, most authors have interpreted these shifts as representing different protonation states for the protein-bound coenzyme, although it has been pointed out that such conclusions are not firm since they are based on "the questionable assumption that the chemical shift in the bound form is identical to that of the corresponding ionic species in the unbound form . . . [Cohn and Rao, 1979]." Both slowly exchanging forms of the pyridoxal phosphate have been observed for all enzyme species where no pH titration effects for the coenzyme have been observed, including the enzyme purified from potato (Klein and Helmreich, 1979). For the *E. coli* enzyme both forms are not observed. Moreover, for this particular enzyme, the resonance shifts with pH (Palm *et al.*, 1979). The general picture emerging from these studies is that in active glycogen phosphorylase, form III is usually observed (see Fig. 8A) on addition of AMP or AMPS or (thio)phosphorylation of the phosphoserine residue, whereas form I prevails in the inactive or inhibited forms of the enzyme, for example, native phosphorylase *b* (Feldmann and Hull, 1977; Withers *et al.*, 1979), glucose- or caffeine-inhibited enzyme (Hoerl *et al.*, 1979; Withers *et al.*, 1979) (Fig. 8B), but surprisingly also after addition of the active substrate analog glucose 1,2-cyclic phosphate (Fig. 8C) (Withers *et al.*, 1981a). Moreover, binding of inhibitor also destabilizes binding of the activator AMP (see Fig. 8B) (Withers *et al.*, 1979, 1983). On the basis of all these studies, the Würzburg group favors a mobile dianionic phosphorus group as the active catalyst (Feldmann and Hull, 1977; Hoerl *et al.*, 1979), whereas the Edmonton group considers a tightly bound and restrained dianionic phosphate or a protonated phosphate in intermediate exchange as more likely (Withers *et al.*, 1981a). However, as indicated, the conclusions about the protonation state may be tenuous; moreover, the mobility has been inferred from the linewidth. Because monomeric, dimeric, and tetrameric forms of the enzyme show the same linewidth for the pyridoxal phosphate (Hoerl *et al.*, 1979; Withers *et al.*, 1981a, 1982e), it seems more likely that the linewidth is not determined by mobility but rather by conformational exchange processes.

Further experiments have shed some more light on the role of the phosphorus moiety in catalysis. Glycogen phosphorylase reconstituted with pyridoxal (5')-diphospho-(1)-$\alpha$-D-glucose is capable of hydrolyzing the glucose from this "transition-state" analog, resulting in an enzyme substituted with a pyridoxal pyrophosphate group as demonstrated by $^{31}$P-NMR studies (Withers *et al.*, 1981b). This has led to the proposal in which the phosphoryl group of the pyridoxal phosphate acts as an electrophile, as depicted in Fig. 9. Further $^{31}$P-NMR studies of pyridoxal pyrophosphate-substituted enzyme as well as of pyridoxal-substituted enzyme with added pyrophos-

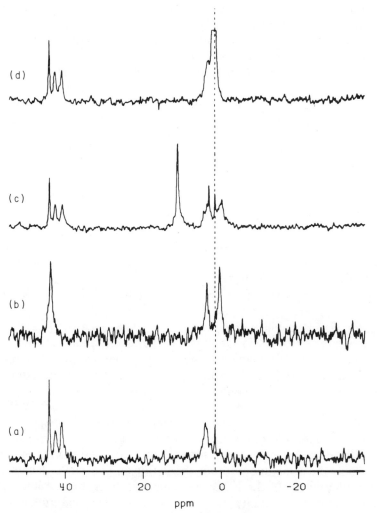

**Fig. 8.** $^{31}$P-NMR spectra (109.3 MHz) of thiophosphorylase *a* in the presence of various ligands. (a) Thiophosphorylase *a* with AMPS. (b) Thiophosphorylase *a* with glucose. (c) Thiophosphorylase *a* with AMPS and glucose 1,2-cyclic phosphate. (d) Thiophosphorylase *a* with AMPS and glucose 1-phosphate. Assignment of the resonances: thiophosphorylase residue, 42.5 ppm, and for (b), 43.4 ppm; bound AMPS, 40.7 ppm; free AMPS, 44.0 ppm; glucose 1,2-cyclic phosphate, 11.3 ppm; glucose 1-phosphate, 1.8 ppm; pyridoxal phosphate form I, 0.3 ppm, and form III, 3.8 ppm. Sharp signals at 1.9 and 3.6 ppm correspond to contaminating phosphate and AMP, respectively. Note the changes for the pyridoxal phosphate resonances with the different additions. From Withers *et al.* (1981a). Copyright 1981 American Chemical Society.

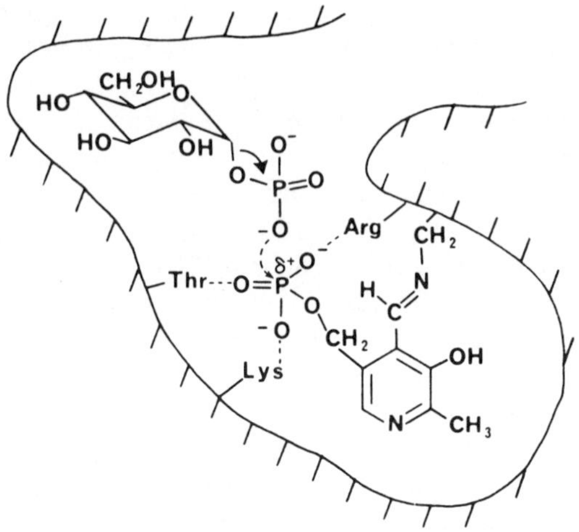

# 4. ³¹P-NMR Studies of Phosphoproteins

**Fig. 9.** Scheme illustrating the proposed role of the coenzyme phosphate as an electrophile in the postulated catalytic mechanism. Note that it is not proposed that a full pyrophosphate bond is formed. Note also that the proposed transition intermediate is similar to UDPglucose, the natural substrate for glycogen synthase catalyzing the opposite reaction. From Withers *et al.* (1981b).

phate are in support of such a model (Withers *et al.*, 1982c). However, a model where the phosphate can act as an acid catalyst still can not be rigorously excluded, although absence of a specific pH dependence argues against it (Withers *et al.*, 1982a,b).

## 2. Other Pyridoxal Phosphate-Containing Enzymes

Aspartate aminotransferase is a typical double-displacement-type enzyme, where the $NH_2$ group from the aspartate is transferred via aldimine and ketimine forms of the pyridoxal phosphate to α-ketoglutarate. Two different forms of the enzyme are known; these can be purified from the cytoplasm or mitochondria of several tissues. The pig heart enzymes are about 50% homologous (Kagiama *et al.*, 1977), and crystallographic studies have revealed a high degree of structural similarity for the two enzymes (Mattingly *et al.*, 1982). However, ³¹P-NMR measurements of the two enzymes gave largely different results (Martinez-Carrion, 1975; Mattingly *et al.*, 1982). Although both model compounds (pyridoxal phosphate and pyridoxamine phosphate) give rise to a titration (Table I), the moiety bound to the cytoplasmic enzyme does not shift with pH. Addition of inhibitors also has no effect (Martinez-Carrion, 1975). Apparently the moiety is inaccessible to solvent or is rigidly held by either an arginine residue or the positive dipolar field generated by the $\alpha$-helix pointing toward the phosphate (Ford et al., 1980). Inorganic phosphate can bind to the apoenzyme of both enzymes at the same position where the pyridoxal phosphate normally resides (Martinez-Carrion, 1975; Mattingly et al., 1982). In sharp contrast to the results obtained with the cytoplasmic enzyme, the mitochondrial isozyme displays sensitivity to pH variations and the effect is different for the various forms of the coenzyme (see Scheme 2B). Values for p$K_a$ of 5.5, 6.3, and 7.6 can be detected for native enzyme, enzyme in the presence of chloride, or after addition of the regulator succinate (Mattingly et al., 1982). It is remarkable that the change in chemical shift is about 1 ppm, which is considerably less than that for the free moiety. Thus it is possible that the observed p$K_a$ values do not represent protonation equilibria for the phosphoryl moiety but they could also be sensitive to the titration of surrounding groups as discussed for the serine proteases (Porubcan et al., 1979).

Another typical pyridoxal phosphate-containing enzyme, serine dehydratase, has also been studied in detail (Schnackerz et al., 1979; Schnackerz and Feldmann, 1980). In the native enzyme the pyridoxal phosphate $^{31}$P-NMR chemical shift is pH dependent, with a p$K_a$ = 6.4. Binding of the competitive inhibitor isoserine gives rise to formation of a transaldimination complex that is now apparently fixed to the enzyme, possibly to an arginine residue, because its resonance does not shift as a function of pH. Apoenzyme reconstituted with pyridoxal phosphate monomethyl ester produced an inactive enzyme, probably because it is incapable of forming the transaldimination complex (Schnackerz et al., 1979). Apoenzyme reconstituted with the 5'-deoxymethylene phosphonate analog results in the formation of an active enzyme, although the p$K_a$ of the phosphonate group is higher than that for the unmodified coenzyme (p$K_a$ = 7.4 versus 6.4). The analog can be used to advantage because this elevated p$K_a$ allows more accurate determination of the titration of the active-site residues, its p$K_a$ being in the region where the enzymes is stable (Schnackerz and Feldmann, 1980).

Other enzymes with pyridoxal phosphate moieties have been studied as well. Tryptophanase behaves similarly to cytoplasmic aspartate transaminase, whereas for serine hydroxymethyl transferase the $^{31}$P-NMR resonance is affected by pH but not by ligands (Schnackerz and Bartholmes, 1983; Quashnock et al., 1983).

## 3. Flavoproteins

Flavodoxins are proteins of low molecular weight that function as electron carriers in low-potential, oxidation–reduction reactions (Mayhew and

## 4. ³¹P-NMR Studies of Phosphoproteins

Ludwig, 1976). They all contain FMN as a prosthetic group. Phosphorus-31 NMR studies of protein purified from *Azotobacter vinelandii* and *Megasphaera elsdenii* have been reported (Edmondson and James, 1979; Moonen and Müller, 1982). For both proteins the chemical shift is independent of pH. The resonance appears slightly downfield from the free dianionic FMN, suggesting that the residue is bound in a dianionic form. From X-ray data the phosphate is known to be buried in the protein (Burnett *et al.*, 1974). Because no positively charged arginine or lysine residues are in the vicinity, the charge on the phosphate is probably stabilized by an α-helix dipole (Hol *et al.*, 1978). Introduction of an electron into the isoalloxazine ring results in a broadening of the phosphorus resonance. From this it could be calculated that the phosphorus atom was 8.5 Å away from the ring (Moonen and Müller, 1982), agreeing nicely with the X-ray data (Burnett *et al.*, 1974). By comparing relaxation measurements obtained for native protein and protein preexchanged in $D_2O$, it appears that several backbone NH groups must be in the vicinity of the phosphorus moiety (Moonen and Müller, 1982).

Studies on glucose oxidase from *Aspergillus niger* showed two resonances for the protein-bound FAD coenzyme. One resonance is shifted remarkably upfield compared to that of the free FAD. Thus the conformation for the enzyme-bound FAD is different from that observed in solution. Only the most upfield resonance is broadened by electrons in the isoalloxazine ring, indicating it to be within 10 Å. The other phosphorus moiety is placed further away. Addition of paramagnetic $Mn^{2+}$ ions does not broaden the resonances, indicating that the pyrophosphate group is buried in the enzyme. Another phosphodiester moiety associated with the enzyme is located on the surface as evidenced by the strongly broadened resonance (James *et al.*, 1981). A similar resonance was observed for the *Azotobacter* flavodoxin. Because the resonance remained associated with the protein on extraction with organic solvents, the authors suggested that a phosphobridge linking two amino acids and comparable to a disulfide linkage exists in these proteins. We have encountered a similar resonance in studies of bacterial pilli (Armstrong *et al.*, 1981). This resonance broadened on $Mn^{2+}$ addition, did not shift with pH, and remained associated with the protein on acid precipitation and detergent treatment, but was dissociated from the protein when the latter was precipitated with acetone or a chloroform/methanol mixture. We subsequently identified the compound as a tightly bound phospholipid, although chemical phospholipid tests previously had given negative results (Armstrong *et al.*, 1981). Be that as it may, the suggestion that *Azotobacter* flavodoxin and *Aspergillus* glucose oxidase contain a bridged phosphoryl group is interesting and deserves confirmation by identification of such a complex by chemical means.

## C. Characterization of Regulatory Phosphorylation Sites

### 1. Muscle Contraction

Several muscle proteins that are involved in the regulation of the contractile state of skeletal muscle can become phosphorylated. $^{31}$P-NMR studies of whole troponin in the presence and absence of $Ca^{2+}$ ions revealed a very narrow resonance with a chemical-shift position and pH titration behavior indistinguishable from that of the phosphoserine standard. No changes were observed after addition of tropomyosin, although the two proteins are known to form a tight complex (Sperling *et al.*, 1979). The phosphorus resonance has been assigned to the N-terminal phosphoserine in the T-subunit of troponin. Later studies showed that it could be readily removed by treatment with alkaline phosphatase (Jahnke and Heilmeyer, 1980). These results suggest that no salt linkage is present in order to stabilize protein–protein interactions and that the phosphate group may actually have the opposite function; namely, to prevent an interaction between protein domains.

Myosin purified from skeletal muscle is the major contractile protein. Each myosin head contains two kinds of light chains, one of which can be phosphorylated and which apparently does not affect contraction (Perry, 1979). $^{31}$P-NMR spectra of purified myosin not only shows a signal corresponding to the phosphoserine of this light chain but also reveals the presence of contaminating ortho- and pyrophosphates and nucleic acids (Koppitz *et al.*, 1980). The phosphoserine resonance has a linewidth of 40 Hz, indicating that the moiety has some flexibility on the surface of the myosin molecule. For the isolated light chain, a linewidth of approximately 10 Hz is measured. The resonance shifts with pH in intact myosin, indicating its exposure to solvent (Koppitz *et al.*, 1980).

Tropomyosin is a long rodlike molecule that can form filaments by association in a head-to-tail fashion. A single phosphorylation site is located at the penultimate serine residue. For both polymerized and depolymerized (addition of 1 $M$ KCl) tropomyosin, a narrow resonance has been observed in the spectra that shifted with pH, as would be expected for an exposed phosphoserine moiety (Vogel and Bridger, 1983a). The observation of a narrow resonance for the polymerized tropomyosin was especially unexpected because the solutions are extremely viscous. However, the tyrosine residues are also known to be mobile in the polymerized state (Edwards and Sykes, 1978). This observation raises questions about the postulated involvement of the phosphoserine in a salt linkage in the head-to-tail overlap region, as had been proposed (Mak *et al.*, 1978). However, further $^{31}$P-NMR analysis of the pH titration behavior (by means of the Hill coefficients)

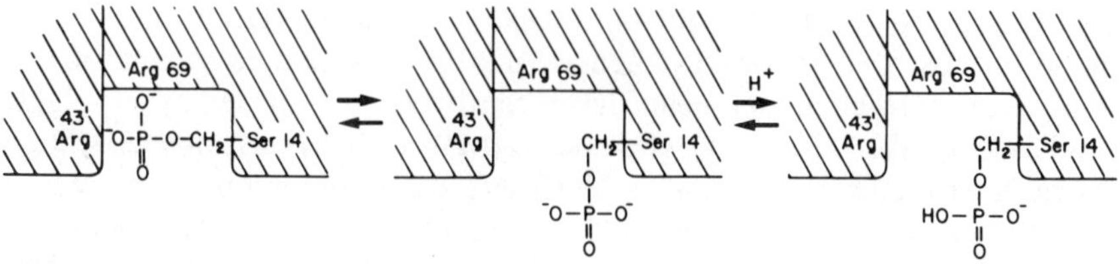

Fig. 10. Model for the regulatory phosphoserine site of glycogen phosphorylase $a$ in the presence of glucose. See text for further explanation. From Vogel and Bridger (1983a).

provided indications that such an interaction may exist (Vogel and Bridger, 1983a).

## 2. Glycogen Phosphorylase

Phosphorylation of serine-14 residue brings about a conformational change that activates the enzyme. The protein phosphatase responsible for the subsequent removal of the phosphate is markedly activated after addition of glucose (Detwiler et al., 1977). The inhibitor glucose is thought to bind to the active site of phosphorylase and induce conformational changes in the region of the phosphoserine. This leads to a higher mobility for this residue than in the absence of glucose, as demonstrated by $^{31}$P-NMR measurements (Hoerl et al., 1979; Withers et al., 1979, 1981a). However, crystals grown in the presence of glucose show that the phosphate moiety is complexed to Arg-69 as well as to Arg-43' of the opposing subunit. In subsequent $^{31}$P-NMR studies, the pH dependence of the phosphoserine resonance was studied in the presence of glucose between pH 6.3 and 8.5, where the enzyme is stable and soluble. The results indicate that the residue could be titrated and that its linewidth decreases with an increase in pH (Vogel and Bridger, 1983a). The existence of more than one conformation for the phosphoserine had been implied from frequency-dependent studies (Vogel et al., 1982). All these different data can be explained by the simple model depicted in Fig. 10. The structure on the left is favored in the absence of glucose. Protonation of the conformation facing the solvent will reduce the affinity and hence will slow down the exchange between the conformations, which is observed in the spectra as a broadening effect.

## 3. Histones

These low-molecular-weight proteins are closely associated with DNA in the nucleosomes. Posttranslational modifications (for example, phosphorylation) of histones may have large effects on the rates of transcription or DNA replication (for review, see Isenberg, 1979). Histones are phosphorylated on hydroxyamino acids; in addition, regulatory phosphorylation of histidine moieties has been reported (Chen *et al.*, 1974). Characterization of this modification by chemical means has recently been confirmed by $^{31}$P-NMR studies (Fujitaki *et al.*, 1981). Histone $H_4$ (MW 11,000) was phosphorylated by rat liver nuclei on N-1 and by Walker-256 carcinosarcomas on the N-3 position of histidine and detected in sodium dodecyl sulfate denatured preparations. Histone $H_4$ can also be phosphorylated chemically at the N-3 position by incubation with phosphoramidate. A $^{31}$P-NMR spectrum of the protein prepared in this way shows one narrow signal at 4.8 ppm and a very broad signal with much higher integrated intensity centered around 7.5 ppm. The authors assigned these peaks on the basis of experiments with proteolytic fragments to histidine-18 and -75, respectively (Fujitaki *et al.*, 1981). From the linewidth they suggested that His-18 is mobile and His-75 is immobilized. However, the linewidth observed for the His-75 residue is such that it far exceeds the value that can be calculated for an immobilized residue on a protein of this size, and therefore the line must be broadened by exchange processes.

### 4. *Other Regulatory Phosphorylations*

The activity of glutamine synthetase from *E. coli* is regulated mainly by covalent adenylation of a specific hydroxyl group of a tyrosine moiety. Thus it has been possible to measure the distance from the catalytic $Mn^{2+}$ or $Co^{2+}$ ion to the phosphorus atom of the AMP moiety bound to the protein (Villafranca *et al.*, 1978). This distance was $\sim 7$ Å. Moreover, it was shown that the phosphoryl moiety is immobilized on the surface of the protein but the adenyl part of the AMP group has considerable flexibility.

Oligosaccharide side chains of glycoproteins play an important role in recognition processes (Hubbard and Ivatt, 1981). Some phosphorylated oligosaccharides have been found to date, and by means of $^{31}$P NMR it has been possible to differentiate among phosphomonoesters, phosphodiesters, and artifacts like cyclic phosphates that are introduced by harsh purification methods (Hashimoto *et al.*, 1980). Further studies should allow insight into the motional characteristics of such protein-bound moieties.

### D. Phosphoamino Acids Involved in Metal-Ion Binding

### *1. Casein*

The first phosphorylated protein to be studied by $^{31}$P NMR was $\alpha$-casein, purified from bovine milk (Ho and Kurland, 1966; Ho *et al.*, 1969). This protein is phosphorylated at several serine side chains and hence detection should be relatively straightforward. Owing to possible ambiguities in the purification of a phosphoamino acid (hydrolysis and migration may occur during acid treatment (Taborsky, 1974)), it was necessary to study the phosphorus linkage in the native protein. Originally no pH dependence of the resonance could be detected and it was concluded that the phosphorus was in a phosphodiester or a pyrophosphodiester linkage (Ho and Kurland, 1966). The authors corrected this assignment in a second paper when further studies revealed a shift in the resonance on changes in pH; hence the phosphorus was part of phosphoserine as was also indicated by the observation of a triplet proton splitting when the protein was dissolved in 8 $M$ urea (Ho et al., 1969). Presence of threonine phosphate would have resulted in a doublet proton splitting. The five different phosphoserine moieties in $\beta$-casein have been resolved and tentatively assigned in 24.1-MHz $^{31}$P-NMR spectra (Humphrey and Jolley, 1982).

## 2. Phosvitin and Phosphodentine Protein

Ho et al. (1966) showed that the phosphorus attached to the protein phosvitin, purified from egg yolk, was in a phosphoserine linkage. Phosvitin is a highly unusual protein in that about 60% of its amino acid residues are phosphorylated. These are probably involved in metal-ion binding (Taborsky, 1974). The protein behaves as a polyelectrolyte and more recent NMR studies have indicated interactions between negatively charged phosphate residues and positively charged amino acid side chains and histidine residues (Krebs and Williams, 1977; Vogel, 1983). Notwithstanding the unusual charge distribution, the majority of the protein's phosphoserine residues have a quite normal p$K_a$ value of 5.8 (determined in a 50 m$M$ Tris buffer), but their titration shows negative cooperativity, as indicated by a very low Hill coefficient of 0.70 (Vogel, 1983). A similarly shallow titration curve was reported for another polyelectrolyte protein, the rat incisor dentine phosphoprotein, although the observed p$K_a$ was somewhat higher (Linde et al., 1980; Cookson et al., 1980). The authors suggested that extensive hydrogen bonding may occur between all the negatively charged side chains. Metal ions are also thought to be able to diffuse across the negatively charged protein surface (Cookson et al., 1980). The mobility that was observed for the phosphate groups in hen phosvitin (Vogel, 1983) also seem to be preserved in the crystalline lipovitellin/phosvitin complex from Xenopus (MW 460,000) (Banaszak and Seelig, 1982). These authors also reported the first solid-state spectra of a phosphoprotein; mobility of the phosphoserine moiety is not observed in pure solid state but becomes readily apparent on addition of small amounts of buffer (Banaszak and Seelig, 1982). In vivo observation of the phosphoserine moieties of phosvitin in intact eggs of Xenopus laevis has been reported (Colman and Gadian,

1976). The resonance disappeared as the development of the embryo progressed, suggesting that the phosphoserine was enzymatically hydrolyzed and that the liberated phosphate was being utilized for growth of the embryo.

## 3. Saliva Proteins

Another class of proteins that may behave somewhat like polyelectrolytes are the acidic proline-rich proteins purified from human saliva. They have a high affinity for calcium and can inhibit formation of hydroxyapatite (for review, see Bennick, 1982). The N-termini of these proteins contains 15 negatively charged amino acids, including two phosphoserines (Wong and Bennick, 1980). All calcium-binding sites are located in this region, and $^{31}$P-NMR experiments have implicated a role for the serine phosphates in calcium-ion binding because presence of saturating amounts of $CaCl_2$ lowers the $pK_a$ for both residues from 6.5 to 5.9 (Fig. 11). Moreover, treatment with alkaline phosphatase removed both residues and reduced the protein's affinity for calcium (Bennick *et al.*, 1981).

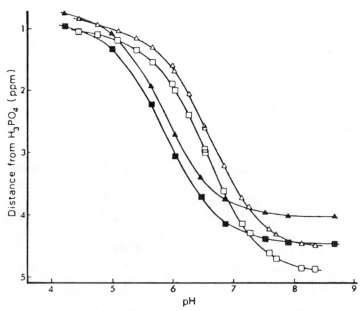

**Fig. 11.** Chemical shifts measured for the two phosphoserine resonances for the proline-rich human saliva acidic protein A as a function of pH. Closed symbols represent values obtained in the presence of 50 m$M$ $CaCl_2$; open symbols refer to those points measured in the absence of metal ion. From Bennick *et al.* (1981).

## 4. Ovalbumin

Another phosphoprotein that can bind metal ions is ovalbumin, the major protein found in hen egg whites (Taborsky, 1974). The protein is phosphorylated at two positions, serine-68 and -344 (Nisbet et al., 1981). These two residues are well resolved in a $^{31}$P-NMR spectrum, both shift as a function of pH, with p$K_a$ values and Hill coefficients identical to those of standards titrated under the same conditions (Vogel and Bridger, 1982c). The residues could be assigned by phosphatase treatment (Fig. 12). The resonance remaining after digestion with phosphatase corresponds to Ser-68, whereas the phosphatase-sensitive (more upfield) peak is SerP-344.

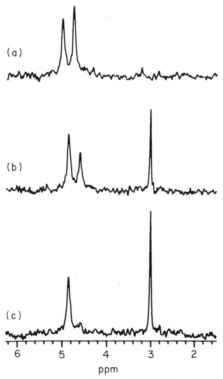

**Fig. 12.** $^{31}$P-NMR spectra (proton-decoupled, 109.3 MHz) of hen egg-white ovalbumin at pH 8.3. Assignments of the resonances: (a) phosphoserine-68, 5.0 ppm, and phosphoserine-344, 4.75 ppm; (b) same as in (a) but 1 h after the addition of 15 m$M$ MgCl$_2$ (causes a small upfield shift) and *E. coli* alkaline phosphatase; (c) same as in (b) but 20 h after the addition of the phosphatase. The new resonance appearing at 3 ppm is inorganic phosphate liberated by the phosphatase treatment. From Vogel and Bridger (1982c). Copyright 1982 American Chemical Society.

$^{31}$P-NMR frequency-dependence measurements of the linewidth indicated that residue 344 has considerable mobility, whereas residue 68 is somewhat more restrained on the surface of the protein (Vogel and Bridger, 1982c). Analysis of the proton–phosphorus spin-coupling constants showed that neither phosphoserine is strained, unlike the active-site group of alkaline phosphatase (Chlebowski *et al.*, 1976). Titrations with metal ions caused shifts for both phosphate resonances. The binding constants deduced from this effect were such that is is unlikely that the phosphorus moieties will contribute significantly to the binding of the metal ion under physiological conditions (Vogel and Bridger, 1982c).

## V. Conclusions

The examples discussed illustrate that assignment of the resonances is usually relatively straightforward. This combined with the 100% natural abundance of the nucleus may explain the popularity of the $^{31}$P-NMR technique for the study of phosphoproteins. The sensitivity is about 5% of $^1$H-NMR sensitivity. For most other nuclei, this disadvantage can be overcome by the use of higher magnetic fields. However, as discussed for the study of phosphoproteins, contributions of the chemical-shift anisotropy relaxation mechanism often prohibit advantageous application of such instruments, much as was observed for the $^{19}$F nucleus (Sykes and Weiner, 1980). Table IV presents an overview of the majority of $^{31}$P-NMR studies presented in this chapter. Some biochemically interesting generalizations can be made on the basis of these results and are discussed further.

### A. Active Sites of Enzymes

The data in Table IV show that the majority of active-side residues do not display a shift on titration, indicating that most are shielded from the solvent or held in position by a salt linkage to a positively charged amino acid side chain or by an $\alpha$-helix dipolar field (Hol *et al.*, 1978). It is noteworthy that the majority of the chemical shifts measured for these resonances are all within 1 ppm of the chemical-shift ranges measured for the comparable model compounds (Table I). Larger deviations in chemical shift have only been observed for alkaline phosphatase, HPr, and possibly the ATPases. Thus a strained phosphate group does not appear to be a general feature common to all these moieties.

The resonances do, however, share another property, namely, that the mobility deduced from the linewidth often shows that the residues are immobilized on the protein. In some cases the linewidth indicates some extra broadening caused by conformational exchange. Using rotational correlation times measured by other spectroscopic techniques (Yguerabide et al., 1970; Hull and Sykes, 1975), it can be calculated that the anisotropy term for the active-site phosphoserine of *E. coli* alkaline phosphatase and the pyridoxal phosphates from aspartate amino transferase, serine dehydratase, and glycogen phosphorylase (forms I and III) all exceed 300 ppm. The data in Table II indicate that the number expected for a phosphomonoester is 120 ppm. Thus it appears that in all these cases, existence of more than one conformation causes exchange broadening. Further studies, including analysis of the frequency dependence of the linewidths, are necessary to indicate if "wobbling" indeed occurs in the active sites of these enzymes.

All the foregoing indicates that the active-site phosphoryl moieties are immobilized or that they are in exchange between two different bound conformations. Exchange can sometimes be induced by the addition of other substrates; for example, addition of coenzyme A to succinyl-CoA synthetase induced the exchange between two different conformations for the phosphohistidine (Figs. 5 and 6); Vogel and Bridger, 1982b). The immobility often observed may facilitate the inline nucleophilic attack known to occur at the phosphorus atom in most phosphoryl-transferring enzymes (Knowles, 1980). In addition to these general conclusions, $^{31}$P-NMR studies have allowed clear insight into the specific mechanisms of alkaline phosphatase, succinyl-CoA synthetase, and glycogen phosphorylase, as discussed.

## B. Regulatory and Metal-Ion Binding Sites

In sharp contrast to the results with active-site residues, the observed linewidths and titration data in Table IV show that all sites that are phosphorylated and dephosphorylated by protein kinases, and phosphatases are mobile and solvent exposed. Such properties must be a necessary requirement for allowing the regulatory proteins to have access ot the sites.

Arginine residues often interact with phosphomonoester moieties of ligands or regulatory phosphoserine residues. If such an interaction is very strong, one would expect that changes in pH would not affect the chemical shift observed in the $^{31}$P-NMR spectra. This predicted behavior has indeed been observed for the 2'-phosphoryl moiety of NADPH or NADP$^+$ bound to *Lactobacillus casei* or *E. coli* dihydrofolate reductase (Feeney et al., 1975; Cayley et al., 1980). These enzymes have a high specificity for the coenzyme NADPH over NADH (Birdsall et al., 1977), suggesting that the 2'-phosphoryl moiety plays a crucial role in the binding of the nucleotide (see also Riordan et al., 1977; Vogel and Bridger, 1981).

## TABLE IV
### $^{31}$P-NMR Parameters Measured for Phosphoproteins

| Protein | Residue | Chemical shift (ppm)[a] | Linewidth (Hz) | Titratable[b] | Frequency (MHz) | Reference[c] |
|---|---|---|---|---|---|---|
| Active-site residue | | | | | | |
| Alkaline phosphatase | SerP | 8.5 | 20 | — | 36.4 | 1 |
| Escherichia coli | | 8.5 | — | — | 40.5 | 2 |
| | | 8.5 | 20 | — | 40.5 | 3 |
| Phosphoglucomutase | SerP | 3.8 | 14 | — | 40.5 | 4 |
| Succinyl CoA synthetase E. coli | $N^3$-HisP | −4.8 | 55 | N.D. | 109.3 | 5 |
| HPr | | | | | | |
| E. coli | $N^1$-HisP | −4.1 | 9 | — | 145.7 | 6 |
| Staphylococcus aureus | $N^1$-HisP | −4.1 | N.D. | — | 40.5 | 7 |
| ATPases | AspP | −17.0 | N.D. | N.D. | 145.7 | 8 |
| Aspartate aminotransferase | | | | | | |
| cytoplasmic | PyrP | 3.7 | 15[d] | — | 40.5 | 9 |
| mitochondrial | PyrP | 3.7 | 14 | + | 40.5 | 10 |
| Serine dehydratase | PyrP | 4.0 | 12 | + | 72.9 | 11 |
| Phosphorylase | PyrP | 3.8/0.6[e] | 60/60[e] | — | 72.9 | 12 |
| Rabbit skeletal muscle | | 3.8/0.6[e] | 160/60[e] | — | 109.3 | 13 |
| Potato | PyrP | 4.5 | N.D. | — | 72.9 | 14 |
| E. coli | PyrP | 3.4 | 40 | + | 72.9 | 15 |
| Flavodoxin | FMN | 5.6 | 3.0[d] | — | 40.5 | 16 |
| | | 4.8 | 2.3 | — | 40.5 | 17 |
| Glucose oxidase | FAD | −11/−13.5 | N.D. | — | 40.5 | 18 |

Regulatory and metal-ion binding-site residues

| Protein | Residue | pH shift | Metal shift | Chemical shift | Ref. |
|---|---|---|---|---|---|
| Phosphorylase | SerP | 3.7/4.8 | N.D. | N.D. | 72.9 | 12 |
| Myosin light chains | SerP | 4.2 | 70 | + | 109.3 | 19 |
| Troponin T | SerP | 4.6 | 40 | + | 72.9 | 20 |
| Ovalbumin | SerP-68 | 4.5 | 3 | + | 72.9 | 21 |
|  | SerP-344 | 5.0 | 4 | + | 109.3 | 22 |
|  | SerP | 4.75 | 2 | + | 109.3 | 22 |
| Tropomyosin | SerP | 4.7 | 5 | + | 109.3 | 19 |
| Glutamine synthetase | 5′-AMP-$O$-Tyr | 4.0 | 22 | N.D. | 40.5 | 23 |
| Histone H$_4$ | $N^3$-HisP | −4.8/−7.3$^f$ | 9/55$^f$ | N.D. | 81.0 | 24 |
| Salivary phosphoproteins | SerP | 4.0/4.5$^f$ | N.D. | + | 145.7 | 25 |
| Casein | SerP | 4.5 | N.D. | + | 40.5 | 26 |
| Phosvitin | SerP | 4.5 | N.D. | + | 40.5 | 26 |
|  |  | 4.5 | N.D. | + | 109.5 | 27 |
| Phosphodentine protein | SerP | 4.5 | N.D. | + | 36.4 | 28 |

[a] All chemical shifts are given with respect to 85% $H_3PO_4$; upfield shifts are given a negative sign.
[b] —, Does not shift with pH; +, shifts with pH; N.D., not determined.
[c] References: 1, Chlebowski et al. (1976); 2, Bock and Sheard (1975); 3, Hull et al. (1976); 4, Ray et al. (1977); 5, Vogel and Bridger (1982b); 6, Dooijewaard et al. (1979); 7, Gassner et al. (1977); 8, Fossel et al. (1981); 9, Martinez-Carrion (1975); 10, Mattingly et al. (1982); 11, Schnackerz et al. (1979); 12, Feldmann and Hull (1977); 13, Withers et al. (1979); 14, Klein and Helmreich (1979); 15, Palm et al. (1979); 16, Edmondson and James (1979); 17, Moonen and Müller (1982); 18, James et al. (1981); 19, Vogel and Bridger (1983c); 20, Koppitz et al. (1980); 21, Sperling et al. (1979); 22, Vogel and Bridger (1982c); 23, Villafranca et al. (1978); 24, Fujitaki et al. (1981); 25, Bennick et al. (1981); 26, Ho et al. (1969); 27, Vogel (1983); 28, Cookson et al. (1980).
[d] Measured from spectra.
[e] Form III and form I, respectively.
[f] Two different residues.

When $^{31}$P-NMR pH titration studies were performed with 2′-AMP bound to dihydrofolate reductase, the resonance shifted with pH (Birdsall et al., 1977). These studies provide a framework for the following interpretation of the $^{31}$P-NMR pH titration studies on regulatory and metal-ion-binding phosphoserine residues. In all cases studied so far, these residues can be titrated, giving rise to pH-dependent shifts in their respective $^{31}$P-NMR spectra. Thus, if a salt linkage exists between the phosphoserine residue and positively charged amino acid side chains in any of these proteins, the strength of this binding must be lower than that of NADP$^+$ bound to Lactobacillus casei dihydrofolate reductase ($K_d = 22$ $\mu M$) and is presumably comparable to the of 2′-AMP bound to this enzyme (270 $\mu M$). On this basis, the fact that all regulatory phosphoserine residues can be titrated suggests that their $K_d$ values for the interaction between the phosphoryl moiety and a positively charged amino acid side chain must be larger than 100 $\mu M$ (i.e., $\Delta G^0 \approx -5.5$ kcal mol$^{-1}$).

Two different models have been proposed to explain how regulatory phosphorylation sites may exert their function. Either the dianionic phosphoserine is salt-linked to an arginine or lysine residue and in doing so may help stabilize one specific protein conformation, or a mobile phosphoryl group may prevent an interaction between protein domains. In the latter case, one would expect the resonance to show titration behavior with a Hill coefficient similar to that of standards, which is indeed observed for some of the proteins. When salt bridges play a role, the energy involved has to be < 5 kcal mol$^{-1}$ in order to observe a titration behavior. A lower p$K_a$ and an increased Hill coefficient ($n > 1$) can provide evidence for the presence of proximal positive charges.

It should finally be stated that the chemical shifts measured for regulatory and metal-ion-binding phosphoamino acids are all very similar to those observed for comparable standards.

## C. Prospects

Although only one report of the study of a solid-state phosphoprotein has appeared to date (Banaszak and Seelig, 1982), application of solid-state NMR techniques to the study of the phosphate groups in the backbone of DNA is a research area of increasing importance (Shindo, Chapter 13). Limited availability of proteins that can be purified in sufficient quantities for these studies may prohibit solid-state NMR research of phosphoproteins, but the study of at least a few with this method will prove worthwhile, because data obtained in this manner are more comparable to what is observed by crystallographic means.

Future solution studies will not only focus on the many phosphoproteins that have not been studied to date but will also go into greater depth on those that have. For instance, frequency-dependent studies will provide more reliable data on the motional characteristics of covalently attached phosphoryl groups (Vogel et al., 1982; Vogel and Bridger, 1982c). Such investigations can be complemented by the utilization of $^{17}O$ isotopically enriched phosphates, which not only provide insight into the mobility of the phosphorus moiety but also minimize problems caused by overlapping resonances in the spectra (Markley et al., 1982). Other isotopes, like the metal $^{113}Cd$, can be used to study the interaction between the phosphate moiety and bound metal ions (Otvos et al., 1979a). Moreover, saturation transfer studies of phosphoamino acid catalytic enzymatic intermediates and substrates (Otvos et al., 1979a) will provide kinetic information that often is not easily available by other techniques.

## Acknowledgments

The author thanks his teachers Drs. George Robillard, William Bridger, Brian Sykes, and Sture Forsén for encouragement and support. The advice, many discussions, and assistance of Drs. William Wolodko and Torbjörn Drakenberg are gratefully acknowledged. This chapter was written while the author was sponsored by a grant from Hässle Pharmaceutical Company. The secretarial help of Ms. Elke Lohmeier is greatly appreciated.

## References

Armstrong, G. D., Frost, L. S., Vogel, H. J., and Paranchych, W. (1981). *J. Bacteriol.* **145**, 1167–1176.
Banaszak, L. J., and Seelig, J. (1982). *Biochemistry* **21**, 2436–2443.
Beard, L. N., and Lenhert, P. G. (1968). *Acta Crystallogr., Sect. B*, **B24**, 1529–1539.
Bendel, P., and James, T. L. (1982). *J. Magn. Reson.* **48**, 76–85.
Bennick, A. (1982). *Mol. Cell. Biochem.* **45**, 83–99.
Bennick, A., McLaughlin, A. C., Grey, A. A., and Madapallimattam, G. (1981). *J. Biol. Chem.* **256**, 4741–4746.
Berden, J. A., Cullis, P. A., Hoult, D. I., McLaughlin, A. C., Radda, G. K., and Richards, R. E. (1974). *FEBS Lett.* **46**, 55–58.
Birdsall, B., Roberts, G. C. K., Feeney, J., and Burgen, A. S. V. (1977). *FEBS Lett.* **80**, 313–316.
Blackburn, B. J., Lapper, R. D., and Smith, I. C. P. (1973). *J. Am. Chem. Soc.* **95**, 2873–2878.
Blomberg, F., Maurer, W., and Rüterjans, H. (1977). *J. Am. Chem. Soc.* **99**, 8149–8159.
Bock, J. L., and Kowalsky, A. (1978). *Biochim. Biophys. Acta* **526**, 135–146.
Bock, J. L., and Sheard, B. (1975). *Biochem. Biophys. Res. Commun.* **66**, 24–30.
Bolton, P. H. (1982). *J. Magn. Reson.* **46**, 91–97.

Brauer, M., and Sykes, B. D. (1981). *Biochemistry* **20**, 6767-6775.
Bridger, W. A. (1973). *PAABS Rev.* **2**, 83-88.
Bridger, W. A. (1974). *In* "The Enzymes" (P. D. Boyer, ed.), 3rd ed., Vol. 10, pp. 581-606. Academic Press, New York.
Bridger, W. A. (1981). *Can. J. Biochem.* **59**, 1-8.
Bridger, W. A., Millen, W. A., and Boyer, P. D. (1968). *Biochemistry* **8**, 3608-3616.
Burnett, R. M., Darling, G. D., Kendall, D. S., LeQuesne, M. E., Mayhew, S. G., Smith, W. W., and Ludwig, M. L. (1974). *J. Biol. Chem.* **249**, 4383-4392.
Busby, S. J. W., Gadian, D. G., Radda, G. K., Richards, R. E., and Seeley, P. J. (1975). *FEBS Lett.* **55**, 14-17.
Cayley, P. J., Feeney, J., and Kimber, B. J. (1980). *Int. J. Biol. Macromol.* **2**, 251-255.
Chauvet-Monges, A. M., Hadida, M., Crevat, A., and Vincent, E. J. (1981). *Arch. Biochem. Biophys.* **207**, 311-315.
Chen, C.-C., Smith, D. L., Bruegger, C. C., Halpern, R. M., and Smith, R. A. (1974). *Biochemistry* **13**, 3785-3789.
Chlebowski, J. F., Armitage, I. M., Tusa, P. P., and Coleman, J. E. (1976). *J. Biol. Chem.* **251**, 1207-1216.
Chlebowski, J. F., Armitage, I. M., and Coleman, J. E. (1977). *J. Biol. Chem.* **252**, 7053-7061.
Cohen, P. (1980). "Protein Phosphorylation in Regulation," Elsevier/North-Holland, Amsterdam.
Cohn, M. (1982). *Annu. Rev. Biophys. Bioeng.* **11**, 23-42.
Cohn, M., and Nageswara Rao, B. D. (1979). *Bull. Magn. Reson.* **1**, 38-60.
Colman, A., and Gardian, D. G. (1976). *Eur. J. Biochem.* **61**, 387-396.
Coleman, J. E., and Chlebowski, J. F. (1979). *Adv. Inorg. Biochem.* **1**, 1-61.
Coleman, J. E., Armitage, I. M., Chlebowski, J. F., Otvos, J. D., and Schoot Uiterkamp, A. J. M. (1979). *In* "Biological Applications of Magnetic Resonance" (R. G. Shulman, ed.), pp. 345-397. Academic Press, New York.
Cookson, D. J., Levine, B. A., Williams, R. J. P., Jontell, M., Linde, A., and deBernard, B. (1980). *Eur. J. Biochem.* **110**, 273-278.
Detwiler, T. C., Gratecos, D., and Fischer, E. H. (1977). *Biochemistry* **16**, 4818-4824.
Dooijewaard, G., Roossien, F. F., and Robillard, G. T., (1979). *Biochemistry* **18**, 2996-3001.
Eckstein, F. (1979). *Acc. Chem. Res.* **12**, 204-210.
Edmondson, D. E., and James, T. L. (1979). *Proc. Natl. Acad. Sci. U.S.A.* **76**, 3786-3789.
Edwards, B. F. P., and Sykes, B. D. (1978). *In* "NMR in Biology" (R. A. Dwek, I. D. Campbell, R. E. Richards, and R. J. P. Williams, eds.), pp. 157-168. Academic Press, New York.
Evans, F. E., and Kaplan, N. O., (1979). *FEBS Lett.* **105**, 11-14.
Evans, F. E., and Sarma, R. H. (1974). *J. Biol. Chem.* **249**, 4754-4759.
Farrar, T. C., and Becker, E. D. (1971). "Pulse and Fourier Transform NMR," pp. 1-111. Academic Press, New York.
Feeney, J., Birdsall, B., Roberts, G. C. K., and Burgen, A. S. V. (1975). *Nature (London)* **257**, 564-566.
Feldmann, K., and Helmreich, E. J. M. (1976). *Biochemistry* **15**, 2394-2401.
Feldmann, K., and Hull, W. E. (1977). *Proc. Natl. Acad. Sci. U.S.A.* **74**, 856-860.
Feramisco, J. R., Glass, D. B., and Krebs, E. G. (1980). *J. Biol. Chem.* **255**, 4240-4245.
Fischer, E. H., Kent, A. B., Snyder, E. R., and Krebs, E. G. (1958). *J. Am. Chem. Soc.* **80**, 2906-2907.
Fletterick, R. J., and Madsen, N. B. (1980). *Annu. Rev. Biochem.* **49**, 31-61.
Fletterick, R. J., Sprang, S., and Madsen, N. B. (1979). *Can. J. Biochem.* **57**, 789-797.
Ford, G. C., Eichele, G., and Jansonius, J. N. (1980). *Proc. Natl. Acad. Sci. U.S.A.* **77**, 2559-2563.
Forsén, S., and Hoffman, R. A. (1963a). *Acta Chem. Scand.* **17**, 1787-1790.

Forsén, S., and Hoffman, R. A. (1963b). *J. Chem. Phys.* **39**, 2892-2895.
Forsén, S., and Hoffman, R. A. (1964). *J. Chem. Phys.* **40**, 1184-1194.
Fossel, E. T., Post, R. L., O'Hara, D. S., and Smith, T. W. (1981). *Biochemistry* **20**, 7215-7219.
Fujitaki, J. M., Fung, G., Oh, E. Y., and Smith, R. E. (1981). *Biochemistry* **20**, 3658-3664.
Gassner, M., Stehlik, D., Schrecker, O., Hengstenberg, W., Maurer, W., and Rüterjans, H. (1977). *Eur. J. Biochem.* **75**, 287-296.
Gettins, P., and Coleman, J. E. (1983). *J. Biol. Chem.* **258**, 408-416.
Gibson, J., Upper, C. D., and Gunsalus, I. C. (1967). *J. Biol. Chem.* **242**, 2474-2477.
Goldman, S. S., and Albers, R. W. (1973). *J. Biol. Chem.* **248**, 867-874.
Gorenstein, D. G. (1975). *J. Am. Chem. Soc.* **97**, 898-900.
Gorenstein, D. G. (1981). *Annu. Rev. Biophys. Bioeng.* **10**, 355-386.
Gorenstein, D. G., and Findlay, J. B. (1976). *Biochem. Biophys. Res. Commun.* **72**, 640-645.
Gorenstein, D. G., and Luxon, B. A. (1979). *Biochemistry* **18**, 3796-3804.
Goss, N. H., Evans, C. T., and Wood, H. G. (1980). *Biochemisry* **19**, 5805-5809.
Gratecos, D., and Fischer, E. H. (1974). *Biochim. Biophys. Acta* **58**, 960-967.
Guéron, M., and Shulman, R. G. (1975). *Proc. Natl. Acad. Sci. U.S.A.* **72**, 3482-3485.
Hall, L. D., and Malcolm, R. B. (1972). *Can. J. Chem.* **50**, 2102-2110.
Hashimoto, C., Cohen, R. E., and Ballou, C. E. (1980). *Biochemistry* **19**, 5932-5938.
Hayman, M. J. (1981). *J. Gen. Virol.* **52**, 1-14.
Helmreich, E. J. M., and Klein, H. W. (1980). *Angew. Chem., Int. Ed. Engl.* **19**, 441-455.
Ho, C., and Kurland, R. J. (1966). *J. Biol. Chem.* **241**, 3002-3007.
Ho, C., Magnuson, J. A., Wilson, J. B., Magnuson, N. S., and Kurland, R. J. (1969). *Biochemistry* **8**, 2074-2082.
Hoerl, M., Feldmann, K., Schnackerz, K. D., and Helmreich, E. J. M. (1979). *Biochemistry* **18**, 2457-2464.
Hol, W. G. J., van Duijnen, P. T., and Berendsen, H. J. C. (1978). *Nature (London)* **273**, 443-446.
Hoult, D. (1978). *NMR Spectrosc.* **12**, 41-77.
Hoving, H., Lolkema, J. S., and Robillard, G. T. (1981). *Biochemistry* **20**, 87-93.
Hubbard, S. C., and Ivatt, R. J. (1981). *Annu. Rev. Biochem.* **50**, 555-584.
Hull, W. E., and Sykes, B. D. (1975). *J. Mol. Biol.* **98**, 121-153.
Hull, W. E., Halford, S. E., Gutfreund, H., and Sykes, B. D. (1976). *Biochemistry* **15**, 1547-1561.
Hultquist, D. E., Moyer, R. W., and Boyer, P. D. (1966). *Biochemistry* **5**, 322-331.
Humphrey, R. S., and Jolley, K. W. (1982). *Biochim. Biophys. Acta* **708**, 294-299.
Isenberg, I., (1979). *Annu. Rev. Biochem.* **48**, 159-191.
Jaffe, E. K., and Cohn, M. (1978). *Biochemistry* **17**, 652-657.
Jahnke, U., and Heilmeyer, L. M. G. (1980). *Eur. J. Biochem.* **111**, 325-332.
James, M. N. G., Delbaere, L. T. J., and Brayer, G. D. (978). *Can. J. Biochem.* **56**, 396-402.
James, T. L., Edmondson, D. E., and Husain, M. (1981). *Biochemistry* **20**, 617-621.
Kagamiyama, H., Sakakibara, R., Wada, H., Tanase, S., and Morino, Y. (1977). *J. Biochem. (Tokyo)* **82**, 291-294.
Kalbitzer, H. R., Deutscher, J., Hengstenberg, W., and Rösch, P. (1981). *Biochemistry* **20**, 6178-6185.
Kalbitzer, H. R., Hengstenberg, W., Rösch, P., Muss, P., Bernsmann, P., Engelmann, R., Dörschug, M., and Deutscher, J. (1982). *Biochemistry* **21**, 2879-2885.
Kearney, E. B. (1960). *J. Biol. Chem.* **235**, 865-877.
Klein, H. W., and Helmreich, E. J. M. (1979). *FEBS Lett.* **108**, 209-214.
Knowles, J. R. (1980). *Annu. Rev. Biochem.* **49**, 877-921.
Koppitz, B., Feldmann, K., and Heilmeyer, L. M. G. (1980). *FEBS Lett.* **117**, 199-202.

Kraut, J. (1977). *Annu. Rev. Biochem.* **46**, 331–358.
Krebs, E. G., and Beavo, J. A. (1979). *Annu. Rev. Biochem.* **48**, 923–961.
Krebs, J., and Williams, R. J. P. (1977). *In* "NMR in Biology" (R. A. Dwek, I. D. Campbell, R. E. Richards, and R. J. P. Williams, eds.), p. 348. Academic Press, New York.
Lazdunski, M. (1972). *Curr. Top. Cell. Regul.* **6**, 267–330.
Linde, A., Bhown, M., and Butler, W. T. (1980). *J. Biol. Chem.* **255**, 5931–5942.
Lowe, G., and Sproat, B. S. (1978). *J. Chem. Soc., Perkins. Trans. 1* pp. 1622–1630.
McCain, D. C., and Markley, J. L. (1980). *J. Am. Chem. Soc.* **102**, 5559–5565.
Mak, A., Smillie, L. B., and Bárńy, M. (1978). *Proc. Natl. Acad. Sci. U.S.A.* **75**, 3588–3592.
Markley, J. L. (1975). *Acc. Chem. Res.* **8**, 70–80.
Markley, J. L. (1979). *In* "Biological Applications of Magnetic Resonance" (R. G. Shulman, ed.), pp. 397–463. Academic Press, New York.
Markley, J. L., Rhyu, G. I., and Ray, W. J. (1982). *Fed. Proc. Fed. Am. Soc. Exp. Biol.* **40**(4), 5069.
Martensen, T. M. (1982). *J. Biol. Chem.* **257**, 9648–9652.
Martinez-Carrion, M. (1975). *Eur. J. Biochem.* **54**, 39–43.
Mattingly, M. E., Mattingly, J. R., and Martinez-Carrion, M. (1982). *J. Biol. Chem.* **257**, 8872–8878.
Maurer, W., Rüterjans, H., Schrecker, O., Hengstenberg, W., Gassner, M., and Stehlick, D. (1977). *Eur. J. Biochem.* **75**, 297–301.
Mayhew, S. G., and Ludwig, M. L. (1976). *In* "The Enzymes" (P. D. Boyer, ed.), 3rd ed., Vol. 12, pp. 57–118. Academic Press, New York.
Meggio, F., Deana, A. D., and Pinna, L. A. (1981). *Biochim. Biophys. Acta* **662**, 1–7.
Mildvan, A. S. (1979). *Adv. Enzymol.* **49**, 103–126.
Moonen, C. T., and Müller, F. (1982). *Biochemistry* **21**, 408–414.
Nanda, R. K., Ribeiro, A., Jardetzky, T. S., and Jardetzky, O. (1980). *J. Magn. Res.* **39**, 119–125.
Nisbet, A. D., Saundry, R. M., Moir, A. J. G., Fothergill, L. A., and Fothergill, J. E. (1981). *Eur. J. Biochem.* **115**, 335–345.
Opella, S. J., Wise, W. B., and Diverdi, J. A. (1981). *Biochemistry* **20**, 284–290.
Osborne, J. B., and Campbell, G. F. (1900). *J. Am. Chem. Soc.* **22**, 422.
Ostrowski, W. (1978). *Biochim. Biophys. Acta* **526**, 147–153.
Otvos, J. D., and Armitage, I. M. (1980). *Biochemistry* **19**, 4031–4043.
Otvos, J. D., Alger, J. R., Coleman, J. E., and Armitage, I. M. (1979a). *J. Biol. Chem.* **254**, 1778–1780.
Otvos, J. D., Armitage, I. M., Chlebowski, J. F., and Coleman, J. E. (1979b). *J. Biol. Chem.* **254**, 4707–4713.
Palm, D., Schächtele, K. H., Feldmann, K., and Helmreich, E. J. M. (1979). *FEBS Lett.* **101**, 403–406.
Parrish, R. F., Uhing, R. J., and Graves, D. J. (1977). *Biochemistry* **16**, 4824–4829.
Perry, S. V. (1979). *Biochem. Soc. Trans.* **7**, 594–616.
Porubcan, M. A., Westler, W. A., Ibañez, I. B., and Markley, J. L. (1979). *Biochemistry* **18**, 4108–4116.
Quashnock, J. M., Chlebowski, J. F., Martinez-Carrion, M., and Shirch, L. V. (1983). *J. Biol. Chem.* **258**, 503–509.
Ray, W. J., Mildvan, A. S., and Grutzner, J. B. (1977). *Arch. Biochem. Biophys.* **184**, 453–463.
Reeck, G. R., Nelson, T. B., Paukstelis, J. V., and Mueller, D. D. (1977). *Biochem. Biophys. Res. Commun.* **74**, 643–649.
Reid, T. W., and Wilson, I. B. (1971). *In* "The Enzymes" (P. D. Boyer, ed.), Vol. 4, pp. 373–416. Academic Press, New York.

Reynolds, W. F., Peat, I. R., Freedman, M. M., and Lyerla, J. R. (1973). *J. Am. Chem. Soc.* **95**, 328–330.
Riordan, J. F., McElvany, K. D., and Borders, C. L. (1977). *Science* **195**, 884–886.
Roberts, J. D., Yu, C., Flanagan, C., and Birdseye, T. R. (1982). *J. Am. Chem. Soc.* **104**, 3945–3949.
Robillard, G. T. (1982). *Mol. Cell. Biochem.* **46**, 3–24.
Rösch, P., Kalbitzer, H. R., Schmidt-Aderjan, U., and Hengstenberg, W. (1981). *Biochemistry* **20**, 1599–1605.
Rose, I. A. (1979). *Adv. Enzymol.* **50**, 361–395.
Rose, Z. B. (1980). *Adv. Enzymol.* **51**, 211–153.
Schliselfeld, L. H., Burt, C. T., and Labotka, R. J. (1982). *Biochemistry* **21**, 317–320.
Schmidt-Aderjan, U., Rösch, P., Frank, R., and Hengstenberg, W. (1979). *Eur. J. Biochem.* **96**, 43–48.
Schnackerz, K. D., and Bartholmes, P. (1983). *Biochem. Biophys. Res. Commun.* **111**, 817–823.
Schnackerz, K. D., and Feldmann, K. (1980). *Biochem. Biophys. Res. Commun.* **95**, 1832–1838.
Schnackerz, K. D., Feldmann, K., and Hull, W. E. (1979). *Biochemistry* **18**, 1536–1539.
Shapiro, B. M., and Stadtman, E. R. (1968). *J. Biol. Chem.* **243**, 3769–3771.
Shindo, H. (1980). *Biopolymers* **19**, 509–522.
Shriver, J. W., and Sykes, B. D. (1981). *Biochemistry* **20**, 1748–1756.
Spector, L. B. (1980). *Proc. Natl. Acad. Sci. U.S.A.* **77**, 2626–2630.
Sperling, J. E., Feldmann, K., Meyer, H., Jahnke, U., and Heilmeyer, L. M. G. (1979). *Eur. J. Biochem.* **101**, 581–592.
Steitz, T. A., and Shulman, R. G. (1982). *Annu. Rev. Biophys. Bioeng.* **11**, 419–444.
Sundralingam, M., and Putkey, E. F. (1970). *Acta Crystallogr., Sect. B* **B26**, 790–800.
Sykes, B. D. (1983). *Can. J. Biochem. Cell Biol.* **61**, 155–164.
Sykes, B. D., and Weiner, J. H. (1980). *In* "Magnetic Resonance in Biology" (J. S. Cohen, ed.), pp. 171–196. Wiley, New York.
Taborsky, G. (1974). *Adv. Protein Chem.* **28**, 1–187.
Titanji, V. P. K., Ragnarson, U., Humble, E., and Zetterqvist, O. (1980). *J. Biol. Chem.* **255**, 11339–11343.
Toyo-Oka, T. (1982). *Biochem. Biophys. Res. Commun.* **107**, 44–50.
Tran Dinh, S., and Roux, M. (1977). *Eur. J. Biochem.* **76**, 245–249.
Van Wazer, J. R., and Letcher, L. H. (1967). *In* "$^{31}$P-NMR" (M. M. Crutchfield, C. H. Dungan, L. H. Letcher, V. Mark, and J. R. Van Wazer, eds.), pp. 75–225. Wiley (Interscience), New York.
Villafranca, J. J., Rhee, S. G., and Chock, P. B. (1978). *Proc. Natl. Acad. Sci. U.S.A.* **75**, 1255–1259.
Vogel, H. J. (1983). *Biochemistry* **22**, 668–674.
Vogel, H. J., and Bridger, W. A. (1981). *J. Biol. Chem.* **256**, 11702–11707.
Vogel, H. J., and Bridger, W. A. (1982a). *Biochemistry* **21**, 394–401.
Vogel, H. J., and Bridger, W. A. (1982b). *J. Biol. Chem.* **257**, 4834–4842.
Vogel, H. J., and Bridger, W. A. (1982c). *Biochemistry* **21**, 5825–5831.
Vogel, H. J., and Bridger, W. A. (1983a). *Can. J. Biochem. Cell Biol.* **61**, 363–369.
Vogel, H. J., and Bridger, W. A. (1983b). *Biochem. Soc. Trans.* **11**, 315–323.
Vogel, H. J., Bridger, W. A., and Sykes, B. D. (1982). *Biochemistry* **21**, 1126–1132.
Walderhaug, M. O., Saccomi, G., Sachs, G., and Post, R. L. (1982). *Fed. Proc., Fed. Am. Soc. Exp. Biol.* **41**(4), 6422.
Walsh, C. (1979). "Enzymatic Reaction Mechanisms." Freeman, San Francisco, California.

Walters, E. D., and Allerhand, A. (1980). *J. Biol. Chem.* **255,** 6200–6204.
Wang, T., Juraszek, L., and Bridger, W. A. (1972). *Biochemistry* **11,** 2607–2610.
Weiner, J. H., and Dickie, P. (1979). *J. Biol. Chem.* **254,** 8590–8593.
Weiner, R. E., Chlebowski, J. F., Haffner, P. H., and Coleman, J. E. (1979). *J. Biol. Chem.* **254,** 9739–9746.
Wimmer, M. J., and Rose, I. A. (1978). *Annu. Rev. Biochem.* **47,** 1031–1078.
Withers, S. G., Sykes, B. D., Madsen, N. B., and Kasvinsky, P. J. (1979). *Biochemistry* **18,** 5342–5348.
Withers, S. G., Madsen, N. B., and Sykes, B. D. (1981a). *Biochemistry* **20,** 1748–1756.
Withers, S. G., Madsen, N. B., Sykes, B. D., Takagi, M., Shimomura, S., and Fukui, T. (1981b). *J. Biol. Chem.* **256,** 10759–10762.
Withers, S. G., Madsen, N. B., Sprang, S. R., and Fletterick, R. J. (1982a). *Biochemistry* **21,** 5372–5382.
Withers, S. G., Shechosky, S., and Madsen, N. B. (1982b). *Biochem. Biophys. Res. Commun.* **108,** 322–327.
Withers, S. G., Madsen, N. B., and Sykes, B. D. (1982c). *Biochemistry* **21,** 6716–6722.
Wong, R. S. C., and Bennick, A. (1980). *J. Biol. Chem.* **255,** 5943–5948.
Yguerabide, J., Epstein, H. F., and Stryer, L. (1970). *J. Mol. Biol.* **51,** 573–590.
Yount, R. G. (1975). *Adv. Enzymol. Relat. Areas Mol. Biol.* **43,** 1–56.

CHAPTER 5

# Paramagnetic Probes of Enzyme Complexes with Phosphorus-Containing Compounds

*Joseph J. Villafranca*
Department of Chemistry
The Pennsylvania State University
University Park, Pennsylvania

| | |
|---|---|
| I. Introduction | 155 |
| II. General Considerations | 156 |
| III. Properties of Cr(III)- and Co(III)-Nucleotide Complexes | 157 |
| IV. Studies with Co(III)-Nucleotide Complexes | 160 |
| V. Studies with Cr(III)-Nucleotide Complexes | 166 |
| VI. Studies with Cr(III)- and Co(III)-Pyrophosphate Complexes | 169 |
| References | 173 |

## I. Introduction

A significant number of biochemically important compounds contain phosphorus, usually in the form of phosphomono- and phosphodiesters. The significant $^{31}$P-NMR properties of such compounds have been covered in early chapters of this volume by Gorenstein. Among the most important NMR properties of phosphorus are the $I = \frac{1}{2}$ nuclear spin, the high natural abundance (100%), reasonable chemical-shift range (~60 ppm for derivatives of phosphoric acid), and easily measurable $T_1$ and $T_2$ relaxation times (0.1–50 s).

The two features that are critical for understanding enzymatic catalysis are structure and mechanism. Several chapters (Rao, Chapter 3; Tsai, Chapter 6; Gerlt, Chapter 7) in this volume deal with mechanistic aspects of catalysis; here I focus on the structural details of substrate binding. The primary spectroscopic methods to be discussed are NMR and EPR (electron paramagnetic resonance). Through advances made in the synthesis of several stable metal-ion–phosphate complexes of biochemical interest, distance relationships among active-site components (metal ions, substrates, etc.) of several ATP-utilizing enzymes have been reported. The experimental design of such experiments and their interpretation comprise the bulk of this chapter. When necessary, the theoretical basis for measurements of distances between nuclei is presented in sufficient detail so the reader can gain an appreciation for the limits of such measurements.

## II. General Considerations

Cr(III)- and Co(III)-nucleotide complexes synthesized by Cleland and co-workers (Cleland, 1982) have provided biochemists with a powerful tool to study the metal-nucleotide environment of enzymes. Previous articles (Cleland, 1982; Cleland and Mildvan, 1979) have summarized the preparation of such nucleotide complexes and their use with several enzyme systems to elucidate stereospecificity of binding and details of catalysis. These stable "substitution-inert" metal-nucleotide complexes have been used to ascertain which coordination isomer of a complex is a substrate for a given enzyme. Combination of the use of Cr(III) and Co(III) complexes of nucleotides, along with appropriate sulfur-substituted nucleotides, has opened a new chapter in the study of the mechanism of phosphoryl transfer processes in these biochemically important problems (Tsai, Chapter 6). Other chapters in this volume discuss the stereospecificity of phosphoryl transfer and the methods used to explore this aspect of enzymic catalysis (Gerlt, Chapter 7).

In this chapter I focus on structural features of metal-nucleotide binding to enzymes and also on other metal-phosphate complexes with individual enzymes. The appropriate theoretical consideration for the NMR and EPR experiments is discussed in each section. The first problem, which has been addressed for several different enzymes, is whether enzymes have a preference for certain stereoisomers of metal-nucleotides. This is fundamentally a structural problem and can be answered by preparation of individual stereoisomers, followed by binding and kinetic studies to detect the preference of an enzyme for one isomer over another. Cleland (1982) has compiled a list of current studies dealing with this question. Other complementary structural problems that can be addressed through the use of these nucleotide complexes is to use them to map distances between the nucleotide site and other substrate or metal-ion sites on enzymes. This is possible because Cr(III) complexes are paramagnetic, whereas Co(III) complexes are diamagnetic. Both types of complexes can be used for distance determinations, and each application is discussed.

## III. Properties of Cr(III)- and Co(III)-Nucleotide Complexes

The preparation and stability of various octahedral complexes of Cr(III)- and Co(III)-nucleotides has been summarized by Cleland (1982). For part of this chapter I focus mainly on the pyrophosphate, ADP and ATP complexes of Cr(III) and Co(III). Whenever necessary, the particular ligands other than the nucleotides or pyrophosphate are given, but for simplicity and aquo-Cr(III) complexes such as $Cr(H_2O)_4ATP$ (bidentate) and $Cr(H_2O_3ATP$ (tridentate) are abbreviated $\beta,\gamma$-CrATP and $\alpha,\beta,\gamma$-CrATP, respectively. If the screw-sense isomer is also important it will be designated $\Delta$ or $\Lambda$ as defined by Cleland. The only Co(III)-nucleotides I discuss are the $\Delta$ and $\Lambda$ isomers of $\beta,\gamma$-Co(NH$_3$)$_4$ATP, abbreviated CoATP and designated accordingly.

As a general rule of thumb, Co(NH$_3$)$_4$ATP complexes are quite stable from pH 3–7 and the individual $\Delta$ or $\Lambda$ isomers do not hydrolyze or racemize appreciably at pH 7 and 25°C. However, $\beta,\gamma$-CrATP is stable at pH 6 but is far less stable at pH 7, hydrolyzing to CrADP-P$_i$ at a rate of 0.3 min$^{-1}$ at 25°. $\alpha,\beta,\gamma$,-CrATP is stable at pH 6–7 but shows appreciable release of ATP at pH 8. Thus studies with both Cr(III)- and Co(III)-nucleotide complexes can be conducted safely at pH values below 6.8, but care must be exercised when using Cr(III)-nucleotides at pH 7 and above.

Octahedral Co(III) complexes are almost exclusively diamagnetic, with very few exceptions (Cotton and Wilkinson, 1972). Certainly any Co(III) complex with a nitrogenous ligand is diamagnetic. It is thus possible to obtain high-resolution $^{31}$-NMR spectra (Cornelius et al., 1977; Granot et al., 1979b) of the individual $\Delta$ and $\Lambda$ diastereomers of $\beta,\gamma$-Co(NH$_3$)$_4$ATP. Figure 1 shows a high-field $^{31}$P-NMR spectrum of a ~60:40 mixture of these two diastereomers with the correct assignment of each isomer in the $\alpha$

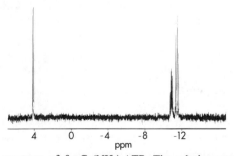

**Fig. 1.** $^{31}$P-NMR spectrum of $\beta,\gamma$-Co(NH$_3$)$_4$ATP. The solution contained 7.0 m$M$ $\beta,\gamma$-Co(NH$_3$)$_4$ATP buffered at pH 7.0. The spectrum was recorded at 20°C on a Brüker WM-360 spectrometer operating at 145.7 MHz for $^{31}$P. Twenty scans were time-averaged.

**TABLE I**

Phosphorus-31 Chemical Shifts for
$\beta,\gamma$-Co(NH$_3$)$_4$ATP[a]

| Position | Isomer | Shift (ppm) |
|---|---|---|
| $P_\alpha$ | $\Delta$ | −11.15 |
|  | $\Lambda$ | −11.01 |
| $P_\beta$ | $\Delta$ | −10.46 |
|  | $\Lambda$ | −10.52 |
| $P_\gamma$ | — | +4.0 |

[a] The Co(NH$_3$)$_4$ATP solution was at pH 7.0, 25°C. The data were collected at a $^{31}$P frequency of 145.7 MHz. Positive chemical shifts are downfield of 85% H$_3$PO$_4$ reference.

and $\beta$ regions of the spectrum. Table I lists the chemical shifts for each of the isomers. Co(III) ligated to the $\beta$- and $\gamma$-phosphoryl groups of ATP produces a downfield shift of ~10 ppm in each multiplet resonance relative to the resonance in free ATP. The $\alpha$-phosphoryl group is shifted upfield slightly by ~0.4 ppm, confirming that this group is not coordinated to the Co(III).

Two types of experiments are possible based on these properties of Co(III)ATP:

1. Enzymatic selectivity for one isomer in a mixture of $\Delta$- and $\Lambda$-$\beta,\gamma$-CoATP could be followed by monitoring the NMR spectrum. Figure 2 shows a spectrum of $\beta,\gamma$-CoATP after reaction with hexokinase. The "unused" isomer is the $\Delta$ isomer (Cornelius and Cleland, 1978) and a small amount of $\Lambda$ isomer remains. This experiment is also quite easily monitored by following the change in CD spectrum (Granot et al., 1979a).

2. Relaxation rates (see Section IV) of $\alpha$-, $\beta$-, and $\gamma$-phosphoryl resonances can be measured for enzyme complexes in which structural information (e.g., distance measurements to a paramagnetic species on an enzyme—Mn$^{2+}$, nitroxide, etc.) is desired. The resolution in the $^{31}$P-NMR spectra is excellent at high magnetic fields, so that a mixture of $\Delta$ and $\Lambda$ can be employed for this experiment as well as the individual isomers.

Cr(III) is paramagnetic ($S = \frac{3}{2}$), and complexes that contain this metal ion cannot be used for high-resolution $^{31}$P- or $^1$H-NMR observation of these complexes. However, these complexes do make ideal stable paramagnetic probes for EPR and NMR experiments. The EPR spectra of several complexes of Cr(III) are given in Fig. 3. All complexes have $g \sim 2$, but the nature of the complex determines the linewidth and line shape. Linewidths vary from 200 to 500 G for simple complexes.

The ground state of a $d^3$ ion is a singlet for octahedral coordination and

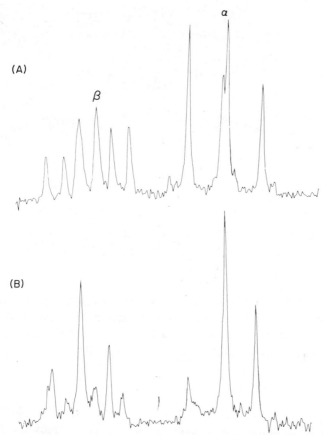

**Fig. 2.** (A) Expanded regions of the $\alpha$- and $\beta$-phosphorus resonances of a 60:40 mixture of the $\Delta$ and $\Lambda$ isomers of $\beta,\gamma$-Co(NH$_3$)$_4$ATP. (B) Spectrum of the $\alpha$ and $\beta$ regions of the $\Delta$ isomer of Co(NH$_3$)$_4$ATP that is *not* a substrate for hexokinase. A mixture of $\Delta$- and $\Lambda$-$\beta,\gamma$-Co(NH$_3$)$_4$ATP was incubated with hexokinase followed by isolation of the unreacted Co(NH$_3$)$_4$ATP. A $^{31}$P-NMR spectrum of the unreacted isomer was obtained as described in Fig. 1. A small amount of the unreacted $\Lambda$ isomer can also be seen in the spectrum.

there is little or no anisotropy found in the $g$ tensor. However, the zero-field splitting can be quite large, producing observable transitions for the $-\frac{1}{2} \rightleftarrows \frac{1}{2}$ and $\frac{1}{2} \rightleftarrows \frac{3}{2}$ electronic transitions. This phenomenon produces the asymmetric shape of the spectrum observed for some complexes. A problem that arises from this is that the linewidth of an EPR transition is quite often used to obtain a lower limit for the correlation time ($\tau_c$) that is used in distance determinations. If the apparent peak-to-peak linewidth is overestimated because of an asymmetric line shape, then $\tau_c$ can be underestimated. This

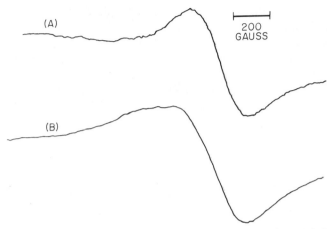

**Fig. 3.** EPR spectra of $Cr^{3+}$-nucleotide complexes obtained at 9 GHz. (A) Spectrum of 0.5 m$M$ $\alpha,\beta,\gamma$-CrATP obtained at 4°C. (B) Spectrum of 0.5 m$M$ $\beta,\gamma$-CrADP obtained at 4°C.

potential source of error should be kept in mind when using EPR spectra of Cr(III) complexes for this purpose. The relation of linewidth and $T_{2e}$, the transverse electron-spin relaxation time for a Lorentzian (Wertz and Bolton, 1972) line shape where $H_{pp}$ is the peak-to-peak linewidth in gauss, is shown by

$$1/T_{2e} = \pi\sqrt{3}(2.8 \times 10^6)H_{pp} \; s^{-1} \tag{1}$$

Most often $T_{2e} \simeq \tau_c$ and the linewidth measurements are suitable for use in distance determinations (see Section V). The CrATP complex in Fig. 2 has a linewidth of 310 G, corresponding to $T_{2e} = 2.1 \times 10^{-10}$ s.

One practical consideration is that the EPR spectrum should be measured on samples that are as dilute as possible (0.5–5 m$M$) but have sufficient signal to noise for accurate linewidth measurements. If samples are too concentrated, substantial dipolar spin–spin broadening occurs that leads to an overestimate of the linewidth.

## IV. Studies with Co(III)–Nucleotide Complexes

This section deals with the use of Co(III)–nucleotides for distance determinations on enzymes. Because these stable diamagnetic complexes are good analogs of metal–nucleotides in several enzyme systems, the assumption is made that the purine or pyrimidine base, the sugar and polyphosphate backbone, still occupy sites identical to those of the normal nucleotides. The metal ion, however, may not.

In some cases where the substrate for the enzyme is the metal–nucleotide, the enzyme may not supply ligands to the metal ion, and therefore the Co(III)–nucleotide may be a nearly identical analog to a Mg–nucleotide complex. In other cases, the enzyme may bind metal ion independently of nucleotide, resulting in nucleotide binding to an enzyme–metal-ion complex. For these enzymes Co(III)–nucleotides may still bind, but the $Co(NH_3)_4$ moiety is not expected to occupy the normal "true" metal-ion site because enzyme ligands would (may) not displace the $NH_3$ ligands. With these considerations in mind, I outline the theory for electron–nuclear dipolar relaxation and limitations of its use in measuring distances from a paramagnetic center to the $^{31}P$ and $^1H$ nuclei of Co(III) complexes. Expanded theoretical discussions have previously been published (Dwek, 1973; James, 1975; Mildvan and Gupta, 1977).

The basic experimental condition required for distance measurements from a paramagnetic center to a ligand nucleus in enzyme systems is that the paramagnetic (p) contribution to the longitudinal ($1/T_1$) or transverse ($1/T_2$) relaxation rates be measurably larger than the diamagnetic (d) contribution of the enzyme itself:

$$1/T_{1p} = 1/T_1(\text{total}) - 1/T_{1d} \tag{2}$$

A similar expression holds for $1/T_{2p}$. The total paramagnetic contributions can arise from two sources: (1) the ligand free in solution can interact with paramagnetic species in solution, or (2) the ligand can interact with enzyme-bound paramagnetic species. Because the measurement of interest deals with the complexes bound to the enzyme, a correction for ligand nuclei being relaxed by paramagnetic species off the enzyme is sometimes necessary. This is expressed in

$$1/T_{1p} = 1/T_{1p}(\text{bound}) - 1/T_{1p}(\text{free}) \tag{3}$$

where "bound" refers to the ligand in an enzyme complex and "free" refers to the ligand off the enzyme. A similar expression holds for $1/T_{2p}$.

If one suspects, or better yet knows, that some of the paramagnetic species is not totally bound to the enzyme, then control experiments must be conducted to assess the paramagnetic contribution to $1/T_1$ and $1/T_2$ of nuclei in binary paramagnetic ion–ligand complexes. Without such a correction, overestimates of relaxation rates are obtained that result in underestimates of distances.

Further discussion focuses on the interpretations of $T_1$ data rather than $T_2$ data, which are sometimes complicated by hyperfine interactions between paramagnetic ion and nucleus and/or exchange contributions between free and bound states. Further discussion of these matters can be found in several references (Dewk, 1973; James, 1975; Mildvan and Gupta, 1977).

Once one is confident that corrections have been made for binary complexes and diamagnetic contributions to $1/T_1$, attention can be paid to one final complication, that is, the lifetime ($\tau_m$) of the ligand in the enzyme complex. The relationship between the experimentally obtained paramagnetic relaxation rate $1/T_{1p}$, the paramagnetic dipolar electron–nuclear relaxation ($T_{1m}$) in the enzyme complex, and the lifetime of this complex $\tau_m$ is described by

$$1/T_{1p} = fq/(T_{1m} + \tau_m) \tag{4}$$

In this expression $\gamma_I$ is the magnetogyric ratio of the nucleus ($2.675 \times 10^4$ rad s$^{-1}$ G$^{-1}$ for $^1$H) and $\mu_{\text{eff}}$ the effective magnetic moment of the electron. For

Three conditions prevail: (1) fast exchange, where $T_{1m} \gg \tau_m$ and $1/T_{1p} = fq/T_{1m}$; (2) slow exchange, where $\tau_m \gg T_{1m}$ and $1/T_{1p} \gg fq/\tau_m$; and (3) intermediate exchange, where $T_{1m} \approx \tau_m$. Only cases (1) and (3) can be used for distance determinations, and case (3) can be used only if $\tau_m$ is known.

The Solomon–Bloembergen equation

$$1/T_{1m} = Cf(\tau_c)r^{-6} \tag{5}$$

relates $1/T_{1m}$ to the distance $r$ between the paramagnetic center and the nucleus (nuclei) being investigated. $f(\tau_c)$ is the correlation function that describes the dipolar electron–nuclear interaction, and $C$ relates the physical properties of the nucleus and the electron(s) according to

$$C = \tfrac{2}{15}\gamma_I^2\mu_{\text{eff}}^2 \tag{6}$$

In this expression $\gamma_I$ is the magnetogyric ratio of the nucleus ($2.675 \times 10^4$ rad s$^{-1}$ G$^{-1}$ for $^1$H) and $\mu_{\text{eff}}$ the effective magnetic moment of the electron. For simple paramagnetic species, the equation

$$\mu_{\text{eff}}^2 = S(S+1)g^2\beta^2 \tag{7}$$

holds, where $S$ is the spin quantum number ($S = \tfrac{1}{2}, 1, \tfrac{3}{2}$, etc. for 1, 2, 3, etc., electrons), $g$ the electronic $g$ factor (usually 2), and $\beta$ the Bohr magneton ($9.274 \times 10^{-21}$ erg G$^{-1}$). The expression in Eq. (7) is too simple for $Cu^{2+}$, where there is anisotropy in $g$ ($g_\parallel$ and $g_\perp$ components) and for ions like $Co^{2+}$, where spin-orbit coupling (Abragam and Bleaney, 1970) generally makes $g \geq 2$ and, worse yet, sometimes gives a "spread" of $g$ values (Abragam and Bleaney, 1970). Equation (7) is valid for "$S$-state" ions like $Cr^{3+}$, $Mn^{2+}$, $Gd^{3+}$, and examples are given for these ions.

For most macromolecular complexes, where $\omega_I$ is the Larmor precession frequency of the nucleus,

$$f(\tau_c) = 3\tau_c/(1 + \omega_I^2\tau_c^2) \tag{8}$$

## TABLE II

Values of $C'$ for Nuclei in Complexes that Have Different Numbers of Electrons[a]

| Number of unpaired electrons | $C'$ | | | | | | | | |
|---|---|---|---|---|---|---|---|---|---|
| | $^1$H | $^{19}$F | $^{31}$P | $^7$Li | $^{13}$C | $^{113}$Cd | $^6$Li | $^{133}$Cs | $^{15}$N |
| 1 | 539 | 528 | 399 | 393 | 341 | 326 | 285 | 274 | 252 |
| 2 | 635 | 622 | 470 | 463 | 401 | 384 | 335 | 323 | 296 |
| 3 | 705 | 691 | 522 | 514 | 445 | 427 | 372 | 358 | 329 |
| 4 | 763 | 748 | 565 | 556 | 481 | 462 | 403 | 388 | 356 |
| 5 | 812 | 796 | 601 | 592 | 512 | 492 | 429 | 413 | 379 |
| 7 | 896 | 878 | 663 | 653 | 565 | 542 | 473 | 455 | 418 |

[a] The values in this table are computed for use in Eq. (9). The value of $g = 2.0$ was used in Eq. (7). The units for $C'$ are cm$^6$ s$^{-2}$.

The nature of $\tau_c$ is discussed later. Rearrangement of Eq. (5) gives

$$r = C' \, [T_{1m} f(\tau_c)]^{1/6} \tag{9}$$

with $C' = C^{1/6}$ when fast-exchange conditions prevail and $T_{1m} = fT_{1p}$. Table II presents $C'$ values for several nuclei. This table was generated using Eqs. (6) and (7) and $g = 2$.

For most paramagnetic species in a macromolecular complex, $\tau_c \simeq \tau_s$, where $\tau_s$ is the longitudinal relaxation time $T_{1e}$ of the electron. $T_{1e}$ can itself be frequency-dependent (for ions with $S > \frac{1}{2}$), that is, the value will change at different magnetic field strengths, but this is usually not important at higher magnetic fields (> 20 kG).

This discussion of the evaluation of $T_{1p}$ and, ultimately, the calculation of $r$ from these data can be reduced in practice to several simple but necessary experiments.

To begin with, the major criterion that has to be met is that there be a believable paramagnetic effect on $T_1$ and $T_2$. With modern Fourier-transform NMR spectrometers, $T_1$ can usually be measured to better than 10% accuracy. Honesty is required of the experimentalist because the quality of the data depends on the signal-to-noise ratio of the individual NMR peaks and how many data points (Mildvan and Gupta, 1977) are taken to determine $T_1$. Realistically, a 50% difference in $T_1$ is a believable change, whereas larger differences are easily measured.

With good $1/T_{1p}$ data in hand, the fraction of ligand in the paramagnetic complex must be known. Knowledge of the dissociation constant under conditions that reasonably approximate the NMR experiment is required. In most cases, the approximation

$$f = [\text{ligand in paramagnetic complex}]/[\text{ligand}]_{\text{total}}$$

**TABLE III**

$\tau_s$ Values for Several Paramagnetic species

| Paramagnetic species | $\tau_s$ Range(s) |
|---|---|
| $Mn^{2+}$ | $10^{-9}$ to $10^{-8}$ |
| $Cu^{2+}$ | $\sim 5 \times 10^{-10}$ to $\sim 5 \times 10^{-9}$ |
| $Fe^{3+}$ (high spin) | $10^{-11}$ to $10^{-9}$ |
| $Fe^{3+}$ (low spin) | $\sim 10^{-11}$ |
| $Co^{2+}$ (high spin) | $\sim 5 \times 10^{-13}$ to $10^{-11}$ |
| $Ni^{2+}$ | $\sim 10^{-12}$ |
| $Fe^{2+}$ | $\sim 10^{-12}$ |
| $Cr^{3+}$ | $\sim 2 \times 10^{-10}$ to $\sim 8 \times 10^{-10}$ |
| Nitroxyl radical | $\sim 10^{-8}$ to $\sim 10^{-6}$ |

can be used safely. A necessary problem to be aware of is that the proper correction be made at this point for the amount of binary ligand–paramagnetic complex that can contribute to $1/T_{1p}$. Neglect of this correction compounds and magnifies errors in the correction to $1/T_{1p}$ and in evaluation of $f$. The ideal case is where all the paramagnetic species is bound to the enzyme and no corrections have to be made.

If one has determined $1/T_{1p}$ and $1/T_{2p}$ and the latter is larger than the former, then fast-exchange conditions prevail and $T_{1m} > \tau_m$. Further discussions of these complications are given elsewhere (Dwek, 1973; James, 1975).

The last "unknown" to be measured is $\tau_c$. There are two direct methods and several indirect methods for determination of $\tau_c$ that I discuss here. One direct method is to determine $1/T_{1p}$ at two or more magnetic fields. Because $\omega_1$ in Eq. (8) is a measure of the frequency of the nucleus at each magnetic field, if $\omega_1^2 \tau_c^2 > 1$, then $1/T_{1p}$ will vary with magnetic field. This condition usually holds if $\tau_c$ is $\tau_s$ of the metal ion or nitroxide spin label and $\tau_s \geq 5 \times 10^{-10}$. This restriction exists because currently available commercial NMR spectrometers are available from 19 to 94 kG. Table III lists $\tau_s$ values for several paramagnetic species.

Another direct method relies on having two magnetically active isotopes of the same nucleus (e.g., $^1H/^2H$, $^6Li/^7Li$, $^{14}N/^{15}N$, etc.). In this case a frequency-dependence study ($\omega_1$ different for each isotope) can be done at one magnetic field. Raushel and Villafranca (1980a,b; Villafranca and Raushel, 1982) described this method and applied it to pyruvate kinase. The obvious advantage of this method is that $\tau_s(\tau_c)$ is directly measured without the added complication of magnetic field dependence of $\tau_s$ itself, a complicating factor of the first method presented to evaluate $\tau_c$.

## 5. Paramagnetic Probes of Enzyme Complexes

**TABLE IV**

Relaxation Data for Co(NH$_3$)$_4$ATP and Carbamoyl-Phosphate Synthetase

| Co(NH$_3$)$_4$ATP nucleus | $\dfrac{1^a}{T_{1d}}$ (s$^{-1}$) | $\dfrac{1^b}{T_{1p}(\text{free})}$ (s$^{-1}$) | $\dfrac{1^c}{T_{1p}(\text{total})}$ (s$^{-1}$) | $r$ (Å) |
|---|---|---|---|---|
| $^{31}$P -$\alpha\beta^d$ | 0.58 | 1.2 | 2.4 | 6.6$^e$ |
| $^{31}$P-$\gamma$ | 0.84 | 2.6 | 4.2 | 6.4$^e$ |

$^a$ Data were obtained on a Brüker WP-200 spectrometer operating at 81 MHz for $^{31}$P. Solution contained enzyme, 5.0 m$M$ Co(NH$_3$)$_4$ATP, and buffer components.

$^b$ Solutions contained 5.0 m$M$ Co(NH$_3$)$_4$ATP and various amounts of Mn$^{2+}$. The concentration of binary Co(NH$_3$)$_4$ATP-Mn$^{2+}$ was determined using a binding constant of 15 m$M$ (Granot et al., 1979b). The value in the table corresponds to the amount of free binary complex present in the solution that has enzyme; Mn$^{2+}$ and CoATP were present.

$^c$ Same as in $^a$ but Mn$^{2+}$ present.

$^d$ The $\alpha$ and $\beta$ resonances overlapped at 81 MHz so a composite $1/T_1$ was measured.

$^e$ Calculated using $\tau_c = 3.2 \times 10^{-9}$ s, determined from the method of Raushel and Villafranca (1980a).

An indirect method to determine $\tau_s(\tau_c)$ is to measure the linewidth of the EPR signal of the bound paramagnetic species as mentioned earlier. This method is a measure of $T_{2e}$ and because $T_{2e}$ may be less than $T_{1e}$, can only give a lower limit of $T_{1e}(\tau_c)$. Another indirect method is to measure the frequency dependence of the interaction of the solvent water protons with the enzyme—paramagnet–ligand complex, but this evaluation of $\tau_c(\tau_s)$ may be complicated by other effects (Burton et al., 1979) and is not discussed.

Several examples of the use of CoATP to map the nucleotide site of enzymes have been presented (Granot et al., 1979b; Balakrishnan and Villafranca, 1978a; Raushel et al., 1983). The data for carbamoyl-phosphate synthetase obtained in our laboratory (Raushel et al., 1983) are presented in Table IV. The correlation time used for these distance calculations is the $\tau_s$ of bound Mn$^{2+}$, which was determined by the method of Raushel and Villafranca (1980a).

Mildvan and co-workers (Granot et al., 1979b) studied the binding of CoATP to bovine heart protein kinase. The enzyme normally binds two Mn$^{2+}$ per ATP at the catalytic site, but the clever use of CoATP as a substitution-inert substrate analog permitted distance determinations from one Mn$^{2+}$ to various nuclei of CoATP. Phosphorus-31 relaxation rates were measured at three frequencies (40, 73, and 146 MHz) to obtain a unique correlation time, and Mn$^{2+}$ to P$_\alpha$ (6.3 Å), P$_\beta$ (6.4 Å), and P$_\gamma$ (5.7 Å) distances were determined. Without the use of CoATP these data could not have been obtained.

## V. Studies with Cr(III)–Nucleotide Complexes

The suitability of Cr(III)–nucleotide complexes as analogs of metal–nucleotides relies on the nucleotide and sugar moieties binding to the same enzyme site as the normal substrate. This appears to be the case from several studies in the literature.

A fundamental property of $Cr^{3+}$ is that it has three unpaired electrons and is therefore paramagnetic. Two uses are apparent from this physical property:

1. $Cr^{3+}$ complexes can be used as paramagnetic probes to measure distances to other substrate or enzyme ligands other than the $Cr^{3+}$ complex.
2. The $Cr^{3+}$ moiety is an EPR probe and can be used to measure distances from $Cr^{3+}$ to other paramagnetic species.

CrATP has been used with several enzymes as a paramagnetic probe (Gupta *et al.*, 1976; Balakrishnan and Villafranca, 1978a; Villafranca, 1980). For some cases a mixture of isomers was used for distance determinations, and one isomer could have bound preferentially to the enzyme that would lead to an error in determining *f*, the mole fraction of paramagnetic species in the enzyme complex. Another problem is that the $\tau_s$ value for $Cr^{3+}$ is $\sim 2 \times 10^{-10}$ s. Thus the paramagnetic influence of $Cr^{3+}$ on the $^1H$ or other nuclei of ligands is quite often small. Distances $> 8$ Å from $Cr^{3+}$ to a nucleus are not well determined and most often must be viewed with caution.

We have demonstrated that $Cr^{3+}$–nucleotides can be used to determine metal–metal distances on glutamine synthetase (Balakrishnan and Villafranca, 1975b; Villafranca *et al.*, 1977). The method relies on the phenomenon of electron–electron dipolar relaxation and has been applied to other enzymes as well (Stein and Mildvan, 1978). Fundamentally, the longitudinal electron-spin relaxation time $T_{1e}$ of the perturbing spin $Cr^{3+}$ must be short compared to the transverse electron-spin relaxation time $T_{2e}$ of the observed spin (e.g., $Mn^{2+}$, nitroxyl radicals, etc.). The phenomenon observed is that there is a diminution of the signal of the observed spin with no appreciable broadening. The assumption used to calculate a distance between these two interacting paramagnetic species is that the Leigh theory (Leigh, 1970) of dipolar electron relaxation prevails with no added complications (Coffman and Buettner, 1979; Eaton and Eaton, 1978). This theory states that if two dissimilar spins are bound to a matrix (the enzyme), the equations

$$\delta H = C''(1 - 3\cos^2 \theta_{R'})^2 + \delta H_0 \qquad (10)$$

$$C'' = g\beta\mu^2 \tau_c / \hbar r^6 \qquad (11)$$

hold for the linewidth $\delta H$ of the observed spin, where $\delta H_0$ is the linewidth in the absence of dipolar effects, $\theta_{R'}$ the angle between the vector joining the two spins and the magnetic field direction, $r$ the distance between the spins, and $\mu$ and $\tau_c(T_{1e})$ are the magnetic moment [see Eq. (7)] and electron-spin relaxation time of $Cr^{3+}$, respectively.

The diminution of signal arises from the angular dependence $\theta_{R'}$. For most angular orientations, the dipolar term $C(1-3\cos^2\theta_{R'})^2$ is large, and the linewidth resulting from this additional spin interaction is large compared to the unperturbed linewidth $\delta H_0$. However, at $\theta_{R'} \simeq 54°$, the dipolar term vanishes and the remaining signal is unbroadened but much lower in amplitude. At saturation with $Cr^{3+}$–nucleotides, the amplitude of the remaining signal ($Mn^{2+}$, nitroxide, etc.) is compared to the signal obtained from that observed with a diamagnetic control (e.g., $Co^{3+}$ ATP). From this $C''$ can be determined from plots given in Dwek (1973) or Villafranca and Raushel (1982), and a distance can be obtained using Eq. (11). Limitations arise in the accuracy of the distance determination when small (<10%) or large (>95%) diminutions in EPR signal amplitude are observed. Under these conditions only upper or lower limits on distances can be determined.

We applied this method to determine the metal–metal distance in *Esche-*

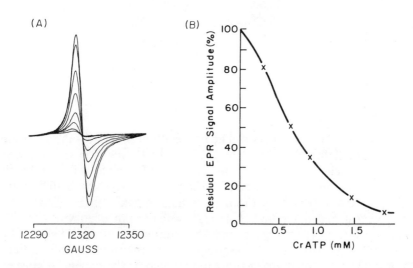

**Fig. 4.** (A) EPR titration experiment demonstrating electron–electron dipolar relaxation. The highest field line of enzyme-bound $Mn^{2+}$ in a sample containing 0.5 m$M$ glutamine synthetase, 10 m$M$ glutamate, and 0.2 m$M$ $Mn^{2+}$ recorded at 35 GHz and 5°C. (B) $\alpha,\beta,\gamma$-CrATP was titrated into this solution at concentrations from 0 to 2 m$M$ and the peak-to-peak signal amplitude replotted. The $Mn^{2+}$ signal is diminished in height by enzyme-bound CrATP without broadening as predicted by the Leigh theory.

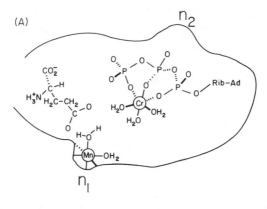

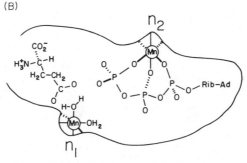

**Fig. 5.** Depiction of the active site of *Escherichia coli* glutamine synthetase showing the $n_1$ and $n_2$ metal-ion sites and the probable location of the nucleotide when $Cr^{3+}$ (A) or $Mn^{2+}$ (B) is the divalent cation.

richia coli glutamine synthetase. The $n_1$ metal-ion site is near the substrate glutamate (Villafranca and Raushel, 1979), whereas the $n_2$ metal-ion site is the metal–nucleotide substrate site. Titration of enzyme–$Mn^{2+}$ ($n_1$) with CrATP ($n_2$) results in diminution of the height of the $Mn^{2+}$ signal (60%) with no appreciable broadening. From these data, a $Mn^{2+}$—$Cr^{3+}$ distance of ~7 Å was calculated. Figure 4 (p. 167) shows a titration of enzyme–$Mn^{2+}$–glutamate by CrATP that results in a 94% diminution of signal height. The $Mn^{2+}$—$Cr^{3+}$ distance becomes shorter (5–6 Å) when glutamate (or a transition-state analog, methionine sulfoximine) binds to the enzyme. It must be pointed out that when a metal–metal distance of less than 7 Å results from use of the Leigh theory, neglect of a contribution from exchange interactions could be serious and the calculated distance could be an underestimate.

Notwithstanding this problem, another problem arises with an enzyme such as glutamine synthetase. The enzyme can bind metal ions to the $n_2$ site in the absence of nucleotide. Therefore, one can anticipate that the $Cr^{3+}$ moiety of $Cr^{3+}$-nucleotides will not bind to the site normally occupied by the divalent metal ion. This is the case for glutamine synthetase. Data from our laboratory (Gibbs, 1980; Gibbs et al., 1983) on $Mn^{2+}$—$Mn^{2+}$ distances between $n_1$ and $n_2$ reveal that these metal ions are 10–12 Å apart. Figure 5 shows this difference in a model of the active site of glutamine synthetase. Thus, whereas the EPR method of determining metal–metal distances using $Cr^{3+}$-nucleotides is attractive, the consideration of whether the $Cr^{3+}$ is in the "proper" site on the enzyme should be kept in mind. This also would apply to using $Cr^{3+}$ as a paramagnetic probe for distance measurements in NMR experiments.

## VI. Studies with Cr(III)- and Co(III)-Pyrophosphate Complexes

Studies of the yeast inorganic pyrophosphatase-catalyzed hydrolysis of pyrophosphate suggest that three divalent cations per enzyme active site are required for catalysis (Spring et al., 1981; Knight and Dunaway-Mariano, 1983). The results from equilibrium dialysis studies of the binding of $Mn^{2+}$ to PPase in the presence and absence of orthophosphate indicate that three divalent cations are bound in the central complex whereas only two divalent cations are bound to the free enzyme (Cooperman et al., 1981; Knight and Dunaway-Mariano, 1983). One of the three divalent cation cofactors is known to coordinate PP to form the active substrate complex, $P^1,P^2$-bidentate $M(H_2O)_4PP$ (M = metal) (Knight et al., 1981). The roles of the remaining two metal cofactors in catalysis have yet to be demonstrated.

In collaboration with D. Dunaway-Mariano, we designed EPR and NMR experiments of these two metal ions relative to the enzyme-bound substrate (Ransom et al., 1983). Earlier NMR studies (Hamm and Cooperman, 1978) showed that the $Mn^{2+}$—P distance from tightly bound $Mn^{2+}$ to P bound in low-affinity phosphate-binding site is 6.2 Å. The weaker $Mn^{2+}$ site was thought to be remote from both sites. At the time these studies were carried out, it was not known that there may be three $Mn^{2+}$ ions bound per enzyme subunit, and thus the presence of the third $Mn^{2+}$ was not taken into consideration.

In our work we have used magnetic-field–dependent $^1$H-NMR and EPR techniques to determine the distance between the two $Mn^{2+}$ sites on the free enzyme, between the two $Mn^{2+}$ sites on the enzyme–$Co(NH_3)_4(P)_2$ and enzyme–$Co(NH_3)_4PNP$ complexes, and between the high-affinity $Mn^{2+}$ site and the Cr(III) site on the enzyme–$Cr(NH_3)_4(P)_2$ and enzyme–$Cr(NH_3)_4PNP$ complexes.

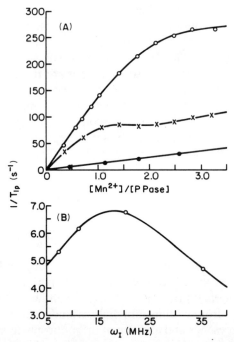

**Fig. 6.** (A) Longitudinal relaxation-rate data plotted as a function of the Mn-PPase ratio. The titration was performed at 20 MHz and 25°C. Key: O, 0.86 m$M$ PPase; X, same but 20 m$M$ Co(NH$_3$)$_4$PNP present; ●, Mn$^{2+}$ titrated into buffer with no PPase present. (B) Frequency dependence of $1/T_{1p}$ data for [Mn$^{2+}$]/[PPase] = 0.26. The shape of this curve is as predicted for enzyme–Mn$^{2+}$ complexes (Dwek, 1973).

Our NMR and EPR data (Figs. 6 and 7) clearly demonstrate that two Mn$^{2+}$ ions (at least) magnetically interact in the absence of substrate or product(s), and that this interaction persists when substrate analogs [e.g., Co(NH$_3$)$_4$PNP or Co(NH$_3$)$_4$(P)$_2$] were also bound to the enzyme. Thus these magnetic resonance data agree with the concept that yeast inorganic pyrophosphatase utilizes three metal ions in substrate binding and catalysis.

Distances between the "tight" Mn$^{2+}$ site and the Cr$^{3+}$ of various complexes were determined using EPR and NMR data. From the diminution elicited by paramagnetic Cr$^{3+}$ on the EPR transitions of bound Mn$^{2+}$, a distance was calculated using the Leigh theory because the results suggested a dipolar relaxation phenomenon similar to that observed with glutamine synthetase. The data that were evaluated were at a Mn:PPase ratio of more than 0.26, where 95% of the Mn$^{2+}$ was in E·Mn complex.

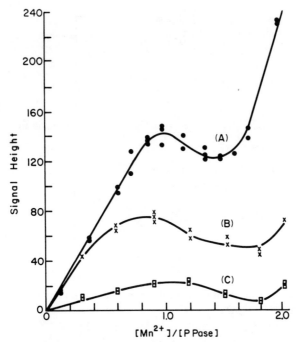

**Fig. 7.** Plot of EPR signal amplitude versus [Mn$^{2+}$]/[PPase] ratio. (A) Titration of 0.86 m$M$ PPase with Mn$^{2+}$. (B) Same as in (A) but in the presence of 17 m$M$ Co(NH$_3$)$_4$PNP. (C) Same as in (A) but in the presence of 19 m$M$ Cr(NH$_3$)$_4$PNP.

**TABLE V**

Metal–Metal Distances Calculated for Various Complexes with Pyrophosphatase[a]

| Complex | Distance (Å) | |
|---|---|---|
| | From NMR | From EPR |
| Mn–Cr(NH$_3$)$_4$(P)$_2$ | 4.8 | 5.2 |
| Mn–Cr(H$_2$O)$_4$(P)$_2$ | — | 5.0 |
| Mn–Cr(NH$_3$)$_4$PNP | 7.0 | 7.2 |
| Mn–Cr(H$_2$O)$_4$PNP | — | 7.5 |
| Mn–Mn(PPase) | 10–12 | 13–16 |
| Mn–Mn[PPase-Co(NH$_3$)$_4$(P)$_2$] | 8–9 | 7–8 |

[a] Data for Cr(NH$_3$)$_4$(P)$_2$ and Cr(NH$_3$)$_4$PNP are at pH 7.5, whereas those for Cr(H$_2$O)$_4$(P)$_2$ and Cr(H$_2$O)$_4$PNP are at pH 6.7. Data from Ransom *et al.* (1983).

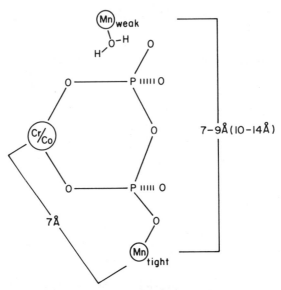

**Fig. 8.** Schematic representation of the distance relationships among the metal ions as determined from the NMR and EPR data.

The method of Gupta (1977) was used to compute distances from the frequency-dependent NMR data and a summary of the Mn—Cr distances computed using both methods is given in Table V (p. 171). Two conclusions are obvious from these data:

1. The distances evaluated by the two magnetic resonance methods are in good agreement.
2. The Mn–Cr distance is shorter in the product complex formed when $Cr(NH_3)_4(P)_2$ and $Cr(H_2O)_4(P)_2$ are used compared to the $Cr^{3+}$-PNP complexes.

This could reflect Mn coordination to $P_i$ in the product complex and a different coordination of $Mn^{2+}$ to PNP in the substrate (analog) complex.

The distance between the two bound $Mn^{2+}$ in various enzyme complexes was evaluated from both the NMR and EPR data and these values are in Table V. The line drawn through the data points in Fig. 6A represents a theoretical curve generated considering the various species present in solution. Both the NMR and EPR data are consistent with a conformational change produced on binding $Co^{3+}$ complexes that results in a shorter Mn—Mn distance. A schematic representation of all the results for distance determinations is presented in Fig. 8.

## Acknowledgments

The author thanks his many co-workers who conducted much of the research discussed in this chapter. Special thanks are extended to F. M. Raushel, P. M. Anderson, M. S. Balakrishnan, E. J. Gibbs, and S. Ransom. W. B. Knight and D. Dunaway-Mariano were enthusiastic collaborators for part of this work. Support from the National Institute of Health (AM-21785 and GM-23529), National Science Foundation (PCM-8108094), and American Heart Association (Established Investigatorship) is gratefully acknowledged.

## References

Abragam, A., and Bleaney, B. (1970). "Electron Paramagnetic Resonance of Transition Ions," p. 446. Oxford Univ. Press (Clarendon), London and New York.
Balakrishnan, M. S., and Villafranca, J. J. (1978a). *Fed. Proc., Fed. Am. Soc. Exp. Biol.* **37**, 830.
Balakrishnan, M. S., and Villafranca, J. J. (1978b). *Biochemistry* **17**, 3531.
Burton, D. R., Forsén, S., Karlström, G., and Dwek, R. A. (1979). *Prog. NMR Spectrosc.* **13**, 1.
Cleland, W. W. (1982). *In* "Methods in Enzymology" 87, (D. L. Purich, ed.), Vol. p. 159. Academic Press, New York.
Cleland, W. W., and Mildvan, A. S. (1979). *Adv. Inorg. Biochem.* **1**, 163.
Coffman, R. E., and Buettner, G. R. (1979). *J. Phys. Chem.* **83**, 2387, 2392.
Cooperman, B. S., Panackal, A., Springs, B., and Hamm, D. J. (1981). *Biochemistry* **20**, 6051–6060.
Cornelius, R. D., and Cleland, W. W. (1978). *Biochemistry* **17**, 3279.
Cornelius, R. D., Hart, P. A., and Cleland, W. W. (1977). *Inorg. Chem.* **16**, 2799.
Cotton, F. A., and Wilkinson, G. (1972). "Advanced Inorganic Chemistry," 3rd ed., p. 884. Wiley (Interscience), New York.
Dwek, R. A. (1973). "NMR and Biochemistry." Oxford Univ. Press (Clarendon), London and New York.
Eaton, S.-S., and Eaton, G. R. (1978). *Coord. Chem. Rev.* **26**, 207.
Gibbs, E. J. (1980). M. S. Thesis, Pennsylvania State University, University Park.
Gibbs, E. J., Ransom, S., and Villafranca, J. J. (1983). In preparation.
Granot, J., Mildvan, A. S., Brown, E. M., Kondo, H., Bramson, H. N., and Kaiser, E. T. (1979a). *FEBS Lett.* **103**, 265.
Granot, J., Kondo, H., Armstrong, R. N., Mildvan, A. S., and Kaiser, E. T. (1979b). *Biochemistry* **18**, 2339.
Gupta, R. K. (1977). *J. Biol. Chem.* **252**, 5183–5185.
Gupta, R. K., Fung, C. H., and Mildvan, A. S. (1976). *J. Biol. Chem.* **251**, 2421.
Hamm, D. J., and Cooperman, B. S. (1978). *Biochemistry* **17**, 4033–4040.
James, T. L. (1975). "Nuclear Magnetic Resonance in Biochemistry." Academic Press, New York.
Knight, W. B., and Dunaway-Mariano, D. (1983). Submitted for publication.
Knight, W. B., Fitts, S. W., and Dunaway-Mariano, D. (1981). *Biochemistry* **20**, 4079–4086.
Leigh, J. S. (1970). *J. Chem. Phys.* **52**, 2608.
Mildvan, A. S., and Gupta, R. K. (1977). *In* "Methods in Enzymology" (C. H. W. Hirs, ed.), Vol. Academic Press, New York.

Ransom, S., Villafranca, J. J., Knight, W. B., and Dunaway-Mariano, D. (1983). Submitted for publication.
Raushel, F. M., and Villafranca, J. J. (1980a). *J. Am. Chem. Soc.* **102**, 6618.
Raushel, F. M., and Villafranca, J. J. (1980b). *Biochemistry* **19**, 5481.
Raushel, F. M., Anderson, P. M., and Villafranca, J. J. (1983). *Biochemistry* **22**, 1872.
Springs, B., Welsh, K. M., and Cooperman, B. S. (1981). *Biochemistry* **20**, 6384–6391.
Stein, P. J., and Mildvan, A. S. (1978). *Biochemistry* **17**, 2675.
Villafranca, J. J. (1980). *In* "Frontiers in Protein Chemistry" (T.-Y. Liu, G. Mamiya, and K. T. Yasunobu, eds.), p. 17. Elsevier/North-Holland, New York.
Villafranca, J. J., and Raushel, F. M. (1979). *Adv. Catal.* **28**, 323.
Villafranca, J. J., and Raushel, F. M. (1982). *Adv. Inorg. Biochem.* **4**, 289.
Villafranca, J. J., Balakrishnan, M. S., and Wedler, F. C. (1977). *Biochem. Biophys. Res. Commun.* **75**, 464.
Wertz, J. E., and Bolton, J. R. (1972). "ESR: Elementary Theory and Practical Application," p. 33. McGraw-Hill, New York.

CHAPTER 6

# Use of Chiral Thiophosphates and the Stereochemistry of Enzymatic Phosphoryl Transfer

*Ming-Daw Tsai*
Department of Chemistry
The Ohio State University
Columbus, Ohio

| | | |
|---|---|---|
| I. Stereochemical Problems Studied with Phosphorothioates | | 175 |
| II. $^{18}O$ Isotope Shifts in $^{31}P$ NMR | | 178 |
| III. $^{17}O$ Quadrupolar Effects in $^{31}P$ NMR | | 181 |
| IV. Prochiral Centers: $^{31}P$ NMR | | 185 |
| V. Pro-Prochiral Centers: $^{31}P(^{18}O)$ NMR or $^{31}P(^{17}O)$ NMR | | 187 |
| VI. Pro-Pro-Prochiral Centers: $^{31}P(^{18}O)$ NMR and $^{31}P(^{17}O)$ NMR | | 190 |
| References | | 196 |

## I. Stereochemical Problems Studied with Chiral Thiophosphates

The use of thiophosphate analogs of biophosphates in studying stereochemical problems was first introduced by Eckstein (1975) and subsequently widely applied to various problems. To illustrate the use of chiral thiophosphates in stereochemistry, consider the phosphoryl transfer reaction catalyzed by hexokinase (Scheme 1).[1] Three types of problems can be studied by

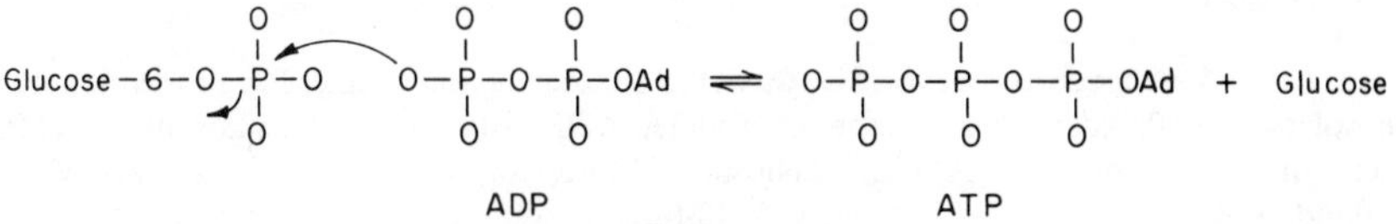

Scheme 1

[1] To avoid confusion, the charges and double bonds on the oxygen and sulfur of phosphoryl group or thiophosphoryl groups are omitted throughout the text. The $R$ and $S$ configurational designations are made according to the Cahn–Ingold–Prelog rule (Cahn *et al.,* 1966), with the assumption that nonbridging oxygens and sulfurs are all singly bonded to phosphorus. According to the priority rules, the atomic number preferences should be applied to exhaustion before the atomic weight preferences are applied.

use of chiral thiophosphates:

1. Stereochemistry of the substitution at phosphorus (retention, inversion, or nonstereospecificity) can be investigated by use of chiral [$^{18}$O]thiophosphoryl groups (Scheme 2).

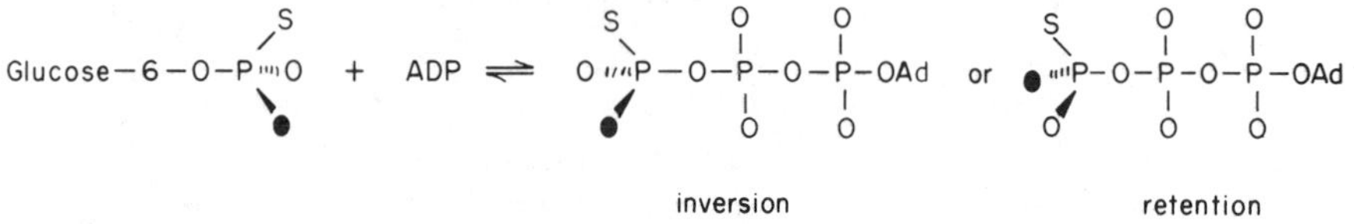

$$\text{Glucose-6-O-P}^{S}_{\bullet}\text{O} + \text{ADP} \rightleftharpoons \underset{\text{inversion}}{\text{O}^{\ominus}\text{P-O-P-O-P-OAd}} \text{ or } \underset{\text{retention}}{\bullet^{\ominus}\text{P-O-P-O-P-OAd}}$$

**Scheme 2**

2. By use of ADP$\beta$S,[2] the P$_\beta$ of ADP becomes a prochiral center. In principle, enzymes can distinguish the two diastereotopic oxygens. Therefore, the phosphorylation may be stereospecific (Scheme 3).

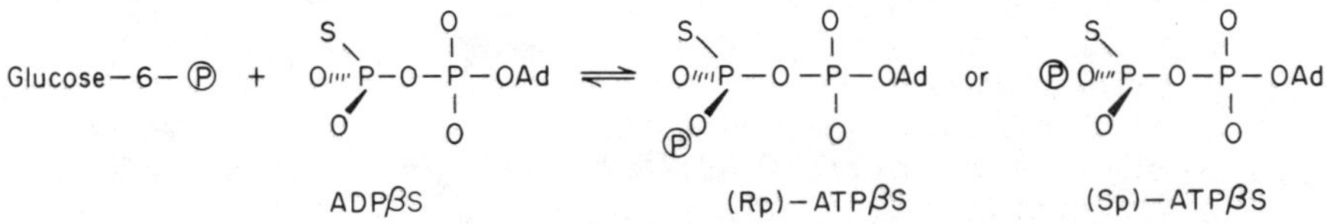

$$\underset{}{\text{Glucose-6-}\textcircled{P}} + \underset{\text{ADP}\beta\text{S}}{\text{O}^{\ominus}\text{P-O-P-OAd}} \rightleftharpoons \underset{(Rp)-\text{ATP}\beta\text{S}}{\text{O}^{\ominus}\text{P-O-P-OAd}} \text{ or } \underset{(Sp)-\text{ATP}\beta\text{S}}{\textcircled{P}\text{O}^{\ominus}\text{P-O-P-OAd}}$$

**Scheme 3**

3. By use of ADP$\alpha$S, the P$_\alpha$ of ADP becomes a chiral center, and there are two possible isomers. Even though the $\alpha$-phosphate of ADP is not directly involved in the reaction, the enzyme may be specific to one of the two isomers (Scheme 4).

The use of thiophosphates also has some disadvantages. They are in most cases poor substrates (0–10% of activity relative to the corresponding natural substrate) of enzymes and are relatively unstable. In addition, it may be argued that the stereochemical results obtained with thiophosphates are

---

[2] Abbreviations used: O, $^{16}$O; ⊖, $^{17}$O; ●, $^{18}$O; P$_i$, inorganic phosphate; P$_{si}$, inorganic thiophosphate; AMP, adenosine 5'-monophosphate; ADP, adenosine 5'-diphosphate; ATP, adenosine 5'-triphosphate, AMPS, adenosine 5'-thiophosphate; ADP$\alpha$S, adenosine 5'-(1-thiodiphosphate); ADP$\beta$S, adenosine 5'-(2-thiodiphosphate); ATP$\alpha$S, adenosine 5'-(1-thiotriphosphate); ATP$\beta$S, adenosine 5'-(2-thiotriphosphate); ATP$\gamma$S, adenosine 5'-(3-thiotriphosphate); CoA, coenzyme A; cAMPS, cyclic AMPS; GTP$\alpha$S, guanosine 5'-(1-thiotriphosphate); GTP$\beta$S, guanosine 5'-(2-thiotriphosphate); U$_p$(S)A, uridinyl (3'→5')-adenosyl-O,O-phosphorothioate; U > pS, uridine 2',3'-cyclic phosphorothioate; UMPS, uridine 5'-thiophosphate; UDP$\alpha$S, uridine 5'-(1-thiodiphosphate); UPT$\alpha$S, uridine 5'-(1-thiotriphosphate); DPP$_S$E, dipalmitoylthiophosphatidylethanolamine; DPP$_S$C, dipalmitoylthiophosphatidylcholine; EDTA, ethylenediaminetetraacetic acid.

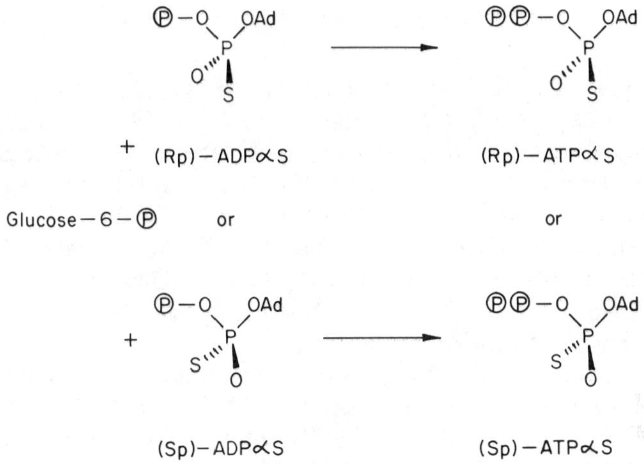

# 6. $^{17}O$ and $^{18}O$ Effects and Chiral Thiophosphates

**Scheme 4**

not authentic. However, so far there has not been an example found in which the stereochemical result obtained with chiral phosphates (see Gerlt, Chapter 7) is different from that obtained with chiral thiophosphates.

The biochemical applications of thiophosphates have been covered in a number of reviews (Eckstein, 1979; Eckstein *et al.*, 1982; Frey, 1982; Frey *et al.*, 1982; Knowles, 1980; Buchwald *et al.*, 1982; Tsai, 1982). The purpose of this chapter is to illustrate the use of $^{31}P$ NMR[3] in these problems, which is based on three factors:

1. When a prochiral center becomes a chiral center because of sulfur substitution, the resultant two diastereomers give different $^{31}P$ chemical shifts.
2. Substitution of an oxygen by $^{18}O$ causes an isotope shift in $^{31}P$ NMR.
3. Substitution of an oxygen by $^{17}O$ causes a quadrupolar broadening on the $^{31}P$-NMR signal.

Phosphorus-31 NMR is not the only method available for some problems, and this chapter does not intend to provide a comprehensive review of all chiral thiophosphate work.

---

[3] The $^{31}P$ chemical shifts described in this chapter are referenced to either 85% $H_3PO_4$ or 1 *M* $H_3PO_4$ as used in original references, which are different by only 0.3 ppm. The + sign always represents a downfield shift in this chapter, although the opposite convention has been used in most of the original references. It should be noted that the $^{31}P$ chemical shifts of most biophosphates are sensitive to pH, concentration, temperature, etc. In many cases the relative shifts are more important than the absolute chemical shifts.

## II. $^{18}O$ Isotope Shifts in $^{31}P$ NMR

Figure 1 shows the $^{31}P$-NMR spectrum of $H_3P^{17}O_4$ (40 atom % $^{17}O$). The spectrum consists of a "broad" signal owing to the $^{31}P-^{17}O$ species and a "sharp" signal owing to the residual non-$^{17}O$-labeled species. Because $^{17}O$-enriched water always contains some $^{18}O$ ($^{18}O/^{17}O = 0.67$ in this case), the sharp signal contains both $^{16}O$ and $^{18}O$ species, as shown by the expanded spectrum in the inset. In this compound, the $^{31}P$-NMR signal of the $^{18}O$-labeled species is shifted upfield by 0.020 ppm per $^{18}O$ atom.

The $^{18}O$ isotope-shift effect in $^{31}P$ NMR ($S_{^{31}P-^{18}O}$) was first reported by Cohn and Hu (1978) and others (Lowe and Sproat, 1978; Lutz et al., 1978). Cohn (1982) and Tsai and Bruzik (1983) have summarized the magnitudes of $S_{^{31}P-^{18}O}$ for a number of biophosphates. Table I lists the $S$ values for commonly used biophosphates, thiophosphates, and some model compounds. Because the data are collected from various reports that have different spectral resolution, some of the data in Table I may have an error of ±10%, unless otherwise specified.

It seems proper to generalize the following statements concerning the

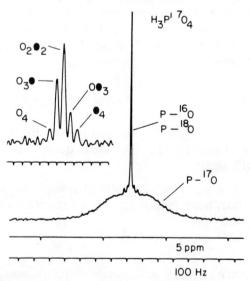

**Fig. 1.** $^{31}P$-NMR spectrum of 50 m$M$ $H_3P^{17}O_4$ (40 atom % $^{17}O$, 27 atom % $^{18}O$) in $D_2O$, pD 1.8 at 81.0 MHz. Spectral parameters: acquisition time 4.1 s, delay 1.0 s, spectra width 2 kHz, 70° pulse, line broadening 2.0 Hz, 1600 scans. The inset shows the expanded spectrum of the sharp peak, processed with Gaussian multiplication (LB = −2, GB = 0.2). Chemical shift 0.09 ppm downfield from 1 $M$ $H_3PO_4$.

## TABLE I
### Magnitude of $^{18}O$ Isotope Shift in $^{31}P$ NMR[a]

| Compound | Labeled position | Condition | $\delta_{^{31}O-^{18}O}$ (ppm) | References[b] |
|---|---|---|---|---|
| $H_4P^{18}OO_3^+ClO_4^-$ | | $0.2M$, $D_2O$ | $0.0188 \pm 0.0007$ | 1 |
| $KH_2P^{18}O_4$ | | pH 2.1 | $0.0201 \pm 0.0007$ | 1 |
| $K_2HP^{18}O_4$ | | pH 8.6 | $0.0218 \pm 0.0007$ | 1 |
| $(PhO)_3P^{18}O$ | | $CDCl_3$ | $0.0391 \pm 0.0029$ | 1 |
| $(PhO)_2P^{18}OO$ | | pD 5.4 | $0.0293 \pm 0.0007$ | 1 |
| $Ph_3P^{18}O$ | | $CDCl_3$ | $0.0399 \pm 0.0007$ | 1 |
| $(CH_3^{18}O)_3P^{18}O$ | P=O | $CDCl_3$ | $0.0392 \pm 0.0029$ | 1 |
| $(CH_3O)_2PO^{18}O^-$ | | | 0.029 | 3 |
| $(CH_3O)PO_2^{18}O^{2-}$ | | | 0.023 | 3 |
| 5′-AMP | | | 0.025 | 5 |
| 2′-Deoxy-AMP | | | 0.032 | 15 |
| ADP | $\alpha$ | | $0.0286 \pm 0.0015$ | 1 |
| | $\alpha\beta(P_\alpha)$ | | 0.0210 | 9 |
| | $\alpha\beta(P_\beta)$ | | 0.0214 | 9 |
| | $\beta$ | | 0.022 | 4,9 |
| ATP | $\alpha\beta(P_\alpha)$ | pH > 7.6 | 0.0172 | 4 |
| | $\alpha\beta(P_\beta)$ | pH > 7.6 | 0.0165 | 4 |
| | $\beta$ | pH > 7.6 | 0.0281 | 4 |
| | $\beta\gamma(P_\beta)$ | pH > 7.6 | 0.0165 | 4 |
| | $\gamma$ | pH > 7.6 | 0.0220 | 4 |
| $Co(NH_3)_4ADP$ | | | | |
| $\Delta$ isomer | $\alpha$, $R_p$ | pH 5.5 | 0.032 | 7 |
| $\Lambda$ isomer | $\alpha$, $R_p$ | pH 5.5 | 0.018 | 7 |
| $Co(NH_3)_4ADP$ | | | | |
| $\Delta$ isomer | $\alpha$, $S_p$ | pH 5.5 | 0.020 | 7 |
| $\Lambda$ isomer | $\alpha$, $S_p$ | pH 5.5 | 0.033 | 7 |
| $Co(NH_3)_4dADP$ | | | | |
| $\Delta$ isomer | $\alpha$, $R_p$ | pH 5.5 | 0.031 | 8 |
| $\Lambda$ isomer | $\alpha$, $R_p$ | pH 5.5 | 0.016 | 8 |
| $Co(NH_3)_4dADP$ | | | | |
| $\Delta$ isomer | $\alpha$, $S_p$ | pH 5.5 | 0.016 | 8 |
| $\Lambda$ isomer | $\alpha$, $S_p$ | pH 5.5 | 0.030 | 8 |
| MgADP | $\alpha$ | pD 8.0 | 0.026 | 1 |
| Ribose-1-P | $P-^{18}O$ | | 0.027 | 11 |
| | $P-^{18}O-C$ | | 0.017 | 11 |
| | $P=^{18}O$ | | 0.041 | 12,14 |
| | $P-^{18}O-CH_3$ | | 0.015 | 12,14 |

*(Continued)*

TABLE I (Continued)

| Compound | Labeled position | Condition | $S_{^{31}O-^{18}O}$ (ppm) | References[b] |
|---|---|---|---|---|
| [cyclohexyl phosphate structure with $^{18}O$] | P—O Axial methyl ester | $H_2O/D_2O$; $CDCl_3$ | 0.026 | 6 |
|  | P—OMe |  | 0.015 | 6 |
|  | P=O |  | 0.040 | 6 |
| [diacylglycerophosphoethanolamine-OSiMe₃ structure] | P=O |  | 0.038 | 10 |
|  | P—O—Si |  | 0.018 | 10 |
| [diacylglycerol oxazaphospholidine structure with $^{18}O$] |  | $CDCl_3$, 30°C | 0.039 ± 0.0029 | 2 |
| [cyclic methyl phosphate structure] | P=O | | 0.043 | 13 |
|  | P—O—$CH_3$ |  | 0.018 | 13 |
| [sugar methyl phosphate structure A] | Methyl ester |  |  |  |
|  | P=O |  | 0.043 | 11,14 |
|  | P—O—$CH_3$ |  | 0.017 | 11,14 |
|  | Ethyl ester |  |  |  |
|  | P=O |  | 0.0418 | 9 |
|  | P—OEt |  | 0.0192 | 9 |
| [sugar cyclic phosphate structure A] | Axial P—O |  | 0.029 | 15 |
|  | Equatorial P—O |  | 0.032 | 15 |
|  | Ethyl ester |  |  |  |
|  | P—OEt |  | 0.014 | 15 |
|  | P=O |  | 0.038 | 15 |
| 5'-AMPS | $\alpha$-$^{18}O$ |  | 0.0331 ± 0.0007 | 1 |
| 2'-UMPS |  |  | 0.032 | 16 |
| 3'-UMPS |  |  | 0.032 | 16 |
| endo-U > PS |  |  | 0.041 | 16 |
| exo-U > PS |  |  | 0.041 | 16 |
| ADP$\alpha$S | $\alpha$-$^{18}O$ |  | 0.037 | 17 |
|  | $\alpha\beta$-$^{18}O$ |  | 0.021 | 17 |
| CNEtO (P) O—P(S)—OAd, $^{18}O$ | $S_p$ isomer | pD 6.4 | 0.0363 ± 0.0045 | 1 |
|  | $S_p$ isomer | pD 6.4 | 0.0363 ± 0.0045 | 1 |

magnitudes of $^{18}$O isotope shifts in $^{31}$P NMR, at least for the compounds in which the phosphorus has an oxidation number of +5:

1. The $S$ value for a P=O double bond is 0.038–0.044 ppm, whereas that for a P—O single bond is 0.015–0.025 ppm. Bonds with partial double-bond character have $S$ values proportional to the bond order (Cohn and Hu, 1980; Lowe et al., 1979).
2. In case of multiple substitution, the magnitude of shift is generally additive.
3. The $S$ values of thiophosphates (in which an O is substituted by an S) are slightly greater than that of the corresponding phosphates.

Experimentally, resolution and quantitation are two important problems that dictate the capability of the $^{18}$O isotope-shift method in solving a specific biochemical problem. Even under optimal conditions both the integrals and the $S$ values may still have an error of ±5%. In general, a medium field (e.g., 81 MHz $^{31}$P) is suitable to resolve most shifts. Aqueous samples are often treated with Chelex-100 or EDTA to remove paramagnetic impurities. Two commonly used techniques for resolution enhancement are Gaussian multiplication and convolution difference (the CD command in the Bruker DISNMR program). Although the Gaussian multiplication is known to change the relative intensity of signals with different line shapes (Clin et al., 1979), it may be safe to assume that the different peaks arising from the $^{18}$O isotope shift have the same line shape. Of course, any manipulation of the free induction decay leading to an improvement in resolution will result in a loss of signal-to-noise ratio.

## III. $^{17}$O Quadrupolar Effects in $^{31}$P NMR

When a dipolar nucleus ($^{31}$P in the present case) is bonded directly to a quadrupolar nucleus ($^{17}$O in the present case), the $^{31}$P nucleus will also be relaxed by virtue of its spin–spin coupling with $^{17}$O. This was termed "scalar relaxation of the second kind" by Abragam (1961). Such a scalar relaxation

---

[a] This table is an extension of the table of Cohn (1982) to cover some more recent data, particularly those from our own laboratory. It is, however, not an exclusive list of all reported data. Unless otherwise specified, the data from our laboratory were obtained at ambient temperature (25–30°C).

[b] References: 1, Sammons et al. (1983); 2, K. Bruzik and M.-D. Tsai (unpublished); 3, Lowe et al. (1979); 4, Cohn and Hu (1980); 5, Lowe and Sproat (1978); 6, Gorenstein and Rowell (1980); 7, Sammons and Frey (1982); 8, Coderre and Gerlt (1980); 9, Sammons (1982); 10, Bruzik and Tsai (1982); 11, Jordan et al. (1981); 12, Jarvest et al. (1980); 13, Buchwald and Knowles (1980); 14, Jarvest et al. (1981); 15, Gerlt and Coderre (1980); 16, Gerlt and Wan (1979); 17, Webb and Trentham (1980); 18, R. L. Van Etten (private communication).

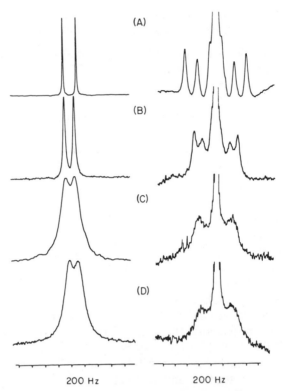

**Fig. 2.** Line shapes of $^{17}O$ NMR (left, at 27.1 MHz) and $^{31}P(^{17}O)$ NMR (right, at 81.0 MHz). (A) $P^{17}OCl_3$, 51 atom % $^{17}O$, in tetrahydrofuran, using [$^2H_6$]acetone for external lock, $\delta =$ 210 ppm for $^{17}O$, 2.5 ppm for $^{31}P$. (B) $(CH_3O)P^{17}O$, 51 atom % $^{17}O$, in $CDCl_3$, $\delta = 73.6$ ppm for $^{17}O$, 2.6 ppm for $^{31}P$. (C) $(PhO)_3P^{17}O$, 51 atom % $^{17}O$, in $CDCl_3$, $\delta = 91.2$ ppm for $^{17}O$, $-17.9$ ppm for $^{31}P$. (D) $(Ph)_3P^{17}O$, 49 atom % $^{17}O$, in $CDCl_3$, $\delta = 43.3$ ppm for $^{17}O$, 28.8 ppm for $^{31}P$. All spectra were run at 31 °C and processed with 5-Hz line broadening. From Sammons et al. (1983).

is dependent on the magnitudes of the longitudinal relaxation time of the quadrupolar nucleus ($T_1$, which is approximately equal to $T_q$ under present conditions) and the spin–spin coupling constant $J_{^{31}P-^{17}O}$ (abbreviated as $J$). When the product $T_q J$ is sufficiently small, the scalar relaxation dominates the relaxation of $^{31}P$ and results in the collapse of the multiplet. Suzuki and Kubo (1964) have calculated the line shape of a dipolar nucleus coupled to a quadrupolar nucleus with $I = \frac{5}{2}$, with different magnitudes of $T_q J$. Figure 2 shows the $^{17}O$- and $^{31}P(^{17}O)$-NMR spectra of $P^{17}OCl_3$ (A), $(CH_3O)_3P^{17}O$ (B), $(PhO)_3P^{17}O$ (C), and $Ph_3P^{17}O$ (D). It can be seen in Fig. 2 that as the $^{17}O$-NMR coupling pattern collapses (decreasing $J$ and/or increasing $^{17}O$

linewidth $\Delta O$), the $^{31}$P-NMR coupling pattern also collapses. In all spectra the strong central peak is due to the residual unlabeled species.

The compounds whose spectra are shown in Fig. 2 are all symmetrical, small molecules with P=O double bonds. These compounds have relatively long $T_q$ and large $J$, thus showing fully or partially resolved $^{17}$O- and $^{31}$P($^{17}$O)-NMR spectra. For biophosphate molecules, $T_q$ is generally shorter (owing to a larger molecule size and a small degree of symmetry) and $J$ generally smaller (owing to a P—O bond with a smaller $\pi$ character). Therefore, the $^{17}$O-NMR signals of biophosphates are broader and less well resolved. Based on Fig. 2, we would expect the $^{31}$P($^{17}$O)-NMR signals of biophosphates to be a "broad singlet." Under this condition ($T_q J < 1$), the scalar relaxation will contribute to the relaxation of the dipolar nucleus according to

$$\frac{1}{T_{1sc}} = \frac{8\pi^2 J^2 I(I+1)}{3} \frac{T_q}{1+(\omega_p - \omega_o)^2 T_q^2} \qquad (1)$$

and

$$\frac{1}{T_{2sc}} = \frac{4\pi^2 J^2 I(I+1)}{3} \left[ T_q + \frac{T_q}{1+(\omega_p - \omega_o)^2 T_q^2} \right] \qquad (2)$$

(Lehn and Kintzinger, 1973; James, 1975; Abragam, 1961), where $I = \frac{5}{2}$, $1/T_{1sc}$ and $1/T_{2sc}$ are the contribution of scalar relaxation to the longitudinal and the transverse relaxations, respectively, of $^{31}$P, $J = J_{31P-17O}$, $T_q$ is the quadrupolar $T_1$ relaxation time of $^{17}$O, and $\omega_p$ and $\omega_o$ are the angular precession frequencies of $^{31}$P and $^{17}$O, respectively.

For small biophosphate molecules at the extreme narrowing limit ($\omega^2 \tau_c^2 \ll 1$), $T_q$ is of the order of $10^{-2}$–$10^{-4}$ s. Because $\omega_p - \omega_o \simeq 10^7$–$10^8$ Hz, $(\omega_p - \omega_o)^2 T_q^2 \gg 1$ and Eqs. (1) and (2) can be reduced to

$$\frac{1}{T_{1sc}} \simeq 0 \qquad (3)$$

and

$$\frac{1}{T_{2sc}} \simeq \frac{35}{3} \pi^2 J^2 T_q \qquad (4)$$

Under this condition $1/T_2 \simeq 1/T_{2sc}$ for $^{31}$P, and $T_1 \simeq T_2 \simeq T_q$ for $^{17}$O, which justifies the approximations $\Delta O \simeq 1/\pi T_q$ and $\Delta P \simeq 1/\pi T_{2sc}$. The following approximate relationship can be obtained from Eq. (4):

$$\Delta P \, \Delta O \simeq \tfrac{35}{3} J^2 \qquad (5)$$

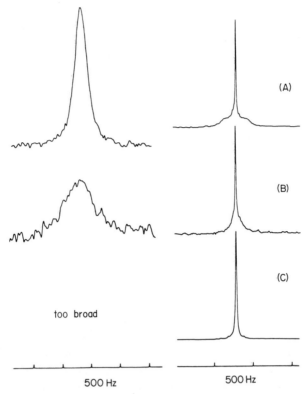

**Fig. 3.** $^{17}$O-NMR spectra (left, at 27.1 MHz) and $^{31}$P($^{17}$O)-NMR spectra (right, at 81.0 MHz) of $H_3P^{17}OO_3$ (50 atom % $^{17}$O) in $D_2O$ (A), $H_2O$/glycerol (1:1 volume ratio) (B), and glycerol (C). All spectra were obtained at 30°C and processed with a line broadening of 20 Hz ($^{17}$O) and 4 Hz ($^{31}$P). From Sammons et al. (1983).

where $\Delta P$ and $\Delta O$ represent the linewidths of $^{31}$P($^{17}$O) and $^{17}$O NMR signals, respectively.

Such an approximate inversely proportional relationship between $\Delta P$ and $\Delta O$, or between $T_q$ (of $^{17}$O) and $T_{2sc}$ (of $^{31}$P), is illustrated by the $^{31}$P($^{17}$O)-NMR signals of $H_3P^{17}OO_3$ in $D_2O$, $H_2O$/glycerol, and glycerol (Fig. 3). As $\Delta O$ increases because of an increased viscosity, $\Delta P$ decreases correspondingly.

The approximate equations [(3)–(5)] derived are valid under the "extreme narrowing limit," and thus should be applicable for small biophosphate molecules in solution. Tsai et al. (1980) have shown that for adenine nucleotides with $^{17}$O labeled at all possible positions, the $^{31}$P($^{17}$O)-NMR signals are "broad." The case of $H_3P^{17}O_4$/glycerol (Fig. 3C), in which the $^{31}$P($^{17}$O) signal has sharpened almost indistinguishably from $^{31}$P($^{16}$O) signal, should not occur in small biophosphates in water. Therefore, $^{17}O$ "quenches" the $^{31}P$-NMR signal of $^{31}P$-$^{17}O$ species, causing an apparent decrease in the intensity of the $^{31}P$-NMR signal.

Such a "line-broadening effect" of $^{17}O$ in $^{31}P$ NMR has been used to locate the position of a $^{17}O$ label (Tsai, 1979) and to calculate the percent enrichment of $^{17}O$ (Huang and Tsai, 1982; Reed and Leyh, 1980). In addition, it has made possible analysis of the configuration of [$^{16}O,^{17}O,^{18}O$]phosphate monoesters and [$^{16}O,^{17}O,^{18}O$]thiophosphates by $^{31}P$ NMR (Section VI).

## IV. Prochiral Centers: $^{31}P$ NMR

By a sulfur substitution, a prochiral phosphorus center (e.g., a phosphodiester) becomes a chiral center with two possible configurations (Scheme 5).

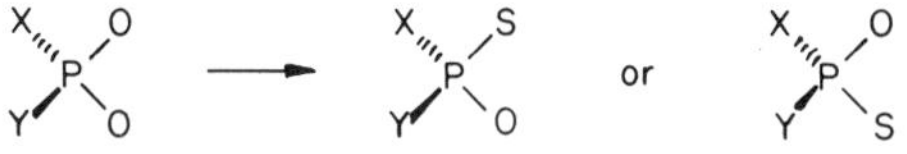

**Scheme 5**

Because most biophosphate molecules contain at least one chiral carbon center, the two isomers **1a** and **1b** are diastereomers and give distinguishable $^{31}P$ chemical shifts. Table II summarizes the chiral thiophosphates that belong to this category and the chemical shifts of the chiral phosphorus of these isomers.

The chiral thiophosphates **1a** and **1b** can be used for two types of studies: stereochemical course of enzymatic substitutions (type 1 in Section I) and stereospecificity of the two isomers as enzyme substrates (type 3 in Section I). A great number of enzyme reactions have been investigated by this approach and have been reviewed as described in Section I. An example in this category is the stereospecific hydrolysis of DPP$_s$C by phospholipases A$_2$ and C (Bruzik *et al.*, 1982). As shown in Scheme 6, when DPPsC(A + B)

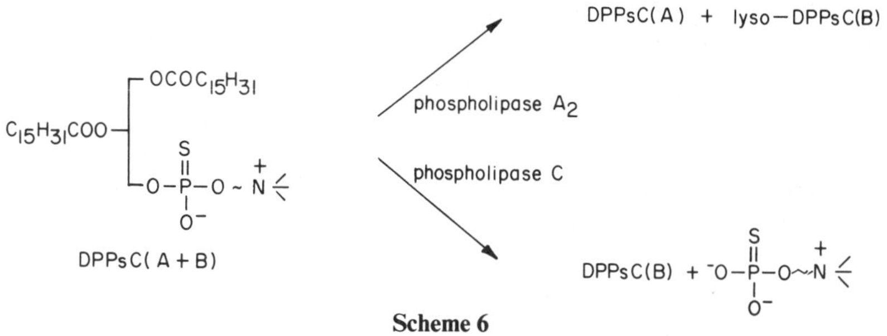

**Scheme 6**

## TABLE II

Phosphorus-31 Chemical Shifts of Diastereomeric Pairs of Chiral Thiophosphates

| Compounds | Solvent | Chemical shifts[a] | | Reference |
|---|---|---|---|---|
| | | $R_p$ | $S_p$ | |
| ADPαS | $H_2O/D_2O$, pH 7.3 | 41.7 | 42.1 | Jaffe and Cohn (1978) |
| | $D_2O$ | 40.61 | 40.96 | Sheu and Frey (1977) |
| ATPαS | $D_2O$ | 42.74 | 42.97 | Sheu and Frey (1977) |
| ATPβS | $H_2O/D_2O$, pH 8.1 | 30.0 | 29.9 | Jaffe and Cohn (1978) |
| UDPαS | $D_2O$ | 40.39 | 40.84 | Sheu et al. (1979) |
| UTPαS | $D_2O$ | 42.62 | 42.39 | Sheu et al. (1979) |
| GTPαS | $D_2O$ | 42.23 | 42.55 | Connolly et al. (1982) |
| GTPβS | $D_2O$ | 28.27 | 28.27 | Connolly et al. (1982) |
| cAMPS | $H_2O$ | 54.27 | 53.22 | Eckstein et al. (1974) |
| | pH 9 | 55.66 | 54.05 | Gerlt et al. (1980) |
| U>pS | | 74.8 (endo) | 76.1 (exo) | Usher et al. (1972) |
| Up(S)A | $D_2O$ | 56.1 | 55.5 | Burgers and Eckstein (1979) |
| DPPsE[b] | $CDCl_3$ | 59.61(A) | 59.47(B) | Orr et al. (1982) |
| DPPsC[c] | $CDCl_3$ | 56.12(A) | 56.07(B) | Bruzik et al. (1983) |
| | $CH_3OD$ | 60.822(A) | 60.801(B) | Bruzik et al. (1983) |
| | $D_2O$/Triton X-100 | 57.133(A) | 57.205(B) | Bruzik et al. (1983) |

[a] Only the chemical shift of the chiral phosphorus is listed. In some of the original references reporting the chemical shift, the absolute configuration was not assigned. The assignments listed are based on some later reports which are not cited.
[b] The absolute configuration is unknown. Isomers A and B are arbitrarily defined.
[c] The absolute configuration is unknown. Isomers A and B are defined as in Bruzik et al. (1982).

(which gives two $^{31}$P-NMR signals in $CDCl_3$, as shown in Fig. 4A; the isomer resonating at lower field was defined as isomer A) was digested with phospholipase $A_2$ from bee venom, only isomer B was specifically hydrolyzed to lyso-DPPsC(B). Figure 4B,C shows the $^{31}$P-NMR spectra of the unreacted DPPsC(A) and the pure DPPsC(B) obtained from reacylation of lyso-DPPsC(B), respectively. On the other hand, phospholipase C is specific to isomer A. Figure 4D shows the $^{31}$P-NMR spectra of DPPsC after partial hydrolysis by phospholipase C from *Bacillus cereus*. The requirement of a specific configuration at phosphorus in the phospholipase C catalysis is to be expected since the reaction involves a P—O bond cleavage. However, the stereospecificity observed for phospholipase $A_2$ is surprising because it hydrolyzes the C-2 ester but not the phosphodiester, and it can tolerate substitution of the choline side chain by other groups.

# 6. $^{17}O$ and $^{18}O$ Effects and Chiral Thiophosphates

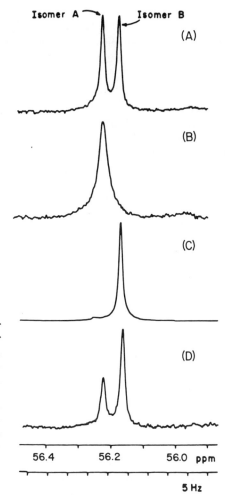

**Fig. 4.** $^{31}$P-NMR spectra (81.0 MHz) of DPP$_s$C (10 m$M$ in CDCl$_3$). (A) Mixture of diastereomers from chemical synthesis. (B) Pure isomer A recovered from hydrolysis by phospholipase A$_2$, (C) Pure isomer B (containing 3% isomer A) obtained from acylation of the product of phospholipase A$_2$ hydrolysis, lyso-DPP$_s$C. (D) DPP$_s$C after partial hydrolysis by phospholipase C. NMR parameters: spectral width 1000 Hz, acquisition time 4.1 s, $^1$H decoupling, line broadening 0.1 Hz, pulse width 12 $\mu$s (90° pulse at 20 $\mu$s). From Bruzik et al. (1982).

## V. Pro-Prochiral Centers: $^{31}P(^{18}O)$ NMR or $^{31}P(^{17}O)$ NMR

By a sulfur substitution, a pro-prochiral phosphorus center (e.g., a phosphomonoester) becomes a prochiral center (Scheme 7).

**Scheme 7**

As described in Section I, when **2** is phosphorylated enzymatically, the reaction is most likely stereospecific because enzymes can differentiate between the two diastereotopic oxygens. Because the prochiral center becomes a chiral center after phosphorylation, the two isomers can be identified based on $^{31}$P chemical shifts.

If **2** is made chiral by an $^{17}$O or $^{18}$O label, the configuration of **2** can be determined by stereospecifically phosphorylating one of the two oxygens, followed by determining whether the labeled oxygen is located at the P—O—P briding position or the P—O nonbridging position. If $^{17}$O is used, a bridging $^{17}$O should cause a broadening (and a decrease in the apparent intensity) of both $^{31}$P-NMR signals, whereas a nonbriding $^{17}$O should exert the effect to only one $^{31}$P signal. If $^{18}$O is used, a nonbridging $^{18}$O should cause a larger isotope shift on the $^{31}$P signal than a bridging $^{18}$O.

The $^{31}$P($^{17}$O)-NMR method has been used to elucidate the steric course of acetyl-CoA synthetase-catalyzed reaction (Scheme 8) (Tsai, 1979). The

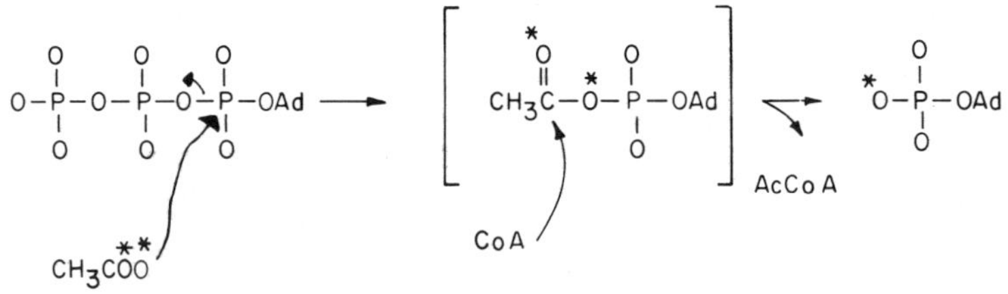

**Scheme 8**

enzyme was found to be specific to $(R_p)$-ATPαS but not to $(S_p)$-ATPαS. As shown in Scheme 9, when $(R_p)$-ATPαS and [$^{17}$O]acetate are used as sub-

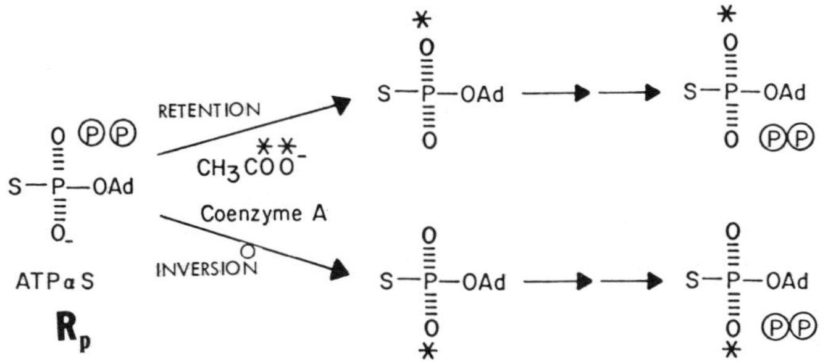

**Scheme 9**

strates, the $^{17}$O from acetate will be incorporated into the pro-$S$ position of AMPS if the reaction proceeds with retention of configuration, or the pro-$R$ position if inversion occurs. To determine the configuration of the $^{17}$O-la-

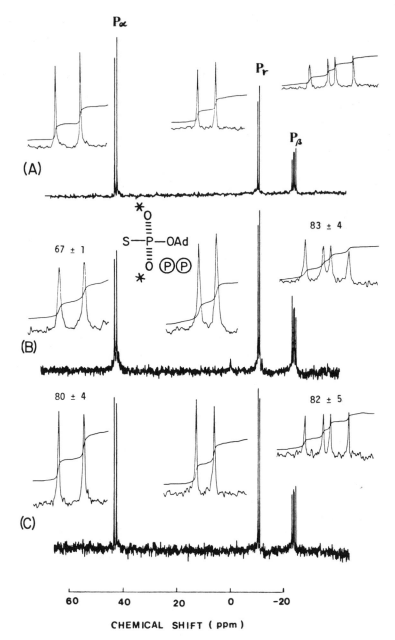

**Fig. 5.** $^{31}$P-NMR spectra (at 32.2 MHz) showing the results of acetyl-CoA synthetase. (A) Unlabeled $(S_p)$-ATPαS. (B) Synthesized $(S_p)$-[α-$^{17}$O, αβ-$^{17}$O]ATPαS. (C) $(S_p)$-ATPαS from [$^{17}$O$_2$]acetate. The insets represent the integrations of the corresponding signals. From Tsai (1979).

beled AMPS, it is converted to $(S_p)$-ATPαS by stereospecific phosphorylation at the pro-$R$ oxygen catalyzed by adenylate kinase, followed by a second phosphorylation catalyzed by pyruvate kinase (Sheu and Frey, 1977; Jaffe and Cohn, 1978). By such a conversion, $^{17}$O should be incorporated into the nonbridging position of $(S_p)$-ATPαS if the step of acetate activation proceeds with retention of configuration. On the other hand, $^{17}$O should be located at the P—O—P bridging position if inversion occurs. A nonbridging $^{17}$O at $P_\alpha$ should cause the $P_\alpha$ signal to broaden and decrease in $^{31}$P NMR, whereas a bridging $^{17}$O should quench both $P_\alpha$ and $P_\beta$ signals.

Figure 5 shows the $^{31}$P-NMR spectra of unlabeled $(S_p)$-ATPαS (A), the synthesized $(S_p)$-[α-$^{17}$O, αβ-$^{17}$O]ATPαS (B), and the $(S_p)$-ATPαS obtained from the enzyme reaction (C). The $^{17}$O isotope used was 20% enriched and the enrichment of [$^{17}$O]acetate was determined as 19%. In Fig. 5B, the $P_\alpha$ signal decreases to 67 ± 1% and the $P_\beta$ signal to 83 ± 4%. In Fig. 5C, the $P_\alpha$ signal decreases to 80 ± 4% and the $P_\beta$ signal to 82 ± 5%. Because both $P_\alpha$ and $P_\beta$ have decreased in Fig. 5C, the results indicate that $^{17}$O must be located at the bridging position, and the reaction catalyzed by acetyl-CoA synthetase must proceed with inversion of configuration (Tsai, 1979).

The same problem could have been solved by use of $^{18}$O isotope shifts in $^{31}$P NMR, which may be advantageous over the $^{17}$O method in terms of quantitation but requires a high-resolution and a higher-field instrument. Indeed, our original idea was to use $^{17}$O NMR to differentiate the bridging and nonbridging $^{17}$O, which was not successful with a low-field spectrometer used earlier; but it has now been shown to be feasible at higher magnetic field and higher temperature (Tsai, 1982; Gerlt et al., 1982; Gerothanassis and Sheppard, 1982). On the other hand, a number of stereochemical studies in this category made use of $^{18}$O-labeling and mass spectroscopy, which gives relatively accurate quantitation but requires derivatization and/or degradation of the product.

There are, however, no alternatives in the configurational analysis of the chiral [$^{16}$O,$^{17}$O,$^{18}$O]thiophosphate ($P_{si}$) discussed in the next section. The only method available is a $^{31}$P-NMR method based on the combined effects of $^{17}$O and $^{18}$O.

## VI. Pro-Pro-Prochiral Centers: $^{31}$P($^{18}$O) NMR and $^{31}$P($^{17}$O) NMR

Hydrolysis of phosphomonoesters generates inorganic phosphate ($P_i$), which contains a pro-pro-prochiral phosphorus center. To make a $P_i$ chiral, it is necessary to make use of all three stable oxygen isotopes ($^{16}$O,$^{17}$O,$^{18}$O) and sulfur, as shown in Scheme 10.

## Scheme 10

The rationale of configurational analysis for chiral $P_{si}$ is illustrated by Scheme 11. The same principle applies to chiral phosphate monoesters

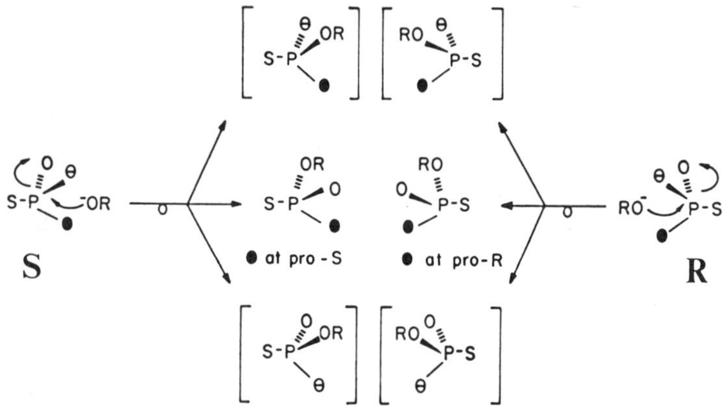

## Scheme 11

where the P—S bond is replaced by P—OR. Displacement of one of the three oxygen isotopes of $(S_p)$-$[^{16}O,^{17}O,^{18}O]P_{si}$ by a nucleophile (RO$^-$) gives a mixture of three inseparable, isotopically different species. Among them, two (those in brackets) contain an $^{17}O$ isotope, which should quench the corresponding $^{31}P$-NMR signals. Only the species that contains only $^{16}O$ and $^{18}O$ ($^{18}O$ at the pro-$S$ position) should give a sharp, unquenched $^{31}P$-NMR signal. Analogously, the $(R)$-$[^{16}O,^{17}O,^{18}O]P_{si}$ should give correspondingly a non-$^{17}O$-containing species with $^{18}O$ at the pro-$R$ position.

Thus, determination of whether $^{18}O$ is at the pro-$R$ or pro-$S$ position would tell the configuration of chiral $P_{si}$ or chiral phosphate monoesters. A general way to achieve this is to derivatize stereospecifically the pro-$R$ or pro-$S$ oxygen. The $^{31}P(^{18}O)$-NMR method can then be used to distinguish the bridging and nonbridging $^{18}O$ on the basis of the different magnitude of isotope shifts. Therefore, two main chemical steps need to be done to convert the chiral phosphoryl group to an analyzable form: a displacement with known stereochemistry and stereospecific derivatization of the pro-chiral oxygens. These two chemical steps vary from compound to compound, but the underlying principles remain the same.

To illustrate the application and configurational analysis of chiral $[^{16}O,^{17}O,^{18}O]$thiophosphate, we describe the stereochemical study of 5'-nucleotidase (Tsai and Chang, 1980; Tsai, 1980). The venom 5'-nucleotidase catalyzes hydrolysis of AMP to adenosine and $P_i$ but does not catalyze transphosphorylation or $P_i \rightleftarrows H_2O$ oxygen exchange (Koshland and Springhorn, 1956). We have first synthesized $(R_p)$-[$^{18}$O]AMPS and $(S_p)$-[$^{18}$O]AMPS of known configuration. Hydrolysis of these two isomers in $H_2^{17}O$ gave two chiral [$^{16}$O,$^{17}$O,$^{18}$O]$P_{si}$ enantiomers with unknown configuration. The two main steps required were available separately in the literature (Eckstein, 1977), as shown in Scheme 12. The stereochemical course of

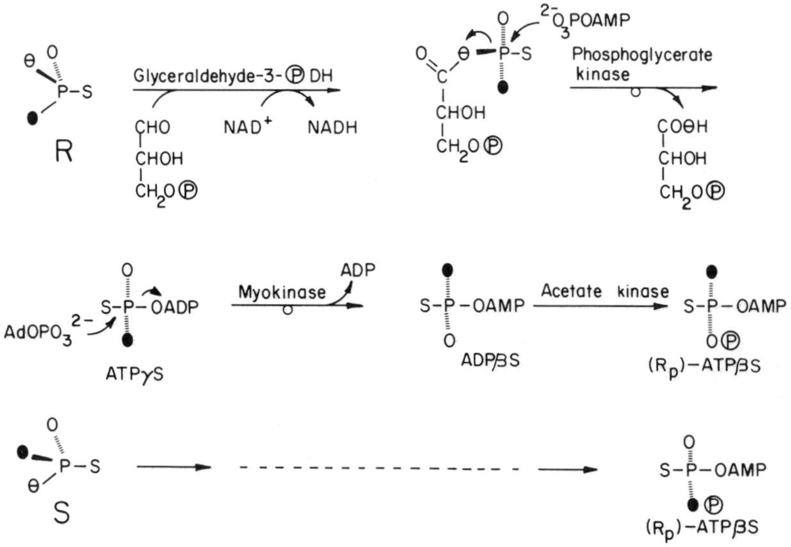

**Scheme 12**

each step in Scheme 12 had been elucidated separately (Richard and Frey, 1978; Richard *et al.*, 1978) except that of phosphoglycerate kinase, which was elucidated by Webb and Trentham (1980) by use of synthesized chiral $P_{si}$ of known configuration on the basis of the same NMR analysis discussed next.

According to Scheme 12, the $(R_p)$-chiral $P_{si}$ should give $(R_p)$-ATP$\beta$S with $^{18}$O located specifically at the $\beta$-nonbridging position. The $(S_p)$-enatiomer should give $(R_p)$-ATP$\beta$S with $^{18}$O at the $\beta\gamma$-bridging position. It is known that a bridging $^{18}$O should cause a smaller isotope shift in $^{31}$P NMR than a nonbridging $^{18}$O. On this basis the configuration can be determined. However, Scheme 12 only shows the species that will give an unquenched $^{31}$P-NMR signal. In reality, each chiral $P_{si}$ species should give a mixture of three $(R_p)$-ATP$\beta$S species (**1a**, **1b**, and **1c** in Scheme 13). In addition, it is impossible to have a chiral $P_{si}$ of 100% purity. A chiral $P_{si}$ sample actually contains up to six isotopic species, as shown in the left column of Scheme 13

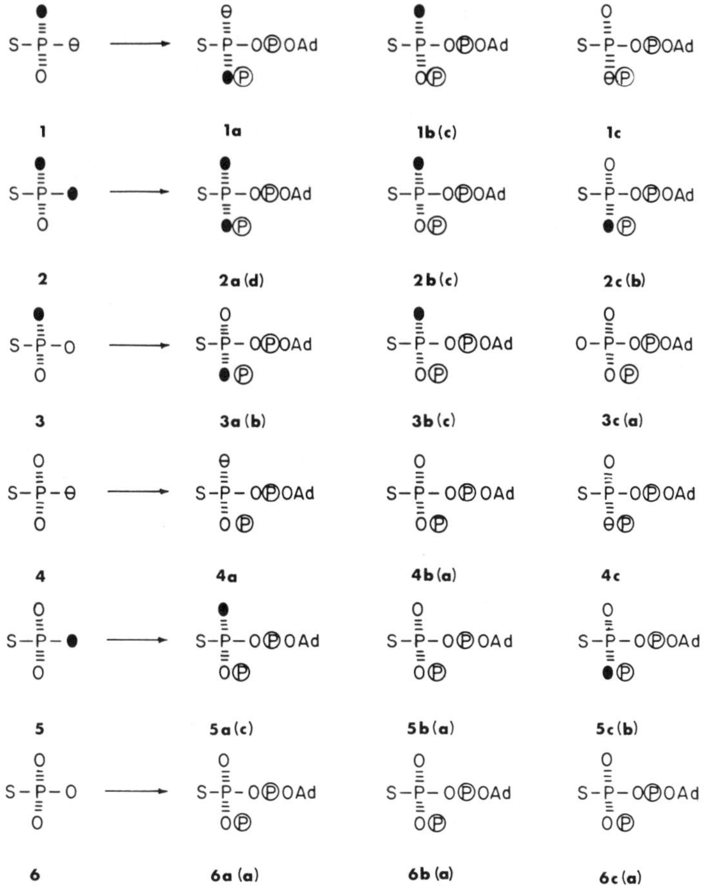

# 6. $^{17}O$ and $^{18}O$ Effects and Chiral Thiophosphates

**Scheme 13**

(two of them are identical species); each of them gives three $(R_p)$-ATPβS species. Fortunately, a careful examination of Scheme 13 reveals that there are only four different non-$^{17}O$-containing species **a**, **b**, **c** and **d**, and that all the nonchirally labeled $P_{si}$ species contribute *equally* to species **b** and **c**. Only the [$^{16}O,^{17}O,^{18}O$]$P_{si}$ species gives specifically **b** or **c**, depending on whether the configuration is $S$ or $R$, respectively. The amounts of species **a** and **d** have to do with isotopic enrichments but not configuration.

Figure 6 shows the $P_\beta$ signals of the $(R_p)$-ATPβS obtained from PS$^{18}O_3^{3-}$ and the two chiral $P_{si}$ enantiomers. The signal contains two overlapping doublets owing to $^{31}P-^{31}P$ coupling. Each half of a doublet contains four lines arising from the four species **a**, **b**, **c**, and **d**. The results are summarized in Table III, where the $F$ value is defined as the ratio **b/c**, the purity refers to the percentage of chirally labeled $P_{si}$ species, and the chirality refers to the

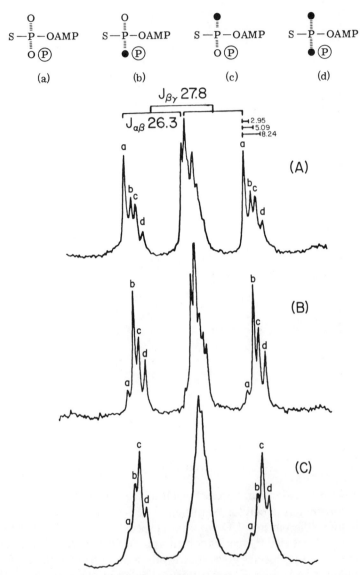

**Fig. 6.** $P_\beta$ signals of the $^{31}$P-NMR spectra of the $(R_p)$-ATP$\beta$S obtained from $[^{18}O_3]P_{si}$ (A) and from the two chiral $P_{si}$ (B, C). The sample (30 μmol) was dissolved in 2.5 ml of $D_2O$ containing 10 m$M$ EDTA and the spectra recorded at 145.7 MHz at ambient temperature. The coupling constants and isotope shifts are epxressed in Hertz. The chemical shift of the $P_\beta$ signal is 29.8 ppm downfield from $H_3PO_4$. From Tsai (1980).

## TABLE III

$^{31}$P-NMR Analysis of the ($R_p$)-ATPβS Derived from Chiral Thiophosphates

| $P_{si}$ samples | Intensity (%)[a] | | | | F value | Configuration |
|---|---|---|---|---|---|---|
| | a | b | c | d | | |
| PS$^{18}$O$_3^{3-}$ | 41.3 ± 1.2 | 24.6 ± 0.1 | 22.1 ± 0.0 | 11.8 ± 1.2 | 1.11 | |
| ($S_p$)-AMPS | 8.8 ± 0.5 | 42.8 ± 0.6 | 28.1 ± 0.5 | 20.3 ± 0.5 | 1.52 | S |
| ($R_p$)-AMPS | 12.2 ± 0.5 | 26.5 ± 1.6 | 38.8 ± 0.1 | 22.4 ± 2.0 | 0.68 | R |
| Calculated[b] | 7.8 | 47.3 | 25.9 | 19.0 | 1.82 | S |
| | 7.8 | 25.9 | 47.3 | 19.0 | 0.55 | R |

[a] Obtained from peak-height measurements for the $P_β$ signal of ATPβS. The errors represent deviations between the two nonoverlapping halves of the two doublets.

[b] Calculated for chiral $P_{si}$ of 47.5% purity and 90% chirality expected based on isotopic enrichments.

optical purity of chirally labeled $P_{si}$ species. The results indicate that 5'-nucleotidase catalyzes the hydrolysis of AMPS with inversion of configuration, as shown in Scheme 14. The causes for the deviations of the observed F values from theoretical values have been discussed (Tsai, 1980).

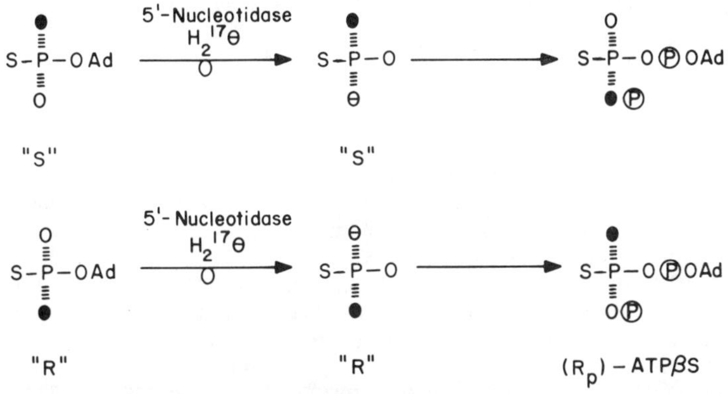

**Scheme 14**

By such a $^{31}$P-NMR analysis, Webb and Trentham have also elucidated the stereochemical course of a number of nucleoside triphosphatases (Webb, 1982). The $^{31}$P-NMR analysis based on the combined effects of $^{17}$O and $^{18}$O is the only method available for configurational analysis of chiral [$^{16}$O,$^{17}$O,$^{18}$O]$P_{si}$, and is also a commonly used method for configurational analysis of chiral [$^{16}$O,$^{17}$O,$^{18}$O]phosphomonoesters, which are covered by Gerlt (Chapter 7.)

## Acknowledgments

The work from the author's laboratory was supported by research grants GM 29041 and GM 30327 from the National Institutes of Health, and PCM 8140443 from the National Science Foundation.

## References

Abragam, A. (1961). "The Principles of Nuclear Magnetism." Oxford Univ. Press, London and New York.
Bruzik, K., and Tsai, M.-D. (1982). *J. Am. Chem. Soc.* **104**, 836–865.
Bruzik, K., Gupte, S. M., and Tsai, M.-D. (1982). *J. Am. Chem. Soc.* **104**, 4682–4684.
Bruzik, K., Jiang, R.-T., and Tsai, M.-D. (1983). *Biochemistry* **22**, 2478–2486.
Buchwald, S. L., and Knowles, J. R. (1980). *J. Am. Chem. Soc.* **102**, 6601–6603.
Buchwald, S. L., Hansen, D. E., Hassett, A., and Knowles, J. R. (1982). *In* "Methods in Enzymology," Vol. 87, pp. 279–301. Academic Press, New York.
Burgers, P. M. J., and Eckstein, F. (1979). *Biochemistry* **18**, 592–596.
Cahn, R. S., Ingold, C. K., and Prelog, V. (1966). *Angew. Chem., Int. Ed. Engl.* **5**, 385–415.
Clin, B., de Bony, J., Palanne, P., Biais, J., and Lemanceau, B. (1979). *J. Magn. Reson.* **33**, 457–463.
Coderre, J. A., and Gerlt, J. A. (1980). *J. Am. Chem. Soc.* **102**, 6594–6597.
Cohn, M. (1982). *Annu. Rev. Biophys. Bioeng.* **11**, 23–42.
Cohn, M., and Hu, A. (1978). *Proc. Natl. Acad. Sci. U.S.A.* **75**, 200–203.
Cohn, M., and Hu, A. (1980). *J. Am. Chem. Soc.* **102**, 913–916.
Connolly, B. A., Romaniuk, P. J., and Eckstein, F. (1982). *Biochemistry* **21**, 1983–1989.
Eckstein, F. (1975). *Angew. Chem. Int. Ed. Engl.* **14**, 160–166.
Eckstein, F. (1977). *Biochim. Biophys. Acta* **483**, 1–5.
Eckstein, F. (1979). *Acc. Chem. Res.* **12**, 204–210.
Eckstein, F., Simonson, L. P., and Bär, H.-P. (1974). *Biochemistry* **13**, 3806–3810.
Eckstein, F., Romaniuk, P. J., and Connolly, B. A. (1982). *In* "Methods in Enzymology" Vol. 87, pp. 197–212. Academic Press, New York.
Frey, P. A. (1982). *Tetrahedron* **38**, 1541–1567.
Frey, P. A., Richard, J. P., Ho, H.-T., Brody, R. S., Sammons, R. D., and Sheu, K.-F. (1982). *In* "Methods in Enzymology" Vol. 87, pp. 213–235. Academic Press, New York.
Gerlt, J. A., and Coderre, J. A. (1980). *J. Am. Chem. Soc.* **102**, 4531–4533.
Gerlt, J. A., and Wan, W. H. Y. (1979). *Biochemistry* **18**, 4630–4637.
Gerlt, J. A., Coderre, J. A., and Wolin, M. S. (1980). *J. Biol. Chem.* **255**, 331–334.
Gerlt, J. A., Demou, P. C., and Mehdi, S. (1982). *J. Am. Chem. Soc.* **104**, 2848–2856.
Gerothanassis, I. P., and Sheppard, N. (1982). *J. Magn. Reson.* **46**, 423–439.
Gorenstein, D. G., and Rowell, R. (1980). *J. Am. Chem. Soc.* **102**, 6165–6166.
Huang, S.-L., and Tsai, M.-D. (1982). *Biochemistry* **21**, 951–959.
Jaffe, E. K., and Cohn, M. (1978). *Biochemistry* **17**, 652–657.
James, T. L. (1975). "Nuclear Magnetic Resonance in Biochemistry," pp. 48–49. Academic Press, New York.
Jarvest, R. L., Lowe, G., and Potter, B. V. L. (1980). *J. Chem. Soc., Chem. Commun.* pp. 1142–1145.

Jarvest, R. L., Lowe, G., and Potter, B. V. L. (1981). *J. Chem. Soc., Perkin Trans. 1* pp. 3186–3195.
Jordan, F., Salamone, S. J., and Wang, A. (1981). *ACS Symp. Ser.* **171,** 585–589.
Knowles, J. R. (1980). *Annu. Rev. Biochem.* **49,** 877–919.
Koshland, D. E., Jr., and Springhorn, S. S. (1956). *J. Biol. Chem.* **221,** 469–476.
Lehn, J. M., and Kintzinger, J. P. (1973). *In* "Nitrogen NMR" (M. Witanowski and G. A. Webb, eds.), pp. 80–161. Plenum, New York.
Lowe, G., and Sproat, B. S. (1978). *J. Chem. Soc., Chem. Commun.,* pp. 565–566.
Lowe, G., Potter, B. V. L., Sproat, B. S., and Hull, W. E. (1979). *J. Chem. Soc., Chem. Commun.* pp. 733–735.
Lutz, O., Nolle, A., and Staschewski, D. Z. (1978). *Z. Naturforsch, A* **33,** 380–382.
Orr, G. A., Brewer, C. F., and Heney, G. (1982). *Biochemistry* **21,** 3202–3206.
Reed, G. H., and Leyh, T. S. (1980). *Biochemistry* **19,** 3472–3480.
Richard, J. P., and Frey, P. A. (1978). *J. Am. Chem. Soc.* **100,** 7757–7758.
Richard J. P., Ho, H. T., and Frey, P. A. (1978). *J. Am. Chem. Soc.* **100,** 7756–7757.
Sammons, R. D., and Frey, P. A. (1982). *J. Biol. Chem.* **257,** 1138–1141.
Sammons, R. D., Frey, P. A., Bruzik, K., and Tsai, M.-D. (1983). *J. Am. Chem. Soc.* **105,** 5455–5461.
Sammons, R. D. (1982). Ph.D. Dissertation, Ohio State University, Columbus.
Sheu, K. F. R., and Frey, P. A. (1977). *J. Biol. Chem.* **252,** 4445–4448.
Sheu, K. F. R., Richard, J.-P., and Frey, P. A. (1979). *Biochemistry* **18,** 5548–5556.
Suzuki, M., and Kubo, R. (1964). *Mol. Phys.* **7,** 201–209.
Tsai, M.-D. (1979). *Biochemistry* **18,** 1468–1472.
Tsai, M.-D. (1980). *Biochemistry* **19,** 5310–5316.
Tsai, M.D. (1982). *In* "Methods in Enzymology" Vol. 87, pp. 235–279. Academic Press, New York.
Tsai, M.-D., and Bruzik, K. (1983). In "Biological Magnetic Resonance" (L. J. Berliner and J. Reuben, eds.), Vol. 5, pp. 129–181. Plenum, New York.
Tsai, M.-D., and Chang, T. T. (1980). *J. Am. Chem. Soc.* **102,** 5416–5418.
Tsai, M.-D., Huang, S. L., Kozlowski, J. F., and Chang, C. C. (1980). *Biochemistry* **19,** 3531–3536.
Usher, D. A., Erenrich, E. S., and Eckstein, F. (1972). *Proc. Natl. Acad. Sci. U. S. A.* **69,** 115–118.
Webb, M. R. (1982). *In* "Methods in Enzymology" Vol. 87, pp. 301–316. Academic Press, New York.
Webb, M. R., and Trentham, D. R. (1980). *J. Biol. Chem.* **255,** 8629–8632.

CHAPTER 7

# Use of Chiral [$^{16}$O,$^{17}$O,$^{18}$O]Phosphate Esters to Determine the Stereochemical Course of Enzymatic Phosphoryl Transfer Reactions

*John A. Gerlt*

Department of Chemistry
Yale University
New Haven, Connecticut

| | |
|---|---:|
| I. Introduction | 199 |
| II. Syntheses of Oxygen Chiral Phosphate Esters | 201 |
|     A. Phosphate Monoesters | 202 |
|     B. Phosphate Diesters | 205 |
| III. Configurational Analyses of Oxygen Chiral Phosphate Esters | 210 |
|     A. [$^{16}$O,$^{18}$O]Phosphate Diesters | 212 |
|     B. Nucleoside [$\alpha$-$^{16}$O,$^{18}$O]Diphosphates | 215 |
|     C. [$^{16}$O,$^{17}$O,$^{18}$O]Phosphate Monoesters | 217 |
| IV. Selected Examples | 220 |
|     A. Adenylate Cyclase | 220 |
|     B. Cyclic Nucleotide Phosphodiesterase | 224 |
|     C. Staphylococcal Nuclease | 227 |
| V. Summary | 230 |
|     References | 230 |

## I. Introduction

A large number of enzymes catalyze nucleophilic substitution reactions at the tetrahedral phosphorus atoms in phosphate esters and anhydrides. The stereochemical consequences of the individual displacement reactions that occur in the active sites of these enzymes might be viewed as potentially diverse when the mechanisms of nonenzymatic reactions at tetrahedral phosphorus atoms are considered, implying that a determination of the overall stereochemical course of an enzyme-catalyzed reaction could well be impossible to interpret. Specifically, associative reactions [$S_N2(P)$] can proceed either with inversion or retention of configuration at phosphorus (Westheimer, 1980). A direct, in-line displacement of a leaving group by the attacking nucleophile results in an inversion of configuration at phosphorus.

However, Westheimer and co-workers have provided evidence for additional complexity in nonenzymic $S_N2(P)$ reactions. Such reactions often occur via formation of a pentacoordinate intermediate, which may undergo a ligand reorganization process termed pseudorotation. Because single pseudorotation results in an inversion of configuration at phosphorus, the overall stereochemical course of an $S_N2(P)$ reaction involving pseudorotation is a retention of configuration at phosphorus. At present, no experimental evidence has been obtained for the occurrence of pseudorotation in the active sites of enzymes. In addition, reactions involving phosphate monoesters can occur by a dissociative mechanism [$S_N1(P)$] that involves the generation of the presumably planar and highly reactive monomeric metaphosphate anion as an intermediate (Westheimer, 1981). In principle, this reaction could be accompanied by racemization at phosphorus, but recent experimental evidence has shown that nonenzymic $S_N1(P)$ reactions proceed with quantitative inversion of configuration at phosphorus (Buchwald and Knowles, 1982): the extreme electrophilicity of the metaphosphate anion demands reaction with a properly positioned accepting nucleophile; otherwise, recombination with the leaving group to reform the original phosphate monoester will occur. Thus stereochemical studies of enzyme-catalyzed phosphoryl and nucleotidyl transfer reactions can be profitably interpreted in terms of the number of nucleophilic displacement reactions [by they either $S_N1(P)$ or $S_N2(P)$] that occur during the course of the overall enzyme-catalyzed reaction: inversion of configuration implying a single bond-breaking reaction, which is most easily explained by the direct transfer of a phosphoryl or nucleotidyl group from a donor to an acceptor, and retention of configuration implying two bond-breaking reactions, which is most easily explained by the necessary participation of a phosphorylated or nucleotidylated enzyme intermediate. More complex interpretations of the stereochemical data are certainly possible, but the enzymologists practicing in this field generally view the simplest explanations as the most likely mechanisms.

This chapter summarizes the techniques that have been developed for the syntheses and configurational analyses of phosphate mono- and diesters that are chiral only by virtue of substitution with the stable isotopes of oxygen: $^{16}O$, 99.759% natural abundance; $^{17}O$, 0.037% natural abundance and a nuclear spin of $\frac{5}{2}$; and $^{18}O$, 0.204% natural abundance. In addition, a few enzyme systems in which these techniques have been used are described. This approach, which dates back only to 1978, was not the first method to ascertain the stereochemical consequences of enzyme-catalyzed reactions at phosphorus. In 1970, Usher and Eckstein reported the seminal experiment in this field (Usher *et al.,* 1970): the stereochemical course of the hydrolysis of the endo isomer of uridine 2′,3′-cyclic phosphorothioate catalyzed by ribonuclease A. Subsequent to this experiment, many stereochemical studies employing chiral phosphorothioate analogs of the natural substrates have been performed (Eckstein, 1979; Knowles, 1980), but these have been plagued by two problems. Some enzymes, notably the mutases, do not accept phosphorothioate analogs as substrates; this problem, of course, is circumvented by the use of oxygen chiral substrates. Perhaps more importantly, many investigators feared that the mechanistic conclusions deduced from studies involving phosphorothioates would be erroneous owing to the low rates at which most enzymes process phosphorothioates; the results that have been obtained with oxygen chiral phosphate esters now strongly suggest that phosphorothioate substitution does not alter the stereochemical course of an enzyme-catalyzed reaction.

The oxygen chiral ester approach to stereochemical studies requires that the syntheses of the substrates be either stereospecific or highly stereoselective and that methods be available to determine unambiguously the configurations of both the substrates and products of the enzymatic reactions. The syntheses must be at least highly stereoselective because it is not possible to physically separate phosphate esters that are enantiomeric or epimeric at phosphorus by virtue of oxygen isotope substitution; the preparation of chiral phosphorothioates often can exploit the ability to separate diastereomers by crystallization, chromatography, or processing by enzymes. It is also desirable to have methods of configurational analysis that are experimentally straightforward and depend on generally available instrumentation; the $^{31}$P-NMR methods discussed in this chapter provide such flexibility.

The methods and results that are described in this chapter have been the subjects of a number of review articles, including much of Volume 87 of *Methods in Enzymology*.

## II. Syntheses of Oxygen Chiral Phosphate Esters

Given the low natural abundance of $^{17}$O and $^{18}$O, chemical methods for the synthesis of phosphate esters that are chiral by virtue of oxygen-isotope substitution must allow for the introduction of any oxygen isotope from (ideally) the commercially available forms of the heavy isotopes $H_2O$, $CO_2$, or $O_2$. In addition, the substrates of the phosphoryl and nucleotidyl transfer reactions include three types of structurally and chemically different phosphates, almost all of which are polyhydroxylic: phosphate monoesters, such as sugar phosphates and mononucleotides, phosphate diesters, such as 3′,5′-cyclic nucleotides and oligonucleotides, and phosphate anhydrides such as ATP. These structural variations have led to different chemical syntheses for each type of compound, and the synthetic strategies developed by the various research groups active in this field are summarized in this section.

Methods for the synthesis of oxygen chiral phosphate esters would be of no practical value without procedures to determine independently the configurations of the synthesized substrates and the products obtained by virtue of enzymatic transformations. This problem was perhaps more challenging than development of the synthetic strategies, and its convenient solution was the key to rapid development of this field. In general, the configurational differences induced by oxygen-isotope substitution do not lead to any directly observable physical property that can be used to ascertain either absolute configuration or chiral purity [although $^{17}$O-NMR spectroscopy can be used to demonstrate the configurational differences in conformationally rigid $^{17}$O-labeled phosphodiesters, that is, the diastereomers of 2'-deoxyadenosine 3',5'-cyclic [$^{17}$O,$^{18}$O]monophosphate (cyclic [$^{17}$O,$^{18}$O]dAMP) (Coderre *et al.*, 1981a)]. The $^{31}$P-NMR techniques that have been used to allow the configurational analyses of phosphate mono- and diesters and anhydrides are described in Section III.

## A. Phosphate Monoesters

Phosphate monoesters are pro-prochiral at phosphorus, so the synthesis of oxygen chiral monoesters demands that all three stable isotopes of oxygen be incorporated into the product. The $^{16}$O position is derived from a mixture of the three stable isotopes present in their natural abundance; the $^{17}$O and $^{18}$O positions are derived from isotopically enriched sources. The two independent syntheses of oxygen chiral monoesters that have been reported utilized enriched water as the source of the heavy oxygen isotopes: $^{17}$O is available commercially in the form of water at enrichments no greater than about 50%, but $^{18}$O is available at enrichments in excess of 99%. The remainder of the oxygen in samples of $^{17}$O is a mixture of $^{16}$O and $^{18}$O, the composition of which depends on the lot of $^{17}$O; the presence of significant amounts of $^{16}$O and $^{18}$O in the enriched $^{17}$O has the consequence that only half the molecules synthesized will be chiral, with the remainder being prochiral at phosphorus. In practice, this mixture of chiral and prochiral molecules demands that the method of configurational analysis be able to ascertain the configuration of the chiral molecules in the presence of a racemic background. Because it can be expected and has been experimentally verified that enzyme-catalyzed reactions will proceed with complete stereospecificity, the imperfect chirality of the phosphate monoesters does not introduce any real experimental problem.

## 1. Hydrolysis and Hydrogenolysis of Cyclic Phosphoramidates

The first account of the synthesis of an oxygen chiral phosphate monoester was reported in 1978 by Knowles and co-workers (Abbott *et al.*, 1978, 1979). This ester, a diastereomer of 1-phospho-($S$)-1,2-propanediol, was selected as the synthetic target because the general method of configurational analysis that Knowles devised for monoesters was based on the ability to assign the configuration of the diastereomers at phosphorus of this particular phosphate monoester. The chemical steps in this synthesis are shown in Fig. 1. Briefly, [$^{17}$O]POCl$_3$, prepared in high yield from H$_2$$^{17}$O and PCl$_5$, was used to phosphorylate (−)-ephedrine, with the products being a 9:1 epimeric mixture of cyclic phosphoramidic chlorides; reaction of this mixture with the protected diol afforded a mixture of cyclic phosphoramidic esters that could be separated chromatographically. The absolute configurations of the phosphoramidic chlorides and esters were known by virtue of the work of Inch and co-workers (Cooper *et al.*, 1977), which provided the basis for Knowles's chemistry. The major ester, having the configuration shown in Fig. 1, was hydrolyzed in H$_2$$^{18}$O, thereby incorporating the second heavy isotope; this process occurs by inversion of configuration at phosphorus. The chiral [$^{17}$O,$^{18}$O]phosphodiester was hydrogenolyzed in the presence

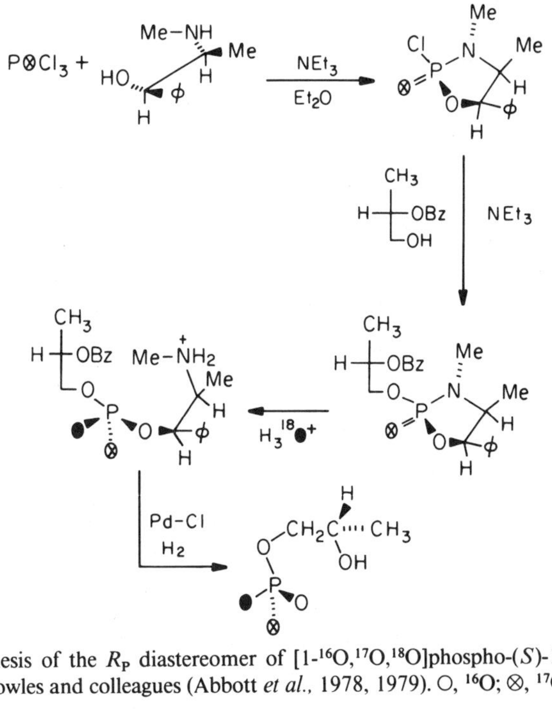

**Fig. 1.** Synthesis of the $R_P$ diastereomer of [1-$^{16}$O,$^{17}$O,$^{18}$O]phospho-($S$)-1,2-propanediol reported by Knowles and colleagues (Abbott *et al.*, 1978, 1979). O, $^{16}$O; ⊗, $^{17}$O; ●, $^{18}$O.

of a Pd catalyst, with C—O bond cleavage of the benzylic ester, yielding the chiral [$^{16}$O,$^{17}$O,$^{18}$O] phospho monoester; in this reaction, the [$^{16}$O]phosphoryl oxygen is derived from the ephedrine moiety in a process that does not alter the configuration of the chiral phosphorus atom. By virtue of the synthetic method, the configuration of the product was predicted to be $R_P$, and this was verified by the methods described in Section III,C.

The same basic synthetic scheme subsequently was utilized by Knowles's laboratory to prepare a number of oxygen chiral phosphate monoesters, including [$\gamma$-$^{16}$O,$^{17}$O,$^{18}$O]ATP (Blättler and Knowles, 1979). In principle, either configuration of the chiral phosphoryl group can be prepared by this chemistry (even thought the synthesis greatly favors one geometric isomer of the cyclic phosphoramidic ester) by simply reversing the order of incorporation of the heavy atom isotopes.

## 2. Hydrogenolysis of Cyclic Hydrobenzoin Triesters

Later in 1978, Cullis and Lowe also reported a synthetic scheme for the preparation of oxygen chiral phosphate monoesters (Cullis and Lowe, 1978, 1981) and the relevant steps in the synthesis of chiral methyl phosphate are shown in Fig. 2. In this chemistry, *meso*-hydrobenzoin, which is chiral by virtue of substitution with a single atom of $^{18}$O, is reacted with [$^{17}$O]POCl$_3$ to afford a single geometric isomer of the cyclic phosphorochloridate; reaction of this material with methanol provided a single geometric isomer of the cyclic methyl triester. Subsequent hydrogenolysis of the cyclic hydrobenzoin ester was accompanied by C—O bond cleavages, resulting in liberation of a chiral sample of methyl phosphate. The original report of this synthesis (Cullis and Lowe, 1978) assigned the cis configuration to both the cyclic

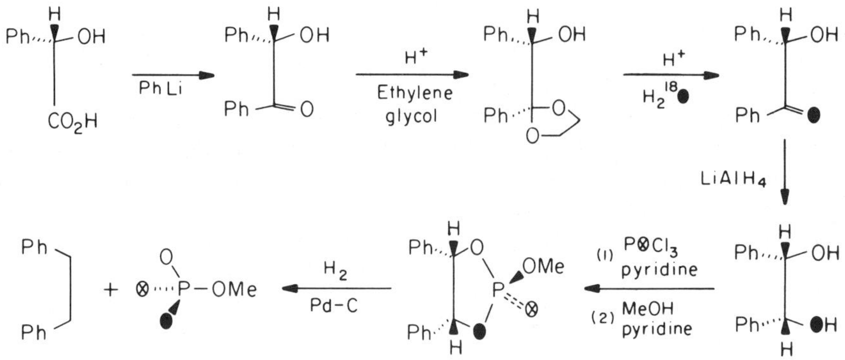

**Fig. 2.** Synthesis of the $S_P$ diastereomer of methyl [$^{16}$O,$^{17}$O,$^{18}$O]phosphate reported by Cullis and Lowe (1978, 1981). O, $^{16}$O; ⊗, $^{17}$O; ●, $^{18}$O.

chloridate and triester, and on this basis claimed that the absolute configuration of the methyl [$^{16}$O,$^{17}$O,$^{18}$O]phosphate was $R_P$; unfortunately, this configuration was not independently determined. Later stereochemical experiments utilizing chiral materials prepared by this route (Jarvest and Lowe, 1980) and analogous experiments performed in the author's laboratory with independently synthesized oxygen chiral esters (Coderre et al., 1981b) were not in agreement, and the reason for the discrepancy was traced to an incorrect assignment of the configurations of Lowe's cyclic chloridate and triester intermediates (Cullis et al., 1981). Thus the synthesis of oxygen chiral monoesters as prescribed by Cullis and Lowe yields esters having the $S_P$ configuration.

This synthesis, which has the admirable feature of not using displacement reactions of potentially imperfect stereospecificity to introduce the heavy oxygen isotopes, has been demonstrated to be a general synthesis because it, too, can be used to synthesize chiral samples of [$\gamma$-$^{16}$O,$^{17}$O,$^{18}$O]ATP (Lowe and Potter, 1981a). Because the method as reported provides only a single isomer of the precursor cyclic hydrobenzoin triester, esters having the $R_P$ configuration can be synthesized by preparing the cyclic chloridate from [$^{18}$O]POCl$_3$ and chiral *meso* [$^{17}$O]hydrobenzoin.

## B. Phosphate Diesters

Phosphate diesters are prochiral at phosphorus, so it is necessary for two of the three stable isotopes of oxygen to be stereospecifically incorporated into the phosphoryl oxygens. Of the three possible combinations chiral [$^{16}$O,$^{18}$O]- and [$^{17}$O,$^{18}$O]phosphodiesters have been reported. The first type can be prepared in a chiral purity approaching 100% whereas the second, as in the case of the monoesters, can be prepared in only about 50% chiral purity.

### 1. Reaction of N-Phenyl Phosphoramidates with Carbonyl Compounds

The first syntheses of chiral [$^{16}$O,$^{18}$O]phosphodiesters were performed independently and reported simultaneously in 1980 by Stec and co-workers (Baraniak et al., 1980) and by this laboratory (Gerlt and Coderre, 1980). The efforts of both laboratories were directed toward the synthesis of oxygen chiral 3',5'-cyclic nucleotides: Stec prepared the $S_P$ diastereomer of cyclic [$^{16}$O,$^{18}$O]AMP, and we reported both diastereomers of cyclic [$^{16}$O,$^{18}$O]dAMP. Both laboratories utilized the Wittig–Staudinger reaction to introduce the heavy oxygen isotope into the precursor N-phenyl phosphoramidates of the cyclic nucleotides. In this chemistry, a primary phosphoramidate is reacted with a strong base, such as NaH, to remove the proton from the nitrogen, and the resulting nitrogen anion reacts with a carbonyl compound to convert stereospecifically the phosphoramidate ester to a phosphate diester; this reaction had been intensively investigated in Stec's laboratory (Stec et al., 1976; Lesiak and Stec, 1978) and was previously applied to the synthesis of a number of chiral phosphorothioate and phosphoroselenate esters, including the diastereomers of adenosine 3′,5′-cyclic phosphorothioate (cyclic AMPS) (Baraniak et al., 1979). In the syntheses of oxygen chiral cyclic AMP and cyclic dAMP, Stec utilized [$^{18}$O]benzaldehyde as the source of $^{18}$O, and we used $C^{18}O_2$, which can be obtained commercially in an enrichment of 99%.

The chemical transformations that we used to synthesize the diastereomers of cyclic [$^{16}$O,$^{18}$O]dAMP are summarized in Fig. 3. 5′-Monomethoxytrityl 2′-deoxyadenosine was reacted with the o-chlorophenyl ester of N-phenyl phosphoramidic chloride, and the product diastereomers of the completely protected 3′-nucleotide were obtained in approximately equal yields, as judged by $^{31}$P NMR. This diastereomeric mixture was easily separated by chromatography, and the trityl groups were removed by treatment with acid. The diastereomerically pure samples of the acyclic nucleotides can be stereospecifically cyclized in the presence of *tert*-butoxide; the product cyclic N-phenyl phosphoramidates are formed with inversion of configuration at phosphorus (Gerlt et al., 1980b). The configurations of the cyclic materials were assigned initially by comparison of their $^{31}$P-NMR chemical shifts with those of several conformationally rigid cyclic phosphoramidates of known configuration at phosphorus (Stec and Okruszek, 1975); later, Saenger and Stec and co-workers reported an X-ray structure of one of the diastereomers (Lesnikowski et al., 1981). Each cyclic N-phenyl phosphoramidate was reacted with NaH and $C^{18}O_2$, and $^{18}$O-labeled samples of cyclic dAMP were isolated; this transformation is predicted and was experimentally verified to occur with retention of configuration at the phosphorus atom. Both diastereomers of cyclic [$^{16}$O,$^{18}$O]dAMP were obtained in approximately equal amounts, since the synthetic procedures produce equal amounts of the phosphoramidate precursors.

Whereas chiral [$^{16}$O,$^{18}$O]phosphodiesters are useful in the study of reactions that convert prochiral substrates to prochiral products, these are not particularly convenient for stereochemical studies of the conversion of prochiral substrates to proprochiral products (i.e., hydrolysis reactions). For studies of the latter type of reaction, we have used chemistry analogous to that depicted in Fig. 3 to prepare the diastereomers of cyclic [$^{17}$O,$^{18}$O]dAMP (Coderre et al., 1981a). For these syntheses, the o-chlorophenyl ester of N-phenyl [$^{17}$O]phosphoramidic chloride was prepared from [$^{17}$O]POCl$_3$ and used H$_2$$^{17}$O as the source of the $^{17}$O label. We have also used Wittig–Stau-

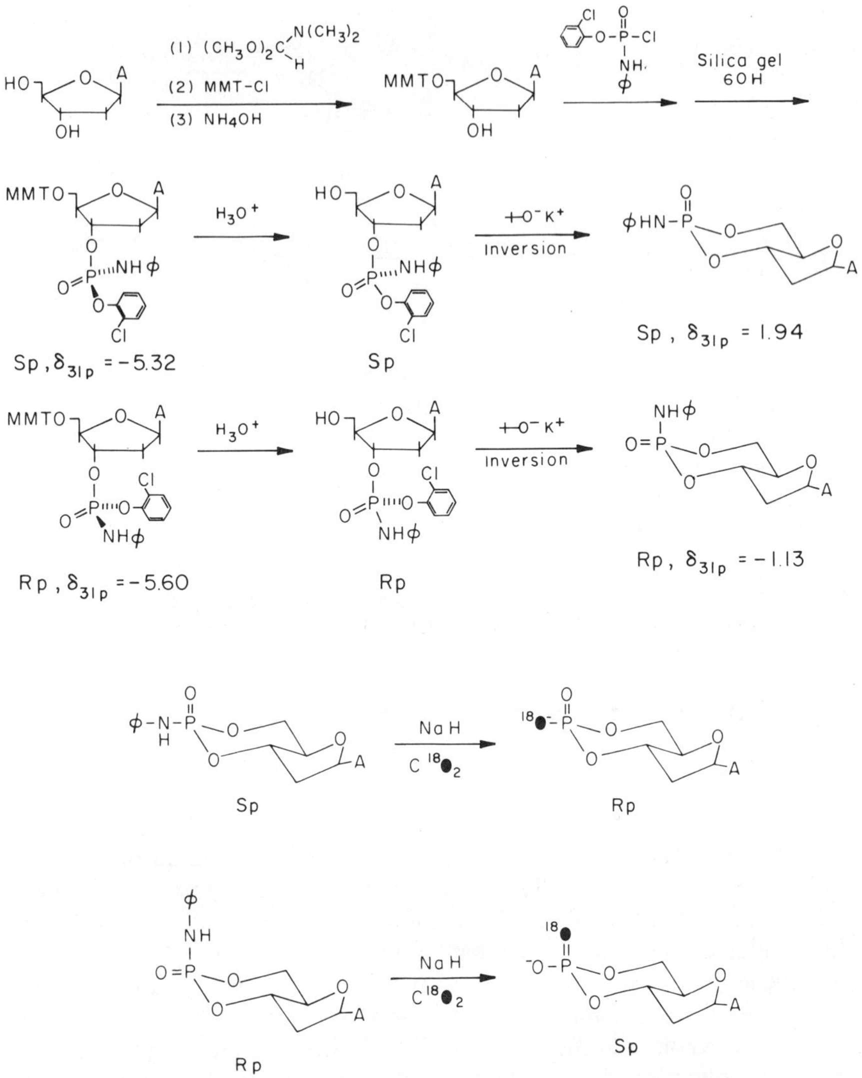

7. Chiral [$^{16}$O,$^{17}$O,$^{18}$O]Phosphate Esters

**Fig. 3.** Synthesis of the diastereomers of cyclic [$^{16}$O,$^{18}$O]dAMP.

dinger chemistry to prepare the diastereomers of the acyclic 4-nitrophenyl esters of thymidine 3′- and 5′-[$^{17}$O,$^{18}$O]phosphates (Mehdi *et al.,* 1981); the sequence of reactions for the preparation of the esters of the 5′-mononucleotide is summarized in Fig. 4.

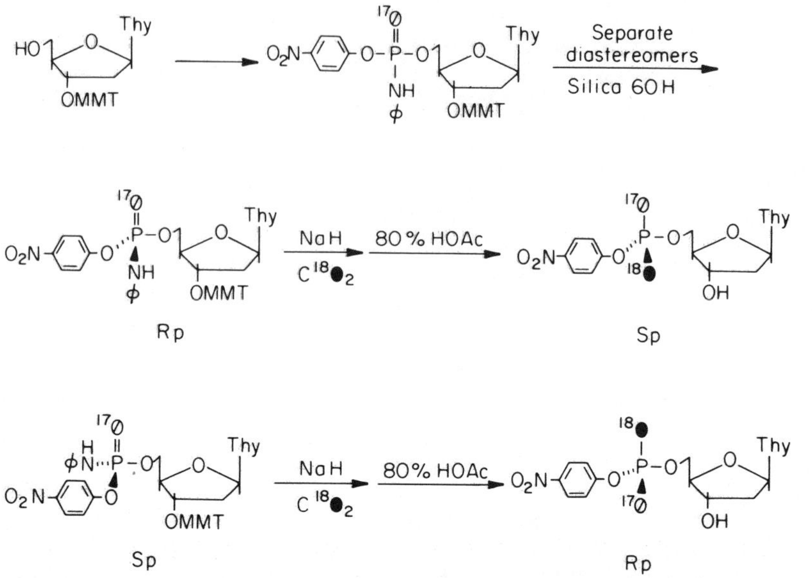

**Fig. 4.** Synthesis of the diastereomers of thymidine 5′-(4-nitrophenyl [$^{17}$O,$^{18}$O]phosphate). From Mehdi et al. (1981). Copyright 1981 American Chemical Society.

## 2. Hydrolysis of Cyclic Phosphoramidates

The vast majority of the stereochemical studies that have been performed to date are those of enzymes that utilize a nucleotide as a substrate. However, Bruzik and Tsai reported the synthesis of the diastereomers of phosphatidylethanolamine that are chiral by virtue of substitution with $^{16}$O and $^{18}$O (Bruzik and Tsai, 1982). Their route to the desired product involves the acid-catalyzed hydrolysis of a five-membered ring cyclic phosphoramidate, which directly yields the ethanolamine moiety on P—N bond cleavage; the steps in the synthesis are shown in Fig. 5. Clearly, this synthetic strategy is relatively inflexible as to the types of phospholipids that are directly accessible, but the known ability of phospholipase D to catalyze a facile transphosphorylation reaction (with retention of configuration) does increase the number of chiral phospholipids that can be synthesized.

## 3. Hydrolysis of Phosphorothioates

The synthetic procedures that are described in the preceding sections are specialized in that those developed for monesters cannot be used to synthesize diesters, and those for diesters cannot be used to synthesize monoesters; in addition, the methods for the synthesis of diesters are not applicable to the

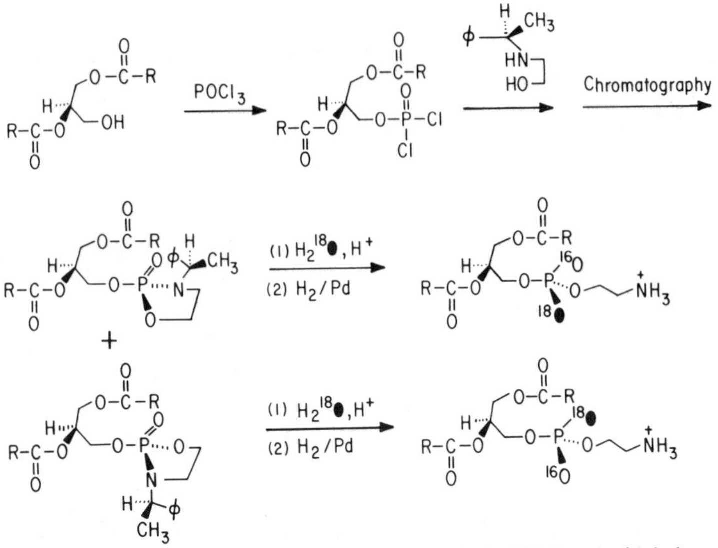

# 7. Chiral [$^{16}$O,$^{17}$O,$^{18}$O]Phosphate Esters

**Fig. 5.** Synthesis of the diastereomers of dipalmitoyl [$^{16}$O,$^{18}$O]phosphatidylethanolamine reported by Bruzik and Tsai (1982).

synthesis of nucleoside polyphosphates chiral at the internal phosphoryl groups. A potentially more general approach for the syntheses of oxygen chiral mono- and diesters was reported independently by the laboratories of Frey (Sammons and Frey, 1982), Eckstein (Connolly et al., 1982), and Lowe et al., 1982). Although the experimental conditions were not identical, all involved the stereospecific replacement of the sulfur atom of a chiral phosphorothioate by an oxygen atom. The applications of these methods to the synthesis of the chiral samples of [$\alpha$-$^{16}$O,$^{18}$O]ADP are summarized in Fig. 6. All the methods activate the sulfur atom such that it becomes a good leaving group displaceable by water in an $S_N2(P)$ reaction; in each case, the reported stereochemical course of the reaction was inversion of configuration of phosphorus. This approach is likely to be very useful in preparing oxygen chiral phosphate esters and anhydrides, especially in view of the synthetic methods Richard and Frey have reported for the preparation of AMPS, ADP$\beta$S, and ATP$\gamma$S with oxygen chiral terminal thiophosphoryl groups (Richard and Frey, 1982). The availability of these compounds and the documented stereospecificities of various kinase reactions in phosphorylating AMPS and ADP$\beta$S, together with a reliable method of stereospecific sulfur removal, will allow the syntheses of any desired oxygen-labeled adenine nucleotide.

Eckstein and Lowe reported that the activation of sulfur in a phosphorothioate diester, such as adenosine 3′,5′-cyclic phosphorothioate, leads to

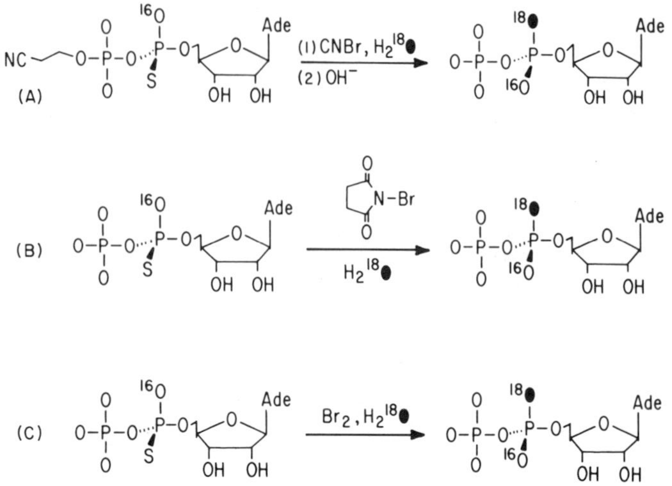

**Fig. 6.** Reported methods used to hydrolyze stereospecifically the $S_P$ diastereomer of ADPαS in $H_2^{18}O$ to form the $R_P$ diastereomer of [α-$^{16}O$,$^{18}O$]ADP: (A) Sammons and Frey (1982); (B) Connolly et al. (1982); (C) Lowe et al. (1982).

formation of oxygen-labeled phosphodiester, but the available evidence suggests that this reaction is regio- but not stereospecific. It is apparent that this methodology may provide a useful complement to the previously described Wittig–Staudinger chemistry (Section II,B,1).

## III. Configurational Analyses of Oxygen Chiral Phosphate Esters

The success of the synthetic methods described in the previous section was tied closely to the availability of techniques to determine the absolute configurations of the synthetic materials. As previously indicated, the configurational differences of phosphate esters that are enantiomeric or diastereomeric at phosphorus are nearly always cyrptic, so that chemical modification reactions must be performed before any physical technique can be used to perform the configurational analysis.

The early experiments reported by Knowles's laboratory (Abbott et al., 1979) utilized a method of configurational analysis that was based on metastable-ion mass spectroscopy and required instrumentation that is not routinely available to most investigators; in addition, this approach involved the chromatographic separation of hydrolytically labile species, which also discouraged routine adoption of the very elegant mass spectral technique.

Stec and co-workers also used a mass spectroscopic procedure to ascertain the stereospecificity of the Wittig–Staudinger reaction in the synthesis of conformationally rigid six-membered cyclic [$^{16}$O,$^{18}$O]phosphates (Baraniak et al., 1980); these experiments include the physical separation of a mixture of benzyl triesters prepared from the chiral diesters, but analogous separations of isomeric derivatives obtained from other chiral esters may not always be compatible with the amounts and identity of the experimental samples. For these reasons, we and others sought an alternative method of configurational analysis for oxygen chiral phosphate esters. The current consensus is that $^{31}$P-NMR methods offer an experimentally straightforward and reasonably sensitive method for the required determinations of configuration.

The ability of $^{31}$P-NMR spectroscopy to provide configurational information is based on the recently documented effects of oxygen-isotope substitution on the resonances of directly bonded $^{31}$P nuclei: $^{17}$O, a quadrupolar nucleus with a nuclear spin of $\frac{5}{2}$, provides an efficient relaxation mechanism for the $^{31}$P nucleus that results in extensive line broadening of the associated $^{31}$P-NMR resonance; and $^{18}$O, which has no nuclear spin, exerts a small shielding effect on a directly bonded $^{31}$P nucleus, when compared to $^{16}$O directly bonded to a $^{31}$P nucleus. Of these, the effect of directly bonded $^{18}$O on the $^{31}$P-NMR chemical shift is the more important because the magnitude of the upfield shift can be used directly to deduce configurations of chiral [$^{16}$O,$^{18}$O]phosphodiesters.

The observation of the shielding effect of $^{18}$O on the resonances of directly bonded $^{31}$P nuclei was independently reported by Cohn (Cohn and Hu, 1978), Lutz (Lutz et al., 1978), and Lowe (Lowe and Sproat, 1978). All three laboratories published spectra of inorganic phosphate that was randomly labeled with approximately 50% enriched $^{18}$O; these revealed that the magnitude of the upfield shift was proportional to the number of bonds between $^{18}$O and $^{31}$P since five resonances having the statistically expected intensities could be resolved. The spectrum published by Cohn and Hu is shown in Fig. 7. The observed additivity of the isotopic perturbations on the $^{31}$P-NMR resonance of inorganic phosphate was further investigated independently by Lowe (Lowe et al., 1979) and Cohn (Cohn and Hu, 1980), both of whom concluded that the magnitude of the shielding caused by directly bonded $^{18}$O nuclei was related to the order of the bond between $^{18}$O and $^{31}$P, with higher bond orders being associated with a greater perturbation of the $^{31}$P-NMR chemical shift. Lowe documented this finding by reporting the $^{18}$O perturbations in samples of inorganic phosphate, monomethyl phosphate, dimethyl phosphate, and trimethyl phosphate that were labeled in the nonesterified oxygens; these were 0.020, 0.024, 0.029, and 0.035 ppm, respectively (Lowe et al., 1979). Cohn's conclusion was based on an exami-

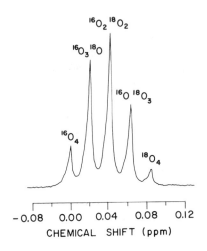

Fig. 7. Phosphorus-31 NMR (145.7 MHz) of randomly $^{18}$O-labeled inorganic phosphate reported by Cohn and Hu (1978).

nation of the $^{18}$O perturbations in a number of labeled adenine nucleotides, with an illustrative example of the relationship of shielding to bond order being the shift perturbations induced by bridging, $\gamma$-nonbridging, and $\beta$-nonbridging $^{18}$O atoms in ATP, 0.0166, 0.0220, and 0.0285 ppm, respectively (Cohn and Hu, 1980). It is this relationship between bond order and the magnitude of the perturbation induced by $^{18}$O that has allowed $^{31}$P-NMR spectroscopy to be used for configurational analyses of oxygen chiral phosphate esters.

## A. [$^{16}$O,$^{18}$O]Phosphate Diesters

Our expectation that $^{18}$O perturbations could be used to determine the configurations of the diasteromers of cyclic [$^{16}$O,$^{18}$O]dAMP was based on the relationships between bond order and shielding described by Lowe and Cohn. Although we observed that no configurational information could be obtained directly by $^{31}$-P-NMR spectroscopy on the samples of oxygen chiral cyclic [$^{16}$O,$^{18}$O]dAMP (Gerlt and Coderre, 1980) with identical upfield shifts of 0.030 ppm being observed for the labeled esters relative to unlabeled material, we reasoned that configurational information should be available from the $^{31}$P-NMR spectra of triesters obtained by alkylation of the diastereotopic phosphoryl oxygens of the diesters. The rationale for this reasoning is summarized in Fig. 8. Alkylation of either the equatorial or axial phosphoryl oxygen will induce single- and double-bond character in the exocyclic P—O bonds, thereby allowing the isotopic identity of the phosphoryl oxygens to be established by virtue of the magnitude of the $^{18}$O perturbation on the $^{31}$P-NMR chemical shift: alkylation of an $^{18}$O will be

# 7. Chiral [$^{16}$O,$^{17}$O,$^{18}$O]Phosphate Esters

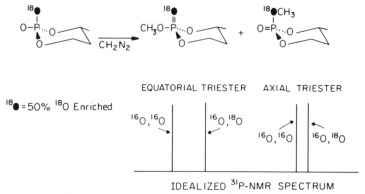

**Fig. 8.** Rationale for the use of $^{18}$O-perturbations on $^{31}$P-NMR chemical shifts to determine the configurations of chiral [$^{16}$O,$^{18}$O]phosphodiesters.

associated with a smaller perturbation than will alkylation of an $^{16}$O. Because the phosphoryl oxygens are chemically nonequivalent, the alkylation reaction can be expected to produce a mixture of the equatorial and axial triesters, and the magnitudes of the $^{18}$O perturbations observed for the pair of triesters are expected to be complementary. Determination of absolute configurations requires that the resonances observed for the isomeric triesters be assignable to a particular ester, and for this reason our early studies utilized alkylation with diazoethane because the chemical shifts of the ethyl (but not methyl) triesters of 3′,5′-cyclic nucleotides had been previously assigned (Engels and Schlaeger, 1977). The spectra that we obtained for samples of the diastereomers of cyclic [$^{16}$O,$^{18}$O]dAMP that had been diluted with an equal amount of unlabeled cyclic dAMP prior to reaction with diazoethane are shown in Fig. 9. In each spectrum the upfield chemical shift measured for the triester with the alkylated $^{18}$O is 0.015 ppm, whereas that measured for the triester with the alkylated $^{16}$O is 0.040 ppm; the isomeric identity of the triester associated with these perturbations permitted the configurational assignments of the samples of cyclic [$^{16}$O,$^{18}$O]dAMP. The configurations that were assigned were in accord with the established configurations of the precursor $N$-phenyl phosphoramidates and the predicted stereochemical course of the Wittig–Staudinger reaction (retention) used to convert them to phosphate esters.

In later work, we reacted the configurationally assigned diasteromers of cyclic [$^{16}$O,$^{18}$O]dAMP with diazomethane and assigned the chemical shifts of the methyl triesters of the cyclic nucleotide by virtue of the magnitudes of the observed $^{18}$O perturbations.

Subsequent to our original report that the magnitudes of $^{18}$O perturba-

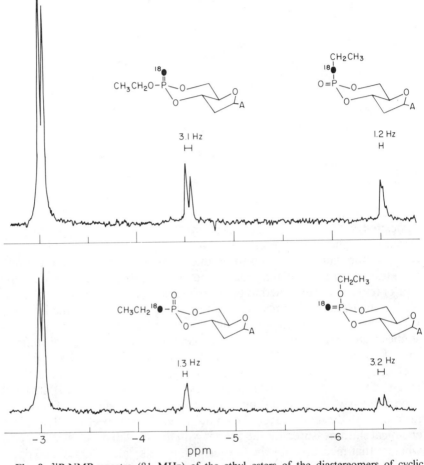

**Fig. 9.** 31P-NMR spectra (81 MHz) of the ethyl esters of the diastereomers of cyclic [$^{16}$O,$^{18}$O]dAMP. From Gerlt and Coderre (1980). Copyright 1980 American Chemical Society.

tions on the 31P-NMR chemical shifts of the triesters of cyclic [$^{16}$O,$^{18}$O]dAMP could be used to establish their absolute configurations, Gorenstein and Rowell independently described an analogous approach to determine the stereochemical course of the base-catalyzed hydrolyses of aryl esters of conformationally rigid six-membered ring cyclic phosphates (Gorenstein and Rowell, 1980). The magnitudes of the $^{18}$O perturbations that they observed were essentially identical to those we reported. Interestingly, their alkylations of the $^{16}$O,$^{18}$O-labeled cyclic diester products with diazomethane in aqueous solution yielded only equatorial triesters rather than the mixture we routinely observe when the reaction is carried out in methanol.

## B. Nucleoside [α-$^{16}$O,$^{18}$O]Diphosphates

Prochiral phosphorus atoms are found not only in phosphodiesters but also in the internal phosphoryl groups of nucleoside di- and triphosphates. Stereochemical studies of reactions occuring at or generating this type of prochiral phosphorus atom require a method for their configurational analysis, and we have shown that $^{18}$O perturbations on $^{31}$P-NMR chemical shifts can be successfully applied to these analyses (Coderre and Gerlt, 1980). At present, relatively few oxygen chiral stereochemical studies have been concerned with reactions involving these prochiral phosphorus atoms, but the synthetic methodology discussed in Section II,B,3 should provide the impetus for a concerted effort in this area.

The approach we have introduced is based on the chemistry that Cleland and co-workers have described for the preparation of Co(III) substitution inert complexes of nucleoside di- and triphosphates (Cornelius et al., 1977). For example, when dADP is reacted with Co(NH$_3$)$_4$(H$_2$O)$_2$$^{3+}$, a diastereomeric mixture of α,β-substitution inert complexes is formed because the reaction involves complex formation with the diastereotopic α-phosphoryl oxygens of the nucleotide; this reaction is illustrated in Fig. 10 using the diastereomers of [α-$^{16}$O,$^{18}$O]dADP as the nucleotide reactants. The complexes formed from dADP (and ADP) can be separated by chromatography

**Fig. 10.** Rationale for the use of substitution inert Co(NH$_3$)$_4$$^{3+}$ complexes to determine the configurations of the diastereomers of [α-$^{16}$O,$^{18}$O]dADP. From Coderre and Gerlt (1980). Copyright 1980 American Chemical Society.

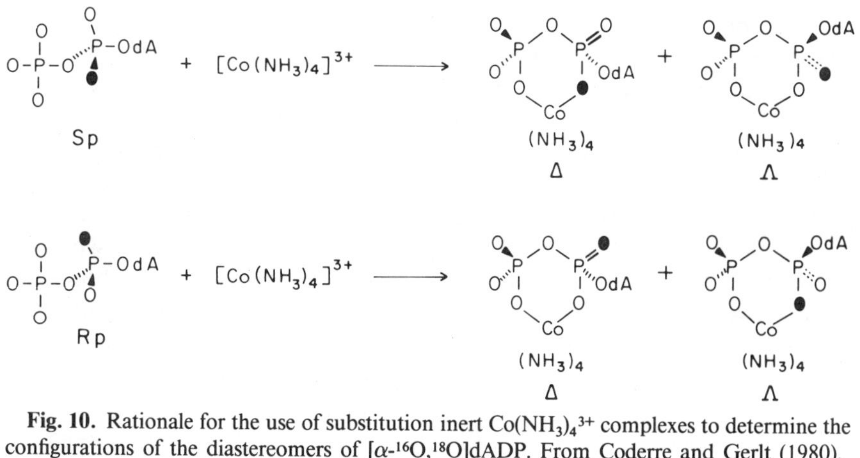

on a column of cross-linked cycloheptaamylose (Dunaway-Mariano and Cleland, 1980). The absolute configurations of the epimeric phosphorus atoms in the complexes have been assigned by using established correlations between configuration and circular dichroism properties (Merritt et al., 1978); using the nomenclature suggested by Cleland, the configurations are described as Δ or Λ. Because Co(III) is diamagnetic, it is possible to obtain high-resolution $^{31}$P-NMR spectra of the complexes, which reveal that the chemical shifts of the resonances associated with the α-phosphoryl nuclei in the dADP (and ADP) complexes have different chemical shifts (Cornelius et al., 1977); using the separated isomers it is possible to assign these resonances.

These well-defined physical and spectral properties are essential to the method of configurational analysis that utilizes the imposition of single- and double-bond character on the α-phosphoryl P—O bonds by complex formation. The Co(III) complexes are reasonably stable, and this stability can be attributed to significant covalent character in the Co(III)—O bond. Coordination of Co(III) to the phosphoryl oxygens of nucleoside di- and triphosphates is thus analogous to alkylation of phosphodiesters. As in the configurational analysis of cyclic phosphodiesters, complex formation will induce single- and double-bond character in the α-phosphoryl P—O bonds of the nucleotide with $^{18}$O coordinated to the Co(III), resulting in a smaller perturbation in the $^{31}$P-NMR chemical shift than when $^{16}$O is coordinated; the structures shown in Fig. 10 summarize this reasoning. Given the absolute configurations of the complexes, the magnitudes of the $^{18}$O perturbations can be used to establish the configurations of the α-phosphorus atoms in the original samples of [α-$^{16}$O,$^{18}$O]dADP (and ADP).

Spectra of the Co(III) complexes of [α-$^{16}$O,$^{18}$O]dADP that demonstrate the validity of this configurational analysis are shown in Fig. 11; the diastereomers of labeled dADP were prepared enzymatically from the diastereomers of cyclic [$^{16}$O,$^{18}$O]dAMP, as described in Section IV,A, and diluted with an equal amount of unlabeled dADP prior to complex formation. In each spectrum the $^{18}$O perturbations measured on the epimeric complexes yield complementary configurational information, and a reversal of the magnitudes of the isotope-induced chemical shift by inversion of the configuration of the chiral α-phosphoryl groups is apparent when the spectra are compared. This technique was subsequently used by both Frey (Sammons and Frey, 1982) and Eckstein (Connolly et al., 1982) to determine the configurations of samples of [α-$^{16}$O,$^{18}$O]ADP synthesized chemically from ADPαS.

At present the diastereomers of [β-$^{16}$O,$^{18}$O]ATP are unknown; however, when their synthesis is accomplished, presumably from the diastereomers of ATPβS using the chemistry described in Section II,B,3, measurements of the

# 7. Chiral [$^{16}$O,$^{17}$O,$^{18}$O]Phosphate Esters

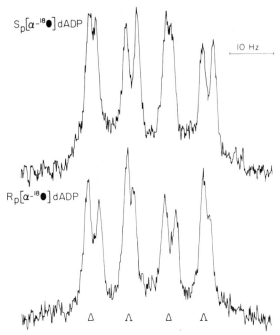

**Fig. 11.** $^{31}$P-NMR spectra (81 MHz) of the α-phosphorus resonances of the Co(NH$_3$)$_4^{3+}$ complexes of the diastereomers of [α-$^{16}$O,$^{18}$O]dADP. From Coderre and Gerlt (1980). Copyright 1980 American Chemical Society.

$^{18}$O perturbations on the chemical shifts of the resonances associated with the β-phosphorus nuclei in β,γ-substitution inert Co(III) complexes should be the method of choice for their configurational analysis.

## C. [$^{16}$O,$^{17}$O,$^{18}$O]Phosphate Monoesters

Although it is likely that $^{17}$O directly bonded to a $^{31}$P nucleus also causes a small upfield change in the $^{31}$P-NMR chemical shift relative to the unlabeled case, this effect has not yet been observed owing to the quadrupolar relaxation that the $^{31}$P nucleus experiences when it is bonded to $^{17}$O (although in principle the $^{17}$O perturbation should be observable in a $^{17}$O-decoupled spectrum). For most phosphates, the line broadening caused by directly bonded $^{17}$O is extensive, and this phenomenon, which was first documented independently by Tsai (1979) and Lowe (Lowe *et al.*, 1979), has been used extensively in the configurational analyses of [$^{16}$O,$^{17}$O,$^{18}$O]phosphate monoesters. A detailed discussion of the line-broadening effect can be found in Chapter 6 by Tsai.

The useful consequence of the extensive line broadening is that the only $^{31}$P-NMR resonances of phosphate esters that are sufficiently narrow to be detected using a narrow sweep width are those without directly bonded $^{17}$O, i.e., those with only $^{16}$O and/or $^{18}$O in the phosphoryl oxygens. For example, in a mixture of the three possible types of double-labeled cyclic phosphodiesters [$^{16}$O,$^{17}$O], [$^{16}$O,$^{18}$O], and [$^{17}$O,$^{18}$O], only the resonance associated with the $^{16}$O,$^{18}$O-labeled ester will have a small linewidth. The configuration of this type of labeled ester can be ascertained using the alkylation and $^{18}$O-perturbation approach described in Section III,A.

The first application of this consideration to the configurational analysis of a [$^{16}$O,$^{17}$O,$^{18}$O]phosphate ester was reported by Buchwald and Knowles (1980); the important chemical transformations required in this analysis are summarized in Fig. 12. Because the configuration of an oxygen chiral phosphate monoester ester cannot be directly established, the strategy underlying this analysis is to convert the monoester to a diester so that the isotopic identity of the exocyclic phosphoryl oxygens can be established. This approach requires that the conversion of the monoester to the cyclic diester be accomplished by a cyclization reaction of known stereochemical course and that a second chiral center be present in the original monoester, thereby establishing the chemical nonequivalence of the exocyclic phosphoryl oxygens. As described by Buchwald and Knowles, the chiral phosphoryl group must be located on chiral 1,2-propanediol, and the oxygen atoms of the monoester are randomly activated for cyclization by reaction with diphenyl phosphorylimidazole; in the presence of a hindered base, the resulting mixed anhydride undergoes cyclization with inversion of configuration at phosphorus, thereby generating a mixture of the three types of doubly labeled cyclic phosphodiester. Methylation with diazomethane produces a mixture of syn and anti methyl esters, the $^{31}$P-NMR chemical shifts of which are resolved and assigned. Measurements of the $^{18}$O perturbations on the resonances for the isomeric $^{16}$O,$^{18}$O-labeled esters allows the configuration of this material and the original chiral monoester to be assigned.

The $^{31}$P-NMR spectra of the mixtures of cyclic methyl esters obtained from the $R_P$ and $S_P$ diastereomers of 1-phospho($S$)-1,2-propanediol are shown in Fig. 13. Unlike the analyses described previously for cyclic [$^{16}$O,$^{18}$O]dAMP and [$\alpha$-$^{16}$O,$^{18}$O]dADP, these spectra are more complex in that four resonances are observed for each ester, even without the addition of unlabeled ester. Because the $^{17}$O presently available is "contaminated" to the extent of approximately 50% with $^{16}$O and $^{18}$O, resonances associated with $^{16}$O,$^{16}$O-, $^{18}$O,$^{18}$O-, and the inverted $^{16}$O,$^{18}$O-labeled diester are also present. However, a simple calculation allows the conclusion that the most intense resonance in the spectrum is that associated with the $^{16}$O,$^{18}$O-labeled ester of interest, with the remaining three resonances providing all of the possible internal standards.

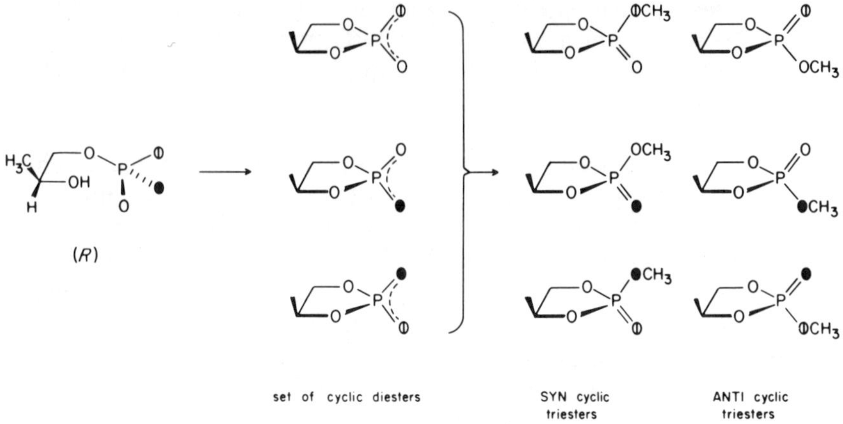

7. Chiral [$^{16}O,^{17}O,^{18}O$]Phosphate Esters

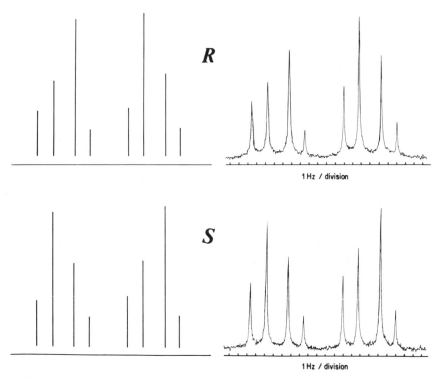

**Fig. 12.** Chemical transformations used in the configurational analysis developed by Knowles and colleagues (Abbott *et al.*, 1979; Buchwald and Knowles, 1980).

**Fig. 13.** Predicted (left) and observed (right) $^{31}$P-NMR spectra (121.5 MHz) of the mixture of methyl esters derived from the diastereomers of [1-$^{16}O,^{17}O,^{18}O$]phospho-(*S*)-1,2-propanediol. From Buchwald and Knowles (1980). Copyright 1980 American Chemical Society.

Given the fact that alkaline phosphatase will catalyze the stereospecific (retention) transfer of a chiral [$^{16}$O,$^{17}$O,$^{18}$O]phosphoryl group from any monoester to the primary hydroxyl group of 1,2-propanediol (Jones et al., 1978), this method of configurational analysis for chiral monoesters can be considered general.

The strategy described by Knowles and Buchwald requires that the chiral [$^{16}$O,$^{17}$O,$^{18}$O]phosphoryl group be located, in general, on a chiral diol. Nucleosides are chiral diols, and this recognition has allowed a convenient configurational analysis of samples of 3′- and 5′-mononucleotides that are chiral at phosphorus. Direct cyclization of the mononucleotides affords 3′,5′-cyclic nucleotides, and the approach to the configurational analysis of $^{16}$O,$^{18}$O-labeled cyclic nucleotides can be used, of course, on these cyclic esters. The required cyclization reactions can be accomplished either by chemical procedures (Jarvest et al., 1981a) or, in the case of 5′-AMP and 5′-dAMP, by enzyme-catalyzed cyclization (Coderre et al., 1981b).

## IV. Selected Examples

At the time of the preparation of this chapter, oxygen chiral phosphate esters and $^{31}$P-NMR methods of configurational analysis had been used to determine the stereochemical consequences of 21 enzyme-catalyzed reactions; the results are summarized in Table I. In addition, Knowles and co-workers had reported the results of five enzymic studies that utilized the metastable-ion mass spectroscopic method of configurational analysis. Also, work from the laboratories of M. D. Tsai and M. R. Webb and D. R. Trentham has resulted in the description of the stereochemistry of six enzymatic reactions in which oxygen chiral inorganic thiophosphate was generated as product; the method of configurational analysis used for these studies is also based on the effects of oxygen-isotope substitution on $^{31}$P-NMR resonances and is discussed in Chapter 6 by Tsai.

Rather than providing a comprehensive description of all the stereochemical studies that have utilized $^{31}$P-NMR methods of configurational analysis, three examples that demonstrate the techniques described in Section III are given. These are chosen from the studies carried out in the author's laboratory.

### A. Adenylate Cyclase

Enzymes that catalyze nucleotidyl transfer reactions and utilize acceptors other than water involve the interconversion of prochiral substrates and products. These enzymes include the nucleotidyl polymerases and cyclases

**TABLE I**

Stereochemical Studies Using Oxygen Chiral Phosphate Esters and $^{31}$P-NMR Methods of Configurational Analysis

| Enzyme | Result | Reference[a] |
|---|---|---|
| Conversion of prochiral substrate to prochiral product | | |
| Adenylate cyclase | Inversion | 1 |
| Guanylate cyclase | Inversion | 2 |
| ATP sulfurylase | Inversion | 3 |
| Isoleucyl-tRNA synthetase (activation step) | Inversion | 4 |
| Phospholipase D (transphosphorylation) | Inversion | 5 |
| Conversion of prochiral substrate to pro-prochiral product | | |
| Cyclic nucleotide phosphodiesterase | Inversion | 6, 7 |
| Cyclic AMP phosphodiesterase | Inversion | 7 |
| Snake venom phosphodiesterase | Retention | 8, 9 |
| Spleen exonuclease | Retention | 10 |
| Staphylococcal nuclease | Inversion | 11 |
| Nuclease $P_1$ | Inversion | 11a |
| Pancreatic DNase I | Inversion | 11b |
| Conversion of proprochiral substrate to pro-prochiral product | | |
| Glycerol kinase | Inversion | 12 |
| Hexokinase | Inversion | 13 |
| Pyruvate kinase | Inversion | 13a |
| $T_4$ polynucleotide kinase | Inversion | 14 |
| Acetate kinase | Inversion | 12 |
| Creatine kinase | Inversion | 15 |
| Glucokinase | Inversion | 16 |
| Phosphofructokinases | Inversion | 17 |
| Phosphoglycerate mutases | Retention | 17a |
| Phosphoglucomutase | Retention | 18 |
| PEP:glucose phosphotransferase (five transfers) | Inversion | 19 |
| Conversion of proprochiral substrate to pro-prochiral product | | |
| Alkaline phosphatase | Retention | 20 |
| Acid phosphatase | Retention | 21 |
| Glucose-6-phosphatase | Retention | 22 |

[a] References: 1, Coderre and Gerlt (1980); 2, Senter et al. (1983); 3, Bicknell et al. (1982); 4, G. Lowe et al. (1983); 5, Bruzik and Tsai (1982); 6, Coderre et al. (1981b); 7, Jarvest et al. (1982); 8, Mehdi and Gerlt (1981b); 9, Jarvest and Lowe (1981b); 10, Mehdi and Gerlt (1981a); 11, Mehdi and Gerlt (1982); 11a, Potter et al. (1983); 11b, S. Mehdi and J. A. Gerlt (unpublished observerations); 12, Blättler and Knowles (1979); 13, Lowe and Potter (1981a); 13a, Lowe et al. (1981); 14, Jarvest and Lowe (1981a); 15, Hansen and Knowles (1981); 16, Pollard-Knight et al. (1982); 17, Jarvest et al. (1981b); 17a, Blättler and Knowles (1980); 18, Lowe and Potter (1981b); 19 Begley et al. (1982); 20, Jones et al. (1978); 21, Saini et al. (1981); 22, Lowe and Potter (1982).

that catalyze the attack of a nucleotide 3′-hydroxyl group on the α-phosphorus atom of a nucleoside triphosphate to form a phosphodiester bond. In the former case (polymerases), the reaction is intermolecular and results in synthesis of internucleotide phosphodiester bonds, and, in the latter case (cyclases), the reaction is intramolecular and results in the formation of a 3′,5′-cyclic nucleotide. To date, no study of a nucleotidyl polymerase reaction using oxygen chiral substrates has been reported (although a number of these enzymes have been examined with phosphorothioate methodology, primarily by F. Eckstein and collaborators; Eckstein et al., 1982). Two studies of nucleotidyl cyclases have been performed with oxygen chiral substrates; the first was performed in our laboratory with a bacterial adenylate cyclase (Coderre and Gerlt, 1980) and the second in Eckstein's laboratory with a bovine guanylate cyclase (Senter et al., 1983).

Few details are available about the chemical mechanisms of the reactions catalyzed by nucleotidyl cyclases since these enzymes can be associated with membranes and therefore difficult to purify and/or are present in very small amounts and therefore difficult to obtain in amounts compatible with many modern enzymological techniques. We have decided to direct our attention to bacterial adenylate cyclases because one, the enzyme produced by *Brevibacterium liquefaciens*, has been purified to homogeneity (Takai et al., 1974) and the gene for another, that from *Salmonella typhimurium*, has been cloned in a multiple-copy plasmid (Wang et al., 1981). Given the limited amounts of enzymes that are presently available, chemical studies designed to ascertain directly if the reaction occurs via formation and breakdown of an adenylated enzyme intermediate are impossible; therefore, we decided to use a stereochemical approach to obtain information regarding the existence of an adenylated enzyme intermediate in the reaction catalyzed by the cyclase isolated from *B. liquefaciens*.

Hayaishi and colleagues, who devised the purification for the *Brevibacterium liquefaciens* enzyme, used it to characterize the reversibility of the adenylate cyclase reaction (Kurashina et al., 1974) and found that the equilibrium constant for the reaction written in the direction of cyclic AMP formation is $0.12\,M$ at pH 7.3; at this pH the rates of the forward and reverse reactions are comparable but about $\frac{1}{25}$ the rate of the forward reaction measured at its pH optimum, pH 9. Our plan for determining the stereochemical course of the reaction is shown in Fig. 14. Since we had synthesized the diastereomers of cyclic [$^{16}O,^{18}O$]dAMP, we would use the cyclase to catalyze their pyrophosphorolysis and form the diastereomers of [$\alpha$-$^{16}O,^{18}O$]dATP. However, the thermodynamics of the cyclase reaction prevents an efficient conversion of cyclic dAMP to dATP, so this reaction was coupled to the glycerol kinase reaction; the kinase reaction utilizes the thermodynamic instability of the $\beta,\gamma$-anhydride bond to displace the overall equilibrium to favor the synthesis of the diastereomers of [$\alpha$-$^{16}O,^{18}O$]dADP. Both the cyclase and glycerol kinase can utilize deoxyadenosine nucleotides as substrates, but only the cyclase reaction can alter the configuration of the chiral phosphorus atoms.

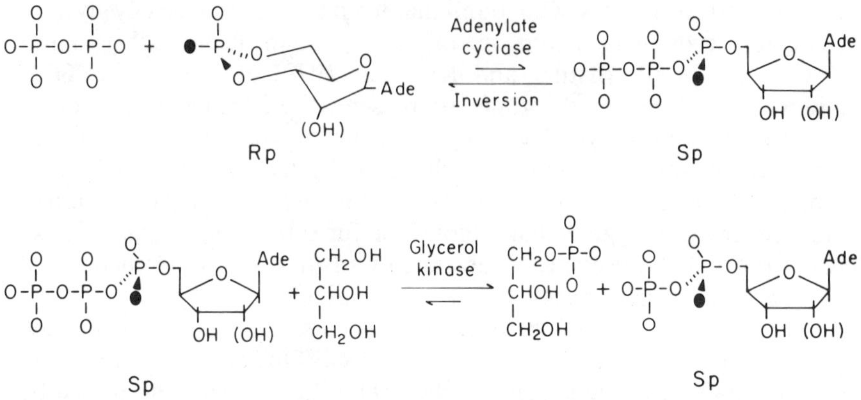

# 7. Chiral [$^{16}O,^{17}O,^{18}O$]Phosphate Esters

**Fig. 14.** Strategy for determining the stereochemical course of the reaction catalyzed by the adenylate cyclase from *Brevibacterium liquefaciens*. From Coderre and Gerlt (1980). Copyright 1980 American Chemical Society.

After carrying out the coupled enzyme reactions with both diastereomers of cyclic [$^{16}O,^{18}O$]dAMP, we isolated $^{18}O$-labeled samples of dADP as judged by measurements of the $^{18}O$-perturbations on the $^{31}P$-NMR resonances of the α-phosphorus atoms. The configurations of these presumably chiral samples were established by formation of their Co(III) α,β-substitution inert complexes, purification by chromatography on Chelex-100, and measurement of the $^{18}O$ perturbations on the $^{31}P$-NMR resonances of the α-phosphorus atoms in the diastereomeric mixture of complexes (Section III,B). The spectra obtained are shown in Fig. 11, and the $^{18}O$ perturbations measured were in accord with the adenylate cyclase reaction occurring with *inversion* of configuration at phosphorus (refer to Fig. 10 for the structures of the complexes formed from the diastereomers of [α-$^{16}O,^{18}O$]dADP that allow prediction of the relative magnitudes of the $^{18}O$ perturbations to be expected in the mixtures of complexes). By microscopic reversibility, the adenylate cyclase must catalyze the intramolecular displacement of pyrophosphate from dATP by its 3'-hydroxyl group to form cyclic dAMP with inversion of configuration.

This result is very persuasive evidence that the mechanism of the adenylate cyclase reaction does not involve the formation of an adenylated enzyme intermediate. In addition, this study provided the second demonstration that oxygen chiral and phosphorothioate substrates are processed by enzymes with the same stereochemical course; we had previously found that the $S_p$ diastereomer of ATPαS is converted to the $R_p$ diastereomer of cyclic AMPS with this cyclase (Gerlt et al., 1980a). The first example of sulfur not altering the stereochemical course of an enzymatic reaction was provided by

Knowles and co-workers, who found that the reaction catalyzed by glycerol kinase proceeds with inversion of configuration using either chiral [$\gamma$-$^{16}$O,$^{17}$O,$^{18}$O]ATP (Blättler and Knowles, 1979) or [$\gamma$-$^{16}$O,$^{18}$O]ATP$\gamma$S (Pliura et al., 1980) as substrate. At present, the stereochemical consequences of the reactions catalyzed by seven enzymes have been investigated with both oxygen chiral and phosphorothioate techniques, and in each comparison the stereochemical results were identical. This accumulation of evidence strongly suggests that although sulfur substitution frequently decreases the rate of processing of substrate by an enzyme, the stereochemical course and many mechanistic details will be unaffected.

Eckstein's study of the stereochemical course of a bovine guanylate cyclase utilized chemically synthesized [$\alpha$-$^{16}$O,$^{18}$O]GTP as substrate, and the configuration of the product cyclic [$^{16}$O,$^{18}$O]GMP was determined by $^{31}$P NMR following methylation, as described in Section III,A. This nucleotidyl cyclase was also found to proceed with inversion of configuration at phosphorus (Senter et al., 1983).

## B. Cyclic Nucleotide Phosphodiesterase

Phosphodiesterases catalyze the conversion of prochiral substrates to proprochiral products, and several of these enzymes have been investigated in our laboratory with oxygen chiral technology. The substrates for our stereochemical studies have been chiral $^{17}$O,$^{18}$O-labeled nucleotide esters, which were prepared according to the procedures described in Section II,B,1.

The cyclic nucleotide phosphodiesterase isolated from bovine heart was the initial phosphodiesterase to be studied with oxygen chiral techniques. Lowe and collaborators were the first to publish the results of stereochemical work on the reaction catalyzed by this enzyme (Jarvest and Lowe, 1980), and they reported that oxygen chiral cyclic AMP [which was prepared by the chemical cyclization of 5′-[$^{16}$O,$^{17}$O,$^{18}$O]AMP synthesized according to the general procedure reported by Cullis and Lowe (Cullis and Lowe, 1978, 1981)] was hydrolyzed by the enzyme with retention of configuration. At the time only Knowles's comparative work on glycerol kinase was known to Lowe, so there was little experimental precedent to cast doubt on this result that was in direct conflict with the finding that Eckstein and Stec and collaborators had reported for the hydrolysis of one of the diastereomers of cyclic AMPS by this enzyme (Burgers et al., 1979). However, subsequent work from this laboratory demonstrated unambiguously that the hydrolysis of cyclic [$^{17}$O,$^{18}$O]dAMP in H$_2$$^{16}$O catalyzed by the same enzyme proceeded with inversion of configuration (Coderre et al., 1981b), thereby providing

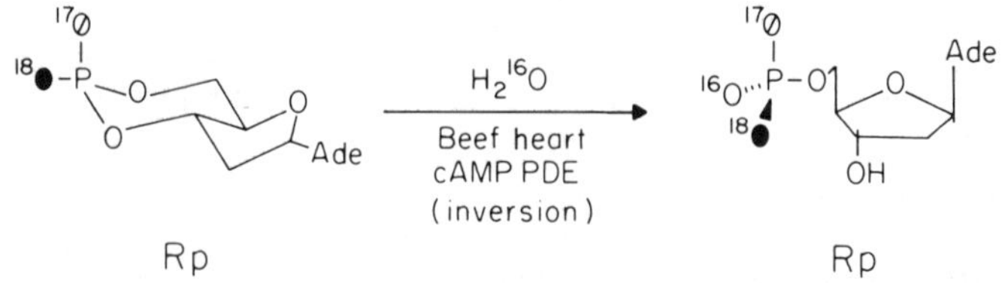

# 7. Chiral [$^{16}$O,$^{17}$O,$^{18}$O]Phosphate Esters

**Fig. 15.** Hydrolysis of cyclic [$^{17}$O,$^{18}$O]dAMP catalyzed by the cyclic nucleotide phosphodiesterase from bovine heart. From Coderre *et al.* (1981b). Copyright 1981 American Chemical Society.

the third example of an enzymatic reaction whose stereochemical course is unaffected by sulfur substitution. Upon reinvestigation of this discrepancy in results (Cullis *et al.*, 1981), Lowe found that the oxygen chiral phosphate monoesters synthesized by his reported procedure have the $S_P$ rather than $R_P$ configuration as was originally claimed (Cullis and Lowe, 1978). This unfortunate error in configurational assignment was the result of an incorrect assignment of the geometric isomerism in the precursor cyclic triester.

We hydrolyzed the $R_P$ diastereomer of cyclic [$^{17}$O,$^{18}$O]dAMP in H$_2$$^{16}$O and isolated a chiral sample of 5′-[$^{16}$O,$^{17}$O,$^{18}$O]dAMP (Fig. 15). Our approach to the configurational analysis of this sample was to use adenylate kinase and pyruvate kinase with their cosubstrates ATP and PEP, respectively, to convert the sample of dAMP to a mixture of the three possible types of triply labeled dATP (Fig. 16). The isolated dATP was then converted to a mixture of the three possible types of doubly labeled cyclic dAMP using the adenylate cyclase from *Brevibacterium liquefaciens*, the reaction of which is accompanied by inversion of configuration at phosphorus (Fig. 16), as was described in Section IV,A. The isolated mixture of diesters was methylated with diazomethane, and the $^{18}$O perturbations on the $^{31}$P-NMR resonances of the equatorial and axial methyl esters were measured. The spectrum obtained is shown in Fig. 17, and this is consistent with the hydrolysis reaction occurring with inversion of configuration at phosphorus. Note that this analysis depends on the presence of directly bonded $^{17}$O-labeled nuclei to extensively broaden the associated $^{31}$P-NMR resonances, so that the $^{31}$P-NMR resonances of phosphorus nuclei bonded only to $^{16}$O and $^{18}$O are observed.

The simplest explanation for the stereochemical result obtained for the cyclic nucleotide phosphodiesterase is that the hydrolysis reaction proceeds by a direct, in-line attack of water on the cyclic ester to yield the acyclic mononucleotide with inversion of configuration at phosphorus, that is, catalysis does not involve the formation of a nucleotidylated enzyme intermediate.

We have investigated the stereochemical course of the base-catalyzed

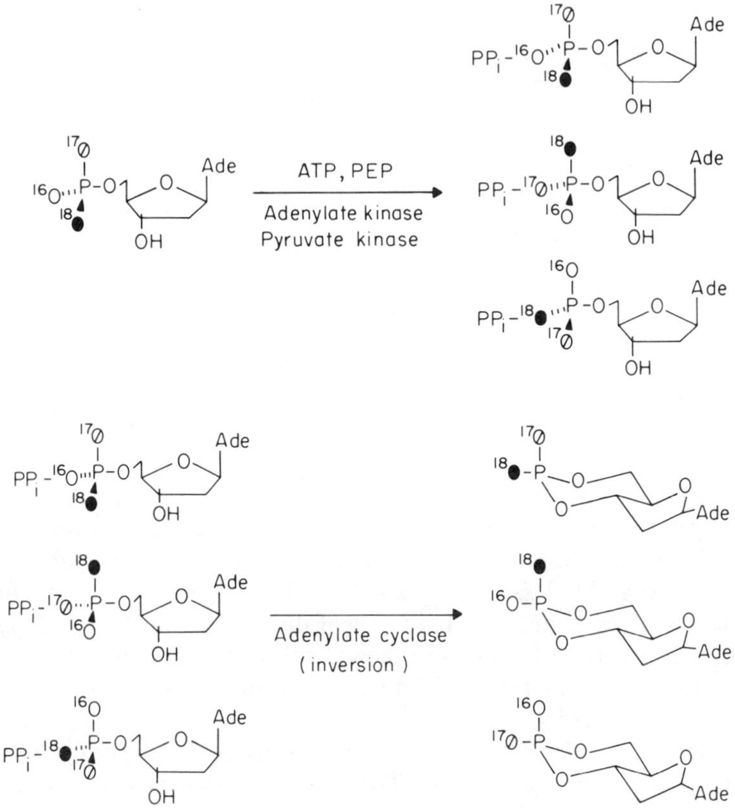

**Fig. 16.** Enzymatic transformations used in the configurational analysis of 5′-[$^{16}$O,$^{17}$O,$^{18}$O]dAMP. From Coderre et al. (1981b). Copyright 1981 American Chemical Society.

hydrolysis of cyclic [$^{17}$O,$^{18}$O]dAMP (S. Mehdi, J. A. Coderre, and J. A. Gerlt, unpublished observations). When heated in 0.2 $N$ Ba(OH)$_2$, cyclic dAMP is hydrolyzed to a 4:1 mixture of 3′- and 5′-dAMP. We have separated and analyzed the configurations of the 3′- and 5′-[$^{16}$O,$^{17}$O,$^{18}$O]dAMPs that were obtained from the analogous hydrolysis of the labeled cyclic ester. The separation was accomplished by subjecting the isolated mixture of hydrolysis products to the coupled action of adenylate kinase and pyruvate kinase in the presence of ATP and PEP, which results in the quantitative conversion of the 5′-dAMP to 5′-dATP; following this reaction, the monophosphate and triphosphates are separated easily by chromatography. The labeled dATP was enzymatically cyclized to the required mixture of the three types of doubly labeled cyclic dAMP, using the bacterial adenylate cyclase as catalyst (Coderre and Gerlt, 1980); the labeled 3′-AMP was chemically cyclized to an analogous mixture of labeled cyclic

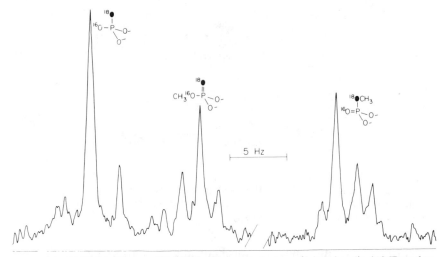

**Fig. 17.** $^{31}$P-NMR spectrum (81 MHz) of the methyl esters of labeled cyclic dAMP used to determine the stereochemical course of the reaction catalyzed by the cyclic nucleotide phosphodiesterase. From Coderre *et al.* (1981b). Copyright 1981 American Chemical Society.

dAMP (Jarvest *et al.*, 1981a). Following methylation and measurements of the $^{18}$O perturbations on the $^{31}$-P-NMR resonances of the mixture of triesters in each sample, the conclusion was clear that both 3'- and 5'-dAMP are obtained from cyclic dAMP with quantitative inversion of configuration at phosphorus. This result is most easily explained by the base-catalyzed hydrolysis occurring by the direct, in-line attack of water on the cyclic ester to yield the monoesters; a mechanism involving obligatory pseudorotation of pentacoordinate intermediates is effectively ruled out by the stereochemical studies. The simplest interpretation of the results describing the enzymatic and nonenzymatic hydrolyses of cyclic dAMP is that the mechanisms are analogous, although the enzyme-catalyzed reaction is perhaps $10^{15}$ times more rapid as a result of factors that may include charge neutralization of the phosphate ester anion, entropic acceleration resulting from the close proximity of the hydroxide ion and cyclic ester in the enzyme active site, and potentiation of the nucleophilicity of the hydroxide ion by a decrease in effective medium polarity in the active site.

## C. Staphylococcal Nuclease

The mechanism by which staphylococcal nuclease catalyzes the hydrolysis of single-stranded RNA and DNA is uncertain, despite the fact that the complete amino acid sequence (Cone *et al.*, 1971) and a 1.5-Å X-ray structure (Cotton *et al.*, 1979) are available. In order to provide additional

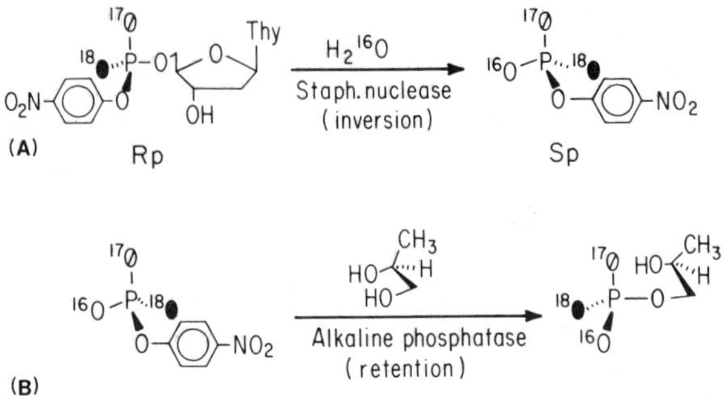

**Fig. 18.** Hydrolysis of thymidine 5′-(4-nitrophenyl [$^{17}O,^{18}O$]phosphate) catalyzed by staphylococcal nuclease (A) and the subsequent transfer of the chiral phosphoryl group to (S)-1,2-propanediol catalyzed by alkaline phosphatase (B). From Mehdi and Gerlt (1982). Copyright 1982 American Chemical Society.

chemical information about the mechanism of this enzymatic reaction, we have determined the stereochemical course of a reaction catalyzed by the nuclease (Mehdi and Gerlt, 1982).

The nuclease requires $Ca^{2+}$ ions for activity, with no other divalent metal ion being able to support catalysis. A large number of other phospodiesterases have been found to be dependent on divalent metal ions for activity, including the restriction endonuclease *Eco*RI (Barton *et al.*, 1982). Thus elucidation of the mechanism of the reaction catalyzed by staphylococcal nuclease may provide important clues to the mechanisms of the other metal-dependent phosphodiesterases. Fortunately, staphylococcal nuclease will catalyze, albeit at a low rate, the hydrolysis of a number of mononucleotide esters (Cuatrecasas *et al.*, 1969), including thymidine 5′-(4-nitrophenyl phosphate); this ester is hydrolyzed to thymidine and 4-nitrophenyl phosphate. We have determined the stereochemical course of the hydrolysis of thymidine 5′-(4-nitrophenyl [$^{17}O,^{18}O$]phosphate) and interpreted the result in terms of the structure of the active site of the enzyme.

Nuclease-catalyzed hydrolysis of the oxygen chiral nucleotide ester yielded oxygen chiral 4-nitrophenyl [$^{16}O,^{17}O,^{18}O$]phosphate (Fig. 18) the configuration of which was analyzed according to Knowles's general method of monoester configurational analysis (Section III,C). The chiral phosphoryl group was transferred to (S)-1,2-propanediol with retention of configuration, using alkaline phosphatase as catalyst (Fig. 18). Following chemical cyclization and methylation with diazomethane, the spectrum reproduced in Fig. 19 was obtained, which allowed the deduction that the nuclease-catalyzed hydrolysis reaction proceeds with inversion of configuration at phosphorus. The simplest interpretation of this result is that the nuclease catalyzes the direct, in-line attack of water on the phosphorus atom

# 7. Chiral [$^{16}$O,$^{17}$O,$^{18}$O]Phosphate Esters

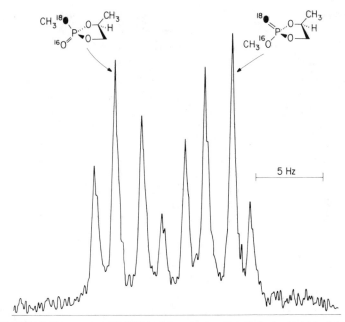

**Fig. 19.** $^{31}$P-NMR spectrum (81 MHz) of the mixture of methyl esters used to determine the stereochemical course of the reaction catalyzed by staphylococcal nuclease. From Mehdi and Gerlt (1982). Copyright 1982 American Chemical Society.

of the substrate to displace thymidine, with no covalent adduct being formed between the enzyme and substrate. This interpretation can be accomodated by the active-site geometry of the nuclease as revealed by the high-resolution structure (Cotton *et al.*, 1979) determined for a complex of the nuclease with Ca$^{2+}$ and thymidine 3',5'-bisphosphate (pdTp), a potent competitive inhibitor of the reaction. The carboxylate of glutamate-43 can be envisaged to act as a general base catalyst to assist in the attack of water on the 5'-phosphorus atom of the substrate.

At present, there are no experimental data to contradict this simplest interpretation; however, the X-ray structure does suggest an alternative explanation that is consistent with the stereochemical study. Nucleophilic attack of glutamate-43 on the 5'-phosphorus atom of a substrate would lead to formation of an ester of glutamyl phosphate, and this mixed anhydride intermediate is likely to hydrolyze by attack of water on the glutamate carboxylate carbon (Kellerman, 1958; DiSabato and Jencks, 1961). Such a mechanism would be accompanied by inversion of configuration at phosphorus because only a single nucleophilic displacement reaction occurs at the phosphorus, despite the fact that an intermediate is formed. Thus, it should be emphasized that the simplest interpretation of an overall inversion of configuration in an enzyme-catalyzed reaction is that *one displacement reaction* occurs at phosphorus and also that an inversion of configuration does not provide *sufficient* evidence to rule out the formation of a covalent intermediate. This potential ambiguity in interpreting stereochemical data can apply only to the phosphohydrolases because in phosphoryl and nucleotidyl transfer reactions that involve an acceptor molecule other than water, the transfer must involve the making and breaking of bonds to the phosphorus atom. An experiment to test this alternative mechanism is to ascertain under single turnover conditions whether the oxygen necessarily incorporated in the hydrolysis reaction is derived from solvent or from some other source (i.e., a carboxylate group in the enzyme). Such an experiment is in progress for staphylococcal nuclease.

## V. Summary

Methods have been developed for the synthesis of essentially any oxygen chiral phosphate ester of biological importance, and the $^{31}$P-NMR methods described and illustrated in this chapter provide a convenient method for determining their configurations and elucidating the stereochemical consequences of enzyme-catalyzed phosphoryl and nucleotidyl transfer reactions.

## Acknowledgments

The research carried out at Yale was performed by Jeffrey A. Coderre and Shujaath Mehdi; I am grateful to them for their enthusiasm, perseverance, and intellectual contributions, which made this research both exciting and successful. I would also like to thank Professors M. Cohn, P. A. Frey, F. Eckstein, J. R. Knowles, and G. Lowe for unpublished data and/or the permission to reproduce their results. The research performed at Yale was supported by a grant (GM-22350) and a Research Career Development Award (CA-00499) from the National Institutes of Health, and by a fellowship from the Alfred P. Sloan Foundation.

## References

Abbott, S. J., Jones, S. R., Weinman, S. A., and Knowles, J. R. (1978). *J. Am. Chem. Soc.* **100**, 2558–2560.
Abbott, S. J., Jones, S. R., Weinman, S. A., Bockhoff, F. M., McLafferty, F. W., and Knowles, J. R. (1979). *J. Am. Chem. Soc.* **101**, 4323–4332.

Baraniak, J., Kinas, R. W., Lesiak, K., and Stec, W. J. (1979). *J. Chem. Soc. Chem. Commun.* pp. 940–941.
Baraniak, J., Lesiak, K., Sochacki, M., and Stec, W. J. (1980). *J. Am. Chem. Soc.* **102**, 4533–4534.
Barton, J. K., Basile, L. A., and Paranawithana, S. R. (1982). *J. Biol. Chem.* **257**, 7911–7914.
Bicknell, R., Cullis, P. M., Jarvest, R. L., and Lowe, G. (1982). *J. Bol. Chem.* **257**, 8922–8927.
Blättler, W. A., and Knowles, J. R. (1979). *Biochemistry* **18**, 3927–3933.
Blättler, W. A., and Knowles, J. R. (1980). *Biochemistry* **19**, 738–743.
Bruzik, K., and Tsai, M. D. (1982). *J. Am. Chem. Soc.* **104**, 863–865.
Buchwald, S. L., and Knowles, J. R. (1980). *J. Am. Chem. Soc.* **102**, 6601–6602.
Buchwald, S. L., and Knowles, J. R. (1982). *J. Am. Chem. Soc.* **104**, 1438–1440.
Burgers, P. M. J., Eckstein, F., Hunneman, D. H., Baraniak, J., Kinas, R. W., Lesiak, K., and Stec, W. J. (1979). *J. Biol. Chem.* **254**, 9959–9961.
Coderre, J. A., and Gerlt, J. A. (1980). *J. Am. Chem. Soc.* **102**, 6594–6597.
Coderre, J. A., Mehdi, S., Demou, P. C., Weber, R. R., Traficante, D. D., and Gerlt, J. A. (1981a). *J. Am. Chem. Soc.* **103**, 1870–1872.
Coderre, J. A., Mehdi, S., and Gerlt, J. A. (1981b). *J. Am. Chem. Soc.* **103**, 1872–1875.
Cohn, M., and Hu, A. (1978). *Proc. Natl. Acad. Sci. U.S.A.* **75**, 200–203.
Cohn, M., and Hu, A. (1980). *J. Am. Chem. Soc.* **102**, 913–916.
Cone, J. L., Cusumano, C. L., Taniuchi, H., and Anfinsen, C. B. (1971). *J. Biol. Chem.* **246**, 3103–3110.
Connolly, B. A., Eckstein, F., and Fuldner, H. H. (1982). *J. Biol. Chem.* **257**, 3382–3384.
Cooper, D. G., Hall, C. R., Harrison, J. M., and Inch, T. D. (1977). *J. Chem. Soc. Perkin Trans. 1* pp. 1969–1980.
Cornelius, R. D., Hart, P. A., and Cleland, W. W. (1977). *Inorg. Chem.* **16**, 2799–2805.
Cotton, F. A., Hazen, E. E., and Legg, M. J. (1979). *Proc. Natl. Acad. Sci. U.S.A.* **76**, 2551–2555.
Cuatrecasas, P., Wilchek, M., and Anfinsen, C. B. (1969). *Biochemistry* **8**, 2277–2284.
Cullis, P. M., and Lowe, G. (1978). *J. Chem. Soc., Chem. Commun.* pp. 512–514.
Cullis, P. M., and Lowe, G. (1981). *J. Chem. Soc., Perkin Trans. 1* pp. 2317–2321.
Cullis, P. M., Lowe, G., Jarvest, R. L., and Potter, B. V. L. (1981). *J. Chem. Soc., Chem. Commun.* pp. 245–246.
DiSabato, G., and Jencks, W. P. (1961). *J. Am. Chem. Soc.* **83**, 4393–4400.
Dunaway-Mariano, D., and Cleland, W. W. (1980). *Biochemistry* **19**, 1496–1505.
Eckstein, F. (1979). *Acc. Chem. Res.* **12**, 204–210.
Eckstein, F., Romaniuk, P. J., and Connolly, B. A. (1982). *In* "Methods in Enzymology, Vol. 87, 197–212. Academic Press, New York.
Engels, J., and Schlaeger, E.-J. (1977). *J. Med. Chem.* **20**, 907–911.
Gerlt, J. A., and Coderre, J. A. (1980). *J. Am. Chem. Soc.* **103**, 4531–4533.
Gerlt, J. A., Coderre, J. A., and Wolin, M. S. (1980a). *J. Biol. Chem.* **255**, 331–334.
Gerlt, J. A., Mehdi, S., Coderre, J. A., and Rogers, W. O. (1980b). *Tetrahedron Lett.* **21**, 2385–2388.
Gorenstein, D. G., and Rowell, R. (1980). *J. Am. Chem. Soc.* **102**, 6165–6166.
Hansen, D. E., and Knowles, J. R. (1981). *J. Biol. Chem.* **256**, 5967–5969.
Jarvest, R. L., and Lowe, G. (1980). *J. Chem. Soc., Chem. Commun.* pp. 1145–1147.
Jarvest, R. L., and Lowe, G. (1981a). *Biochem. J.* **199**, 273–276.
Jarvest, R. L., Lowe, G. (1981b). *Biochem. J.* **199**, 447–451.
Jarvest, R. L., and Lowe, G., and Potter, B. V. L. (1981a). *J. Chem. Soc., Perkin Trans. 1* pp. 3186–3195.
Jarvest, R. L., Lowe, G., and Potter, B. V. L. (1981b). *Biochem. J.* **199**, 427–432.

Jarvest, R. L., Lowe, G., Baraniak, J., and Stec, W. J. (1982). *Biochem. J.* **203**, 461–470.
Jones, S. R., Kindman, L. A., and Knowles, J. R. (1978). *Nature (London)* **275**, 564–565.
Kellerman, G. M. (1958). *J. Biol. Chem.* **231**, 427–443.
Knowles, J. R. (1980). *Annu. Rev. Biochem.* **49**, 877–919.
Kurashina, Y., Takai, K., Suzuki-Hori, C., Okamoto, H., and Hayaishi, O. (1947). *J. Biol. Chem.* **249**, 4824–4828.
Lesiak, K., and Stec, W. J. (1978). *Z. Naturforsch., B: Anorg. Chem., Org. Chem.* **33B**, 782–785.
Lesnikowski, Z. J., Stec, W. J., Zielinski, W. S., Adamiak, D., and Saenger, W. (1981). *J. Am. Chem. Soc.* **103**, 2862–2863.
Lowe, G., and Potter, B. V. L. (1981a). *Biochem. J.* **199**, 227–233.
Lowe, G., and Potter, B. V. L. (1981b). *Biochem. J.* **199**, 693–698.
Lowe, G., and Potter, B. V. L. (1982). *Biochem. J.* **201**, 665–668.
Lowe, G., and Sproat, B. S. (1978). *J. Chem. Soc., Chem. Commun.* pp. 565–566.
Lowe, G., Potter, B. V. L., Sproat, B. S., and Hull, W. E. (1979). *J. Chem. Soc., Chem. Commun.* pp. 733–735.
Lowe, G., Cullis, P. M., Jarvest, R. L., Potter, B. V. L., and Sproat, B. S. (1981). *Philos. Trans. R. Soc. London. Ser. B* **293**, 75–92.
Lowe, G., Tansley, G., and Cullis, P. M. (1982). *J. Chem. Soc., Chem. Commun.* pp. 595–598.
Lowe, G., Sproat, B. S., Tansley, G., and Cullis, P. M. (1983). *Biochemistry* **22**, 1229–1236.
Lutz, O., Nolle, A., and Staschewski, D. (1978). *Z. Naturforsch., A* **33A**, 380–382.
Mehdi, S., and Gerlt, J. A. (1981a). *J. Am. Chem. Soc.* **103**, 7018–7020.
Mehdi, S., and Gerlt, J. A. (1981b). *J. Biol. Chem.* **256**, 12164–12166.
Mehdi, S., and Gerlt, J. A. (1982). *J. Am. Chem. Soc.* **104**, 3223–3225.
Mehdi, S., Coderre, J. A., and Gerlt, J. A. (1981). *ACS Symp. Ser.* **171**, 109–114.
Merritt, E. A., Sundaralingam, M., Cornelius, R. D., and Cleland, W. W. (1978). *Biochemistry* **17**, 3274–3278.
Pliura, D. H., Schomburg, D., Richard, J. P., Frey, P. A., and Knowles, J. R. (1980). *Biochemistry* **19**, 325–329.
Pollard-Knight, D., Potter, B. V. L., Cullis, P. M., Lowe, G., and Cornish-Bowden, A. (1982). *Biochem. J.* **201**, 421–423.
Potter, B. V. L., Connolly, B. A., and Eckstein, F. (1983). *Biochemistry* **22**, 1369–1377.
Richard, J. P., and Frey, P. A. (1982). *J. Am. Chem. Soc.* **103**, 3476–3481.
Saini, M. S., Buchwald, S. L., Van Etten, R. L., and Knowles, J. R. (1981). *J. Biol. Chem.* **256**, 10453–10455.
Sammons, R. D., and Frey, P. A. (1982). *J. Biol. Chem.* **257**, 1138–1141.
Senter, P. D., Eckstein, F., Mülsch, A., and Böhme, E. (1983). *J. Biol. Chem.* **258**, 6741–6745.
Stec, W. J., and Okruszek, A. (1975). *J. Chem. Soc., Perkin Trans. 1* pp. 1828–1832.
Stec, W. J., Okruszek, A., Lesiak, K., Uznanski, B., and Michalski, J. (1976). *J. Org. Chem.* **41**, 227–233.
Takai, K., Kurashina, Y., Suzuki-Hori, C., Okamoto, H., and Hayaishi, O. (1974). *J. Biol. Chem.* **249**, 1965–1972.
Tsai, M. D. (1979). *Biochemistry* **18**, 1468–1472.
Usher, D. A., Richardson, D. I., and Eckstein, F. (1970). *Nature (London)* **228**, 663–665.
Wang, J. Y. J., Clegg, D. O., and Koshland, D. E. (1981). *Proc. Natl. Acad. Sci. U.S.A.* **78**, 4684–4688.
Westheimer, F. H. (1980). *In* "Rearrangements in Ground and Excited States" (P. de Mayo, ed.), Vol. 2, pp. 229–271. Academic Press, New York.
Westheimer, F. H. (1981). *Chem. Rev.* **81**, 313–326.

CHAPTER 8

# DNA and RNA Conformations

*Chi-Wan Chen*
*Jack S. Cohen*

Developmental Pharmacology Branch
National Institute of Child Health and Human Development
National Institutes of Health
Bethesda, Maryland

| | |
|---|---:|
| I. Introduction | 233 |
| II. Helix–Coil Transitions | 234 |
|     A. RNA | 235 |
|     B. DNA | 237 |
| III. Sequence Dependence of Double-Stranded DNA Conformations | 239 |
|     A. B-DNA | 240 |
|     B. Alternating B-DNA | 245 |
|     C. C-DNA | 246 |
|     D. A-DNA | 249 |
|     E. Z-DNA | 250 |
| IV. Conformational Transitions of Double-Stranded DNA | 253 |
|     A. Salt-Induced Transitions | 253 |
|     B. Ethanol-Induced B-to-A Transitions | 255 |
| V. Dynamic Behavior of RNA and DNA | 255 |
| VI. Nucleosomal DNA | 256 |
| VII. Biological and Genetic Significance | 259 |
|     References | 260 |

## I. Introduction

$^{31}$P-NMR spectroscopy has proved to be a very useful technique in probing the secondary structures of DNA and RNA in solution. It owes its success to four factors:

1. The convenient NMR properties of the $^{31}$P nucleus (i.e., spin $\frac{1}{2}$ and 100% natural abundance)
2. The sensitivity of $^{31}$P-NMR chemical shifts toward changes in

O—P—O bond angles and P—O torsional angles (Gorenstein, 1981)
3. The development of Fourier-transform NMR methods, which to some extent make up for the low sensitivity of the $^{31}$P nucleus
4. The increasing availability of synthetic DNA and RNA of defined sequences in varying lengths

Thus, one can conveniently use $^{31}$P-NMR spectroscopy to follow the helix–coil transitions of RNA and DNA, as well as to examine the various secondary structures of DNA, to monitor the transitions between these structures, to evaluate the dynamic properties of RNA and DNA, and, it is hoped, to extract some information about the biological and/or genetic importance of the different base sequences and their corresponding conformations.

Polymorphism of double-stranded DNA has been observed in fibers (Leslie *et al.*, 1980) and in solutions (Ivanov *et al.*, 1973) for some time. Four major forms of DNA secondary structures were characterized, namely, A, B, C, and D (for a review, see Zimmerman, 1982). Whereas models of DNA secondary structures might differ depending on the base sequences, the counterion, and the water content, their double helices were all considered to be right-handed and held together through classical Watson and Crick base-pairing with a uniform sugar phosphate backbone. More recent circular-dichroism (CD) studies of poly(dGdC)·poly(dGdC) (Pohl and Jovin, 1972) and poly(dG-m$^5$dC)·poly(dG-m$^5$dC) (Behe and Felsenfeld, 1981), X-ray studies of d(CGCG) (Drew *et al.*, 1980), d(CGCGCG) (Wang *et al.*, 1979, 1981), and d(ATAT) (Viswamitra *et al.*, 1978) crystals, and $^{31}$P-NMR (Patel *et al.*, 1979; Simpson and Shindo, 1979, 1980; Cohen *et al.*, 1981; Cohen and Chen, 1982; Chen and Cohen, 1983; Chen *et al.*, 1983) and $^{13}$C-NMR (Shindo, 1981; Chen *et al.*, 1983) studies of polydeoxynucleotides have significantly altered our view of DNA structures. It has been shown that not only can DNA exist with both right- and left-handedness, but also the secondary structure along the sugar phosphate backbone can be nonuniform, depending on the base sequence and the environment. A majority of the studies to be discussed in this chapter have utilized synthetic polymers of defined sequences because these polymers enable the sequence-dependent structural variations to be studied most effectively.

## II. Helix–Coil Transitions

Gorenstein (see review, 1981) has shown that $^{31}$P-NMR chemical shifts in phosphate esters can serve as a direct probe for O—P—O bond angles and P—O torsional angles $\omega$, $\omega'$ (for definition, see Sundaralingam, 1969).

Molecular orbital calculations (Gorenstein and Kar, 1975; Prado et al., 1979) predicted that the $^{31}$P chemical shift of a phosphate diester in a gauche-gauche (g,g) conformation should be 3-6 ppm upfield from that in a trans-gauche (t,g) conformation. Furthermore, it has been found through molecular orbital calculations and comparisons with X-ray structures (Gorenstein et al., 1976a; Perahia and Pullman, 1976) that a reduction in O—P—O bond angles is coupled to a change from a gauche to a nongauche or trans conformation. Although the molecular flexibility is more restricted and the variation in bond angles and torsional angles is more limited in polynucleotides and polydeoxynucleotides, a similar correlation between $^{31}$P chemical shifts and phosphodiester geometry can be established. Therefore, $^{31}$P-NMR spectroscopy provides an alternate means to monitor the helix-coil transition of RNA and DNA. The melting temperatures this method yields are generally consistent with those obtained by UV spectroscopy (Fig. 1) (Gorenstein et al., 1982; Patel, 1979; Patel and Canuel, 1979; Shindo et al., 1980a) and $^1$H-NMR spectroscopy (which monitors the helical base-pairing and base-stacking interactions; Patel, 1977; Patel and Canuel, 1979) or other conventional techniques.

## A. RNA

A 0.7- to 1.3-ppm downfield shift has been observed for a variety of oligonucleotides and polynucleotides with increasing temperatures (Gorenstein et al., 1976b). This corresponds to a change from the all g,g phosphodiester conformation in the stacked helical structure (Sundaralingam, 1969; Day et al., 1973) to a mixture of g,g and t,g conformations in the unstacked, random-coil structure.

In a 1:1 or 1:2 mixture of poly(A) and oligo(U) (Fig. 1) (Gorenstein, et al., 1982, Figs. 2 and 6; Gorenstein and Goldfield, Chapter 10, Fig. 3), three $^{31}$P signals are observed: a sharp peak corresponding to oligo(U), a moderately sharp peak [0.2 ppm upfield from oligo(U)] corresponding to poly(A), and a broad peak 0.2-0.5 ppm further upfield assigned to poly(A)·[oligo(U)]$_n$ and poly(A)·2[oligo(U)]$_n$. As the temperature increases, the signal intensity of the double/triple helix drops, and a sharp, cooperative transition is observed at 50°C. The resonances of the single helices then shift downfield noncooperatively by ~1 ppm when temperature reaches 85°C. The observation of separate signals representing both the single helices and the double/triple helices suggests a slow-exchange process on the NMR time scale between the two states with a rate $< 125^{-1}$ (rate $< 2\pi\Delta v$, $\Delta v = 20$ Hz at 32.4 MHz) or a lifetime $\tau > 8$ ms [$\tau > 1/(2\pi\Delta v)$].

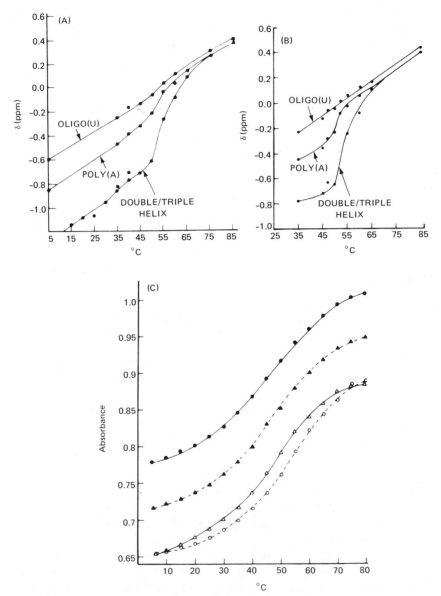

**Fig. 1.** Comparison of $^{31}$P-NMR and UV melting curves for mixtures of poly(A) and oligo(U) in 0.2 $M$ NaCl, 10 m$M$ cacodylate, and 1 m$M$ EDTA at pH 7.0. (A, B) $^{31}$P-NMR melting curves for 1:1 and 1:2 mixtures of poly(A) and oligo(U), respectively. (C) UV (260 nm) melting curves of 1:1(▲), 1:2(△), 1:3(○), and 2:1(●) mixtures of poly(A) and oligo(U). Reproduced with permission from Gorenstein *et al.*, 1982. Copyright 1982 American Chemical Society.

Breslauer et al. (1975) have also reported a lifetime greater than 14 ms for the melting of double-stranded $A_7U_7$ in 1 $M$ NaCl.

Phosphorus-31 signals of self-complementary dimers (e.g., ApU and GpC) are 0.1–0.2 ppm upfield from those of their corresponding homodimers (i.e., ApA, UpU, GpG, CpC) (Gorenstein et al., 1976b). These self-complementary dimers are apparently more constrained to the g,g conformation as a result of miniature double-helix formation, although the chemical-shift difference is small owing to the small population of the double helix present (probably < 10% at 0°C; Krugh et al., 1976).

## B. DNA

The temperature dependence of $^{31}$P-NMR chemical shifts has also been monitered for oligodeoxynucleotide duplexes, for example, d(CCGG), d(GGCC), (Patel, 1977), d(GCGC) (Patel, 1979a), d(CGCGCG) (Patel, 1976), d(GGAATTCC) (Patel and Canuel, 1979), and d(CGCGAATTCG-CG) (Patel et al., 1982a); polydeoxynucleotides, for example, poly(dAdT)· poly(dAdT) (Patel and Canuel, 1976; Simpson and Shindo, 1979), and other alternating purine–pyrimidine sequences (Cohen et al., 1981); and natural DNA (Hanlon et al., 1976; Yamada et al., 1978; Mariam and Wilson, 1979; Temussi et al., 1979; Shindo et al., 1980a). In general, the $^{31}$P signals sharpen with increasing temperature and shift downfield by up to 1 ppm.

Patel et al. (1982a) have observed that the 11 phosphodiesters in d(CGCGAATTCGCG) duplex exhibit eight partially resolved $^{31}$P resonances that spread over a 0.45-ppm chemical-shift range at 31°C, indicating a distribution of O—P—O bond angles and P—O torsional angles along the sequence. During melting at 71°C, the eight signals shift downfield and collapse into three major peaks spreading over a 0.25-ppm range, suggestive of an increase in the population of t,g conformations and a narrower distribution of O—P—O bond angles and P—O torsional angles.

Synthetic poly(dAdT)·poly(dAdT) (Patel and Canuel, 1976; Patel, 1979b), which apparently had a wide distribution of long-chain molecules [perhaps >800 base pairs (bp)], showed a $^{31}$P signal that was too broad (linewidth ~600 Hz at 146 MHz) below 50°C to be useful for helix–coil transition study. Simpson and Shindo (1979), on the other hand, have demonstrated that a 145-bp poly(dAdT)·poly(dAdT) prepared from semisynthetic chromatin core particles gives rise to a sharp, partially resolved, resonance (linewidth ~64 Hz at 109 MHz) even at 21°C (Fig. 2). The significant reduction in the linewidth of 145-bp poly(dAdT)·poly(dAdT) can be attributed to a reduction in the molecular size, which results in a

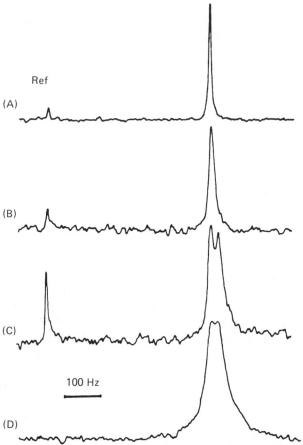

**Fig. 2.** Temperature dependence of the $^{31}$P-NMR spectrum at 109.3 MHz of 145-bp poly(dAdT)·poly(dAdT) in 10 m$M$ Tris-HCl and 1 m$M$ EDTA pH 8.0 and at (°C): A, 53; B, 39; C, 30; D, 21. Chemical shifts are upfield from internal reference trimethyl phosphate (TMP). Reproduced with permission from Simpson and Shindo, 1979. Copyright 1979 IRL Press.

more rapid molecular tumbling. Hanlon et al. (1976) have shown that the $^{31}$P linewidth of calf thymus DNA (~14,000 bp) at 36 MHz is reduced significantly when sonicated (~260 bp).

Several alternating purine–pyrimidine polydeoxynucleotides, for example, poly(dAdT), poly(dA-br$^5$dU), and poly(dIdC) duplexes, exhibit a splitting in their $^{31}$P-NMR signals at room temperature (Shindo et al., 1979; Cohen et al., 1981). The doublets observed suggest that these sequences adopt a so-called alternating B-form, in which two distinct phosphodiester conformations alternate along the helical backbone (details discussed in Section III). In some cases, the doublet shifts upfield on cooling and collapses to a broad singlet, indicating a double helix with a uniform phosphodiester backbone and an increase in the g,g population. Conversely, when heated above the melting temperature, the doublet shifts downfield and collapses to a sharp singlet, indicative of a random coil with an increase in the t,g population.

Mariam and Wilson (1979) have shown that both the helical and coil forms of calf thymus and salmon sperm DNA are observed in their $^{31}$P-NMR spectra in the melting region ($\sim 70°C$), which suggests a slow-exchange process between the two forms with an interconversion rate $\ll 36$ s$^{-1}$. At least three different signals are also detected at temperatures above $T_m$, indicating the coexistence of three conformational states.

Patel et al. (1982b,c,d) have used $^1$H- and $^{31}$P-NMR spectroscopies and differential scanning calorimetry to investigate structural and energetic features of the helix–coil transitions of d(CGTGAATTCGCG) duplex, which contains a dG·dA mismatch, of d(CGCAGAATTCGCG) duplex with an extra noncomplementary dA residue, and of d(GAATTCGCG) duplex, which contains a d(GAATTC) hexanucleotide duplex flanked by d(GCG) trinucleotide ends. They were able to show, for example, that the adenosine of the extra dA residue in d(CGCAGAATTCGCG) duplex is stacked into the helix and the base-pairing on either side of the extra stacked dA residue remains intact; that the $^{31}$P resonance of the extended phosphodiester linkage opposite the extra stacked dA residue is shifted downfield by $\sim 0.7$ ppm from the rest of the $^{31}$P resonances; and that the melting temperature is reduced by 19° compared to the corresponding fully base-paired dodecamer without the extra dA residue.

## III. Sequence Dependence of Double-Stranded DNA Conformations

X-Ray diffraction studies of natural (Fuller et al., 1965) and synthetic (Leslie et al., 1980) DNA fibers have shown that depending on its base sequence, water content, and counterion, DNA can adopt at least four different conformations: A, B, C, and D forms. Since the early 1970s it has become evident through CD spectroscopy (Pohl and Jovin, 1972; Ivanov et al., 1973) that DNA secondary structure also varies in solution. However, owing to the limited resolution of X-ray fiber diffraction, the DNA models thus derived were sequence-averaged and each had a uniform backbone conformation with a mononucleotide repeat unit (Fig. 3). Circular dichroism also has the drawback that detailed structures of DNA in solution cannot be

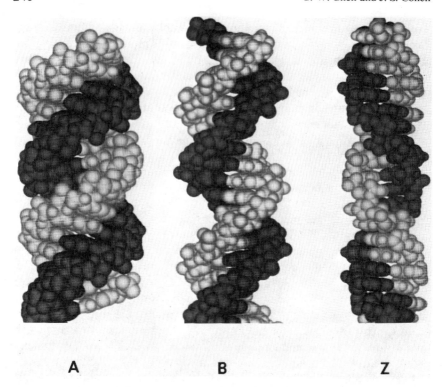

**Fig. 3.** Space-filling molecular models of A-, B-, and Z-DNA. The coordinates used for A- and B-DNA were derived from the fiber X-ray diffraction study by Arnott *et al.* (1980), and those for Z-DNA were determined from single-crystal X-ray analysis by Wang *et al.* (1981).

established by this method. X-Ray studies of oligonucleotide crystals (Viswamitra *et al.*, 1978; Wang *et al.*, 1979; Crawford *et al.*, 1980; Drew *et al.*, 1980; Wing *et al.*, 1980; Dickerson and Drew, 1981; Dickerson *et al.*, 1982) and $^{31}$P-NMR (Patel, 1979; Patel *et al.*, 1979; Shindo *et al.*, 1979; Cohen *et al.*, 1981; Chen and Cohen, 1982; Chen and Cohen, 1983; Chen *et al.*, 1983) and $^{13}$C-NMR (Shindo, 1981; Chen *et al.*, 1983) studies of oligo- and polydeoxynucleotides have provided evidence for nonuniform sugar phosphate backbone conformations of DNA and their dependence on base sequence.

## A. B-DNA

Simpson and Shindo (1980) have compared the $^{31}$P-NMR spectra at 109.3 MHz of three 145-bp double-stranded DNAs (Fig. 4), namely, poly(dGdC)·poly(dGdC), poly(dAdT)·poly(dAdT), and random-sequence chicken erythrocyte DNA obtained from nucleosome core particles. In

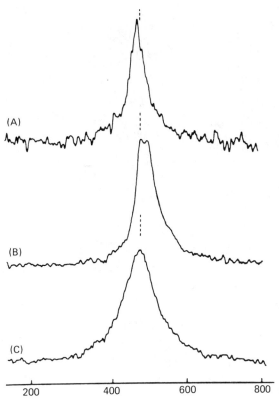

**Fig. 4.** ³¹P-NMR spectra at 109.3 MHz of 145-bp length (A) poly(dGdC)·poly(dGdC), (B) poly(dAdT)·poly(dAdT), and (C) chicken erythrocyte DNA in 10 m$M$ Tris-HCl and 1 m$M$ EDTA at pH 8 and 20°C. Chemical shifts in Hertz are upfield positive from internal TMP. Reproduced with permission, from Simpson and Shindo, 1980. Copyright 1980 IRL Press.

low-salt solution both poly(dGdC)·poly(dGdC) and random-sequence DNA give rise to a symmetrical single peak at about −4.26 ppm [relative to trimethyl phosphate (TMP); to convert to 85% $H_3PO_4$, add 3.71 ppm]; however, the linewidth for the random-sequence DNA (103 Hz) is much larger than that for poly(dGdC)·poly(dGdC) (41 Hz). Because it is known that the two DNAs are of the same length and are both of the B-form family, the broader linewidth observed for the random-sequence DNA is clearly a result of chemical-shift dispersion along the phosphodiester backbone. An analysis of the ³¹P resonance linewidth of the 145-bp chicken erythrocyte DNA as a function of frequency (Shindo, 1980) revealed that there is 0.5-ppm chemical-shift dispersion contribution to the linewidth. An essentially linear relationship has been found between the frequency of observation and the linewidth of the broad ³¹P resonance of the nucleosomal DNA from chicken erythrocytes (Fig. 5; Shindo et al., 1980a). Shindo et al.

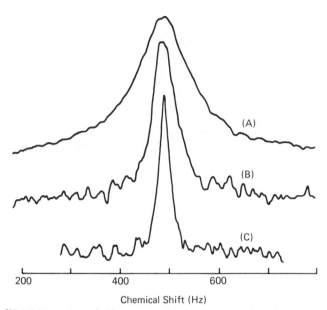

**Fig. 5.** $^{31}$P-NMR spectra of chicken erythrocyte nucleosomal DNA in the core particles (~1 mg ml$^{-1}$) in 5 m$M$ Tris-HCl and 0.1 m$M$ EDTA at pH 8 and 19°C at three observing frequencies: A, 109.3 MHz; B, 40.3 MHz; C, 24.3 MHz. The linewidths are: A, 130 Hz; B, 45 Hz; C, 25 Hz, with an approximately linear dependence on field strength (Shindo et al., 1980a). Chemical shifts in Hertz are upfield positive from internal TMP. Reproduced with permission, from Cohen and Chen, 1982. Copyright 1982 American Chemical Society.

(1980a) attributed this effect to chemical-shift dispersion, which suggests the presence of heterogeneity of the phosphodiester conformations in the DNA backbone. This backbone conformational variance was also observed in the $^{31}$P spectra of B-form DNA fibers (Shindo et al., 1980b). In contrast, the $^{31}$P resonance of poly(dAdT)·poly(dAdT) appears as a doublet with a total linewidth of 64 Hz and a separation of 24 Hz. The average linewidth for each phosphorus peak of this doublet is then 40 Hz, which is nearly identical to the linewidth for poly(dGdC)·poly(dGdC).

As mentioned, poly(dAdT), poly(dA-br$^5$dU), and poly(dIdC) duplexes all exhibit a partially resolved doublet at room temperature (Figs. 2 and 6, Table I), corresponding to an alternating structure (Cohen et al., 1981). Poly(dAdU)·poly(dAdU) and poly(dGdC)·poly(dGdC), on the other hand, show only a singlet at room temperature (Fig. 6, Table I; Cohen et al., 1981). The singlet resonance suggests that although poly(dAdU) and poly(dGdC) duplexes contain alternating purine–pyrimidine sequences, they exist in a uniform B form because either the two types of phosphodiesters are equivalent or are in rapid exchange on the NMR time scale. This is in con-

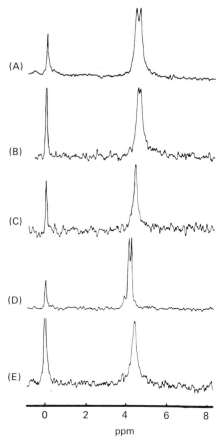

**Fig. 6.** $^{31}$P-NMR spectra at 109.3 MHz of sonicated alternating purine–pyrimidine deoxynucleotide duplexes: A, poly(dAdT)·poly(dAdT) at 24°C; B, poly(dA-br$^5$dU)·poly(dA-br$^5$dU) at 30°C; C, poly(dAdU)·poly(dAdU) at 31°C; D, poly(dIdC)·poly(dIdC) at 39°C; E, poly(dGdC)·poly(dGdC) at 23°C; all in 0.1 $M$ NaCl and 0.05 m$M$ EDTA, pH 6–7, except (B) in 5 m$M$ Tris and 0.1 m$M$ EDTA. Chemical shifts are upfield positive from internal TMP. Reproduced with permission, from Cohen *et al.*, 1981. Copyright 1981 American Chemical Society.

trast to the alternating B form Lomonossoff *et al.* (1981) proposed for poly(dGdC)·poly(dGdC) based on their DNase I digestion study.

The homopolymer poly(dA)·poly(dT) also gives a singlet $^{31}$P resonance (Table I; Cohen *et al.*, 1981) similar to those of random-sequence DNA, poly(dGdC)·poly(dGdC) (Simpson and Shindo, 1980), and poly(dAdU)·poly(dAdU) (Cohen *et al.*, 1981) but at slightly higher field,

## TABLE I

### $^{31}$P-NMR Parameters of Sonicated Polydeoxynucleotides

| Polydeoxynucleotide | T (°C) | Salt or organic solvent present[a] Type | Concentration | Chemical shifts[b] (ppm) | $\Delta\nu^c$ (Hz) | Reference[d] |
|---|---|---|---|---|---|---|
| (dAdT)·(dAdT) | 24 | NaCl | 0.1 M | $\begin{cases} -4.346 \\ -4.542 \end{cases}$ | 21 | 1 |
|  | 24 | NaCl | 4.0 M | $\begin{cases} -4.041 \ (-4.317) \\ -4.394 \ (-4.670) \end{cases}$ | 39 | 1 |
|  | 37 | NaCl | 0.1 M | $\begin{cases} -4.273 \\ -4.477 \end{cases}$ | 22 | 2 |
|  | 37 | CsF | 3.9 M | $\begin{cases} -4.233 \ (-4.339) \\ -4.831 \ (-4.937) \end{cases}$ | 65 | 2 |
|  | 37 | CsCl | 5.0 M | $\begin{cases} -4.342 \ (-4.278) \\ -4.747 \ (-4.683) \end{cases}$ | 44 | 2 |
|  | 37 | LiCl | 5.0 M | $\begin{cases} -4.342 \ (-4.278) \\ -4.747 \ (-4.672) \end{cases}$ | 41 | 2 |
|  | 37 | MgCl$_2$ | 1.0 M | $\begin{cases} -4.683 \ (-4.220) \\ -4.884 \ (-4.422) \end{cases}$ | 22 | 2 |
|  | 37 | EtOH | 50% | $-3.724 \ (-4.248)$ |  | 2 |
| (dA-br$^5$dU)·(dA-br$^5$dU) | 32 | Tris | 5 mM | $\begin{cases} -4.456 \\ -4.594 \end{cases}$ | 15 | 1 |
| (dAdU)·(dAdU) | 30 | NaCl | 0.1 M | $-4.467$ |  | 1 |
| (dIdC)·(dIdC) | 38 | NaCl | 0.1 M | $\begin{cases} -4.091 \\ -4.212 \end{cases}$ | 13 | 1 |
| (dGdC)·(dGdC) | 23 | NaCl | 0.1 M | $-4.264$ |  | 1 |
|  | 23 | NaCl | 4.0 M | $\begin{cases} -2.817 \\ -4.282 \end{cases}$ | 160 | 1 |
| (dG-m$^5$dC)·(dG-m$^5$dC) | 50 | Tris | 5 mM | $\begin{cases} -4.087 \\ -4.329 \end{cases}$ | 26 | 3 |
|  | 50 | NaCl | 1.0 M | $\begin{cases} -2.865 \\ -4.134 \end{cases}$ | 139 | 3 |
|  | 50 | CsF | 1.2 M | $\begin{cases} -3.018 \\ -4.318 \end{cases}$ | 142 | 3 |
|  | 50 | MgCl$_2$ | 1.5 mM | $\begin{cases} -3.139 \\ -5.177 \end{cases}$ | 223 | 3 |
|  | 50 | Co(NH$_3$)$_6$Cl$_3$ | 0.5 mM | $\begin{cases} -2.807 \\ -4.134 \end{cases}$ | 145 | 3 |
| (dA)·(dT) | 37 | NaCl | 0.1 M | $-4.467$ |  | 4 |
|  | 37 | EtOH | 50% | $-3.710 \ (-4.234)$ |  | 4 |

[a] All samples also contained Tris and EDTA.
[b] Upfield from internal reference trimethyl phosphate (TMP). To convert to external 85% H$_3$PO$_4$, add 3.71 ppm. Actual values from the computer output are quoted. Braces indicate doublet resonances. Values in parentheses are corrected for the ionic–hydration effect by using (dT)$_9$ as a control.
[c] Separation of the doublet to the nearest Hz.
[d] References: 1, Cohen et al. (1981); 2, Chen and Cohen (1983); 3, Chen et al. (1983).

presumably owing to a higher population in the g⁻,g⁻ conformation. It might correspond to the B' form, a minor variant of B form, that Leslie et al. (1980) observed in the fibers of poly(dA)·poly(dT). No ³¹P-NMR spectra have yet been reported for poly(dG)·poly(dC) because of its extremely low solubility in water.

## B. Alternating B-DNA

Viswamitra et al. (1978, 1982) have obtained the crystal structure of the tetradeoxynucleotide d(ATAT) that contains right-handed, double-helical segments with dA and dT sugar puckers in the C-3'-endo and C-2'-endo conformations, respectively, and dTpdA and dApdT phosphodiesters in the t,g⁻ and g⁻,g⁻ conformations, respectively. This prompted Klug et al. (1979) to propose an alternating B form for poly(dAdT)·poly(dAdT), in which every second phosphodiester linkage is different from that of the classical B form. Shindo et al. (1979) and Simpson and Shindo (1979) observed a partially resolved doublet of equal area in the ³¹P-NMR spectra of 145-bp poly(dAdT)·poly(dAdT) (Fig. 2). The same phenomenon was reported for sonicated poly(dAdT)·poly(dAdT) by Cohen et al. (1981). The separation of the doublet (24 Hz) is too large to be accounted for by ring-current effects alone, which predicts a 4-Hz difference between the two phosphodiester signals in the classical B form (Shindo et al., 1979). The doublet can only be explained by the presence of two distinct phosphodiester conformations that differ in their P—O torsional angles $\omega$ and $\omega'$ (Gorenstein, 1981). The upfield component at −4.47 ppm is thus assigned to dApdT phosphodiester in the g⁻,g⁻ conformation and the downfield component at −4.27 ppm to dTpdA phosphodiester in the t,g⁻ conformation (Chen and Cohen, 1983). A ³¹P doublet was also observed for poly(dAdT)·poly(dAdT) fibers when oriented parallel to the magnetic field, whereas only a broad singlet was observed for salmon sperm DNA fibers (Shindo and Zimmerman, 1980).

Several other synthetic DNAs with alternating purine–pyrimidine sequences, for example, poly(dA-br⁵dU)·poly(dA-br⁵dU), poly(dIdC)·poly(dIdC) (Fig. 6, Table I; Cohen et al., 1981), and poly(dG-m⁵dC)·poly(dG-m⁵dC) (Fig. 7A,B, Table I; Chen et al., 1983) also exhibit two partially resolved ³¹P signals of equal area. This again implies the existence of two distinct alternating phosphodiester backbone conformations for these DNAs in solution. Of particular interest is poly(dG-m⁵dC)·poly(dG-m⁵dC) that affords a doublet ($\Delta\nu = 23-32$ Hz at 109.3 MHz) in low- and medium-salt solutions (Fig. 7A,B) whereas its unmethylated analog, poly(dGdC)·poly(dGdC), of the same molecular size gives a singlet under similar conditions (Figs. 4 and 6, Table I).

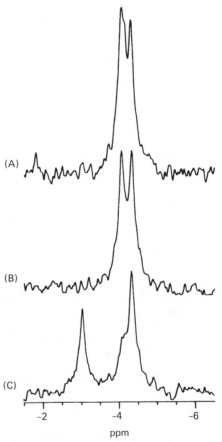

**Fig. 7.** Effect of CsF concentration on $^{31}$P-NMR spectra at 109.3 MHz of poly(dG-m$^5$dC)·(dG-m$^5$dC) in 5 m$M$ Tris-HCl and 0.1 m$M$ EDTA, at pH 8 and 50°C. CsF concentration ($M$): A, 0; B, 0.6; C, 1.2. The copolymer obtained from P. L. Biochemicals had been sonicated for 3 h before use. Chemical shifts are upfield negative from internal TMP (Chen *et al.,* 1983).

## C. C-DNA

Although Patel and Canuel (1976) did not detect a resolved $^{31}$P doublet for poly(dAdT)·poly(dAdT) in low-salt solution because of its large molecular size (discussed previously), their later study of the same copolymer at high CsF concentration did result in a well-resolved doublet (Patel *et al.,* 1981). They attributed the spectral change to a noncooperative transition from a regular B form in low-salt solution to an alternating B form at high CsF concentration. Kypr and co-workers (Kypr *et al.,* 1981; Vorlickova *et*

# 8. DNA and RNA Conformations

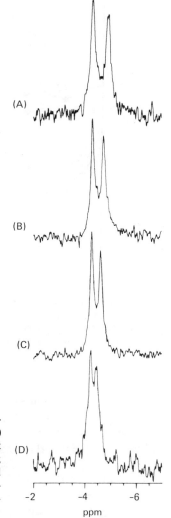

**Fig. 8.** Effect of CsF concentration on $^{31}$P-NMR spectra at 109.3 MHz of sonicated poly(dAdT)·poly(dAdT) in 0.1 $M$ NaCl, 50 m$M$ Tris-HCl, and 1 m$M$ EDTA, at pH 7 and 37°C. CsF concentration ($M$): A, 4; B, 2; C, 1; D, 0. Chemical shifts are upfield negative from internal TMP (Chen and Cohen, 1983.) Reproduced, with permission, from Cohen and Chen, 1982. Copyright 1982 American Chemical Society.

*al.,* 1983) also observed a significant change in the $^{31}$P spectra of sonicated poly(dAdT)·poly(dAdT) from a partially resolved to a completely resolved doublet on addition of CsF. A systematic $^{31}$P-NMR study of salt effects on the secondary structure of sonicated polydeoxynucleotides has been carried out by Chen and Cohen (1983). Addition of monovalent salts [e.g., NaCl, CsF (Fig. 8), LiCl, and CsCl] causes the upfield component of the doublet of poly(dAdT)·poly(dAdT) to shift upfield (Fig. 9, Table I) relative to a single-stranded oligonucleotide control (dT)$_9$. The oligonucleotide (dT)$_9$ is

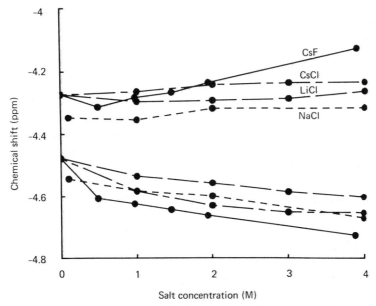

**Fig. 9.** $^{31}$P-NMR chemical shifts of poly(dAdT)·poly(dAdT) doublet as a function of salt content and concentration. Experimental conditions were identical to those in Fig. 8, except that curves for NaCl were obtained at 22°C. All chemical shifts have been corrected for the ionic effect observed in (dT)$_9$ (Chen and Cohen, 1983).

used to estimate the nonselective (i.e., sequence-independent), ionic, and/or hydration effects that are assumed to be the same for all polydeoxynucleotides. The net chemical-shift changes should then reflect the sequence-dependent conformational effects. The downfield component of the doublet of poly(dAdT)·poly(dAdT) shifts downfield when compared to (dT)$_9$, with increasing CsF concentration, but remains essentially constant in LiCl, NaCl, and CsCl (Fig. 9). These changes indicate a fast, noncooperative transition for the AT copolymer from an alternating B form to another alternating structure with a higher g$^-$,g$^-$ population and thus an increased helical winding angle (Gorenstein, 1981), presumably a variant of the so-called C form.

There have been a number of studies (Ivanov *et al.*, 1973; Zimmer and Luck, 1974) that attribute the linear changes in CD spectra of DNA on increasing LiCl and NaCl concentrations to a continuous change of the DNA structure toward a C form, although this interpretation has been disputed by Zimmerman and Pheiffer (1980). Other studies have shown that the CD spectral changes can be related to a change in the number of base pairs per turn of the double helix (Basse and Johnson, 1979) or a change in the helical winding angle (Chan et al., 1979). Wilson and Jones (1982) have demonstrated that there is a linear correlation between the downfield shifts of the $^{31}$P signal of DNA caused by intercalating drugs and the unwinding angles these drugs produce (see Fig. 8, Gorenstein and Goldfield, Chapter 10). It is therefore reasonable to assume that the selective upfield shifts of the dApdT signal correspond to an increase in the helical winding angle, which in turn corresponds to a C-like form. In a fiber diffraction study, Mohendrasingham et al. (1983) found that fibers of poly(dAdT)·poly(dAdT) gave D-type patterns over a wide range of conditions but found C-type patterns only at low salt and low humidity. However, it is difficult to simply correlate fiber diffraction studies with NMR solution studies given the sensitivity of these patterns to humidity effects.

Poly(dA)·poly(dT), on the other hand, shows little change in its $^{31}$P-NMR spectra relative to control (dT)$_9$ under similar conditions. This suggests that the AT homopolymer remains in the B form (or B' form) in the presence of monovalent cations. Poly(dGdC)·poly(dGdC) exhibits spectral and conformational changes in high-salt solution that are in marked contrast to the two AT-containing polymers, and they are discussed in Section III,E.

### D. A-DNA

On addition of ethanol, the upfield component of the doublet of poly(dAdT)·poly(dAdT) shifts downfield and collapses into the downfield component, while the downfield component remains unchanged relative to the control (dT)$_9$ (Fig. 10, Table I; Chen and Cohen, 1983). Because the downfield shifts are indicative of a decrease in the helical winding angle (Wilson and Jones, 1982), this copolymer apparently approaches a uniform A-like form with increasing ethanol concentration. This agrees well with the findings from X-ray (Zimmerman and Pheiffer, 1979) and CD (Ivanov et al., 1973) studies that show an ethanol-induced A form with a winding angle smaller than that in a B form. Although the copolymer precipitated at the concentration required for NMR studies (1–2 mg ml$^{-1}$) when ethanol concentration reached 55% and thus the transition could not be followed to completion, there is no doubt that the structure does indeed shift toward an A form within the range studied.

The $^{31}$P singlet resonance of poly(dA)·poly(dT) shifts downfield relative to (dT)$_9$ with added ethanol (Table I). Therefore, one can conclude that the AT homopolymer, like its copolymer counterpart, also exists in an A form in ethanol.

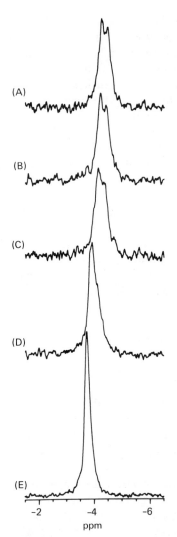

**Fig. 10.** Dependence of $^{31}$P-NMR spectra at 109.3 MHz of sonicated poly(dAdT)·poly(dAdT) on ethanol at 37°C. The concentration of ethanol is expressed in percent by volume: A, 0; B, 10; C, 20; D, 40; E, 50. All other experimental conditions were identical to those in Fig. 8 (Chen and Cohen, 1983).

## E. Z-DNA

Mitsui et al. (1970) reported an unusual X-ray diffraction pattern of ethanol-precipitated fiber of poly(dIdC)·poly(dIdC) and proposed a left-handed helical structure for this copolymer. Pohl and Jovin (1972) found that poly(dGdC)·poly(dGdC) underwent a salt-induced conformational transition in solution to a new form with an inverted CD spectrum and suggested that a left-handed helix could be a possible explanation. Subsequently, Pohl (1976) also observed the same CD spectral change for

## TABLE II

Salt Concentrations[a] at the Midpoints of the B-to-Z Transition for Poly(dGdC)·Poly(dGdC) and Poly(dG-m⁵dC)·Poly(dG-m⁵dC) from ³¹P-NMR Measurements[b]

| Salt | Poly(dGdC)·poly(dGdC) | Poly(dG-m⁵dC)·poly(dG-m⁵dC) |
|---|---|---|
| NaCl | 3000 (3000) | 700 |
| CsF | ≫ 3590 (~4800) | 800 |
| MgCl$_2$ | 900 (600) | 0.5 |
| Co(NH$_3$)$_6$Cl$_3$ | — | 0.1 (~1) |

[a] All concentrations are in millimolar units. Values in parentheses are salt concentrations at which precipitation occurred.
[b] Chen et al. (1983).

poly(dGdC)·poly(dGdC) on addition of ethanol. X-ray structural elucidation of d(CGCG) and d(CGCGCG) crystals (Wang et al., 1979; Crawford et al., 1980; Drew et al., 1980; Wing et al., 1980; Dickerson and Drew, 1981) has presented firm evidence that these oligomers can form a left-handed helix with a zigzag phosphodiester backbone, thus termed Z-DNA (Fig. 3). It was realized (Arnott et al., 1980) that such a left-handed Z form could exist in fibers of poly(dGdC)·poly(dGdC), as well as other polydeoxynucleotides containing alternating purine–pyrimidine sequence, for example, poly(dAdC)·poly(dGdT) and poly(dA-s⁴dT)·poly(dA-s⁴dT). CD studies of poly(dCdA)·poly(dTdG) seemed to suggest that this alternating polymer can undergo a CsF-induced (Vorlickova et al., 1982) or ethanol-induced (Zimmer et al., 1982) conformational transition from a B form to a Z form.

Patel et al. (1979) observed two well-resolved ³¹P resonances ($\Delta v = 1.5$ ppm) of approximately equal area for d(GCGCGCGC) in 4 $M$ NaCl solution but at that time attributed it to an alternating B-DNA conformation. Simpson and Shindo (1980) also detected a similar ³¹P doublet for 145-bp poly(dGdC)·poly(dGdC) in high-salt solution that was considered to correspond to the two distinct phosphodiester environments in the Z form. Cohen et al. (1981) demonstrated that sonicated poly(dGdC)·poly(dGdC) gave rise to the same characteristic doublet in high NaCl solution (Table I), of which the upfield component was assigned to dGpdC in the g,g conformation and the downfield component to dCpdG in the t,g conformation in the Z form. Although other salts are known to cause the B-to-Z transition in poly(dGdC)·poly(dGdC) (Pohl and Jovin, 1972; Vorlickova et al., 1980), only the LiCl-induced Z form has been characterized by ³¹P-NMR spectroscopy (Patel et al., 1982e). Attempts to monitor the B-to-Z transition for poly(dGdC)·poly(dGdC) in MgCl$_2$ or CsF solution by ³¹P-NMR spectroscopy always result in undesirable aggregation and thus line broadening before the transition takes place (Table II; Chen et al., 1983) owing to the relatively high concentration of the polymer (1 mg ml⁻¹) required.

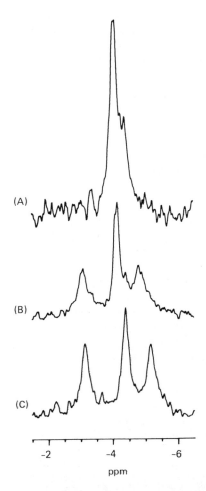

**Fig. 11.** Effect of $MgCl_2$ concentration on $^{31}$P-NMR spectra at 109.3 MHz of poly(dG-m$^5$dC)·poly(dG-m$^5$dC) in 5 mM Tris-HCl and 0.1 mM EDTA, at pH 8 and 50°C. $MgCl_2$ concentration (mM): A, 0.3; B, 0.6; C, 1.5. The copolymer prepared in this laboratory contained an impurity, possibly poly(dIdC)·poly(dG-m$^5$dC) hybrid, which appeared at about $-4$ to $-4.2$ ppm and constituted one-third of the total $^{31}$P-NMR intensity. Chemical shifts are upfield negative from internal TMP (Chen et al., 1983).

Behe and Felsenfeld (1981) have shown that the B-to-Z transition for poly(dG-m$^5$dC)·poly(dG-m$^5$dC) can be induced at much lower salt concentrations (close to physiological conditions) than those required for the unmethylated copolymer. This finding not only brings the B-to-Z transition within the realm of physiological significance but also allows extensive studies of the Z form in solution to be performed by NMR spectroscopy. Thus addition of NaCl, CsF, $MgCl_2$, or $Co(NH_3)_6Cl_3$ to poly(dG-m$^5$dC)·poly(dG-m$^5$dC) in solution causes its $^{31}$P spectra to change from a partially resolved doublet ($\Delta v = 23-32$ Hz at 109.3 MHz), indicating a right-handed alternating B form, to a well-separated doublet ($\Delta v = 140-220$ Hz at 109.3 MHz), indicating a left-handed, alternating Z form (Figs. 7 and 11, Table I; Chen et al., 1983). The upfield and downfield components correspond to g,g$^+$ and t,g$^+$ phosphodiester conformations (Wang et al., 1981), respectively. The midpoints of the transition with these salts derived from the NMR study (Table II) are in good agreement with those reported by Behe and Felsenfeld (1981) from CD measurements. Patel et al. (1982e) have also reported similar $^{31}$P spectra of poly(dG-m$^5$dC)· poly(dG-m$^5$dC) in low salt and 1.5 $M$ NaCl solutions.

## IV. Conformational Transitions of Double-Stranded DNA

It is apparent from Section III that various secondary structures are accessible to double-stranded DNAs, depending on their base sequences and their environment. Almost all of these conformations have their own characteristic $^{31}$P-NMR spectra in solution (Fig. 12). Reversible transitions can occur between certain conformations and can be brought about by changing the salt concentration or the water content (achieved by changing the relative amount of organic solvents, e.g., ethanol). With a few exceptions these processes can be successfully monitored by $^{31}$P-NMR spectroscopy as discussed. Some of the results are summarized in Table III.

### A. Salt-Induced Transitions

*1. B-to-C Transition*

Poly(dAdT)·poly(dAdT) undergoes a fast, noncooperative transition from a right-handed, alternating B form to a right-handed, alternating C form (Fig. 12, Table III) with increasing monovalent salt concentration (Figs. 8 and 9; Chen and Cohen, 1983). The $^{31}$P signal corresponding to dTpdA remains relatively unchanged with NaCl, LiCl, and CsCl and shifts slightly downfield with CsF, whereas the signal corresponding to dApdT shifts linearly upfield with all four salts. The linear dependence of the dApdT upfield shift on salt concentration strongly suggests a continuous increase in the helical winding angle for this copolymer and a transition through a continuum of structures within the C-form family. This is consistent with the results obtained by CD spectroscopy (Ivanov et al., 1973).

Divalent salts (e.g., MgCl$_2$) have no significant effect on the conformation of poly(dAdT)·poly(dAdT) (Chen and Cohen, 1983; see also Table I). Poly(dA)·poly(dT) assumes a right-handed B or B' form and does not appear to be affected by either monovalent or divalent salts (Table II; Chen and Cohen, 1983).

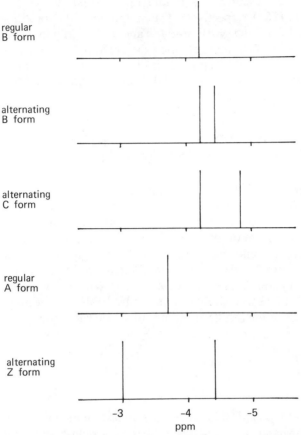

**Fig. 12.** Schematic presentation of $^{31}$P-NMR spectra of various DNA secondary structures. Typical examples are: regular B form, poly(dGdC)·poly(dGdC), in low salt solution; alternating B form, poly(dAdT)·poly(dAdT), in low salt solution; alternating C form, poly(dAdT)·poly(dAdT), in high monovalent salt solution; regular A form, poly(dAdT)·poly(dAdT), in high ethanolic solution; alternating Z form, poly(dGdC)·poly(dGdC), in high salt solution or poly(dG-m$^5$dC)·poly(dG-m$^5$dC) in relatively low to intermediate salt solution. Chemical shifts are upfield negative from internal TMP.

## 2. B-to-Z Transition

On addition of NaCl (Cohen *et al.*, 1981) or LiCl (Patel *et al.*, 1982e), poly(dGdC)·poly(dGdC) undergoes a slow cooperative transition from a right-handed B form to a left-handed Z form (Tables II and III), as evidenced by the replacement of half of the B form resonance at about −4.3 ppm by a new signal at −3 ppm. The methylated analog poly(dG-m$^5$dC)·poly(dG-m$^5$dC), which adopts a right-handed alternating B form in low-salt solution,

**TABLE III**

Comparison of DNA Conformations[a] among Different Base Sequences as Determined from $^{31}$P-NMR Studies[b]

| Base sequence | Low-salt conformation | Conformational transition | High-salt conformation | High-ethanol conformation |
|---|---|---|---|---|
| (dAdT)·(dAdT) | RaB | Noncooperative Cooperative[c] | Ra"C" | RrA |
| (dGdC)·(dGdC) | RrB | Cooperative | LaZ | |
| (dG-m⁵dC)·(dG-m⁵dC) | RaB | Cooperative | LaZ | |
| (dA)·(dT) | RrB′ | None Cooperative[c] | RrB′ | RrA |

[a] Notations used in this table: R, right-handed; L, left-handed; a, alternating; r, regular; B, B form; "C," C-like form; A, A form; Z, Z form; B′, B′ form.
[b] Chen and Cohen (1983).
[c] The transition was not followed to completion because of precipitation.

also undergoes the same type of salt-induced transition to a left-handed Z form (Figs. 7, 11 and 12, Table III) but at much lower salt levels (Table II; Chen et al., 1983).

### B. Ethanol-Induced B-to-A Transition

Addition of ethanol causes both poly(dAdT)·poly(dAdT) and poly(dA)·poly(dT) to change to an A form with decreased helical winding angle (Chen and Cohen, 1983). The two phosphodiesters in poly(dAdT)·poly(dAdT) that are of two distinct conformations in the absence of ethanol become equivalent in ~50% ethanol (Figs. 10 and 12).

## V. Dynamic Behavior of RNA and DNA

Several reports have investigated the dynamic behavior of RNA and DNA in solution through $^1$H-NMR (Early and Kearns, 1979; Hogan and Jardetzky, 1979, 1980), $^{31}$P-NMR (Klevan et al., 1979; Hogan and Jardetzky, 1979, 1980; Bolton and James, 1979, 1980; Keepers and James, 1982; Shindo, 1980; Shindo et al., 1980a; Opella et al., 1981 Hart, Chapter 17; James, Chapter 12), and $^{13}$C-NMR (Bolton, and James, 1979; Hogan and Jardetzky, 1980; Levy et al., 1981; Shindo, 1981; Keepers and James, 1982) relaxation measurements. It is generally agreed that large-amplitude internal motions in the nanosecond range are present in RNA and DNA, contrary to the previous belief that RNA and DNA behave like rigid rods. This was confirmed by Lipari and Szabo (1981) in a detailed analysis of the NMR relaxation data with a variety of models. However, they also concluded that the relaxation data are consistent with several physical pictures of the internal motions. Therefore, no single model can yet satisfactorily describe the local backbone flexibility and the conformational fluctuations existing in RNA and DNA.

It is pointed out in Section III that in low-salt solution, several alternating purine–pyrimidine polydeoxynucleotides exhibit an alternating B form whereas others exhibit a regular B form. The separation (10–32 Hz) in the $^{31}$P doublet (Table I; Cohen *et al.*, 1981), which represents the two distinct phosphodiester conformations in the alternating B form, clearly suggests that the interconversion between the two conformations is either nonexistent or slower than $60-190$ s$^{-1}$. The failure to observe a doublet for sequences such as poly(dGdC)·poly(dGdC) and poly(dAdU)·poly(dAdU) in low-salt solution indicates that their phosphodiester backbone is uniform or that the energy barrier between the two phosphodiester conformations is so small as to allow rapid interconversion at a rate of more than 180 s$^{-1}$ (calculated from the linewidth) (Cohen *et al.*, 1981). The well-resolved doublet ($\Delta v = 140-220$ Hz) for poly(dGdC)·poly(dGdC) and poly(dG-m$^5$dC)·poly(dG-m$^5$dC) in high-salt solution (Cohen *et al.*, 1981; Chen *et al.*, 1983) reveals that the two phosphodiester conformations in the Z form exchange at a rate much slower than $0.9-1.4$ ms$^{-1}$ or do not exchange at all. In view of the two distinct phosphodiester conformations observed in the X-ray structures of both d(ATAT) (Viswamitra *et al.*, 1978) and (dCdG)$_{2-3}$ (Drew *et al.*, 1980; Wang *et al.*, 1979) crystals, it is most probable that these distinct conformations are not interconverting in solution. But this does not preclude each of these conformations from exhibiting rapid local motions as indicated here and in Chapters 11 (Hart) and 12 (James) of this book.

## VI. Nucleosomal DNA

The nucleosome core particle, isolated from chromatin by treatment with nucleases, consists of 145-bp DNA wrapped around an octameric histone core and is the main structural unit of the genetic material. Several, $^{31}$P-NMR studies (Cotter and Lilley, 1977; Kallenbach *et al.*, 1978; Klevan *et al.*, 1979; Shindo *et al.*, 1980a) have shown that nucleosomal DNA yields a broad symmetrical resonance. The absence of any detectable asymmetry in

8. DNA and RNA Conformations 257

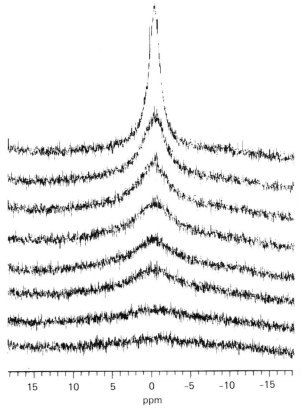

**Fig. 13.** $^{31}$P-NMR spectra at 109.3 MHz of chicken erythrocyte nucleosome core particles in 5 m$M$ Tris-HCl, pH 8, alone (upper spectrum) and in the presence of increasing concentrations of MnCl$_2$ (see Fig. 14).

the $^{31}$P resonance tends to rule out a model that invokes DNA backbone "kinks" at a frequency of 1 in 10 bp as a result of DNA bending. The broad linewidths observed for the 145-bp DNA, either complexed in the core particle or free in solution, are largely caused by chemical-shift dispersion as a result of conformational heterogeneity present in the phosphodiester backbone (Shindo *et al.*, 1980a; Shindo, 1980; Simpson and Shindo, 1980), as discussed in Section II,A. Shindo *et al.* (1980a) and Klevan *et al.* (1979) have compared the relaxation parameters of DNA in the core particles and of histone-free DNA in solution and concluded that there is considerable local mobility in both complexed and free DNA and that packing of DNA into nucleosomes has little effect on its local motions.

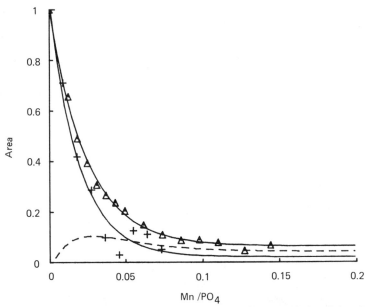

**Fig. 14.** Effect of $Mn^{2+}$ on the integrated area between the same limits of the $^{31}$P-NMR signal of nucleosome core particles ($\triangle$) and the naked DNA derived from them (+). The difference at high $Mn^{2+}$ concentrations is of the same order as the error in the area measurements, about $\pm 0.05$, indicating that phosphates of nucleosomal DNA are ~95 $\pm$ 5% accessible to $Mn^{2+}$ ions in solution.

In order to gain some insights into the DNA – protein interaction in the nucleosome core particles, J. S. Cohen, J. D. McGhee, and G. Felsenfeld (unpublished results, 1982) examined the effect of the paramagnetic ion $Mn^{2+}$ on the $^{31}$P-NMR signal. They found that $Mn^{2+}$-induced line broadening for the complexed DNA (Fig. 13) was essentially the same as that for the naked DNA (Fig. 14). This clearly suggests that most of the phosphates in DNA bound to histones are as accessible to $Mn^{2+}$ ions (or solvent molecules) as those in free DNA. Furthermore, in the presence of excess $Mn^{2+}$, only 5% of the $^{31}$P signal intensity of the bound DNA remained (Fig. 14), indicating that at most 1 phosphate in 20 is relatively inaccessible. This is consistent with the findings of McGhee and Felsenfeld (1980) through melting-temperature analysis. In view of these results, the periodicities evident in the DNase I digestion study (Rhodes and Klug, 1980) of DNA adsorbed to a flat surface are likely to arise mainly from the steric hindrance to the enzyme sites rather than from the inaccessibility of phosphates protected by specific histone binding.

## VII. Biological and Genetic Significance

Many drugs and carcinogens that interact with DNA, through either intercalation or ionic/covalent binding, have specificities toward certain base sequences (for reviews of NMR studies, see Krugh et al., 1979; Patel, 1982; Gorenstein and Goldfield, Chapter 10). Because it has never been determined whether the specificities of these compounds are a result purely of recognition of the base sequences or of the conformations inherent for these sequences, one cannot rule out either possibility. Santella et al. (1981) reported that poly(dGdC)·poly(dGdC) modified with $N$-acetyl-$N^2$-acetylaminofluorene (AAF) at the guanine residues to an extent of 28% showed a CD spectrum that is characteristic of a Z-DNA. It is conceivable that if a stretch of dCdG sequence in the B form were modified with an alkylating agent such as AAF *in vivo*, its transition to the Z form would be greatly facilitated.

Long tracts of alternating dCdG have not yet been discovered in nature. However, a 7-bp length of this sequence does exist in a hairpin region of four related parvovirus genomes (Wells et al., 1980). A number of related $(dCdG)_2$ sequences have also been found (Nishioka and Leder, 1980). Klysik et al. (1981) have prepared several recombinant DNAs containing oligomers of dCdG within the restriction fragments and demonstrated by CD and $^{31}$P-NMR methods that dCdG segments can exist in the Z form in high-salt concentrations while being flanked by long segments of B form. The coexistence of conformations as different as right-handed B form and left-handed Z form may have significant implications for the role of DNA structures in protein or ligand recognition as well as gene regulation (Wells et al., 1977).

The most interesting discovery that suggests the existence of Z-DNA in nature was made by Rich and co-workers (Lafer et al., 1981; Nordheim et al., 1981). These authors have successfully produced antibodies that are specific to the Z-DNA conformation, as proved by their binding *in vitro* to three different DNA polymers capable of forming Z-DNA. These antibodies are found to bind *in vivo* to the interband regions of polytene chromosomes of *Drosophila melanogaster* as visualized by fluorescent staining, thus indicating that these regions consist of Z-DNA.

Before the discoveries by Rich and co-workers (Lafer et al., 1981; Nordheim et al., 1981) and by Behe and Felsenfeld (1981), the main objection to the idea that Z-DNA occurs in nature was that it requires unphysiologically high salt concentrations. The observation that methylated poly(dGdC)·poly(dGdC) can exist in the Z form in the normal physiological salt range has changed our perceptions. Moreover, there is increasing evidence that the methylation of dC at its 5-position in $(dCdG)_n$ sequences in eukaryotic DNA is tied to gene expression (Razin and Riggs, 1980; Ehrlich and Wang, 1981; Felsenfeld and McGhee, 1982), suggesting that Z-DNA may be involved in gene regulation. Another alternating sequence, that is, $(dCdA)_n \cdot (dTdG)_n$, has been found to exist in nature and to convert from a right-handed to a left-handed form under mild conditions. Hamada *et al.* (1982a, 1982b) first reported that stretches of $(dCdA)_n \cdot (dTdG)_n$ are widely found in evolutionarily diverse eukaryotic genomes ranging from yeast to human. It has also been shown by two-dimensional gel electrophoresis (Haniford and Pulleyblank, 1983) that $(dCdA)_n \cdot (dTdG)_n$ readily undergoes a transition to a Z conformation when subjected to unwinding torsional stress in ionic conditions that are close to physiological. Concomitantly, Nordheim and Rich (1983) have demonstrated, by using Z-DNA specific antibodies, the formation of left-handed Z-DNA in sequences of $(dCdA)_{32} \cdot (dTdG)_{32}$ in negatively supercoiled plasmids in the absence of high salt concentration or chemical modifications.

Although stretches of alternating dAdT sequence are not common in nature, a so-called TATA box has been found to exist in several cellular and viral genes centered ~25 bp upstream from the mRNA starting points (Corden *et al.*, 1980). *In vitro* genetic study revealed that the TATA box participates in the initiation of transcription and possibly may play an essential role as a promoter sequence *in vivo*. Whether or not the alternating B-DNA conformation is crucial in this recognition process is not known.

## References

Arnott, S. Chandrasekaran, R., Birdsall, D. L., Leslie, A. G. W., and Ratliff, R. L. (1980). *Nature (London)* **283**, 743-745.
Baase, W. A., and Johnson, W. C., Jr. (1979). *Nucleic Acids Res.* **6**, 797-814.
Behe, M., and Felsenfeld, G. (1981). *Proc. Natl. Acad. Sci. U.S.A.* **78**, 1619-1623.
Bolton, P. H., and James, T. L. (1979). *J. Phys. Chem.* **83**, 3359-3366.
Bolton, P. H., and James, T. L. (1980). *J. Am. Chem. Soc.* **102**, 25-31.
Breslauer, K. J., Sturtevant, J. M., and Tinoco, J. (1975). *J. Mol. Biol.* **99**, 549-565.
Chan, A., Kilkuskie, R., and Hanlon, S. (1979). *Biochemistry* **18**, 84-91.
Chen, C.-W., and Cohen, J. S. (1983). *Biopolymers* **22**, 879-893.
Chen, C.-W., Cohen, J. S., and Behe, M. (1983). *Biochemistry* **22**, 2136-2142.
Cohen, J. S., and Chen, C.-W. (1982). *ACS Symp. Ser.* **191**, 249-267.
Cohen, J. S., Wooten, J. B., and Chatterjee, C. L. (1981). *Biochemistry* **20**, 3049-3055.
Corden, J., Wasylyk, B., Buchwalder, A., Sassone-Corsi, P., Kedinger, C., and Chambon, P. (1980). *Science* **209**, 1406-1414.
Cotter, R. I., and Lilley, D. M. J. (1977). *FEBS Lett.* **82**, 63-68.
Crawford, J. L., Kolpak, F. J., Wang, A. H.-J., Quigley, G. J., van Boom, J. H., van der Marel, G., and Rich, A. (1980). *Proc. Natl. Acad. Sci. U.S.A.* **77**, 4016-4020.

Day, R. O., Seeman, N. C., Rosenberg, J. M., and Rich, A. (1973). *Proc. Natl. Acad. Sci. U.S.A.* **70**, 849–853.
Dickerson, R. E., and Drew, H. R. (1981). *J. Mol. Biol.* **149**, 761–786.
Dickerson, R. E., Drew, H. R., Conner, B. N., Wing, R. M., Fratini, A. V., and Kopka, M. L. (1982). *Science* **216**, 475–485.
Drew, H., Takano, T., Tanaka, S., Itakura, K., and Dickerson, R. E. (1980). *Nature (London)* **286**, 567–573.
Early, T. A., and Kearns, D. R. (1979). *Proc. Natl. Acad. Sci. U.S.A.* **76**, 4165–4169.
Ehrlich, M., and Wang, R. Y.-H. (1981). *Science* **212**, 1350–1357.
Felsenfeld, G., and McGhee, J. D. (1982). *Nature (London)* **296**, 602–603.
Fuller, W., Wilkins, M. H. F., Wilson, H. R., and Hamilston, L. D. (1965). *J. Mol. Biol.* **12**, 60–80.
Gorenstein, D. G. (1981). *Annu. Rev. Biophys. Bioeng.* **10**, 355–386.
Gorenstein, D. G., and Kar, D. (1975). *Biochem. Biophys. Res. Commun.* **65**, 1073–1080.
Gorenstein, D. G., Kar, D., Luxon, B. A., and Momii, R. K. (1976a). *J. Am. Chem. Soc.* **98**, 1668–1673.
Gorenstein, D. G., Findlay, J. B., Momii, R. K., Luxon, B. A., and Kar, D. (1976b). *Biochemistry* **15**, 3796–3803.
Gorenstein, D. G., Luxon, B. A., Goldfield, E. M., Lai, K., and Vegeais, D. (1982). *Biochemistry* **21**, 580–589.
Hamada, H., and Kakunaga, T. (1982a). *Nature (London)* **298**, 396–398.
Hamada, H., Petrino, M. G., and Kakunaga, T. (1982b). *Proc. Natl. Acad. Sci. U.S.A.* **79**, 6465–6469.
Haniford, D. B., and Pulleyblank, D. E. (1983). *Nature (London)* **302**, 632–634.
Hanlon, S., Glonek, T., and Chan, A. (1976). *Biochemistry* **15**, 3869–3875.
Hogan, M. E., and Jardetzky, O. (1979). *Proc. Natl. Acad. Sci. U.S.A.* **76**, 6341–6345.
Hogan, M. E., and Jardetzky, O. (1980). *Biochemistry* **19**, 3460–3468.
Ivanov, V. I., Minchenkova, L. E., Schyolkina, A. K., and Poletayev, A. I. (1973). *Biopolymers* **12**, 89–110.
Kallenbach, N. R., Appleby, D. W., and Bradley, C. H. (1978). *Nature (London)* **272**, 134–138.
Keepers, J. W., and James, T. L. (1982). *J. Am. Chem. Soc.* **104**, 929–939.
Klevan, L., Armitage, I. M., and Crothers, D. M. (1979). *Nucleic Acids Res.* **6**, 1607–1616.
Klug, A., Jack, A., Viswamitra, M. A., Kennard, O., Shakked, Z., and Steitz, T. A. (1979). *J. Mol. Biol.* **131**, 669–680.
Klysik, J., Stirdivant, S. M., Larson, J. E., Hart, P. A., and Wells, R. D. (1981). *Nature (London)* **290**, 672–677.
Krugh, T. R., Liang, J. W., and Young, M. A. (1976). *Biochemistry* **15**, 1224–1228.
Krugh, T. R., Hook, J. W., III, Lin, S., and Chen, F.-M. (1979). In "Stereodynamics of Molecular Systems" (R. H. Sarma, ed.), pp. 423–435. Pergamon, Oxford.
Kypr, J., Vorlickova, M., Budesnsky, M., and Sklena, V. (1981). *Biochem. Biophys. Res. Commun.* **99**, 1257–1264.
Lafer, E. M., Moller, A., Nordheim, A., Stollar, B. D., and Rich, A. (1981). *Proc. Natl. Acad. Sci. U.S.A.* **78**, 3546–3550.
Leslie, A. G. W., Arnott, S., Chandrasekaran, R., and Ratliff, R. L. (1980). *J. Mol. Biol.* **143**, 49–72.
Levy, G. C., Hilliard, P. R., Jr., Levy, L. F., and Rill, R. L. (1981). *J. Biol. Chem.* **256**, 9986–9989.
Lipari, G., and Szabo, A. (1981). *Biochemistry* **20**, 6250–6256.
Lomonossoff, G. P., Butler, P. J. G., and Klug, A. (1981). *J. Mol. Biol.* **149**, 745–760.

Mahendrasingam, A., Rhodes, N. J., Goodwin, D. C., Nave, C., Pigram, W. J., and Fuller, W. (1983). *Nature (London)* **301**, 535–537.
McGhee, J. D., and Felsenfeld, G. (1980). *Nucleic Acids Res.* **8**, 2751–2769.
Mariam, Y. H., and Wilson, W. D. (1979). *Biochem. Biophys. Res. Commun.* **88**, 861–866.
Mitsui, Y., Langridge, R., Shortle, B. E., Cantor, C. R., Grant, R. C., Kodama, M., and Wells, R. D. (1970). *Nature (London)* **228**, 1166–1169.
Nishioka, Y., and Leder, P. (1980). *J. Biol. Chem.* **255**, 3691–3694.
Nordheim, A., Pardue, M. L., Lafer, E. M., Moller, A., Stollar, B. D., and Rich, A. (1981). *Nature (London)* **294**, 417–422.
Nordheim, A., and Rich, A. (1983). *Proc. Natl. Acad. Sci. U.S.A.* **80**, 1821–1825.
Opella, S. J., Wise, W. B., and DiVerdi, J. A. (1981). *Biochemistry* **20**, 284–290.
Patel, D. J. (1974). *Biochemistry* **13**, 2388–2396.
Patel, D. J. (1976). *Biopolymers* **15**, 533–558.
Patel, D. J. (1977). *Biopolymers* **16**, 1635–1656.
Patel, D. J. (1979a). *Biopolymers* **18**, 553–569.
Patel, D. J. (1979b). *Acc. Chem. Res.* **12**, 118–125.
Patel, D. J. (1982). *In* "Nucleic Acid Geometry and Dynamics" (R. H. Sarma, ed.), pp. 185–231. Pergamon, Oxford.
Patel, D. J., and Canuel, L. L. (1976). *Proc. Natl. Acad. Sci. U.S.A.* **73**, 674–678.
Patel, D. J., and Canuel, L. L. (1979). *Eur. J. Biochem.* **96**, 267–276.
Patel, D. J., Canuel, L. L., and Pohl, F. M. (1979). *Proc. Natl. Acad. Sci. U.S.A.* **76**, 2508–2511.
Patel, D. J., Kozlowski, S. A., Suggs, J. W., and Cox, S. D. (1981). *Proc. Natl. Acad. Sci. U.S.A.* **78**, 4063–4067.
Patel, D. J., Kozlowski, S. A., Marky, L. A., Broka, C., Rice, J. A., Itakura, K., and Breslauer, K. J. (1982a). *Biochemistry* **21**, 428–436.
Patel, D. J., Kozlowski, S. A., Marky, L. A., Rice, J. A., Broka, C., Dallas, J., Itakura, K., and Breslauer, K. J. (1982b). *Biochemistry* **21**, 437–444.
Patel, D. J., Kozlowski, S. A., Marky, L. A., Rice, J. A., Broka, C., Itakura, K., and Breslauer, K. J. (1982c). *Biochemistry* **21**, 445–451.
Patel, D. J., Kozlowski, S. A., Marky, L. A., Rice, J. A., Broka, C., Itakura, K., and Breslauer, K. J. (1982d). *Biochemistry* **21**, 451–455.
Patel, D. J. Kozlowski, S. A., Nordheim, A., and Rich, A. (1982e). *Proc. Natl. Acad. Sci. U.S.A.* **79**, 1413–1417.
Perahia, D., and Pullman, B. (1976). *Biochim. Biophys. Acta* **435**, 282–289.
Pohl, F. M. (1976). *Nature (London)* **260**, 365–366.
Pohl, F. M., and Jovin, T. M. (1972). *J. Mol. Biol.* **67**, 375–396.
Prado, F. R., Giessner-Prettre, C., Pullman, B., and Daudey, J.-P. (1979). *J. Am. Chem. Soc.* **101**, 1737–1742.
Razin, A., and Riggs, A. D. (1980). *Science* **210**, 604–610.
Rhodes, D., and Klug, A. (1980). *Nature (London)* **286**, 573–578.
Santella, R. M., Grunberger, D., Weinstein, I. B., and Rich, A. (1981). *Proc. Natl. Acad. Sci. U.S.A.* **78**, 1451–1455.
Shindo, H. (1980). *Biopolymers* **19**, 509–522.
Shindo, H. (1981). *Eur. J. Biochem.* **120**, 309–312.
Shindo, H., and Zimmerman, S. B. (1980). *Nature (London)* **283**, 690–691.
Shindo, H., Simpson, R. T., and Cohen, J. S. (1979). *J. Biol. Chem.* **254**, 8125–8128.
Shindo, H., McGhee, J. D., and Cohen, J. S. (1980a). *Biopolymers* **19**, 523–537.
Shindo, H., Wooten, J. B., Pheiffer, B. H., and Zimmerman, S. B. (1980b). *Biochemistry* **19**, 518–526.
Simpson, R. T., and Shindo, H. (1979). *Nucleic Acids Res.* **7**, 481–492.

## 8. DNA and RNA Conformations

Simpson, R. T., and Shindo, H. (1980). *Nucleic Acids Res.* **8**, 2093–2103.
Sundaralingam, M. (1969). *Biopolymers* **7**, 821–860.
Temussi, P. A., Guidoni, L., Ramoni, C., and Podo, F. (1979). *Physiol. Chem. Phys.* **11**, 445–452.
Viswamitra, M. A., Kennard, O., Jones, P. G., Shelbrick, G. M., Salisbury, S., Falvello, L., and Shakked, Z. (1978). *Nature (London)* **273**, 687–688.
Viswamitra, M. A., Shakked, Z., Jones, P. G., Shelbrick, G. M., Salisbury, S. A., and Kennard, O. (1982). *Biopolymers* **21**, 513–533.
Vorlickova, M., Kypr, J., Kleinwachter, V., and Palecek, E. (1980). *Nucleic Acids Res.* **8**, 3965–3973.
Vorlickova, M., Kypr., J., Stokrova, S., and Sponar, J. (1982). *Nucleic Acids Res.* **10**, 1071–1080.
Vorlickova, M., Kypr, J., and Sklenar, V. (1983). *J. Mol. Biol.* **166**, 85–92.
Wang, A. H.-J., Quigley, G. J., Kolpak, F. J., Crawford, J. L., van Boom, J. H., van der Marel, G., and Rich, A. (1979). *Nature (London)* **282**, 680–686.
Wang, A. H.-J., Quigley, G. J., Kolpak, F. J., van der Marell, G., van Boom, J. H., and Rich, A. (1981). *Science* **211**, 171–176.
Wells, R. D., Blakesley, R. W., Hardies, S. C., Horn, G. T., Larson, J. E., Selsing, E., Burd, J. F., Chan, H. W., Dodgson, J. B., Jensen, K. F., Nes, I., F., and Wartell, R. M. (1977). *CRC Crit. Rev. Biochem.* **4**, 305–340.
Wells, R. D., Goodman, T. C., Hillen, W., Horn, G. T., Klein, R. D., Larson, J. E., Muller, U. R., Neuendorf, S. K., Panayotatos, N., and Stirdivant, S. M. (1980). *Prog. Nucleic Acid Res. Mol. Biol.* **24**, 167–267.
Wilson, W. D., and Jones, R. L. (1982). *Nucleic Acids Res.* **10**, 1399–1410.
Wing, R., Drew, H., Takano, T., Broka, C., Tanaka, S., Itakura, K., and Rich, A. (1980). *Nature (London)* **287**, 755–758.
Yamada, A., Kaneko, H., Akasaka, K., and Hatano, H. (1978). *FEBS Lett.* **93**, 16–18.
Zimmer, C., and Luck, G. (1974). *Biochim. Biophys. Acta* **361**, 11–32.
Zimmer, C., Tymen, S., Marck, C., and Guschlbauer, W. (1982). *Nucleic Acids Res.* **10**, 1081–1091.
Zimmerman, S. B. (1982). *Annu. Rev. Biochem.* **51**, 395–427.
Zimmerman, S. B., and Pheiffer, B. H. (1979). *J. Mol. Biol.* **135**, 1023–1027.
Zimmerman, S. B., and Pheiffer, B. H. (1980). *J. Mol. Biol.* **142**, 315–330.

CHAPTER 9

# High-Resolution $^{31}$P-NMR Spectroscopy of Transfer Ribonucleic Acids

### David G. Gorenstein

Department of Chemistry
University of Illinois at Chicago
Chicago, Illinois

| | |
|---|---|
| I. Introduction | 265 |
| II. Spectral Comparison of Different Acceptor tRNAs | 268 |
|    A. $^{31}$P Spectral Differences | 268 |
|    B. Structural Basis for $^{31}$P Differences | 273 |
| III. $Mg^{2+}$ Dependence of $^{31}$P Spectra of tRNA$^{Phe}$ | 276 |
| IV. $Mn^{2+}$ Effects on $^{31}$P Spectra of tRNA$^{Phe}$ | 280 |
| V. Assignment of $^{31}$P Signals | 281 |
|    A. Stereoelectronic Origin of $^{31}$P Chemical Shifts | 281 |
|    B. Assignment of Scattered Signals | 283 |
| VI. Conformational Transitions | 287 |
| VII. Spermine Effects | 291 |
| VIII. $^{31}$P NMR of Yeast tRNA$^{Phe}$ · *E. coli* tRNA$^{Glu}_2$ Complex: Spectral Changes | 292 |
| IX. Conclusions | 294 |
|    References | 295 |

## I. Introduction

This chapter describes the application of $^{31}$P-NMR spectroscopy to the study of the conformation of transfer ribonucleic acids (tRNAs). The $^{31}$P-NMR spectrum of pure acceptor tRNA species has been shown to contain considerable fine structure (Figs. 1 and 2). High-resolution $^{31}$P-NMR spectra by Guéron and Shulman (1975) and later by Salemink *et al.* (1979) and Gorenstein and Luxon (1979) revealed ~16 individual phosphate resonances spread over 7 ppm that were not observed in earlier $^{31}$P tRNA spectra (Guéron, 1971; Weiner *et al.*, 1974; Gorenstein and Kar, 1975).

As discussed in Chapter 1, Gorenstein and co-workers have proposed that $^{31}$P chemical shifts of phosphate esters are sensitive to stereoelectronic

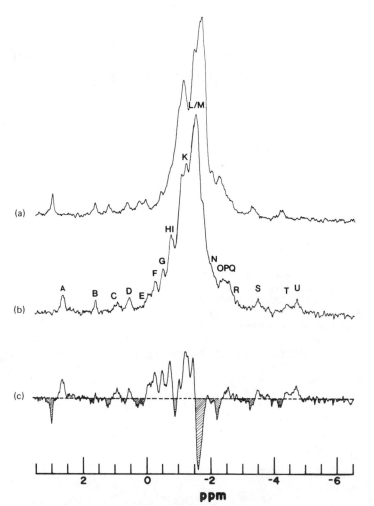

**Fig. 1.** $^{31}$P-NMR spectra of (a) yeast tRNA$^{Phe}$, (b) *E. coli* tRNA$^{Tyr}$, and (c) tyrosine–phenylalanine difference $^{31}$P-NMR spectrum at 35°C, pH 7, 10 m$M$ MgCl$_2$, 100 m$M$ NaCl, 10 m$M$ cacodylate, 20% D$_2$O, 1 m$M$ EDTA, 80.9 MHz ($^{31}$P), and 2-Hz line broadening applied to free-induction decay (FID).

effects (Gorenstein and Kar, 1975, 1977; Gorenstein, 1975, 1977, 1978, 1981; Gorenstein and Goldfield, 1982a,b), and model-system studies on single- and double-stranded nucleic acids (Patel, 1976; Shindo *et al.*, 1980; Cohen and Chen, 1982; Gorenstein and Luxon, 1979; Gorenstein *et al.*, 1976, 1981, 1982) suggested that a phosphate diester in a gauche, gauche (g,g; see definition in Gorenstein, Chapter 1, Fig. 5) conformation should

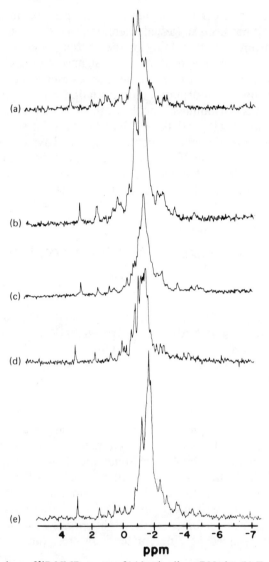

**Fig. 2.** Comparison of $^{31}$P-NMR spectra of (a) bovine liver tRNA$^{Asp}$, (b) *E. coli* tRNA$^{fMet}$, (c) tRNA$^{Tyr}$, (d) tRNA$_2^{Glu}$, and (e) yeast tRNA$^{Phe}$ at pH 7, 20% D$_2$O, 10 m$M$ MgCl$_2$, 0.2 $M$ NaCl, 10 m$M$ cacodylate, 1 m$M$ EDTA, 31°C (except tRNA$_2^{Glu}$ at 35°C), and 32.4 MHz with 0.5-Hz spectral line broadening.

resonate several parts per million upfield from a diester in a non-gauche conformation. It has been suggested, therefore, that stereoelectronic differences arising from tertiary folding of the phosphate ester backbone are responsible for the 7-ppm spread in the $^{31}P$ signals of the tRNA.

Phosphorus-NMR spectroscopy alone among other spectroscopic probes can provide detailed conformational information on the phosphate ester bonds in the transfer ribonucleic acids. In this Chapter, an attempt first is made to assign some of these tRNA $^{31}P$-NMR signals. These assigned signals are then used to probe the solution conformation and dynamics of the tRNA ribose phosphate backbone. Finally, conformational changes resulting from the interaction of the tRNA with metal ions, polyvalent cations, and a complementary nucleic acid are analyzed.

## II. Spectral Comparison of Different Acceptor tRNAs[1]

### A. $^{31}P$ Spectral Differences

A comparison of the $^{31}P$-NMR spectra of yeast tRNA$^{Phe}$ and *Escherichia coli* tRNA$_2^{Glu}$, tRNA$^{Tyr}$, tRNA$^{fMet}$, and bovine liver tRNA$^{Asp}$ in 10 m$M$ Mg$^{2+}$ is shown in Figs. 1 and 2. As Guéron and Shulman (1975), Salemink *et al.* (1979), and Gorenstein and Luxon (1979) have shown for yeast tRNA$^{Phe}$, the other tRNAs also show a number of similar spectral features. Thus between +2.6 and +3.0 ppm[2] is the 5′-terminal phosphate that integrates for a single phosphate residue and is the only signal which is pH-sensitive (being a monoester with p$K \sim 6$). Between −0.5 and −2.0 ppm is a main cluster of signals, which, as will be shown, represent the undistorted phosphate diesters in the double-helical stems and hairpin loops. The main cluster peaks L/M for the helical stems integrate for 35–37 phosphates in all the tRNAs. Upfield and downfield of the main cluster, spread over 6–7 ppm, are ~16 scattered signals, a number of which are well resolved at 30°. As shown in Figs. 3 and 4 for two of the tRNA species, other scattered signals become better resolved at different temperatures. Individually resolved

---

[1] Abbreviations used: tRNA$^{Phe}$, phenylalanine tRNA; tRNA$_2^{Glu}$, glutamic acid tRNA; tRNA$^{Tyr}$, tyrosine tRNA; tRNA$^{fMet}$, formylmethionine tRNA; EDTA, ethylenediaminetetraacetic acid.

[2] All $^{31}P$ chemical shifts are referenced to an external sample of 85% $H_3PO_4$, and positive chemical shifts are to low field. Note that our earlier tRNA $^{31}P$ spectra were referenced to 15% $H_3PO_4$, which is 0.453 ppm upfield of 85% $H_3PO_4$. All figures from previous publications reproduced in this chapter have been corrected to reflect the 85% $H_3PO_4$ standard.

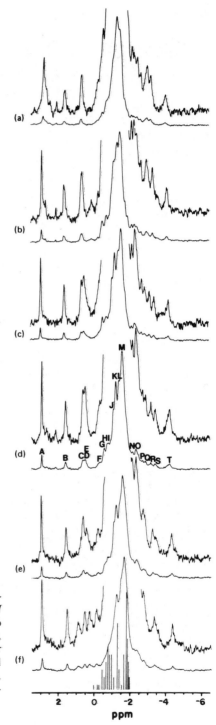

**Fig. 3.** $^{31}$P-NMR spectra of yeast phenylalanine tRNA (33 mg/ml) in 0.1 $M$ NaCl, 10 m$M$ cacodylate, 10 m$M$ MgCl$_2$, 1 m$M$ EDTA, 10% D$_2$O, pH 7, and at temperatures (°C): a, 66; b, 60; c, 54; d, 49; e, 44; f, 34. Expanded scale for the scattered peaks is shown above the normal spectrum. Number of acquisitions 8000 FIDs, 1.86-s acquisition time, 145.8 MHz. A simulated stick figure is also shown at the bottom.

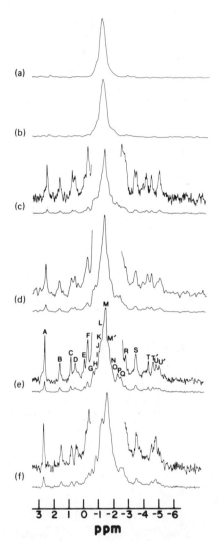

**Fig. 4.** $^{31}$P-NMR spectra of *E. coli* tRNA$^{Tyr}$ (2 mg/ml) under conditions similar to those in Fig. 1: 80.9 MHz, 4000–16,000 FIDs, 0.5-Hz line broadening. Temperatures (°C): a, 70; b, 65; c, 50; d, 45; e, 35; f, 25.

signals (such as B, C, D, and T) integrate in all tRNAs for approximately one phosphate.

The temperature dependence of the $^{31}$P chemical shifts of the labeled signals in Figs. 3 and 4 is shown in Figs. 5 and 6. Other spectra (not shown) behave similarly. Thus between 20 and 60°C most of the scattered and main cluster signals shift very little with temperature. As shown earlier (Guéron and Shulman, 1975; Salemink et al., 1979; Gorenstein and Luxon, 1979; Gorenstein and Goldfield, 1982a,b), this lack of temperature sensitivity to

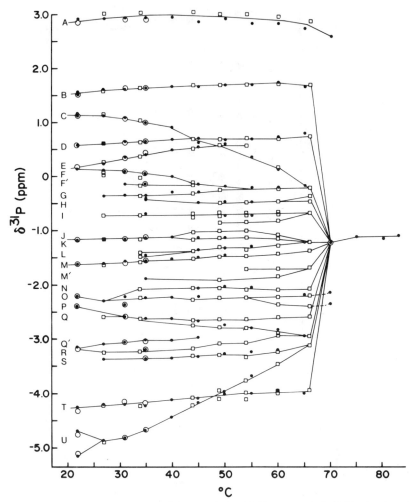

**Fig. 5.** Temperature dependence of the chemical shift $\delta$ in the peaks identified in Fig. 3 for tRNA$^{Phe}$: □, 145.8 MHz (from spectra in Fig. 3); ●, 32.4 MHz; ⊙, repeat spectra at 32.4 MHz.

most features in the tRNA $^{31}$P spectra in 10 m$M$ Mg$^{2+}$ suggests that the tRNAs (and the backbone phosphates) for the most part retain their native conformation throughout this temperature range (see also Gorenstein, Chapter 1). Eventually, at $T > T_m$ ($\sim 60-75°C$), all of the diester peaks merge into a single signal with the tRNA melting into a random-coil conformation. The melting of the secondary and tertiary structure around 60–75°C is confirmed in the UV melting curve of tRNA$^{Tyr}$ (see Fig. 6).

Although the main cluster signals and most of the scattered peaks show

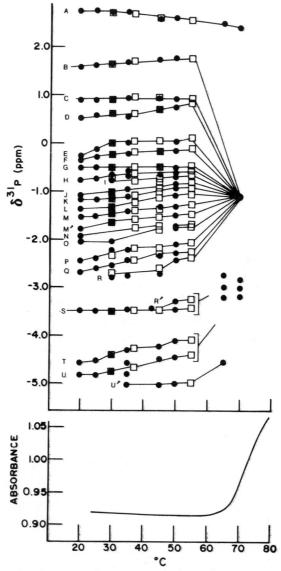

**Fig. 6.** Temperature dependence of the chemical shift $\delta$ in the peaks identified in Fig. 4 for tRNA$^{Tyr}$. UV (260 nm) melting curve for sample shown at bottom.

only small shifts with temperature, several important exceptions are notable (see plots in Figs. 5 and 6). For tRNA$^{Phe}$ (Figs. 2 and 5), peak C shifts 1.3 ppm upfield and broadens from 22 to 65°C whereas peak E shifts $\sim 0.6$ ppm downfield; D shifts very little over this temperature range and does not broaden. Peaks F, P, T, and U are the only other peaks that display a large shift with temperature, shifting 0.3–1.7 ppm over this temperature range.

The temperature dependence of the chemical shifts for tRNA$^{Tyr}$ (Figs. 3 and 6) differs somewhat from tRNA$^{Phe}$. Peak C shows almost no temperature dependence of its chemical shift between 20 and 60°C (melting temperature $\sim 65$°C). Peak D shifts slightly upfield at higher temperature and peaks E and F both shift upfield with increasing temperature (as opposed to F shifting downfield in tRNA$^{Phe}$). The E and F resonances also appear further upfield than in the tRNA$^{Phe}$ spectra; peak E is at $-0.2$ ppm at 25°C and peak F is at $-0.45$ ppm. There are three peaks far upfield, T, U, and U′, that are all temperature-dependent.

The temperature dependence of the $^{31}$P signals for bovine liver tRNA$^{Asp}$, E. coli tRNA$^{fMet}$, and tRNA$_2^{Glu}$ also generally follows that of either tRNA$^{Phe}$ or tRNA$^{Tyr}$ (Gorenstein and Goldfield, 1982a,b; Gorenstein et al., 1982). For example, tRNA$^{fMet}$ is the most stable of the tRNAs because it has more G–C base pairs than the other tRNAs, and its melting temperature at high Mg$^{2+}$ is 75°C. Therefore at 70°C, considerable tertiary structure remains, although the peaks have broadened substantially. The temperature dependence of the chemical shifts resembles that of tRNA$^{Phe}$ (Gorenstein and Goldfield, 1982a,b; Gorenstein et al., 1982). However, peak C is located further upfield at $\sim 1.5$ ppm at 20°C and merges with peak D at 65°C, never moving as far upfield as it does in tRNA$^{Phe}$ (at least as a distinguishable peak). The behavior of peaks E, F, and F′ very much resemble that of tRNA$^{Phe}$. Peak U at $-4.75$ ppm at 20°C moves downfield to $-4.2$ ppm at 60°C, after which it disappears. There is no resonance corresponding to peak T of tRNA$^{Phe}$ but there are two temperature-dependent resonances (S and S′) that begin appearing as one peak at 25°C at $-3.45$ ppm and move downfield to $-2.75$ ppm (S) and $-3.0$ ppm (S′) at 60°C, after which they also disappear.

## B. Structural Basis for $^{31}$P Differences

The origin of these $^{31}$P spectral differences between the tRNA species is not apparent from their primary and secondary structures. All tRNAs are now known to fit the cloverleaf secondary structure (Fig. 7a,b), and it is becoming increasingly clear that the tRNAs also have a tertiary structure quite similar to that found for yeast tRNA$^{Phe}$. Thus it is known from

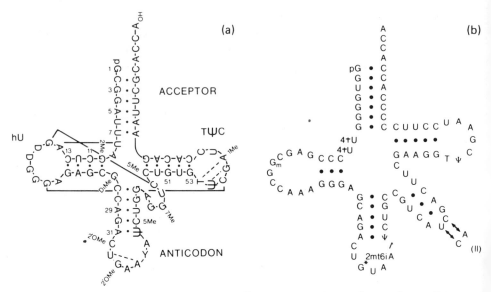

Fig. 7. (a) Cloverleaf model for yeast tRNA$^{Phe}$. (b) Cloverleaf model for *E. coli* tRNA$^{Tyr}$.

sequence studies that most of the conserved or semiconserved bases in different tRNAs are involved in tertiary structure hydrogen bonding (Kim *et al.*, 1974). Low-resolution X-ray studies of tRNA$^{Gly}$ (Wright *et al.*, 1979), tRNA$^{fMet}$ (Schevitz, *et al.*, 1979; Woo *et al.*, 1980), and tRNA$^{Asp}$ (Moras *et al.*, 1980) show that these molecules have structures similar to tRNA$^{Phe}$.

The yeast phenylalanine acceptor molecule is a class 1 (or $D_4V_5$) tRNA with 4 base pairs in the dihydrouridine stem and 5 bases in the variable loop and a total of 76 nuclcotides. *Escherichia coli* tRNA$^{Tyr}$ is in class 2 (or $D_3V_{13}$) with a total of 85 bases; *E. coli* tRNA$_2^{Glu}$ is in class 1 ($D_3V_4$) with 76 nucleotides; tRNA$^{fMet}$ is in class 1 ($D_4V_5$) with 77 nucleotides; and bovine liver tRNA$^{Asp}$ is in class 1 ($D_4V_5$) with 76 nucleotides. $^1$H-NMR studies (Kearns, 1976) confirm that in most tRNAs there are $\sim 20$ (20 in tRNA$^{Phe}$ and 23 in tRNA$^{Tyr}$) base pairs stabilizing the secondary structure and $7 \pm 1$ tertiary structure base pairs (Reid and Hurd, 1979). The tRNA$^{Tyr}$ differs from tRNA$^{Phe}$ and other class 1 tRNAs by the presence of the larger variable arm (Fig. 7B) that presumably forms a helical hairpin loop protruding from the center of the molecule (Brennan and Sundaralingam, 1976). As shown by the $^{31}$P difference spectra in Fig. 1, most of the additional nine phosphates in tRNA$^{Tyr}$ relative to the above class 1 tRNAs are found in the main cluster spectral region (0 to $-1.5$ ppm), rather than in the scattered signals.

## 9. High-Resolution $^{31}$P-NMR Spectroscopy of tRNAs

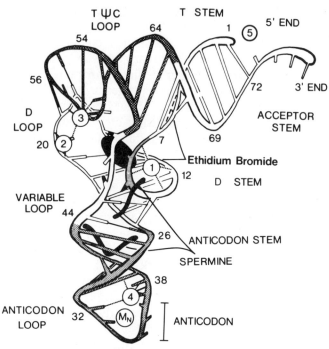

**Fig. 8.** Schematic model for three-dimensional structure of yeast tRNA$^{Phe}$ showing sugar–phosphate backbone, base pairs, five magnesium-binding sites (numbered circles), manganese site, ethidium-binding site, and spermine-binding site (dark curved shape in AC stem). Partially derived from Holbrook *et al.* (1978) and Kim (1979).

In tRNA$^{Tyr}$ there are eight extra phosphates (relative to tRNA$^{Phe}$) in the variable arm, six of which can base-pair and which are presumably in a double-helical $-g,-g$ conformation. The D loop in tRNA$^{Tyr}$, however, has one less base pair than tRNA$^{Phe}$, so only four more phosphates might be expected in the $-g,-g$ conformation. The other five extra phosphates in the tRNA$^{Tyr}$ structure are in the hairpin loops of the D and variable arms. In the $^{31}$P difference spectra in Fig. 1 are found extra phosphates in signals J, K, and L (and less in M) in tRNA$^{Tyr}$ relative to tRNA$^{Phe}$.

It is especially significant that the general features of the scattered $^{31}$P signals are retained in the different acceptor tRNAs. The good agreement between the shifts of these scattered signals associated with the tertiary structures in the different tRNAs provide further support to the belief that the tertiary structure of all tRNAs (even classes 1 and 2) are quite similar (such as that of tRNA$^{Phe}$, Fig. 8).

## III. Mg$^{2+}$ Dependence of $^{31}$P Spectra of tRNA$^{Phe}$

Magnesium stabilizes the functional, native conformation of tRNAs through stabilization of the loops and sharp turns in the tertiary structure (Kim, 1979). This explains the higher melting temperature of tRNAs in the presence of Mg$^{2+}$ and the magnesium dependence of the $^{31}$P signals.

At lower temperature ($T < 30\,°C$) the $^{31}$P-NMR spectra of tRNA$^{Phe}$ in the absence or in the presence of 10 m$M$ Mg$^{2+}$ are basically quite similar, as concluded previously by Guéron and Shulman (1975) and Gorenstein and Luxon (1979), Gorenstein *et al.* (1981), and Salemink *et al.* (1981). Apparently, even without Mg$^{2+}$ but in 0.2 $M$ total Na$^+$ (0.1 $M$ added NaCl) the native secondary and tertiary structures are similarly stabilized at lower temperatures. However, as shown in Figs. 9 and 10, a number of differences do exist even at 19 °C. By titrating the tRNA$^{Phe}$ solution with added Mg$^{2+}$, it is possible to identify which signals in the absence of Mg$^{2+}$ correspond to the signals in the presence of Mg$^{2+}$, and these signals are labeled in Figs. 9 and 10. At 19°C without Mg$^{2+}$, peak C of tRNA$^{Phe}$ is shifted 1.0 ppm *upfield* relative to the high-Mg$^{2+}$ (10 m$M$ dialyzed) sample whereas peak E is shifted 1.0 ppm *downfield* relative to the high-Mg$^{2+}$ sample. Salemink *et al.* (1981) have found a similar reversal of peak C and E positions at 30°C, as has Müller *et al.* (1980). With the additional exception of peaks P and T/U, other peaks experience only small shifts (<0.3 ppm).

Significantly, the integrated intensity for the scattered peaks (B–F and N–O) of tRNA$^{Phe}$ (and roughly the same for tRNA$^{fMet}$) is only ~ 3 in the absence of Mg$^{2+}$ compared to 14 at 39° in 30 m$M$ Mg$^{2+}$. At 19°, peaks B–E and N–U integrate for a total of 9 phosphates and 14 phosphates at 0 and 20 m$M$ Mg$^{2+}$, respectively (Fig. 8; Gorenstein *et al.*, 1981).

These $^{31}$P spectra at 19°C without added Mg$^{2+}$ suggest that insofar as the phosphate ester backbone is concerned, both partially denatured (but certainly not cloverleaf) and native (tertiary) structures are present. All of the scattered signals (B–E and N–U) integrate for less than one phosphate each, even at 19°C (no Mg$^{2+}$). Utilizing the integrated intensity of the scattered signals (B–E and N–U) as an indicator of tertiary structure, 64% of the molecules exist in the native conformation at 19°C (integrated total of 9 phosphates are observed whereas 14 are expected in the fully native structure). Extra intensity (12–13 phosphates) is observed in the downfield portion of the main cluster (F–K), where signals for the unstrained hairpin-loop phosphates in the cloverleaf structure are expected to be found. About six of the double-helix phosphate signals normally observed in the native structures (35 total) are also missing in the 19°C, no magnesium spectrum. Some melting of the stems (fraying at the ends of the double helices) is also indicated. At 39°C and no Mg$^{2+}$, only 21% of the native structure remains

# 9. High-Resolution $^{31}$P-NMR Spectroscopy of tRNAs

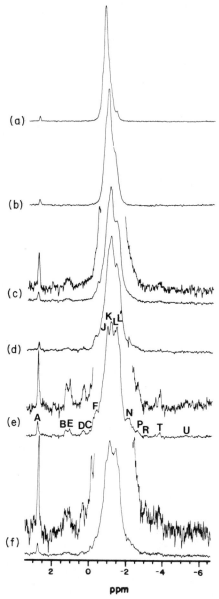

**Fig. 9.** $^{31}$P-NMR spectra (145.8 MHz) of tRNA$^{Phe}$ (33 mg/ml) in 100 m$M$ NaCl, 10 m$M$ cacodylate, 1 m$M$ EDTA, no Mg$^{2+}$, and at pH 7.0, at temperatures (°C): a, 71; b, 61; c, 51; d, 46; e, 41; f, 36.

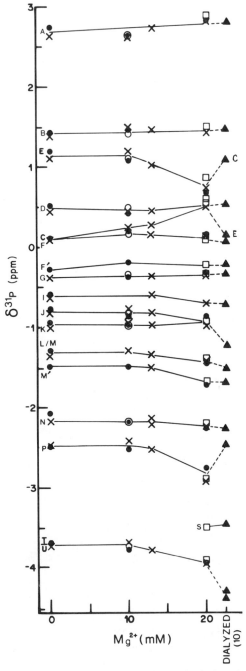

**Fig. 10.** Magnesium dependence of the $^{31}$P chemical shifts $\delta$ of tRNA$^{Phe}$ peaks; 19°C. ●, Sample 1, 145.8 MHz; □, sample 1, 32.4 MHz; △, sample 1 redialyzed against 10 m$M$ MgCl$_2$, 32.4 MHz. ○, Sample 2; ×, sample 3, both at 145.8 MHz.

(total of three scattered peaks, spectra not shown). Again, the additional intensity (21 phosphates) is found in the region (peaks J/K) that is associated with the unstrained hairpin loop structures.

Little change in the 19°C spectra is observed until $\geq 10$ mM $Mg^{2+}$ has been added. Both the chemical shifts (Fig. 10) and integrated intensities of the signals are unperturbed in 0–10 mM $Mg^{2+}$. Further addition of $Mg^{2+}$ (to 20 mM) produces a $^{31}P$ spectrum that looks quite similar to the spectrum of a sample dialyzed against 10 mM $Mg^{2+}$. Thus the integrated intensity of the scattered signals B–E and N–U in the 20 mM sample is 13 (versus 14 in the 10 mM dialyzed sample spectrum), and 35 phosphates in the double-helical signals L/M are found under both conditions.

The $Mg^{2+}$ dependence of the $^{31}P$ chemical shifts at 19°C (Fig. 10) is consistent with the requirement for 10–15 mM $Mg^{2+}$ ($\sim$ 6–10 $Mg^{2+}$ ions per tRNA) to stabilize the native structure. However, even in 20 mM $Mg^{2+}$, at 19°C, the tRNA is not in an identical conformational state as found in the 10 mM dialyzed sample (high-$Mg^{2+}$ state). Assuming no $Mg^{2+}$, the low-temperature spectrum corresponds to a different (low-magnesium) conformation, and using the $^{31}P$ chemical shifts of peaks C, E, P, and T/U as a monitor of the conformational state, in 20 mM $Mg^{2+}$ ([$Mg^{2+}$]/[tRNA] $\sim$ 14) the two conformations are populated about equally. Further addition of $Mg^{2+}$ ([$Mg^{2+}$]/[tRNA] > 28) shifts the scattered signals into their high-$Mg^{2+}$ native state (Salemink et al., 1981; Müller et al., 1980).

The requirement for at least 10 mM added $Mg^{2+}$ to produce significant spectral changes in tRNA[Phe] is consistent with other studies that show $Mg^{2+}$ binds cooperatively to the nonnative structure (Sander and Ts'o, 1971; Schreier and Schimmel, 1975). Apparently at least a $Mg^{2+}$/tRNA ratio of 4 is required to stabilize the tertiary structure.

The signals in the 10 mM $Mg^{2+}$ dialyzed sample showing the largest temperature sensitivity are C, E, F, P, T, and U (Fig. 5). Gorenstein et al. (1981) and Salemink et al. (1981) have demonstrated that the signals showing the largest magnesium sensitivity are A, C, E, F, P, and T/U (peak U appears to be underneath peak T in low-magnesium buffers (Salemink et al., 1981). Gorenstein et al. (1981) have suggested that the $Mg^{2+}$ dependence of the chemical shifts of the scattered diester signals A, C, E, F, P, and T/U is likely the result of a conformational change of the tRNA involving the tightly bound $Mg^{2+}$ ions. X-Ray studies on yeast tRNA[Phe] have located the four strong magnesium-binding sites shown in Fig. 8. Magnesium in two of the sites (2 and 3) cross-links the D and TΨC loops, which is likely responsible for the cooperative stabilization of the tertiary structure by magnesium. The temperature dependence of peaks C, E, F, P, T, and U in the 10 mM $Mg^{2+}$ dialyzed sample probably reflects partial disruption of the binding site at higher temperature for one or more of these four tightly bound, tertiary-structure-stabilizing magnesium ions. This high-temperature (or low-magnesium) conformation must retain most of the tertiary structure because UV and $^1$H-NMR melting studies (Reid and Hurd, 1979) under these conditions show no change in 3° structure. As discussed later, chemical-exchange line-broadening ($1/T_2$) effects for peaks C–F, T, and U also provide evidence for this premelting conformational transition.

## IV. $Mn^{2+}$ Effects on $^{31}P$ Spectra of tRNA$^{Phe}$

Addition of up to 0.12 $Mn^{2+}$ ions per yeast tRNA$^{Phe}$ molecule in 10 m$M$ $Mg^{2+}$, 0.1 $M$ NaCl, 10 m$M$ cacodylate, *no* EDTA, and at 32°C produces negligible chemical-shift (<0.1 ppm) and signal-integral changes (Fig. 11). However, as shown in Fig. 10, selective line-broadening effects are observed for signals, A, D, E, F, and U below $Mn^{2+}$/tRNA ratios of 0.01 (0.013 m$M$ $Mn^{2+}$). The phosphate monoester signal A experiences the largest paramagnetic broadening, from 2 Hz at no $Mn^{2+}$ to 12.7 Hz at a $Mn^{2+}$/tRNA ratio of 0.053. The signal is broadened below detection at a $Mn^{2+}$/tRNA ratio of 0.12. Signals D and E broaden from 3–5 Hz at low $Mn^{2+}$ to 8–10 Hz at $Mn^{2+}$/tRNA ratio of 0.053. Signal U increases from 6 Hz to 17 Hz with increasing $Mn^{2+}$. Other signals, such as peak B, Fig. 11, show little broadening below $Mn^{2+}$/tRNA ratios of 0.05.

The broadening of a given phosphate signal by $Mn^{2+}$ will depend on the phosphorus–manganese distance and the residence time of the ion at the binding site. The greatest paramagnetic broadening will thus occur at tight binding sites, with direct phosphate–oxygen–manganese association. Guéron and Shulman (1975), Guéron and Leroy (1979), and Gorenstein *et al.* (1981) have studied the effect of added paramagnetic ions ($Co^{2+}$ and $Mn^{2+}$) on the $^{31}$P-NMR spectrum of tRNA$^{Phe}$. Gorenstein *et al.* (1981) have concluded from the selective line broadening for signals A, D, E, and U at low $Mn^{2+}$/tRNA ratios (<0.010) that there are *specific* $Mn^{2+}$-binding sites. However, Guéron and Leroy (1979) have suggested that much of the $Mn^{2+}$-induced broadening arises from nonspecific binding of the ion to the tRNA acting as a highly anionic polyelectrolyte.

Of the specifically $Mn^{2+}$-broadened signals, only the signals A, E, and U are also $Mg^{2+}$-dependent, and certainly the dianionic phosphate (signal A) is a likely candidate for binding divalent metal ions (site 5, Fig. 8). It is surprising that peaks C, P, and T which are $Mg^{2+}$-sensitive and likely reflect that the specific divalent-metal-ion tight binding sites are not especially $Mn^{2+}$-sensitive. Gorenstein *et al.* (1981) have made the assumption that the phosphates of peaks A, D, E, and U, are associated with the manganese tight binding sites identified in $^1$H-NMR and X-ray studies (largely proximate to or identical to the five $Mg^{2+}$ binding sites labeled in Fig. 8).

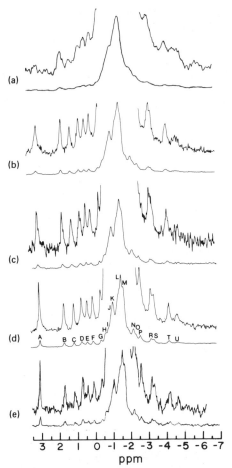

**Fig. 11.** ³¹P-NMR spectra of tRNA^Phe in 100 m$M$ NaCl, 10 m$M$ cacodylate, 10 m$M$ MgCl$_2$, no EDTA, and at pH 7.0, 32°C, and various MnCl$_2$/tRNA ratios: a, 0.12; b, 0.025; c, 0.012; d, 0.004; e, 0. Number of acquisitions 30,000–40,000. (a) 2-Hz exponential multiplication line broadening; (b)–(e), 0.5-Hz line broadening.

## V. Assignment of ³¹P Signals

### A. Stereoelectronic Origin of ³¹P Chemical Shifts

Gorenstein and Luxon (1979) attempted to simulate the ³¹P spectra of tRNA^Phe utilizing only the correlation established by Gorenstein and Kar (1975) between phosphorus chemical shifts and torsional angles (described by Gorenstein, Chapter 1, Fig. 6). Based on this and model-system (Gorenstein, Chapter 1; Gorenstein and Goldfield, Chapter 10) studies, the main upfield signal at $-1.6$ ppm in Fig. 2 is assigned to the $-g,-g$ double-helical stem phosphates. The torsional angles for the 75 phosphate diesters from the X-ray structure (Sussman *et al.*, 1978) of tRNA$^{Phe}$ and the chemical-shift contour map (Chapter 1, Fig. 6) were used to simulate the tRNA $^{31}$P spectrum (Fig. 3, stick figure). Although the central cluster shape is approximately reproduced in the computed spectrum, the fit to the scattered peaks is hopelessly inadequate. In the main cluster, the helical $-g,-g$ phosphate peaks L and M between $-1.3$ and $-2.0$ ppm integrate for 35 phosphates, whereas the X-ray structure indicates that there should be 32 phosphates having $^{31}$P signals falling within this chemical-shift range. If the model systems (Gorenstein *et al.*, 1976, 1982; Gorenstein *et al.*, 1982; Chapters 8 and 10) and $^{31}$P chemical-shift calculations are at least qualitatively correct, then conformational effects are responsible for a spread of at most 2.6 ppm (g,g relative to t,t). Because the actual spread in the scattered peaks is nearly 7 ppm, it would appear as though some factor other than torsional effects was responsible for the additional spread. It is quite likely that this second effect is ester RO—P—OR' bond-angle changes because $^{31}$P chemical shifts are quite sensitive to RO—P—OR bond-angle changes as well, as discussed by Gorenstein (Chapter 1, Fig. 2; see also Gorenstein, 1975). Thus reduction of ester O—P—O bond angles by $10°$ results in a downfield shift of $20-30$ ppm so that a $2-3$ ppm shift per degree of bond-angle distortion might be expected from this correlation. Alteration of the ester bond angle through the tertiary interactions in the tRNA could readily explain the scattered peak shifts. The scattered downfield peaks would represent phosphates with bond angles several degrees smaller than normal, whereas the scattered upfield peaks would represent phosphates with bond angles several degrees larger than normal. Unfortunately, these bond-angle differences are beyond the resolution of present X-ray diffraction studies.

The observed coalescence of all the diester signals of the tRNAs at $\sim 70°$C in the presence of 10 m$M$ Mg$^{2+}$ is consistent with this stereoelectronic origin (bond/torsional angles) for these $^{31}$P shift differences because at this temperature the native structure melts into a random-coil state. Except for some of the scattered signals (discussed later), the overall constancy of the $^{31}$P spectra in the temperature range $22-66°$C is also consistent with the hypothesis that $^{31}$P chemical shifts are sensitive to phosphate ester torsional angles and bond angles. The torsional angles are presumably constrained to fixed values in the native tRNA, and therefore the $^{31}$P signals shift little with temperature prior to disruption of this structure.

Without added Mg$^{2+}$ in the tRNA$^{Phe}$ sample, the main cluster upfield signals L' and M at $-1.5$ ppm, shift downfield by a total of $0.8-0.9$ ppm with increasing temperature. The $^{31}$P chemical-shift changes of peak M in the high-salt (no $Mg^{2+}$) tRNA$^{Phe}$ sample is very similar to changes in the upfield (multistrand) signal in a poly(A) · oligo(U) double- and triple-strand complex (under similar buffer conditions; see Fig. 11 in Gorenstein, Chapter 1). The similarity in the melting transition ($T_m \sim 48°$) is largely coincidental though the disruption of 20 A · U (or A · U · U) base pairs or triplets in poly(A) · oligo(U) and is probably similar energetically to the melting of the four stems with 20 base pairs in native tRNA. The sharp melting of the native tRNA structure in the absence of $Mg^{2+}$ around 50°C is consistent with $^1$H-NMR (Kearns, 1976; Robillard and Reid, 1979) and optical melting studies (Romer et al., 1969). This and other evidence (see Gorenstein et al., 1981) suggest that the *disappearance of the scattered peaks is associated with loss of tertiary structure.*

## B. Assignment of Scattered Signals

Partial assignment of the $^{31}$P scattered signals can also be attempted based on the metal-binding experiments described before and the chemical modification studies of Salemink et al. (1979). As shown in Table I and discussed previously, peaks A, C, E, F, P, and T/U are all sensitive to magnesium concentration (particularly with regard to their temperature-dependent behavior). It is likely that the phosphates that are $Mg^{2+}$-sensitive are associated with the conformation of phosphates near the four magnesium tight binding sites. It is possible of course that magnesium binding produces a conformational change and $^{31}$P perturbation that is transmitted to a phosphate some distance from the actual $Mg^{2+}$ binding site. Without additional information to the contrary, however, we take the former, simpler position. Similarly, we assume that the phosphates of peaks A, D, E, and U are associated with the manganese tight binding sites identified by $^1$H-NMR and X-ray studies (Fig. 8).

We know from earlier $^{31}$P studies that peaks B–H and N–U are associated with the tertiary structure, and peak A is the terminal 5'-phosphate monoester. Salemink et al. (1979) have shown that peak F is lost on removal of the Y base in the anticodon (AC) loop. Similarly, bovine pancreatic RNase A treatment of tRNA$^{Phe}$ cleaves the ACCA terminus and the phosphodiester bond at $U_{33}$ in the AC loop. As shown in Table I, peaks C, F, and U are lost in this partial hydrolysis of the native structure. In addition, RNase $T_1$ treatment of tRNA$^{Phe}$ apparently cleaves the D and TΨC loops but does not affect the AC loop. Salemink et al. (1979) have indicated that the only scattered signals remaining after RNase $T_1$ treatment are B, C, D, E, U, and some in the N–P region.

As shown in Table I, we can conclude from these experiments that peak C

TABLE I

Effect of Perturbants and Modifications on $^{31}P$ Spectrum of Yeast tRNA$^{Phe}$ and Tentative Identification of Signals

| Peak[a] | Tertiary[b] | Temperature[c] | $T_2$[d] | Et[e] | Mg$^{2+}$[f] | Mn$^{2+}$[g] | $-Y$[h] | RNase A[i] | RNase T$_1$[j] | Location | Tentative identification |
|---|---|---|---|---|---|---|---|---|---|---|---|
| A |   |   |   |   |   |   |   |   | Retained | AA arm | Terminal 5'-phosphate |
| B | + | + |   |   |   |   |   |   | Retained |   |   |
| C | + |   | + |   | + |   |   | Lost | Retained | AC loop, site 4 | $Y_{37}$ |
| D | + |   | + |   | + | + |   |   |   |   |   |
| E | + | + | + | + | + | + |   |   | Retained | AC loop, or AC/D arm, Site 1 | $U_8$, $A_9$, $C_{11}$, or $U_{17}$ / $U_{33}$ (or U) |
| F | + | + |   |   |   |   | Lost | Lost |   |   |   |
| G |   |   |   |   |   |   |   |   |   |   |   |
| H |   |   |   |   |   |   |   |   |   |   |   |
| I |   |   |   |   |   |   |   |   |   |   |   |
| J |   |   |   |   |   |   |   |   |   |   |   |
| K |   |   |   |   |   |   |   |   |   |   |   |
| L |   |   |   |   |   |   |   |   |   |   |   |
| M |   |   |   |   |   |   |   |   |   |   |   |

| Peak | | | | | | | | Status | Location | Assignment |
|---|---|---|---|---|---|---|---|---|---|---|
| N | + | + | | | | | | Retained } Some | TΨC or D arm, site 2/3 | $G_{19}$, $G_{20}$, or $A_{21}$ |
| O | + | + | | | | | | | | |
| P | + | + | + | | | | | | | |
| Q | + | + | + | + | | | | | | |
| R | + | + | + | | | | | | | |
| S | + | + | + | | | | | | | |
| T | + | + | | | + + | + + | | Lost One } Lost | TΨC or D arm, site 2/3 AC loop, Mn site | $G_{19}$, $G_{20}$, or $A_{21}$ $U_{33}$ (or E/F) |
| U | + | + | | | + + | (+) | + | Retained | | |

[a] See Fig. 1 for peak identification.
[b] Peaks associated with tertiary structure.
[c] Temperature-dependent peaks.
[d] Peaks that show chemical exchange line broadening (other than denaturation effects).
[e] Peaks perturbed by ethidium ion.
[f] Peaks affected by $Mg^{2+}$ ions.
[g] Peaks affected by $Mn^{2+}$ ions.
[h] Peaks lost on removal of Y base (Salemink et al., 1979).
[i] Peaks lost on partial hydrolysis by RNase A (Salemink et al., 1979).
[j] Peaks still remaining after partial hydrolysis by RNAse $T_1$ (Salemink et al., 1979).

is in the AC loop (from the RNase A and $T_1$ data) and is likely one of the $Mg^{2+}$- but not one of the $Mn^{2+}$-binding sites. This description fits best with the phosphate of $Y_{37}$, which is the only $Mg^{2+}$-binding site in the AC. It is also not a $Mn^{2+}$-binding site.

Either peak E, F, or U may be tentatively identified as the phosphate of $U_{33}$ because they are in or near the AC loop based on the RNase A and $T_1$ results (Table I), show chemical-exchange line-broadening effects (see below), are affected by $Mg^{2+}$ binding and temperature but are likely associated with the only $Mn^{2+}$-binding site ($U_{33}$) in the AC loop (at least E and U). Salemink *et al.* (1979) have argued that peak U (labeled $j_2$ in their paper) is likely the hydrogen-bonded phosphate $P_{36}$. However, as discussed by Gorenstein (Chapter 1), hydrogen-bonding interactions to phosphates produce only small perturbations (Gorenstein, 1978, 1981; <0.5 ppm) and probably are not responsible for the large upfield shift (>3.0 ppm) of peak U. Stereoelectronic effects resulting from tertiary folding of the polynucleotide backbone, however, can readily explain these large shifts. Because peak E and not U is perturbed on binding the drug ethidium bromide to tRNA$^{Phe}$, it is likely that E is not in the anticodon loop (see Gorenstein and Goldfield, Chapter 10, for further discussion). It is therefore more likely that peak U is associated with $U_{33}$.

Peak F is magnesium-sensitive and is lost when treated with RNase A and RNAse $T_1$. It, like peak E, could thus be associated with the magnesium-binding site 1 that is proximate to both the AC and D arms.

Peaks U and T show evidence for chemical-exchange line broadening at temperatures below $T_m$ (Gorenstein *et al.*, 1979). The peaks are narrow at low temperature, broaden at intermediate temperature (because of an intermediate rate of chemical exchange), and then narrow again at higher temperature. As discussed later, the broadening and chemical-shift changes for peak U are probably associated with a conformational change of the AC loop as the binding site for $Mg^{2+}$ is disrupted at higher temperatures (but still below $T_m$). It is significant that peak U in *E. coli* tRNA$^{Tyr}$ shows similar chemical-exchange line-broadening effects. At low temperature (<25°C), the signal is too broad to be observed and only appears as it sharpens at $T > 35°C$ (Fig. 4), presumably owing to fast exchange removing any chemical-exchange line broadening. This provides further support that the tertiary structures of the two tRNAs are quite similar.

Peaks P and T are phosphates in the hinge region of the molecule where the TΨC and D loops fold together and where two of the bound magnesiums are found (Fig. 8, sites 2 and 3). Direct interaction of the phosphates of $G_{19}$, $G_{20}$, and $A_{21}$ with the magnesiums is likely responsible for the sensitivity of these peaks to $Mg^{2+}$. These phosphates are not affected by $Mn^{2+}$ (at low concentrations), and it is believed that these $Mg^{2+}$-binding sites are not one of the three $Mn^{2+}$ tight binding sites.

## VI. Conformational Transitions

Previous optical (Sprinzl *et al.,* 1974) and $^1$H-NMR studies (Kan *et al.,* 1977; Reid and Hurd, 1979) have established that tRNA$^{Phe}$ retains its tertiary and secondary structures up to 65–70°C in the presence of 10 m$M$ Mg$^{2+}$. The overall constancy of the $^{31}$P spectra in the temperature range 22–66°C is consistent with this conclusion, although some conformational changes are detected by $^{31}$P NMR in this range. As shown in Figs. 3 and 4, some of the scattered peaks show moderate upfield or downfield shifts with temperature. In addition, line-broadening effects and analysis of the integrated intensities of the main cluster and scattered peaks suggest that a second lower temperature transition may be monitored with $^{31}$P spectral changes.

The simplest model to analyze these results is in terms of low-temperature (<50°C) and high-temperature (>60°C) transitions. We suggest that transition I at lower temperatures is associated with the interconversion of two different conformations, both of which retain the main features of the secondary and tertiary structures found in the X-ray structure. Transition II at higher temperature represents the cooperative melting of the secondary and tertiary structures into a random coil.

Looking first at transition II, the spectral changes suggestive of thermal denaturation can be summarized as follows. The scattered signals B–F and T–U start broadening at ~50°C (except for D at 146 MHz). Intensity decreases are also observed for the well-resolved scattered peaks B–E, T, and U at higher temperatures (Gorenstein and Luxon, 1979).

Except for C, E, and U, only small changes in the chemical shifts of the signals are observed between 50 and 60°C. Suddenly, between 60 and 70°C almost all the scattered peaks disappear and the integrated intensity of the main cluster (J–M) increases to 73 phosphates. At temperatures greater than 70°C, all the diester peaks merge into a single signal.

The broadening and intensity changes and lack of major chemical-shift effects with temperature between 50 and 66°C are consistent with slow chemical exchange[3] (Pople *et al.,* 1959) between the native structure and the random coil. The chemical shift of the random coil is similar to that of peaks G–J, and as the population of the denatured random-coil conformation increases between 60 and 70°C, the signal intensities for the native structure are transferred into this region. By 66°C, a total of about 6 of the 16 scattered phosphates in the native structure are lost, representing 36% in the melted state. The estimated melting temperature ($T \sim 68°C$) from Fig. 5 agrees

---

[3] On the NMR time scale, two signals are in slow exchange if $1/\tau \ll 2\pi(\Delta\nu)$, where $\Delta\nu$ is the chemical-shift difference between the two sites and $\tau$ the lifetime of the state. For fast exchange where only one averaged NMR signal is observed, $1/\tau \gg 2\pi(\Delta\nu)$.

with the single cooperative thermal transition between 65 and 75°C estimated by optical and $^1$H-NMR techniques under similar conditions (Romer et al., 1970; Robillard et al., 1977).

In contrast to the slow chemical exchange observed for the thermal denaturation (transition II), the $^{31}$P spectral changes for the lower temperature transition I are consistent with more rapid chemical exchange between two (or possibly more) different tertiary conformations.

For most of the tRNA species, peaks C, E, F, P, T, and U shift with temperature between 22 and 60°C. The linewidths for the resolvable peaks, especially T and U, are similarly temperature-sensitive. For tRNA$^{Phe}$ at 32 MHz, the linewidth of peak T increases from 10 to 14 Hz, and U increases from 8 to 24 Hz between 22 and 27°C. With further increase in temperature, both signals resharpen to linewidths of 5 Hz by ~45°C (at 32 MHz). At 146 MHz, U broadens from ~70 Hz at 27°C to 100 Hz at higher temperatures. Peaks C and U in tRNA$^{fMet}$ undergo similar linewidth changes with temperature (Fig. 9B; Gorenstein and Goldfield, 1982b). Peak U in E. coli tRNA$^{Tyr}$ shows similar chemical-exchange line-broadening effects. At low temperature (35°C) the signal is too broad to be observed and only appears as it sharpens at $T > 35$°C, presumably because of fast exchange removing any chemical-exchange line broadening.

The line-broadening maximum with temperature for T and U and the frequency dependence of the line broadening are strong evidence for chemical-exchange effects in transition I. The slow-exchange broadening component of the linewidth $\delta v_{ex}$ relates to the lifetime $\tau_I$ of the signal in this transition, with $1/\tau_I = \pi(\delta v_{ex})$, where $\delta v_{ex} = \delta v_{obsd} - \delta v_0$ and $\delta v_{obsd}$ is the observed linewidth and $\delta v_0$ the intrinsic linewidth in the absence of chemical exchange (Pople et al., 1959). The term $\delta v_0$ can be estimated from the linewidths of signals B–E, which show no evidence of any (or remaining) chemical-exchange broadening (at least at 32 MHz and 35–50°C). Thus for peak T at 30°C, $\delta v_{ex} \simeq 14 - 5 = 9$ Hz or $\tau_I \sim 35$ ms. For U, $\tau_I$ is estimated at 8 ms (tRNA$^{Phe}$) and 39 ms (tRNA$^{fMet}$) (Gorenstein and Goldfield, 1982b).

The lifetime can also be estimated from the chemical-shift difference between the two sites $\Delta v$: $\tau \sim \sqrt{2}/(2\pi\Delta v)$. This expression is only valid in the intermediate-exchange region (where maximum line broadening occurs). With a chemical-shift difference between the two conformations for the T peak of 0.3 ppm, $\tau$ is estimated at 23 ms, in reasonable agreement with the 35-ms lifetime at lower temperature in the slow-exchange region. Because of the field dependence of $\Delta v$, at 146 MHz maximum broadening would occur at a rate of exchange 4.5 times faster than at 32 MHz (146:32) or $\tau_I \sim 5$ ms. Maximum broadening for peak T occurs 20°C higher at 146 MHz than at 32 MHz. This temperature dependence is consistent with an increase of a factor of ~2 in the rate of exchange per 10°C rise in temperature for a transition with a "moderate" activation energy. Also, the linewidth at maximum broadening is larger at higher field strength (for T and U, 14 and 24 Hz at 32 MHz, respectively, and 42 and >100 Hz at 146 MHz, respectively).

The decrease in tRNA$^{Phe}$ linewidths for peaks C–F at 32 MHz (and C for tRNA$^{fMet}$) from 22 to 36°C (32.4 MHz) is also a chemical-exchange broadening effect. The increase in linewidth for C–F at 146 MHz from 26 to 36°C is certainly consistent with this because we again expect a field dependence to the chemical-exchange broadening effect as described above. Below 36°C and at 32 MHz, these signals experience fast exchange, whereas at 146 MHz they are in slow exchange. The minimum 12- to 13-Hz exchange broadening for C–F at 34°C indicates $\tau_1 \sim 24$ ms, consistent with the relaxation lifetime measured from line broadening of T. The anomalously small linewidth for B and its uniquely slow spin-lattice relaxation rate (Gorenstein and Luxon, 1979) suggest that B is in a special environment. (Note that it is also not Mg$^{2+}$-sensitive).

The estimated millisecond time scale for exchange cannot be due to a helix-to-coil transition because for hairpin loops lifetimes of 10–100 $\mu$s are expected (no Mg$^{2+}$ in 30 m$M$ Na$^+$). Robillard et al. (1977), Romer et al. (1970), and Coutts et al. (1975) have shown that disruption of the teritary structure is associated with a longer (2–23 ms) relaxation time. More recently, Labuda and Porschke (1980, 1982), using temperature-jump Y-base fluorescence relaxation kinetics, have identified a conformational transition in the anticodon loop of tRNA$^{Phe}$ in a similar Mg$^{2+}$ buffer. Their measured relaxation time of 1 ms at 7°C is, however, shorter than our rate process.

Most significantly, as shown in Table I, the peaks that are most temperature- and magnesium-sensitive for this early transition I conformational change are largely in the anticodon loop (particularly C, F, and E/U). Transition I therefore represents a conformational change in the anticodon loop, supporting the earlier suggestions (Labuda and Porschke, 1980, 1982; Urbanke and Maass, 1978). Although highly speculative, we present one possible interpretation for this conformational change.

Besides their temperature sensitivity, peaks C and E/U are also sensitive to magnesium ion (we still are not sure whether either E or U or both are involved in the AC-loop conformational transition). It is potentially significant that temperature and magnesium ion have roughly an *equal and opposite* effect on the $^{31}$P chemical shifts of peaks C and E/U. Thus C shifts $\sim 1$ ppm upfield with increasing temperature (20–65°C) or decreasing magnesium (from 10 m$M$ Mg$^{2+}$ to none), whereas E/U shifts 0.6–1.5 ppm downfield under similar changes. Recall that peak C is assigned to the phosphate of Y$_{37}$ and E/U to the phosphate of U$_{33}$. They are both involved in the divalent metal-ion-binding site(s) that must contribute to the stabilization of the anticodon-loop conformation. The ~1.0-ppm changes in the chemical shifts of C and E/U in this AC-loop transition are about the right magnitude for rotation about one of the diester bonds from a gauche to a trans conformation or the reverse. Thus $P_{37}$ of $Y_{37}$ (Peak C) is in a $-g,t$ conformation (Sussman *et al.*, 1978) and is directly stabilized by coordination to $Mg^{2+}$ (site 4, Fig. 8). Disruption of this structure by increasing the temperature or decreasing the magnesium ion concentration could alter the conformation about this phosphate to the lower energy $-g,-g$ (Gorenstein and Kar, 1977). Phosphate-33 of $U_{33}$ (peak E/U) is in a $-g,-g$ conformation at high magnesium concentration and low temperature (at least from the X-ray structures; Sussman *et al.*, 1978). Loss of magnesium (either by increasing the temperature or decreasing the magnesium concentration could allow $P_{33}$ to assume a $-g,t$ conformation, stabilized by the hydrogen bond between the $N_3H$ of $U_{33}$ and $P_{35}$ (Sussman *et al.*, 1978; Quigley *et al.*, 1975; Jack *et al.*, 1976; Stout *et al.*, 1978). This hydrogen-bond arrangement is observed in the X-ray structures and helps provide for the "U turn" in the phosphate ester backbone of the AC loop.

This magnesium-dependent AC-loop conformational change could be biochemically significant. It is well-known that magnesium ion is essential for the proper function of the tRNA in the translation process (Gassen, 1980). Urbanke and Maass (1978) have also noted a similar low-temperature conformational change in the AC loop as monitored by $Y_{37}$ fluorescence changes. Proton-NMR chemical shift changes of base Y methyl signals are also consistent with a change in the stacking arrangement of the anticodon triplet ($mG_{34}A_{35}A_{36}$) and $Y_{37}$ (Davanloo *et al.*, 1979). This AC loop conformational change could represent a shift from the AC triplet 3'-stacked conformation to a 5'-stacked conformation, as first suggested by Fuller and Hodgson (1967). Only two bases would remain stacked on the 3' side in the 5'-stacked conformation (the triplet and two other bases are stacked on the 5' side). The reverse is true for the 3'-stacked conformation. Because the major conformational flexibility in the tRNA backbone is about the phosphate ester bond (Stout *et al.*, 1978), it is reasonable that the two "pivot points" for this conformational change are two phosphates. They simply switch between $-g,-g$ and $-g,t$ conformations.

This analysis gains additional support from the X-ray structure of *E. coli* initiator tRNA$^{fMet}$ (Woo *et al.*, 1980). Overall, the yeast tRNA$^{Phe}$ and *E. coli* tRNA$^{fMet}$ structures are quite similar. A major difference, however, exists in the anticodon loop where in the initiator tRNA, $U_{33}$ has an almost opposite orientation to that found in tRNA$^{Phe}$. In the latter, $U_{33}$ folds into the AC loop, with $N_3$ of $U_{33}$ forming a hydrogen bond with $P_{36}$. In tRNA$^{fMet}$, $U_{33}$ is rotated into the solvent region and a hydrogen bond is perhaps made between the 2'-OH of ribose-33 and $P_{36}$. The oppositely rotated $U_{33}$ in tRNA$^{fMet}$ could represent the high-temperature, low-magnesium, anticodon-loop conformation. No divalent-metal-ion sites have yet been described in the X-ray structure for tRNA$^{fMet}$, and it would be significant if one were either not found or were located in a different position than in tRNA$^{Phe}$.

The conformational change in the anticodon loop has been suggested to be an important event in protein biosynthesis (Woese, 1970; Lake, 1977). Thus in the Lake (1977) model, the anticodon conformation switches from a 5' stack to a 3' stack, after the correct recognition of the tRNA in the ribosome site. This would presumably allow the aminoacyl group to move toward the ribosome-bound peptidyl-tRNA.

Our results suggest that initiator tRNA and chain-elongator tRNAs can exist in both anticodon-loop conformations, although some of the elongator tRNAs (such as tRNA$^{Tyr}$; Fig. 6) do not appear to undergo a low-temperature conformational transition. Aminoacylated initiator tRNA goes only into the ribosomal A (aminoacyl) site and then during chain elongation later, translocates to the P site.

If the tRNA$^{fMet}$ existed in only one AC-loop conformation, we would have a simple explanation for the ability of a tRNA to discriminate between the P and A ribosomal sites. However, it has recently been shown that initiator tRNA likely exists in a least two conformation states when bound to the 70-S ribosomal peptidyl site (Eckhardt and Luhrmann, 1981). Obviously the conformational flexibility of both initiator and elongator tRNAs is both complex and potentially significant.

## VII. Spermine Effects

The most striking effect of addition of spermine on the $^{31}$P-NMR spectrum of tRNA$^{Phe}$ is the broadening of the upfield peak U at $-4.1$ ppm (Gorenstein and Goldfield, 1982b). This line broadening varies with the spermine/tRNA ratio and reaches a maximum at the 1:1 ratio at 31°C. Previous spectra on different tRNA$^{Phe}$ samples have shown variable linewidths for peak U at $\sim 30$°C. Salemink et al. (1979) had noted a sharpening of peak U on phenol extraction of tRNA$^{Phe}$. Residual quantities of spermine in the dialyzed tRNA samples may have been responsible for these effects. Cohen (1972) has shown that tRNAs bind 10 molecules of spermine and that 2 or 3 spermine molecules are tightly bound (similarly reported by Schreier and Schimmel, 1975). Takeda and Ohnishi (1975) have found that spermine is, in fact, more tightly bound than Mg$^{2+}$. Either dicationic magnesium or tetracationic spermine can convert tRNA from an inactive conformation to

$$NH_3^+-(CH_2)_3H_2N^+-(CH_2)_4H_2N^+-(CH_2)_3-NH_3^+$$

an active one (Cohen, 1972), at least as monitored by the ability of the tRNA to be aminoacylated (Tabor and Tabor, 1976; Kayne and Cohn, 1972). Two of these bound spermine molecules have been resolved in the X-ray structure of tRNA$^{Phe}$ (Quigley et al., 1978). One spermine binds in the deep groove of the anticodon double-helix stem, apparently hydrogen-bonded to four different phosphate residues. The positively charged spermine appears to draw the phosphates on opposite sides of the groove 3 Å closer together than might be expected in its absence. The second spermine binds at the beginning of the D stem near the top of the AC stem. It wraps around phosphate-10 and neutralizes the charge interaction between phosphates-9, -10, and -11 of one chain and phosphates-47, -46, and -45 of the other. Together with the two bound magnesium ions in the D and AC arms, the spermine immobilizes the AC end of the tRNA.

The linewidth and chemical-shift changes of the scattered signals that we have observed in the presence of spermine are consistent with the X-ray results. As discussed previously, the low-temperature conformational transition I is associated with a change in the anticodon-loop conformation. Peak U has been assigned to the anticodon loop (Salemink et al., 1979; Gorenstein et al., 1981), and at the spermine/tRNA ratio of 3:1 peak U narrows at higher temperatures. At higher temperature, the narrowing of peak U as the temperature is increased is to be expected if the broadening of this peak is due to chemical exchange between the two conformations of the anticodon loop.

## VIII. $^{31}$P NMR of Yeast tRNA$^{Phe}$ · *E. coli* tRNA$_2^{Glu}$ Complex: Spectral Changes

The anticodons of yeast tRNA$^{Phe}$ (G$^{2'OMe}$AA) and *E. coli* tRNA$_2^{Glu}$ (s$^2$UUC) are complementary, and the two form a strong complex through hydrogen-bonding interaction of the anticodons (Eisinger, 1971; Grosjean et al., 1976). Comparison of the $^{31}$P-NMR spectra of the tRNA$^{Phe}$ and tRNA$_2^{Glu}$ alone and in a 1:1 complex at 55°C is shown in Fig. 12.

The most striking features of the dimer spectra are (1) the complete absence of intensity of peak C, the temperature-variable peak (which is absent also in tRNA$_2^{Glu}$ and assigned to Y$_{37}$ in tRNA$^{Phe}$) and (2) an apparent gain in intensity in the main cluster. This region accounts for 50–53 phosphates in each of tRNA$_2^{Glu}$ and tRNA$^{Phe}$ and at first inspection ~124 resonances in the dimer. However, the dimer main cluster is broader than in

9. High-Resolution $^{31}$P-NMR Spectroscopy of tRNAs

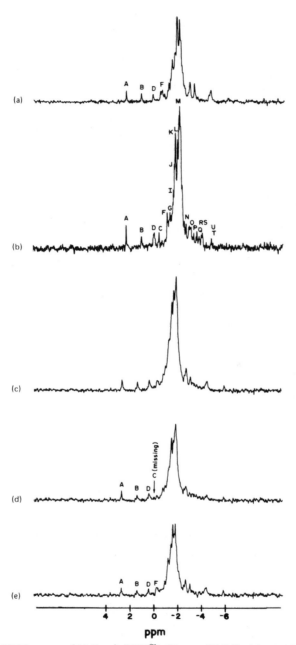

**Fig. 12.** $^{31}$P-NMR spectra of (a) *E. coli* tRNA$_2^{Glu}$, (b) yeast tRNA$^{Phe}$, (c) 1:1, (d) dimer minus tRNA$_2^{Glu}$, and (e) dimer minus tRNA$^{Phe}$; 55°C, 10 m$M$ Mg$^{2+}$ buffer, 32.4 MHz (25,000–35,000 scans).

either tRNA$_2^{Glu}$ or tRNA$^{Phe}$, so that it covers up some of the upfield resonances between $-1.0$ and $-1.4$ that are resolved in the individual tRNA monomer spectra and that integrate in total for 6 resonances. Therefore, the main cluster peaks actually account for $\sim 118$ phosphates, still larger than the expected $100-106$. This increase in the main cluster peak area is largely due to an increase in peak I that at 55°C is at $-0.7$ ppm in tRNA$_2^{Glu}$ and at $-0.77$ ppm in tRNA$^{Phe}$, or the left-most shoulder of the main cluster. These peaks in the dimer account for $\sim 10$ more phosphates than expected. The other striking difference is the reduction in the intensity of the upfield scattered peaks. Although the dimer main cluster is broad and obscures some of the resolved peaks in the tRNA monomers at $-1.0$ to $-1.4$ ppm, it has resolved upfield structure at $-2.0$ ppm and $-2.1$ ppm that integrates for 4 phosphates, whereas the tRNA$_2^{Glu}$ and the tRNA$^{Phe}$ monomer spectra contribute about 7 phosphates in this region. Further upfield, in the region from peaks P and R (spectral window $-2.5$ ppm to $-3.2$ ppm), there is also lost intensity. The tRNA$_2^{Glu}$ has a peak at $-2.5$ ppm that accounts for 2 phosphates, and the tRNA$^{Phe}$ has 5 peaks in this window that integrate for $6 \pm 1$ phosphates. However, the dimer contains *at most* 5 phosphates in this region, and the tRNA$^{Phe}$ peak at $-3.2$ ppm has either greatly diminished in intensity or has disappeared. Peaks (T/U) at $-3.85$ to $-3.95$ ppm account for 6 phosphates in tRNA$_2^{Glu}$ and 2 in tRNA$^{Phe}$ but only 2 in the dimer. Thus it appears that at least 10 phosphates have disappeared from the upfield spectral window. The loss of intensity in this region could be caused either by equilibrium intensity loss, line broadening owing to chemical exchange, or a shift of intensity downfield into the central cluster region. Finally, a new far-upfield peak that integrates for $\frac{1}{2}$ phosphate has appeared in the dimer at $-5.3$ ppm. No peak has ever been observed so far upfield in any tRNA monomer spectra at 55°C, although this signal likely arises from the phosphate assigned to peak U ($U_{33}$?).

## IX. Conclusions

These features of the dimer spectra and previous discussion argue strongly that there are at least three conformations for tRNA$^{Phe}$ at high Mg$^{2+}$ (below $T_m$). The disappearance (or perhaps upfield shift) of peak C in the dimer strongly suggests that the AC loop has undergone a conformational change on codon binding, and that this conformation is different from either of the AC-loop conformations described previously. Furthermore, the loss of upfield intensity indicates loss or modification of tertiary structure because these peaks are associated with tertiary structure. This loss is similar to the spectral changes observed on treatment with RNase $T_1$, which cleaves in the TΨC and D loops of tRNA disrupting the tertiary structure: the upfield resonances decrease from a total of 11 phosphates to 3 (Salemink *et al.*, 1979). Also, the growth of the downfield portion of the main cluster, which is where we find unstrained hairpin loop g,g phosphates, indicates that the phosphates that were once in the tertiary folds of the tRNA are now in relatively free hairpin loops. Thus the tertiary structure has been considerably opened up. This conclusion is supported by equilibrium binding studies that have suggested that codon–anticodon recognition induces a partial unfolding of the tertiary structure, making part of the TΨC loop (TΨCG sequence) available for complementary base-pairing to CGAA (Schwarz *et al.*, 1974; Moller *et al.*, 1978). Chemical modification (Wagner and Garrett, 1978) and fluorescence (Robertson *et al.*, 1977) experiments have supported this conclusion, whereas $^1$H-NMR studies (Davanloo *et al.*, 1979; Geerdes *et al.*, 1978) have suggested that no (or very minor) changes occur on codon–anticodon complexation.

Phosphorus-31 NMR appears to be a uniquely sensitive magnetic resonance probe for these subtle conformational changes in tRNA structure. Unlike other spectroscopic probes that largely monitor the base structure of tRNA, $^{31}$P NMR can provide conformation and dynamics on the phosphate ribose backbone.

## Acknowledgments

Support of this research by the National Science Foundation, the National Institutes of Health, and the Alfred P. Sloan Foundation for a fellowship is gratefully acknowledged. Support of the Purdue Biological NMR facility by the NIH (RRO 1077) is also appreciated. The collaboration with Roulhwai Chen, John Findlay, Evelyn Goldfield, Dave Kar, Ken Kovar, Bruce Luxon, and Kofen Lai has, of course, been crucial to these studies.

## References

Brennan, T., and Sundaralingam, M. (1976). *Nucleic Acid Res.* **3**, 3525–3251.
Cohen, J. S., and Chen, C.-W. (1982). *ACS Symp. Ser.* **191**, 249–266.
Cohen, S. S. (1972). *Adv. Enzyme Regul.* **10**, 207–233.
Coutts, S. M., Riesner, D., Romer, R., Rabl, D. R., and Maass, G. (1975). *Biophys. Chem.* **3**, 275–289.
Davanloo, P., Sprinz, M., and Cramer, F. (1979). *Biochemistry* **18**, 3189–3199.
Eckhart, H., and Luhrmann, R. (1981). *Biochemistry* **20**, 2075–2080.
Eisinger, J. (1971). *Biochem. Biophys. Res. Commun.* **43**, 854–861.

Fuller, W., and Hodgson, A. (1967). *Nature (London)* **215,** 817–821.
Gassen, H. G. (1980). *Prog. Nucleic Acid Res. Mol. Biol.* **24,** 57–86.
Geerdes, H. A. M., Van Boom, J. H., and Hilbers, C. W. (1978). *FEBS Lett.* **88,** 27–32.
Gorenstein, D. G. (1975). *J. Am. Chem. Soc.* **97,** 898–900.
Gorenstein, D. G. (1977). *J. Am. Chem. Soc.* **99,** 2254–2258.
Gorenstein, D. G. (1978). *Jerusalem Symp. Quantum Chem. Biochem.* **11,** 1–15.
Gorenstein, D. G. (1981). *Annu. Rev. Biophys. Bioeng.* **10,** 355–386.
Gorenstein, D. G., and Goldfield, E. M. (1982a). *J. Mol. Cell. Biochem.* **46,** 97–120.
Gorenstein, D. G., and Goldfield, E. M. (1982b). *Biochemistry* **21,** 5839–5849.
Gorenstein, D. G., and Kar, D. (1975). *Biochem. Biophys. Res. Commun.* **65,** 1073–1080.
Gorenstein, D. G., and Kar, D. (1977). *J. Am. Chem. Soc.* **99,** 672–677.
Gorenstein, D. G., and Luxon, B. A. (1979). *Biochemistry* **18,** 3796–3804.
Gorenstein, D. G., Findlay, J. B., Momii, R. K., Luxon, B. A., and Kar, D. (1976). *Biochemistry* **15,** 3796–3803.
Gorenstein, D. G., Goldfield, E. M., Chen, R., Kovar, K., and Luxon, B. A. (1981). *Biochemistry* **20,** 2141–2150.
Gorenstein, D. G., Luxon, B. A., Goldfield, E. M., Lai, K., and Vegeais, D. (1982). *Biochemistry* **21,** 580–589.
Grosjean, H., Soll, D. G., and Crothers, D. M. (1976). *J. Mol. Biol.* **103,** 499–519.
Guéron, M. (1971). *FEBS Lett.* **19,** 264–266.
Guéron, M., and Leroy, J. L. (1979). *In* "ESR and NMR of Paramagnetic Species in Biological and Related Systems" (I. Bertini and R. S. Drago, Eds.), pp. 327–368. Reidel, Holland.
Guéron, M., and Shulman, R. G. (1975). *Proc. Natl. Acad. Sci. U.S.A.* **72,** 3482–3485.
Holbrook, S. R., Sussman, J. L., Warrant, R. W., and Kim, S.-H. (1978). *J. Mol. Biol.* **123,** 631–660.
Jack, A., Ladner, J. E., and Klug, A. (1976). *J. Mol. Biol.* **108,** 619–649.
Kan, L. S., Ts'o, P. O. P., Sprinzl, M., van der Haar, F., and Cramer, F. (1977). *Biochemistry* **16,** 3143–3154.
Kayne, M. S., and Cohn, M. (1972). *Biochem. Biophys. Res. Commun.* **46,** 1285–1291.
Kearns, D. R. (1976). *Prog. Nucleic Acid Res. Mol. Biol.* **18,** 91–149.
Kim, S.-H. (1979). *In* "Transfer RNA: Structure, Properties, and Recognition" (P. R. Schimmel, P. Soll, and J. R. Abelson, eds.), pp. 83–100. Cold Spring Harbor Lab., Cold Spring Harbor, New York.
Kim, S.-H., Suddath, F. L., Quigley, G. J., McPherson, A., Sussman, J. L., Wang, A. H. J., Seeman, N. C., and Rich, A. (1974). *Science* **185,** 435–440.
Labuda, D., and Porschke, D. (1980). *Biochemistry* **19,** 3799–3805.
Labuda, D., and Porschke, D. (1982). *Biochemistry* **21,** 49–53.
Lake, J. A. (1977). *Proc. Natl. Acad. Sci. U.S.A.* **74,** 1903–1907.
Möller, A., Schwarz, U., Lipecky, R., and Gassen, H. G. (1978). *FEBS Lett.* **89,** 263–266.
Moras, D., Comarmond, M. B., Fischer, J., Weiss, R., and Thierry, J. C. (1980). *Nature (London)* **288,** 669–674.
Müller, A., Guéron, M., and Leroy, J. L. (1980). *EMBO-FEBS tRNA Workshop, 16–21 July 1980.*
Patel, D. J. (1976). *Biopolymers* **15,** 533–558.
Pople, J., Schneider, W. G., and Bernstein, H. J. (1959). "High Resolution Nuclear Magnetic Resonance." McGraw-Hill, New York.
Quigley, G. J., Wang, A. H. J., Seeman, N. C., Suddath, F. C., Rich, A., Sussman, J. L., and Kim, S. H. (1975). *Proc. Natl. Acad. Sci. U.S.A.* **72,** 4866–4870.
Quigley, G. J., Teeter, M. M., and Rich, A. (1978). *Proc. Natl. Acad. Sci. U.S.A.* **75,** 64–68.
Reid, B. R., and Hurd, R. E. (1979). *In* "Transfer RNA: Structure, Properties, and Recogni-

tion" (P. R. Schimmel, D. Soll, and J. N. Abelson, eds.), pp. 177–190. Cold Spring Harbor Lab., Cold Spring Harbor, New York.

Robertson, J. M., Kahan, M., Wintermeyer, W., and Zachau, H. G. (1977). *Eur. J. Biochem.* **72,** 117–125.

Robillard, G. T., and Reid, B. R. (1979). *In* "Biological Applications of Magnetic Resonance" (R. G. Shulman, ed.), pp. 45–112. Academic Press, New York.

Robillard, G. T., Tarr, C. E., Vosman, F., and Reid, B. R. (1977). *Biochemistry* **16,** 5261–5274.

Romer, R., Riesner, D., and Maass, G. (1970). *FEBS Lett.* **10,** 352–357.

Romer, R., Reisner, D., Maass, G., Wintermeyer, W., Thiebe, R., and Zachau, H. G. (1969). *FEBS Lett.* **5,** 15–19.

Salemink, P. J. M., Swarthof, T., and Hilbers, C. W. (1979). *Biochemistry* **18,** 3477–3485.

Salemink, P. J. M., Reijerse, E. J., Mollevanger, L., and Hilbers, C. W. (1981). *Eur. J. Biochem.* **115,** 635–641.

Sander, C., and Ts'o, P. O. P. (1971). *J. Mol. Biol.* **55,** 2.

Schevitz, R. W., Podjarny, A. D., Krishnamachari, N., Hughes, J. J., and Sigler, P. B. (1979). *In* "Transfer RNA: Structure, Properties and Recognition" (P. R. Schimmel, D. Soll, and J. N. Abelson, eds.), pp. 133–143. Cold Spring Harbor Lab., Cold Spring Harbor, New York.

Schreier, A. A., and Schimmel, P. R. (1975). *J. Mol. Biol.* **93,** 323–329.

Schwarz, U., Luhrmann, R., and Gassen, H. G. (1974). *Biochem. Biophys. Res. Commun.* **56,** 807–814.

Shindo, H., McGhee, J. D., and Cohen, J. S. (1980). *Biopolymers* **19,** 523–537.

Sprinzl, M., Kramer, E., and Stehlik, D. (1974). *Eur. J. Biochem.* **49,** 595–605.

Stout, C. D., Mizuno, H., Rao, S. T., Swaminathan, P., Rubin, J., Brennan, T., and Sundaralingam, M. (1978). *Acta Crystallogr., Sect. B* **B34,** 1529–1544.

Sussman, J. L., Holbrook, S. R., Warrant, R. W., Church, G. M., and Kim, S. H. (1978). *J. Mol. Biol.* **123,** 607–630.

Tabor, C. W., and Tabor, H. (1976). *Annu. Rev. Biochem.* **45,** 285–306.

Takeda, Y., and Ohnishi, T. (1975). *Biochem. Biophys. Res. Commun.* **63,** 611–617.

Urbanke, C., and Maass, G. (1978). *Nucleic Acids Res.* **5,** 1551–1560.

Wagner, R., and Garrett, R. A. (1978). *FEBS Lett.* **85,** 291–294.

Weiner, L. M., Backer, J. M., and Rezvukhin, A. I. (1974). *FEBS Lett.* **41,** 40–42.

Woese, C. (1970). *Nature (London)* **226,** 817–820.

Woo, N. H., Roe, B. A., and Rich, A. (1980). *Nature (London)* **286,** 346–351.

Wright, H. T., Manor, P. C., Beurling, K., Karpel, R. L., and Fresco, J. R. (1979). *In* "Transfer RNA: Structure, Properties, and Recognition" (P. R. Schimmel, D. Soll, and N. I. Abelson, eds.), pp. 145–160. Cold Spring Harbor Lab., Cold Spring Harbor, New York.

# CHAPTER 10

# Phosphorus-31 NMR of Drug–Nucleic Acid Complexes

*David G. Gorenstein*
*Evelyn M. Goldfield*

Department of Chemistry
University of Illinois at Chicago
Chicago, Illinois

| | |
|---|---|
| I. Introduction | 299 |
| II. Drug–Double-Helical Nucleic Acid Complexes: Intercalation | 300 |
|   A. Small Duplexes | 300 |
|   B. Intercalating Drug–Double-Helical Polynucleic Acid Complexes | 301 |
| III. Electrostatic Mode of Interaction | 310 |
| IV. tRNA·Ethidium $^{31}$P Spectra | 311 |
|   A. $^{31}$P Perturbations | 311 |
|   B. Structural Interpretation of Effects of Ethidium Ion on tRNA | 311 |
| References | 315 |

## I. Introduction

As described in Chapters 1 and 8, $^{31}$P-NMR chemical shifts in nucleic acids have been shown to provide a direct probe of P—O ester bond torsional angles. The interaction of drugs with nucleic acids is believed to perturb the conformation of the sugar–phosphate backbone, and hence structural and dynamic information on these complexes is potentially readily accessible through $^{31}$P-NMR spectroscopy.

It is now believed that much of the pharmacological activity of these drugs derives from their direct interaction with double-helical deoxyribonucleic acids (Gale *et al.*, 1972; Lown, 1977). By perturbing DNA structure, these drugs may inhibit the synthesis of nucleic acids and interfere with cellular mitosis. The antitumor activity of many of these drugs likely involves this inhibition of DNA synthesis.

Two of the mechanisms for drug binding to double-helical nucleic acids are:

1. Intercalation of the drug (such as ethidium bromide and actinomycin D) between stacked base pairs of the nucleic acid
2. Electrostatic association of positively charged drugs (such as netropsin) to the negatively charged phosphate–ribose backbone of DNA

In this chapter we explore the ability of $^{31}$P NMR to distinguish between these two modes of drug–nucleic acid interaction and it is hoped, provide more detailed structural and dynamic information on various double-helical DNA and RNA and tRNA complexes.

## II. Drug–Double-Helical Nucleic Acid Complexes: Intercalation

### A. Small Duplexes

In one of the earliest clear demonstrations of $^{31}$P spectral perturbations on drug binding, Patel (1974, 1976a,b) and Reinhardt and Krugh (1977) showed that actinomycin D (Act D) shifted several phosphate diester signals up to 2.6 ppm downfield from the double-helical signal on binding to oligonucleotide duplexes containing dCdC base pairs. Thus downfield $^{31}$P shifts of 1.6 and 2.6 ppm in the d-GpGpCpG·Act D (2:1) complex (Fig. 1) and 1.6 ppm in the d-pGpC·Act D(2:1) complex at 8°C have been observed. Reinhardt and Krugh (1977) showed that at even lower temperature (−18°C) in methanol–water, the $^{31}$P signal for the two phosphates in duplex d-pGpC is split into two signals and shifted 1.7 and 2.4 ppm downfield on complexation with Act D.

These shifts are consistent with the Jain and Sobell (1972) model for these intercalated complexes: partial unwinding of a specific section of the double helix allows these planar, heterocyclic drugs such as Act D to stack between two base pairs (Fig. 2). X-Ray studies of various intercalating drug·duplex complexes (Reddy *et al.*, 1979; Shieh *et al.*, 1980) suggest that the major backbone deformation of the nucleic acid on intercalation of the drug involves the C-5′—O-5′ torsional angle. However, in several complexes (see Reddy *et al.*, 1979), the $\omega,\omega'$ torsional angles are altered from the normal values of 290°,290° (−g,−g) to values such as 273°,323°. According to the $^{31}$P chemical-shift contour map of Fig. 6 in Gorenstein (Chapter 1; see also Gorenstein and Kar, 1975), such torsional angle perturbations of 20–30° can result in $^{31}$P deshielding of ~0.5–1.0 ppm. Either the actual solution $\omega,\omega'$ torsional-angle changes for these drug complexes are some-

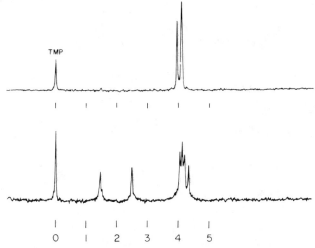

**Fig. 1.** 145.7-MHz $^{31}$P-NMR spectrum of d(CpGpCpG) (top) and d(CpGpCpG)·Act-D(2:1) (bottom) in 0.1 $M$ cacodylate, 0.01 $M$ EDTA, D$_2$O, and at pH 6.5 and 30°C. Chemical-shift scale relative to trimethyl phosphate and increasing field, increasing chemical shift. From Patel (1976a). Copyright 1976 John Wiley & Sons, Inc.

what larger than indicated by the X-ray studies or some other effect (such as bond-angle distortion or the C-5'—O-5' torsional-angle changes) also contributes to the larger observed shift perturbations (Gorenstein, 1975, 1978, 1981; see also Chapter 1 of this volume).

### B. Intercalating Drug–Double-Helical Polynucleic Acid Complexes

#### 1. Poly(A)·Oligo(U)·Ethidium Complex

Similar downfield shifts as in the Act D·GC complexes are observed in the poly(A)·oligo(U) (1:1) complex with ethidium ion (Et) (Goldfield et al., 1983). $^{31}$P-NMR spectra of poly(A)·oligo(U) at 1:1 ratio are shown in Fig. 3. At high temperature the spectra correspond to the superposition of the individual nucleic acid components. The downfield signal at −0.053 ppm at 60°C is the oligo(U). The upfield signal corresponds to poly(A). The smaller peaks are due to heterogeneity of the samples as discussed in Gorenstein et al. (1976, 1982). At temperatures below the $T_m$ of 48°C, a new, broad signal appears 0.4–0.6 ppm upfield from the single-strand nucleic acids. As discussed by Chen and Cohen (Chapter 8), this upfield signal has been assigned to the resonance of phosphates in double/

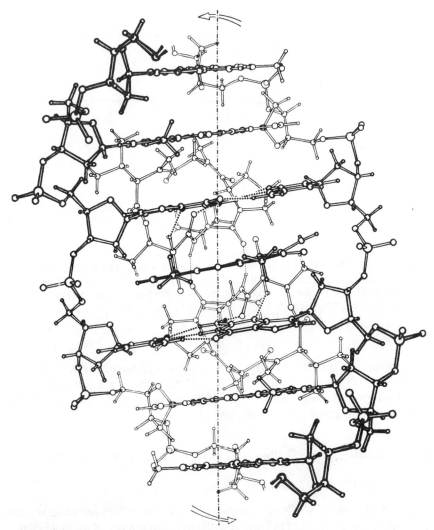

**Fig. 2.** Jain and Sobell model for intercalation of actinomycin D to double-helical (dApTpGpCpApT)$_2$. Only the planar nitrogen heterocycle ring of Act D is shown. From Jain and Sobell (1972).

triple helices (Patel and Canuel, 1979; Shindo *et al.*, 1980; Cohen and Chen, 1982; Gorenstein, 1978; Gorenstein *et al.*, 1982).

In Fig. 4, we show the $^{31}$P spectra of a (1:1:3) sample of poly(A)·oligo(U)·Et. A new, broad signal at ~0.7 ppm appears ~2 ppm downfield from the multistrand helix signal at −1.45 ppm. This signal arises

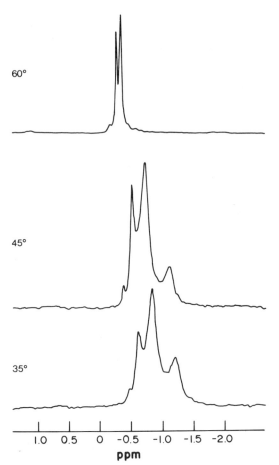

**Fig. 3.** ³¹P-NMR spectra for poly(A)·oligo(U) (1:1) at various temperatures (°C) in 0.2 $M$ NaCl, 10 m$M$ cacodylate, 1 m$M$ EDTA, and at 20% $D_2O$, pH 7.0; 32.4 MHz. Total nucleotide concentration, 24 mg/ml. Exponential line broadening, 1 Hz.

from phosphates that are structurally perturbed because of intercalated ethidium ion. The ³¹P melting curve for the signals in Fig. 4 is also shown in Fig. 5. The melting temperature as given by these data is ~62°C, considerably higher than the melting temperature of the double helix in the absence of Et (Gorenstein et al., 1982; see Chen and Cohen, Chapter 8, Fig. 1).

The new downfield signal that we observe in the poly(A)·oligo(U)·Et complex is similar to the downfield signals of the Act D complexes. This deshielding for the drug complex signal is also entirely consistent with the intercalation perturbation of the phosphate ester geometry observed in the

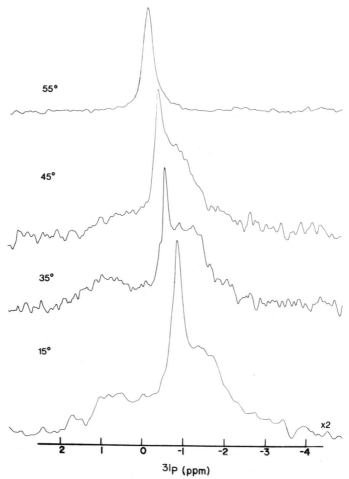

**Fig. 4.** $^{31}$P-NMR spectra of poly(A)·oligo(U)·ethidium ion (1:1:3) at various temperatures (°C) under the same conditions as Fig. 3.

Act D complexes. The unperturbed helix signals at −1.45 ppm in these complexes likely represent undistorted phosphates in regions adjacent to the intercalation site. Hogan and Jardetzky (1980), in an NMR study of Et binding to DNA, found that the effects of the intercalation are localized to a 2-base-pair-long DNA region, and that internal motions outside the binding site are nearly unaffected by the drug (see also James, Chapter 12)

Surprisingly, however, other Et complexes of nucleic acids show smaller $^{31}$P shift changes (Patel and Canuel et al. 1977; Reinhardt and Krugh, 1977). In fact, except for the results shown in Figs. 4 and 5 and the studies of Patel

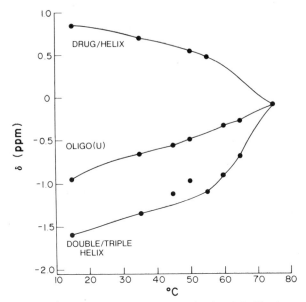

**Fig. 5.** Phosphorus-31 melting curve for signals in Fig. 4.

and Krugh on Act D, all other drug·duplex complexes show only small changes in the $^{31}$P-NMR spectra (Patel and Canuel, 1977; Reinhardt and Krugh, 1977; Wilson and Jones, 1982). Thus Reinhardt and Krugh (1977) have shown that the drugs Et and 9-aminoacridine produce small (<0.2 ppm) upfield and downfield shifts on complexation with complementary and noncomplementary deoxydinucleotides at 3–6°C.

## 2. *$^{31}$P NMR of Poly(A)·Oligo(U)·Ethidium Ion in 10 mM MgCl$_2$*

Phosphorus-31 spectra of poly(A)·oligo(U) with and without added Et in 10 m$M$ MgCl$_2$ are shown in Fig. 6. In the absence of Et in Mg$^{2+}$ buffer at 31°C, we have two peaks: the downfield peak is due to oligo(U) and the upfield peak is due to poly(A) superimposed on the double-helix phosphates. In the presence of 10 m$M$ Mg$^{2+}$, the oligo(U) and poly(A) resonances are shifted 0.3–0.5 ppm upfield so that the poly(A) peak obscures the double-helix peak. The longitudinal relaxation times $T_1$ measured for a 1:1 sample of poly(A)·oligo(U) at 40°C are 1.97 ± 0.11 s for oligo(U), 1.005 ± 0.069 s for the double helix, and 1.69 ± 0.081 s for poly(A). The $T_1$ for poly(A) is less certain than the others because the poly(A) and double helix are superimposed, but the number agrees roughly with other measurements on poly(A) alone (Akasaka, 1974; see also James, Chapter 12).

In the presence of Et ion and Mg$^{2+}$, we again have a downfield peak at 0.55

ppm. As the temperature is raised, the downfield peak first broadens and then suddenly shifts upfield to merge with the upfield double-helix peak at 57.5°C. On further heating, the broad resonance narrows, but double helix is still visible at 65°C. (Compare to 54°C spectra in the absence of EtBr). In Fig. 7 we show a melting curve for these resonances that indicates a melting temperature of 57–58°C.

## 3. Stabilization of Double Helix by Ethidium Ion

$^{31}$P-NMR spectroscopy demonstrates the stabilization of the double helix by Et. The rise in $T_m$ with increased concentration of Et (compare Chen and Cohen, Chapter 8, Fig. 1, and Fig. 5, this chapter) illustrate this effect. In the 31°C $^{31}$P-NMR spectra in the presence of 10 m$M$ Mg$^{2+}$, we estimate that the

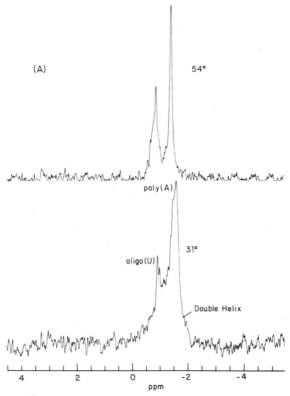

**Fig. 6.** (A) $^{31}$P-NMR of poly(A)·oligo(U) (1.32:1) in 10 m$M$ MgCl$_2$ at the temperatures (°C); 80.99 MHz. Total nucleotide concentration, 8.4 m$M$. Exponential line broadening,

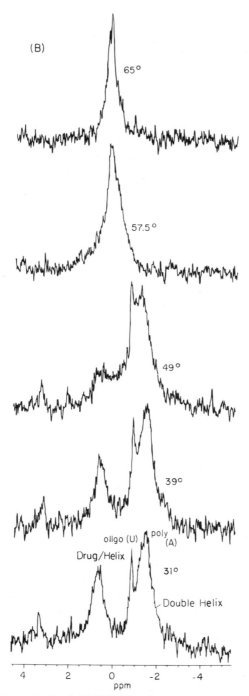

2 Hz. Other conditions as in Fig. 3. (B) $^{31}$P NMR of poly(A)·oligo(U)·Et (1.32:1.00:0.62) at various temperatures (°C). Conditions same as in (A).

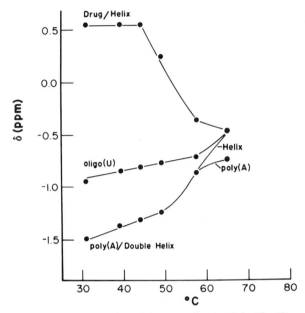

Fig. 7. Phosphorus-31 melting curve for signals in Fig. 6B.

downfield drug–helix peak accounts for ~30% of the integrated intensity. According to the nearest-neighbor exclusion hypothesis, Et binds to every other base pair under saturating conditions (Breslof and Crothers, 1975). If 50% of the double-helix phosphates are represented by this peak, then 60% of the phosphates are in double-helix conformations in this poly(A)·oligo(U) system. In the absence of Et under the same conditions, we estimate that ~31% of the potential base pairs are in double-helix conformations.

Both the $^{31}$P spectra (and results discussed by Goldfield et al., 1983) in 10 m$M$ Mg$^{2+}$ demonstrate that the effects of Mg$^{2+}$ on the $T_m$ are minimal. Waring (1965) noted that the binding of Et to DNA and RNA is reduced in the presence of Mg$^{2+}$ and attributes this to an increase in the dissociation rate constant and not to a decrease in binding sites. It is likely that in solutions with no Mg$^{2+}$, both the electrostatic and intercalation effects of Et are used to stabilize the double helix.

The melting of poly(A)·oligo(U)·Et as monitored by $^{31}$P-NMR spectroscopy shows the result of chemical-exchange broadening. The narrower peaks in the 10 m$M$ Mg$^{2+}$ samples at low temperature are probably mainly due to the higher spectrometer frequency because the slow-exchange region, which gives rise to narrow, separate peaks, is determined by $\tau/2\pi\Delta\nu \gg 1$,

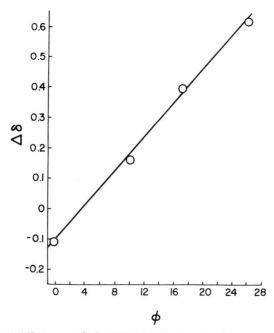

**Fig. 8.** Chemical-shift change $\Delta\delta$ of the DNA $^{31}$P signal on titration to saturation, versus the unwinding angle $\phi$ for ethidium ($\phi = 26°$), quinacrine ($\phi = 17°$), daunorubicin ($\phi = 10°$), and tetralysine ($\phi \sim 0°$).

where $\tau$ is the lifetime of a given state. At 32 MHz, $\Delta\nu$ will be $2\frac{1}{2}$ times smaller than at 81 MHz, so the slow-exchange region will occur at lower temperatures at lower spectrometer frequency. Figure 6 illustrates nicely the effects of chemical exchange. The linewidths remain narrow and there is little change in the separation of the signals until ~49°C, when the upfield peak broadens dramatically. At 57°C, the peaks merge and begin to narrow. The region of maximal line broadening occurs at intermediate exchange, between these two temperatures. In this region we can estimate the lifetime of the state from the formula $\tau = \sqrt{2}/2\pi(\Delta\nu)$ as between 1 and 3 ms.

In their $^{31}$P-NMR work on Et·DNA complexes, Wilson et al. (1981) and Wilson and Jones (1982) observed downfield shifts and line broadening as a function of the Et/DNA ratio, but they did not observe a separate drug/helix signal. Under Wilson et al.'s conditions and magnetic field strength, the DNA and drug are in fast chemical exchange based on temperature-jump kinetic studies (Bresloff and Crothers, 1975; Wakelin and Waring, 1980). However, the relaxation times of a Et·poly(A)·poly(U) system are from 5 to 10 times larger than those for calf thymus DNA (Bresloff and Crothers,

1975). At 81 MHz and low temperature, our poly(A)·oligo(U)·Et system is clearly in slow chemical exchange.

Wilson and Jones (1982) have measured the $^{31}$P chemical-shift change induced in DNA by various drugs: ethidium, quinacrine, daunorubicin, and tetralysine. They found a linear correlation between the change in chemical shift and the unwinding angle of the DNA double helix measured for each drug (Fig. 8). In all cases exchange is fast because a separate $^{31}$P signal for the drug–DNA complex is not observed.

## III. Electrostatic Mode of Interaction

It is significant that the large Et-induced $^{31}$P shift shown in Figs. 4–7 is *not* observed in low-salt poly(A)·oligo(U) solution (Gorenstein, 1978). It is possible that in these low-salt solutions the main interaction between the drug and the nucleic acids are electrostatic. Consistent with this interpretation, the purely electrostatic association between the antibiotic netropsin and the self-complementary duplex d-GpGpApApTpTpCpC produces only small upfield $^{31}$P shifts (Patel, 1979d). This (and other data) supports a drug·duplex model involving binding at D-ApT base pairs in the minor groove, with the two charged guanidine ends of the drug electrostatically associating with the phosphates.

Patel and Canuel (1979) have shown that small $^{31}$P chemical-shift changes occur on complexation of a steroidal diamine to the duplex poly(d-ApT). Presumably this drug does not completely intercalate between the stacked base pairs but rather electrostatically associates with the anionic phosphates, resulting in little change in the phosphate geometry.

Davanloo and Crothers (1979) have noted a 1-ppm *upfield* shift of the $^{31}$P signal of d-ApApApGpCpTpTpT at 30°C on complex formation with the oligopeptide tetralysine. This probably results from a stabilization of the double helix at this temperature by the oligopeptide and hence an increase in the population of $-g,-g$ phosphates. At 4°C, the d($A_3GCT_3$) duplex shifts only 0.5 ppm upfield on complex formation. Note that tetralysine produces only a very small upfield shift in DNA (Fig. 8). The tetralysine is not expected to induce any significant alteration in the double-helix $-g, -g$ phosphate conformation, in contrast to the short duplex in Davanloo and Crothers' study.

These small upfield shifts are similar to the nonspecific $Mg^{2+}$ ion effects on the duplex signals in tRNA (Guéron and Shulman, 1975; Gorenstein *et al.*, 1981). This further supports the hypothesis that large (>2 ppm) downfield shifts are only observed for the intercalation mode of drug binding.

## IV. tRNA·Ethium ³¹P Spectra

### A. ³¹P Perturbations

The ³¹P spectra of tRNA plus Et ion at 31°C are shown in Fig. 9. The chemical shifts of the scattered peaks (see discussion of tRNA spectra in Gorenstein, Chapter 9) are virtually unaffected with one possible exception. However, a number of peaks undergo significant line broadening as a function of added Et (Fig. 10). The only peak in the downfield portion of the spectrum (+ 3.3 to −0.453 ppm) that broadens is peak E at ~0.245 ppm. The linewidth of this peak goes from ~15 Hz at an Et/tRNA ratio of 0 to >25 Hz when the Et/tRNA ratio is increased to 1.32. At an Et/tRNA ratio of 2.0, the peak cannot be detected as a separate peak. It is possible that the chemical shift of peak E moves upfield to coincide with peak F at ~0.077 ppm. However, the intensity of peak F with added Et increases only slightly. This can be explained by the superposition of a broadened peak E under the sharper peak F. The entire upfield portion of the spectrum (− 1.85 to −4.53 ppm) broadens and loses resolution as the Et/tRNA ratio increases. Peaks N, O, P, and Q between − 1.85 ppm and − 2.65 ppm broaden and merge with one another. Peaks R/S and T also broaden as a function of added Et (Fig. 10). These upfield peaks and peak E narrow as the temperature is raised to 49°C. Between 49 and 55°C, peaks R/S and T again begin to broaden as the temperature is increased. Peak U (Gorenstein and Luxon, 1979) is not visible in these spectra.

The other downfield peaks show no significant line broadening nor changes in intensity within experimental error. Peak A especially remains narrow, which contrasts with its linewidth behavior because of added $Mn^{2+}$ (Gorenstein *et al.*, 1981), where A shows a rapid linear increase in linewidth at low $Mn^{2+}$/tRNA ratios. This result gives us confidence that the line-broadening effects are not due to paramagnetic metal-ion impurities in the Et solution.

At Et/tRNA ratios of 1.32 and above, there possibly is a small new peak at 0.847 ppm that appears as a shoulder to peak D (0.7 ppm).

### B. Structural Interpretation of Effects of Ethidium Ion on tRNA

Fluorescence depolarization studies indicate that in the presence of $Mg^{2+}$ there is one strong binding site for Et to $tRNA^{Phe}$ (Tao *et al.*, 1970; Wells and Cantor, 1977) and perhaps a weaker site (Sturgill, 1978). The binding constant for the stronger site is $4-5 \times 10^5 \, M^{-1}$ and for the weaker site is $\sim 4 \times 10^4 \, M^{-1}$ at temperatures of 22–25°C (Sakai and Cohen, 1976; Stur-

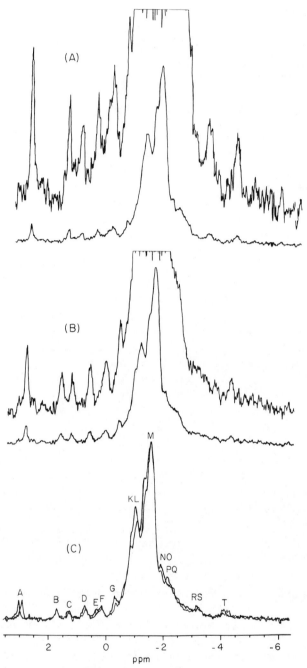

**Fig. 9.** $^{31}$P-NMR spectra of tRNA$^{Phe}$ at various molar ratios of Et in 10 m$M$ MgCl$_2$, 0.1 $M$ NaCl, 10 m$M$ cacodylate, 1 m$M$ EDTA, 20% D$_2$O, and at pH 7.0, 31°C; 80.99 MHz. Exponential line broadening, 2 Hz. Et/tRNA: A, 0.44; B, 1.32; C, 2.0. Thin curve, drug; thick curve; no drug.

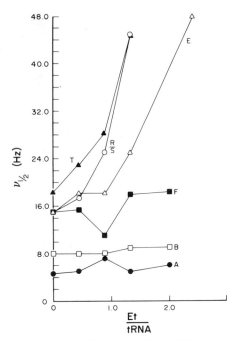

**Fig. 10.** Corrected linewidths at half-height at 80.99 MHz versus molar ratio of Et for selected signals from Fig. 9, spectrum C. ●, Peak A; □, peak B; ■, peak F; △, peak E; ○, peak R/S; ▲, peak T.

gill, 1978; Torgerson *et al.*, 1980). Lifetime-measurement studies indicate that at low EtBr/tRNA ratios, the fluorescence decays with a single lifetime of 28 ms, and at higher ratios the decay contains shorter-lived components as well (Jones *et al.*, 1978). Wells and Cantor (1977) have determined that the binding site of Et/tRNA is between 33 and 40 Å from the 3' end of the tRNA (see Fig. 8 in Gorenstein, Chapter 9, for the X-ray structure of tRNA$^{Phe}$). Tao *et al.* (1970) have shown that in the presence of Mg$^{2+}$ the binding of Et is very far away from the Y base (anticodon loop), whereas in the absence of Mg$^{2+}$ it binds closely to the Y base.

Based on relaxation studies (Tritton and Mohr, 1971), fluorescence lifetime, steady-state fluorescence studies (Jones *et al.*, 1978), and, especially, $^1$H-NMR studies (Jones and Kearns, 1975; Jones *et al.*, 1978), the major mode of binding of Et to tRNA$^{Phe}$ is believed to be intercalation between base pairs AU$_6$ and AU$_7$ (or AU$_5$–AU$_6$) of the acceptor stem. The interpretation has received wide acceptance among many authors. However, the X-ray diffraction studies of crystals of tRNA into which Et have been allowed to diffuse indicate another mode of binding. The ethidium is found to lodge in a cavity at the mouth of the $P_{10}$ loop of the tRNA tertiary base pair $U_8-A_{14}$ (Liebman et al., 1977; Gorenstein, Chapter 9, Fig. 8). Liebman et al. have reinterpreted the fluorescence and NMR data in light of this mode of binding.

These results do not resolve the controversy but do lend some support to the view that the Et lodges itself in the tertiary pocket. The tRNA[Phe] $^{31}$P spectrum consists of a major cluster at $-0.433$ to $-1.453$ ppm with both upfield and downfield scattered signals. It has been suggested that bond- and torsional-angle differences arising from tertiary interactions account for the 7-ppm spread in the $^{31}$P signals of tRNA (Gorenstein, Chapter 9; Gorenstein and Luxon, 1979; Guéron and Shulman, 1975; Salemink et al., 1979).

Based on the poly(A) · oligo(U) results, one would expect that the major result of intercalation in the $^{31}$P-NMR spectra would be a downfield shift of one or two phosphates from the main cluster region to 0.5–0.8 ppm. It is possible that the small peak (less than one phosphate) at 0.847 ppm is due to phosphates at the intercalation site. However, we do not see any major (one to two phosphate) growth in intensity in this spectral region.

The most striking effect of the addition of Et is the major broadening of the upfield scattered peaks that have been associated with tertiary interactions of the TΨC and D stems and of peak E (Gorenstein and Goldfield, 1982a,b; see also Chapter 9). Based on its temperature and $Mg^{2+}$ and $Mn^{2+}$ dependence, we have previously suggested that this peak is associated with the AC loop or the loop between the acceptor and D stems (Gorenstein et al., 1981; Gorenstein and Goldfield, 1982a,b; see also Chapter 9). However, peak E, as the broadened peaks in the upfield spectral region, is likely associated with tertiary interactions near the Et binding site. Because the Et binding is not near nor in the AC loop, we may conclude that peak E is likely not in this loop.

Our interpretation of the broadening of the various $^{31}$P resonances is that the resonances of phosphates near the Et binding site(s) are affected because the phosphates are held more rigidly when Et is bound. Thus if the Et rests in the tertiary pocket described previously or if it intercalates in the acceptor stem, it is likely that the phosphates' motions near where the acceptor stem joins the D stem will be slowed down, whereas the overall tumbling of the molecule will be affected only slightly. Therefore, the linewidths of the terminal phosphate, those in the AC loop, and others removed from the Et will remain narrow. Bolton and James (1979) and others (see James, Chapter 12) have shown that to account for the $^{31}$P relaxation behavior of nucleic acids at least two correlation times are required: a slow (milliseconds) time associated with long-range bending (or tumbling) and a faster (0.3–0.5 ns) time associated with rotational wobbling about P—O ester bonds. If the Et restricts the local flexibility of phosphates involved in tertiary interactions near the binding site, then the differential broadening of the scattered peaks could be explained.

## Acknowledgments

We wish to acknowledge the contributions of Bruce Luxon, Kofen Lai, Vickie Bowie, Donna Vegeais, and Rouhlwai Chen to these studies. Supported by research grants from the National Institutes of Health and the National Science Foundation. Purchase of the Bruker WP-80 spectrometer was assisted by a National Science Foundation Departmental Equipment Grant. Suport of the Purdue Biological NMR facility by NIH (RRO 1077) is acknowledged.

## References

Akasaka, K. (1974). *Biopolymers* **13,** 2273–2280.
Bolton, P. H., and James, T. L. (1979). *J. Phys. Chem.* **83,** 3359–3366.
Bresloff, J. L., and Crothers, D. M. (1975). *J. Mol. Biol.* **95,** 103–123.
Cohen, J. S., and Chen, C.-W. (1982). *ACS Symp. Ser.* **191,** 249–267.
Davanloo, P., and Crothers, D. M. (1979). *Biopolymers* **18,** 2213–2231.
Gale, E. F., Cundliff, E. F. Reynolds, P. E. Richmond, M. H., and Waring, M. J. (1972). "The Molecular Biology of Antibiotic Action." Wiley, New York.
Goldfield, E. M., Luxon, B. A., Bowie, V., and Gorenstein, D. G. (1983). Biochemistry 22.
Gorenstein, D. G. (1975). *J. Am. Chem. Soc.* **97,** 898–900.
Gorenstein, D. G. (1978). *Jerusalem Symp. Quantum Chem. Biochem.* **11,** 1–15.
Gorenstein, D. G. (1981). *Annu. Rev. Biophys. Bioeng.* **10,** 355–386.
Gorenstein, D. G., and Kar, D. (1975). *Biochem. Biophys. Res. Commun.* **65,** 1073–1080.
Gorenstein, D. G., and Goldfield, E. M. (1982a). *J. Mol. Cell. Biochem.* **46,** 97–120.
Gorenstein, D. G., and Goldfield, E. M. (1982b). *Biochemistry* **21,** 5839–5849.
Gorenstein, D. G., and Luxon, B. A. (1979). *Biochemistry* **18,** 3796–3804.
Gorenstein, D. G., Findlay, J. B., Momii, R. K., Luxon, B. A., and Kar, D. (1976). *Biochemistry* **15,** 3796–3803.
Gorenstein, D. G., Goldfield, E. M., Chen, R., Kovar, K., and Luxon, B. A. (1981). *Biochemistry* **20,** 2141–2150.
Gorenstein, D. G., Luxon, B. A., Goldfield, E. M., Lai, K., and Vegeais, D. (1982). *Biochemistry* **21,** 580–589.
Guéron, M., and Shulman, R. G. (1975). *Proc. Natl. Acad. Sci. U.S.A.* **72,** 3482–3485.
Hogan, M. E., and Jardetzky, O. (1980). *Biochemistry* **19,** 2079–2085.
Jain, J. C., and Sobell, H. M. (1972). *J. Mol. Biol.* **68,** 21–34.
Jones, C. R., and Kearns, D. R. (1975). *Biochemistry* **14,** 2660–2665.
Jones, C. R., Bolton, P. H., and Kearns, D. R. (1978). *Biochemistry* **17,** 601–607.
Liebman, M., Rubin, J., and Sundaralingam, M. (1977). *Proc. Natl. Acad. Sci. U.S.A.* **74,** 4821–4825.
Lown, J. W. (1977). *Bioorg. Chem.* **3,** 95–121.
Patel, D. J. (1974). *Biochemistry* **13,** 2388–2402.

Patel, D. J. (1976a). *Biopolymers* **15,** 533–558.
Patel, D. J. (1976b). *Biochim. Biophys. Acta* **442,** 98–108.
Patel, D. J. (1979a). *Acc. Chem. Res.* **12,** 118.
Patel, D. J. (1979b). *In* "Stereodynamics of Molecular Systems" (R. H. Sarma, ed.), pp. 397–472. Pergamon, Oxford.
Patel, D. J. (1979c). *Eur. J. Biochem.* **96,** 267–276.
Patel, D. J. (1979d). *Eur. J. Biochem.* **99,** 369–378.
Patel, D. J., and Canuel, L. L. (1977). *Proc. Natl. Acad. Sci. U.S.A.* **73,** 3343–3347.
Patel, D. J., and Canuel, L. L. (1979). *Proc. Natl. Acad. Sci. U.S.A.* **76,** 24–28.
Reddy, B. S., Seshadri, T. P., Sakore, T. D., and Sobell, H. M. (1979). *J. Mol. Biol.* **135,** 787–812.
Reinhardt, C. G., and Krugh, T. R. (1977). *Biochemistry* **16,** 2890–2895.
Sakai, T. T., and Cohen, S. S. (1976). *Prog. Nucleic Acid Res. Mol. Biol.* **17,** 15–42.
Salemink, P. J. M., Swarthof, T., and Hilbers, C. W. (1979). *Biochemistry* **18,** 3477–3485.
Shieh, H-S, Berman, H. M., Dabrow, M., and Neidle, S. (1980). *Nucleic Acids Res.* **8,** 85–97.
Shindo, H., McGhee, J. D., and Cohen, J. S. (1980). *Biopolymers* **19,** 523–537.
Sturgill, T. W. (1978). *Biopolymers* **17,** 1793–1810.
Tao, T., Nelson, J. H., and Cantor, C. R. (1970). *Biochemistry* **9,** 3514–3524.
Torgerson, P. M., Drickamer, H. G., and Weber, G. (1980). *Biochemistry* **19,** 3960–3966.
Tritton, T. R., and Mohr, S. C. (1971). *Biochem. Biophys. Res. Commun.* **45,** 1240–1249.
Wakelin, L. P. G., and Waring, M. (1980). *J. Mol. Biol.* **144,** 183–214.
Waring, M. J. (1965). *J. Mol. Biol.* **13,** 269–282.
Wells, B. D., and Cantor, C. R. (1977). *Nucleic Acids Res.* **4,** 1667–1680.
Wilson, W. D., and Jones, R. L. (1982). *Nucleic Acids Res.* **10,** 1399–1410.
Wilson, W. D., Keel, R. A., and Mariam, R. H. (1981). *J. Am. Chem. Soc.* **103,** 6267–6269.

CHAPTER 11

# Phosphorus Relaxation Methods: Conformation and Dynamics of Nucleic Acids and Phosphoproteins

*Phillip A. Hart*
School of Pharmacy
University of Wisconsin
Madison, Wisconsin

| | |
|---|---:|
| I. Introduction | 317 |
| II. General Considerations | 319 |
| III. The Phosphorus–Proton Nuclear Overhauser Effect | 322 |
| IV. Phosphorus Relaxation Times | 328 |
|    A. Oligonucleotides | 328 |
|    B. Double-Stranded DNA | 334 |
|    C. Phosphoproteins | 341 |
| V. Summary | 344 |
|    References | 345 |

## I. Introduction

The phosphodiester is the fundamental phosphorus-containing functional unit of oligonucleotides as well as of single- and double-stranded nucleic acids. The conformation of the phosphodiester group is a major structural determinant in these systems as well as the functional group that prescribes their polyelectrolyte behavior. The phosphomonoester is a common functional group found in sugar phosphates, mononucleotides, phosphoproteins, and other phosphorylated substances and in the context of those systems is not so much a primary structure determinant as it is a determinant of interaction with the polar milieu and is, through that interaction, a determinant of structural and physical state. This chapter focuses attention on $^{31}$P-NMR studies of these phosphorus-containing functional groups, studies that have been undertaken to solve problems of conformation and dynamics that arise from the chemical or physical properties of those groups. This chapter emphasizes approaches to the solution of conformation problems. Questions of dynamics are considered because they are

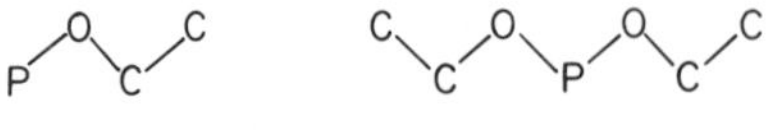

Fig. 1. Skeletal representations of the phosphomonoester **1** and the phosphodiester **2**.

unavoidable; however, they are covered in greater detail elsewhere in this volume (James, Chapter 12). I hope the reader is struck by the beautiful complexity of the subject rather than discouraged by the lack of complete analytical methods.

The most accessible and the most definitive conformational parameters of the phosphomonoester group symbolized in **1** are the rotameric distributions about the O—C and C—C bonds (Fig. 1). The most accessible parameters of the phosphodiester symbolized in **2** are the distributions about the C—C and O—C bonds, and these parameters are definitive as well. However, the rotameric distribution about the O—P bonds are definitive but not so accessible. Placing these functional groups in the specific context of a mononucleotide and of a dinucleoside monophosphate as in **3** and **4** allows a specific discussion of the means available to determine the conformational distributions of interest (Fig. 2). The bonds of interest, indeed those for which torsion-angle values are required for detailed conformational analyses, are those labeled in **3** and **4**. Table I gives the two principal conventions used to label those bonds, the atoms required to formally describe the torsion angles about them, and the common rotamer designations that arise from Newman projections involving those basis atoms.

A large body of work reports experimental probes of most of the torsion angles of Table I. This work is far too extensive to quote here. Lee *et al.* (1976; Lee and Tinoco, 1977), Alderfer and T'so (1977), Davies (1978), and Sarma (1980) have summarized most of it. Those bonds that involve phosphorus are the salient ones for the purposes of this volume. Those cases where phosphorus relaxation methods have been useful or *show promise* are emphasized in this chapter.

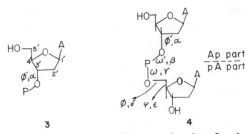

Fig. 2. General representation of a nucleoside monophosphate **3** and a specific representation of adenylyl-(3' → 5')-adenosine **4**, showing typical bond and group designations.

**TABLE I**

Torsion-Angle Conventions Referenced to the Phosphodiester Backbone

| Angle[a] | Atom basis[b] | Typical specific values[c] | | |
|---|---|---|---|---|
| $\phi'/\alpha$ | P—O-3'—C-3'—C-4' | 60° (g⁺) | 180° (t) | −60° (g⁻) |
| $\omega'/\beta$ | O—P—O-3'—C-3' | 60° | 180° | −60° |
| $\omega/\gamma$ | O—P—O-5'—C-5' | 60° | 180° | −60° |
| $\psi/\delta$ | O-5'—C-5'—C-4'—C-3' | 60° (g⁺) | 180° (t) | −60° (g⁻) |
|  | H-5',5"—C-5'—C-4'—H-4' | gg | gt | tg |
| $\phi/\epsilon$ | P—O-5'—C-5'—C-4' | 180° | −60° | 60° |
|  | P—O-5'—C-5'—H-5',5" | g'g' | t'g' | g't' |

[a] Most common designations.
[b] See Fig. 2, structure **4**.
[c] Alternative designation shown parenthetically. The numerical values are assigned according to Sundaralingam et al. (1973).

The phosphorus relaxation methods that are routinely available are the phosphorus–hydrogen nuclear Overhauser effect (NOE) and the various relaxation times, the spin–lattice relaxation time $T_1$, the spin–spin relaxation time $T_2$, and the spin–lattice relaxation time in the rotating frame $T_{1\rho}$. Under ideal circumstances for interpretation, both geometric and dynamic information can be gotten from these experimental values; however, it is almost inevitable that the ideal is seldom encountered. Instead, various assumptions and approximations are required before any interpretation can be made. I partition the following discussion along the lines of the relaxation methods listed here and I show what interpretation can be placed on the data. As well, I try to clarify directions for further work that I consider potentially fruitful.

## II. General Considerations

An interpretation of experimental observations made by relaxation methods requires a knowledge of the mechanism(s) responsible for the time dependence of the net magnetization. This subject has been widely reported and extensively reviewed by Abragam (1961), Farrar and Becker (1971), and Noggle and Schirmer (1971). For the purposes of this chapter, only the dipole–dipole interaction and the modulated anisotropic chemical shift need be considered. The dipolar interaction receives the major emphasis.

The dipole–dipole interaction depends on the correlated motion of magnetic dipoles, usually assumed to be executing independent and random motion. In the present treatment, the interaction is between phosphorus and some other magnetic nucleus or nuclei (usually hydrogen). The central objects in the expression of this interaction are the autocorrelation function and its Fourier transform — the spectral density. The formulation of appropriate correlation functions and their corresponding spectral density functions have been widely discussed in the primary literature and have been well reviewed. Abragam (1961), Slichter (1963), Redfield (1965), Werbelow and Grant (1977), Vold and Vold (1978), and Lenk (1977) are selected authors of general treatments of nuclear magnetic relaxation; and Woessner (1962), Tsutsumi (1979), Tsutsumi et al. (1979), London and Avitable (1978), Levine et al. (1973), Tropp (1980), King and Jardetzky (1978), King et al. (1978), Wittebort and Szabo (1978), Keepers and James (1982), Allison et al. (1982), Hart et al. (1981), and Lipari and Szabo (1982) have published specific spectral density formulations relevant to specific models of molecular motion.

The specific form of the autocorrelation function [Eq.(1)] that is relevant to phosphorus-proton dipolar interactions and from which the spectral densities used in this chapter were derived is an adaptation of the Tsutsumi (1979) formulation derived originally from the work of Woessner (1962). In

$$G_k(t) = \sum_{i,j} f_k(i) P[i(t), j(o)] f_k(j) \tag{1}$$

$i$ and $j$ are different conformations, $P$ is a two-time joint probability. The $k$ are the five orientation factors required by the symmetry of the dipole–dipole interaction and have the form

$$f_1 = (1 - 3n^2)/r^3$$
$$f_2 = (l^2 - m^2)/r^3$$
$$f_3 = mn/r^3$$
$$f_4 = nl/r^3$$
$$f_5 = lm/r^3$$

in which $l$, $m$, and $n$ are direction cosines of the internuclear vector relative to the molecular frame and $r$ is the length of the internuclear vector. This formulation is sufficiently general to include any mode of overall motion and any mode of internal motion that satisfies the conditions of randomness and independence from overall motion. By including $r$ in the geometric factor, one automatically accounts for its dependence on conformation and time. The specific form of $P$, the time-dependent conformational probability, depends on the model chosen for internal conformational change and the time dependence of that change. The spectral densities derived from Eq. (1) are of the general form

$$J(\omega) = \sum_{k=1}^{5} [\mathbf{f}_{ik}][P][\mathbf{f}_{jk}]\left(\frac{\tau_R}{1+\omega^2\tau_R^2}\right)$$
$$+ [\mathbf{f}_{ik}][P'][\mathbf{f}_{jk}]\left(\frac{\tau_{RC}}{1+\omega^2\tau_{RC}^2}\right) \quad (2)$$

in which $\tau_{RC} = (\tau_R^{-1} + \tau_{int}^{-1})^{-1}$; $P$ and $P'$ are matrices of equilibrium probabilities, the dimensions of which depend on the number of internal conformational states chosen; $\mathbf{f}_i$ is a row vector and $\mathbf{f}_j$ a column vector where $i,j$ assumes the values 1 to the number of conformational states; $\tau_R$ the overall correlation time and will be single-valued in the case of isotropic overall motion but will be multivalued in the case of anisotropic overall motion; $\tau_{int}$ an internal motion parameter, the complexity of which depends on the model chosen for internal motion; and $\omega$ the appropriate Larmor precession frequency function. An application of this formulation has been published by Hart et al. (1981). I find this formulation useful because it is possible for me to see at once how geometry, equilibrium populations, and dynamics participate in the spectral density expressions. Furthermore, because the geometric factors are partitioned, they can be computed separately and the often very complex geometry problems can be considered separately from the time dependence of the geometry. A more detailed rationale for the form of Eq. (2) is given in Section IV,A. The formulation of Eq. (2) can be compared with the "model independent" formulation of Lipari and Szabo (1982), which is similar in form to that of King and Jardetzky (1978) and King et al. (1978). The Lipari and Szabo spectral density expression is

$$J(\omega) = \frac{2}{5}\left[\frac{S^2\tau_M}{1+(\omega\tau_M)^2} + \frac{(1-S^2)\tau}{1+(\omega\tau)^2}\right] \quad (3)$$

in which $\tau_M$ is the correlation time for overall motion and $\tau \equiv (\tau_M^{-1} + \tau_e^{-1})^{-1}$; $\tau_e$ is a time constant for internal motion; and $S^2$ an order parameter, more specifically "a model independent measure of the degree of spatial restriction of the [internal] motion [Lipari and Szabo, 1982, p. 4549]." Lipari and Szabo (1982) have shown that it is possible to fit spectral densities exemplified by Eq. (3) to relaxation data by treating $S^2$ and $\tau_e$ as adjustable parameters. The success of the fit depends on the range of $\tau_e$ relative to $\tau_M$, and the authors found that when internal motion (but not necessarily overall motion) obeys the extreme narrowing condition, relaxation rates can be rationalized by more than one model for internal motion. I am intrigued by the possibility that the model-independent parameters $S^2$ and $\tau_e$ of Eq. (3) might be related to the geometric and equilibrium coefficients of Eq. (2).

By whatever means the spectral densities are calculated, they enter the calculation of phosphorus dipolar relaxation involving P and H spins:

$$1/T_1 = \tfrac{9}{2}\gamma_P^2\gamma_H^2\hbar^2[J(\omega_P - \omega_H) + 18J(\omega_P) + 9J(\omega_P + \omega_H)] \quad (4)$$

$$1/T_2 = 9\gamma_P^2\gamma_H^2\hbar^2[4J(0) + J(\omega_P - \omega_H) + 18J(\omega_P) \\ + 36J(\omega_H) + 9J(\omega_P + \omega_H)] \quad (5)$$

$$\text{NOE} = 1 - \frac{\gamma_H}{\gamma_P}\frac{J(\omega_P - \omega_H) - 9J(\omega_P + \omega_H)}{J(\omega_P - \omega_H) + 18J(\omega_P) + 9J(\omega_P + \omega_H)} \quad (6)$$

where $\gamma_P$ and $\gamma_H$ are the magnetogyric ratios and $\omega_P$ and $\omega_H$ the Larmor precession frequencies. This particular NOE formulation assumes that cross-relaxation terms have been reduced to zero by simultaneously saturating all interacting nuclei other than the one ($^{31}$P in this case) being observed. The relaxation of that requirement is discussed in the following section. Note that the factor $r_{PH}^{-6}$ does not appear in Eqs. (4) and (5) because it is absorbed into the spectral densities by way of the geometric factors $f_{ik}$.

It should be clear from the expressions for the spectral densities that even given experimental values for $T_1$, $T_2$, and NOE, the prospect of fitting specific geometries (or linear combinations of geometries) to the experimental relaxation rates is problematical because a priori the problem is overparameterized. Furthermore, Lipari and Szabo (1982) have contended that when internal motion satisfies the extreme narrowing condition, relaxation rates are not uniquely sensitive to specific models for internal motion. In addition, Hart *et al.* (1981) have shown that for certain values of overall motion, no sensitivity to internal motional rates is manifested.

Given these rather sobering potential restrictions, let us examine specific systems and discover what information about conformation and dynamics can be extracted from single relaxation measurements or combinations thereof.

## III. The Phosphorus–Proton Nuclear Overhauser Effect

The nuclear Overhauser effect (NOE) arises when dipole–dipole interaction-mediated populations are perturbed by irradiation of one (or more) of the interactive nuclei. The experimental procedure and many qualitative interpretations obscure the fact that the NOE is an expression of relaxation processes, a fact made clear by Eq. (6). Although Eq. (6) is a valid expression for the NOE when cross-relaxation has been obviated as mentioned previously, a far more useful application of the experiment for conformational analysis permits cross-relaxation by leaving some interacting spins unperturbed, that is, unirradiated. When cross-relaxation is permitted and when specific conditions are met and assumptions made, fairly sophisticated conformational analyses can be accomplished. The homonuclear version of the NOE has been widely applied to conformational analysis beginning with the work of Hart and Davis (1969, 1972), Schirmer et al. (1970), Noggle and Schirmer (1971), Son et al. (1972), and Son and Chachaty (1973). The heteronuclear version, specifically the phosphorus–hydrogen NOE, has been applied to conformational analyses of ATP (Hart, 1976), thiamin pyrophosphate (P. A. Hart, unpublished), and nucleoside monophosphates (Hart, 1978). It is the work on nucleoside monophosphates that is reviewed here. Emphasis is placed on the method as well as on restrictions on its use in more complex phosphorus-containing systems.

Noggle and Schirmer (1971) have shown that the fractional change of the integrated intensity of nucleus I (spin $\frac{1}{2}$, magnetogyric ratio $\gamma_I$), when the nuclei S (all spin $\frac{1}{2}$, all having the same magnetogyric ratio $\gamma_S$) are successively saturated, can be written

$$f_I(S) = \left[ \frac{\gamma_S}{2\gamma_I} \sum_S \rho_{IS} - \frac{\gamma_n}{2\gamma_I} \sum_{n \neq I,S} \rho_{In} f_n(S) \right] \Big/ R_I \qquad (7)$$

in which $\rho_{IS} = \gamma_I^2 \gamma_S^2 h^2 \tau_c / r_{IS}^6$ (the dipolar relaxation rate of I owing to the single spins S, and $R_I = \Sigma_k \rho_{Ik} + \rho_I^*$ the total relaxation rate of I). In these expressions, $\rho_I^*$ is the spin–lattice relaxation rate of I contributed by independent relaxation mechanisms other than dipole–dipole interactions, $\tau_c$ the time constant for overall molecular reorientation, and $r_{IS}$ the distance between I and S in angstroms. The quantity $f_n(S)$ that appears in Eq. (7) is defined entirely analogously to $f_I(S)$. Equation (7) is a general expression for the NOE in multispin systems, given the following assumptions:

1. The extreme narrowing condition is satisfied that is, $\omega_0 \tau_c \ll 1$, where $\omega_0$ is the Larmor frequency of the nucleus having the larger magnetogyric ratio.
2. All spins are loosely coupled, that is, $J < \Delta\delta$.
3. All dipole–dipole interactions involving the nucleus I are modulated by the same correlation time $\tau_c$.
4. Other relaxation mechanisms comprising $\rho_I^*$ are characterized by the same correlation time that modulates the dipolar interactions.

Although it is not obvious from Eq. (7), the second term arises because of cross-relaxation, and Noggle and Schirmer (1971) have shown in detail how this term contributes to successful conformational analyses. When $I \equiv {}^{31}P$ and all other interacting atoms are $^1H$ (when the extreme narrowing condition is met) and when $\rho^* = 0$, then $f_P(H) = \gamma_H/2\gamma_P = 1.24$, and this maximum enhancement will be found in a multispin system when all hydrogens that can contribute to phosphorus relaxation via dipolar coupling are saturated. If the conditions $\rho^* = 0$ and $\omega_0\tau_c \ll 1$ are not met, the maximum observable P{H} NOE will be attenuated. It is convenient to express that modulation by modifying Eq. (7) to read

$$f_I(S) = \Phi(\omega_I, \omega_S, \tau_C)\frac{1}{\beta}\left[\frac{\gamma_S}{\gamma_I}\sum_S \rho_{IS} - \frac{\gamma_n}{\gamma_I}\sum_{n\neq I,S}\rho_{In}f_n(S)\right]\bigg/ R_I^{dd} \quad (8)$$

in which

$$\Phi(\omega_I, \omega_S, \tau_C) = \frac{9[1 + (\omega_I + \omega_S)^2\tau_C^2]^{-1} + [1 + (\omega_I - \omega_S)^2\tau_C^2]^{-1}}{9[1 + (\omega_I + \omega_S)^2\tau_C^2]^{-1} + [1 + (\omega_I - \omega_S)^2\tau_C^2]^{-1} + 18[1 + \omega_I^2\tau_C^2]^{-1}}$$

where $R_I^{dd}$, the total dipolar relaxation rate of I, and $\beta$ are defined as follows: $R_I^{dd} \equiv R_I - \rho_I^*; \beta \equiv R_I/R_I^{dd}$. By measuring the total P{H} NOE via saturation of all interacting protons, the combined effects of failure of extreme narrowing and the incursion of other relaxation mechanisms can be discerned. Note that the coefficients that appear in the function $\Phi$ are different from those in the work of Hart (1978). They were chosen to be consistent with Eq. (6), which adopts the spectral density coefficients of Woessner (1962) rather than those of Kuhlman et al. (1970).

If the assumption that a single $\tau_C$ modulates all contributions to the spin–lattice relaxation rate is maintained, so that $\rho^*$ and the effects of extreme narrowing can be factored as shown in Eq. (8), then $\tau_C$ can be eliminated from the expressions for $\rho_I^{dd}$ and the NOE will depend only on $r_{IS}^{-6}$ over the various weighted conformations. Equation (8) then will take the form

$$f_I(S) = \Phi(\omega_I, \omega_S, \tau_C)\frac{1}{\beta}\left[\gamma_I\gamma_S^3\sum_S \langle r_{IS}^{-6}\rangle - \gamma_I\gamma_n^3\sum_{n\neq I,S}\langle r_{In}^{-6}\rangle f_n(S)\right]\bigg/ R_I^{dd} \quad (9)$$

where $R_I^{dd} = \Sigma_k \langle \gamma_I^2\gamma_S^2 r_{Ik}^{-6}\rangle$. The $\gamma$ factors can be factored much more pletely if it is assumed that n or k is always hydrogen. In some cases (ATP, thiamin pyrophosphate), it is necessary to include phosphorus in the group of spins with which the observed phosphorus interacts. The average of $r^{-6}$ over the various conformations is used based on the assumption that rates of conformational change are faster than $1/T_1$. Noggle and Schirmer (1971) have given, a detailed rationale for this averaging process.

The application of Eq. (9) to small-molecule conformational analyses is straightforward. A set of single-frequency P{H} NOEs is developed by standard methods; internuclear distances are computed from X-ray-derived bond lengths and bond angles for various dihedral angles corresponding to the test conformations; a set of equations having the form of Eq. (9) corresponding to all necessary interactions is solved iteratively. Various

## 11. Conformation of Nucleic Acids

conformations or linear combinations of conformations are tried until an adequate fit to the experimental enhancements is found. This procedure is given in greater detail by Hart (1978) for P{H} NOEs and a more sophisticated fitting procedure is given for H{H} NOEs by Schirmer et al. (1970).

Hart (1978) has applied this procedure to a conformational analysis of 2′,3′-cyclic cytidine monophosphate (cCMP), as well as 3′- and 5′-adenosine monophosphate (3′-AMP and 5′-AMP). Because proton chemical-shift dispersion is limited at the fields used for these experiments (2.1 T), proton irradiation power was less than optimum to minimize multiple irradiation, and correction factors were applied to the computed enhancement profiles before comparing them to the experimental profiles. The corrections also take into account the observation that total NOEs are less than the theoretical maximum of 1.24.

For each of the three mononucleotides examined, Fig. 3 shows experimental single-frequency NOEs on one diagram and best-fit conformations on the other. These conformations must be regarded as approximate because of the difficulty of the experiments, because some of the assumptions may not be justified in these systems and because exhaustive computer fits were impractical. However, it is gratifying that the results for cCMP are in qualitative agreement with the coupling-constant analyses of Lavalee and Coulter (1973) and of Lapper and Smith (1973) and that the profile of phosphorus enhancements is reasonable in light of the moderately fixed cyclic nucleotide conformation and the reasonably predictable location of the phosphorus nucleus.

Application of the method to 5′-AMP and 3′-AMP gave the results indicated in the remainder of Fig. 3. There were two computed distributions in the 5′-AMP case as shown. The second one is unlikely, however, because it is not consistent with four-bond P–H coupling constants as discussed by Hart (1978). The first computed conformation distribution is reasonably consistent with four-bond coupling constants and with other P–H and P–C coupling-constant analyses reported by Sarma et al. (1973), Evans and Sarma (1974), Lee et al. (1976), Cozzone and Jardetzky (1976), and Schleich and Smith (1976).

Turning to the experiments on 3′-AMP, the computed distribution shown in Fig. 3 was difficult to find because it required an arbitrary (albeit reasonable) restriction on the sugar ring conformation and because the small negative enhancement produced on irradiation of the two C-5′ protons had to be considered in the simulations. Negative enhancements arise by way of the cross-relaxation term (the negative term) in Eq. (9). Nonetheless, good agreement was found with the phosphorus–hydrogen coupling-constant analysis of Cozzone and Jardetsky (1976) but not with the P–C coupling-constant analysis of Schleich and Smith (1976).

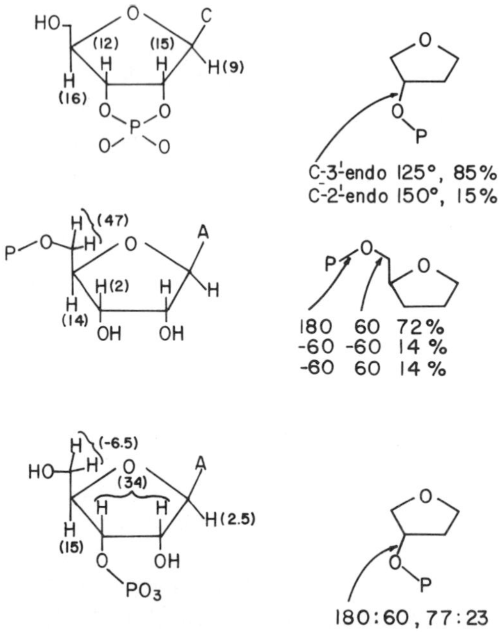

**Fig. 3.** Observed P{H} NOEs (left) and best-fit angles (right).

The work summarized here on mononucleotides, while crude, was encouraging. Considerable effort was needed to refine the analyses (both theory and experiment); however, the relevant systems that hold the potential for information about nucleic acid structure are dinucleotides and oligonucleotides, both single- and double-stranded. Initial work by Hart (1978) on adenosylyl-(3′→5′)-adenosine (ApA) symbolized in **4**, showed a complex but potentially informative phosphorus–proton enhancement profile (Table II). Of particular interest was the substantial phosphorus enhancement in response to irradiation of pA–H-2′,3′, the 2′ and 3′ hydrogens of the adenylic acid portion of the dimer. No such enhancement was detected in the 5′-AMP case (the closest monomer analog); thus ApA must manifest a different conformation distribution about the bonds $\delta$ and/or $\epsilon$ than is seen in 5′-AMP. The ApA data are qualitative only for several reasons. The primary reason is the substantial proton resonance overlap at 2.1 T, making unique (single-proton) irradiations rare (only the 1′ protons can be irradiated separately). As well, however, because of the total NOE value of 78%, considerably below the theoretical maximum, the system might be in substantial violation of the extreme narrowing condition. Nevertheless, it was my opinion and so it remains, that relevant

## TABLE II

P{H} NOE and $T_1$ Values for ApA and $d_3$ApA

| Compound | $T_1$ (s)[a] | Proton(s) saturated | NOE (%)[b] |
|---|---|---|---|
| ApA | 7.1 | All | 78 |
|  |  | Ap-1' | 0 |
|  |  | pA-1' | 0 |
|  |  | pA-2',3' | 20 |
|  |  | Ap-2',3' | 30 |
|  |  | Ap-5',5" | 1 |
|  |  | Ap-4'; pA-4',5',5" | 32 |
| $d_3$ApA | 8.2 | All | 78 |
|  |  | Ap-1' | 2 |
|  |  | pA-1' | 0 |
|  |  | Ap-2',3'; pA-2' | 17 |
|  |  | pA-3' | 14 |
|  |  | Ap-4' | 12 |
|  |  | Ap-5',5" | −1 |

[a] Measured by inversion–recovery; conc = 20 mg/ml; $T$ = 26°C; samples degassed.
[b] $T$ = 26°C; less than optimum proton irradiation power; samples degassed.

conformational analyses must be accomplished at the dimer and the oligomer level in order to provide a means for studying the correlation between conformation and environmental parameters.

If it were possible to work at higher fields, at least some of the overlap in the proton spectrum of, for example, ApA would be mitigated and less ambiguous irradiations could be accomplished. Indeed, at 6.3 T the proton spectrum is sufficiently dispersed so that virtually all ambiguity would vanish. Unfortunately, at that field extreme narrowing is violated to such an extent that the phosphorus–proton NOE is nearly completely attentuated (P. A. Hart, unpublished; heteronuclear NOEs drop to zero; only homonuclear NOEs become negative when extreme narrowing is violated). Thus it is necessary to work at the lower-field values and simplify the proton spectrum by deuterium substitution. Yang (1980) has completed preliminary work of the kind required. Yang accomplished the synthesis of 4',5',5"-trideuteroadenosine and coupled that with the appropriately blocked adenosine 3'-phosphite to give, after oxidation and deprotection, trideuterated ApA labeled at pA-4',5',5". Because of the synthetic route chosen, this substance was difficult to make and very difficult to purify, thus only a small amount was available. Nevertheless, it was possible to get the P{H} NOE data of Table II. Note that some proton degeneracy still exists; however, it was possible to irradiate pH–H-3′ unequivocally and to establish the distinct difference on that basis between the pA portion of ApA and 5′-AMP as noted. Differences in the Ap region of the dimer were noted relative to 3′-AMP as well and the important point is, these subtle differences appear to be inconsistent with analyses of Lee *et al.* (1976). However, more deuterium-substituted oligomers are required to allow a more thorough analysis. The difficulty does not end there. The conformational distribution about the $\beta,\gamma$ bonds (the phosphodiester bonds) cannot be determined by the NOE method. Rotation about those bonds does not change phosphorus–proton distances. Therefore, not only must further work be completed to specify the $\alpha,\delta,\epsilon$-rotamer distributions but a different method entirely must be used to determine the $\beta,\gamma$-rotamer distribution. It is ironic that these bonds are thought to be the most important sites of conformational change in oligonucleotides.

## IV. Phosphorus Relaxation Times

### A. Oligonucleotides

#### 1. General Considerations

Section III establishes how the NOE method could be used to analyze the rotamer distribution about the $\alpha,\delta,\epsilon$ bonds (and to some extent, the bonds of the sugar ring), but that the method is not at all applicable to $\beta,\gamma$-bond analysis in oligonucleotides. In this section, I show how it is possible to approach the $\beta,\gamma$ distribution problem by interpretation of phosphorus relaxation times, and I show how this method can be used in conjunction with the phosphorus chemical-shift correlations of Gorenstein (1981).

The salient feature of Eq. (2), given in general form in Section II, that motivates this section is the appearance in the equation of separate factors that depend on internuclear vector orientation and equilibrium conformer populations. These factors persist into the extreme narrowing regime for any mode(s) of overall motion, so long as conformational change is sufficiently fast to preclude exchange effects on relaxation rates. This is true even though conformational change might *not* be fast enough to contribute to the harmonic average of correlation times in the second term of Eq. (2). Exploitation of these geometric factors for conformational analysis is difficult not only because of the number of adjustable parameters (as mentioned earlier) but also for more fundamental reasons that are revealed in the sequel.

## 2. The Two-State Jump Model

Hart et al. (1981) and P. A. Hart and C. F. Anderson (unpublished) have treated this problem using an approach analogous to that of Tsutsumi (1979). The very simplest model for internal motion that allows specification of conformational probabilities is the two-state jump model in which independent internal motion is manifested as a transition between two states populated unequally. While this model has not been applied to small oligonucleotides, it has found use in the rationalization of DNA restriction fragment relaxation times. The details of that work are discussed following the completion of this section. The two-state model is included here because it leads easily and naturally to the general form of Eq. (2).

For the present analysis, it suffices to adopt a model for internal motion that contains the motion as an equilibrium between two states — the conformational parameters of which can be specified. It is required, as usual, that this motion be independent from overall motion. If these states are labeled 1 and 2 and the conformational change is assumed to follow first-order kinetics, time-dependent probabilities are

$$P_1(t) = P_1^E + C_1 \exp[-k_2/P_1^E t] \tag{10a}$$

$$P_2(t) = P_2^E + C_2 \exp[-k_2/P_2^E t] \tag{10b}$$

in which $P^E$ is an equilibrium probability (or population), $k_1$ and $k_2$ are the relevant rate constants, and $C_1$ and $C_2$ constants that are determined by the boundary conditions. For the sake of constructing the correlation function mentioned earlier, it is necessary to have the two-time joint probability $P(i, j)$, which is the probability that at time $t$ the system is in the state $i$ and at an earlier time it was in the state $j$. This joint probability can be expressed in terms of conditional and a priori (or equilibrium) probabilities, $P(i, j) = P(i|j)P_j^E$, an expression that can be evaluated by imposing boundary conditions on the time-dependent expressions [Eq. (10)]. Thus, when

$$P_1(0) = 1, \quad C_1 = 1 - P_1^E$$

and when

$$P_1(0) = 0, \quad C_1 = -P_1^E$$

and similarly for $C_2$. Furthermore, by the equilibrium conditions, $k_1/P_2^E = k_2/P_1^E$, so we may write for the required conditional probabilities,

$$P(1|1) = P_1^E + (1 - P_1^E) \exp[-k_1/P_2^E t] \tag{11a}$$

$$P(1|2) = P_1^E - P_1^E \exp[-k_1/P_2^E t] \tag{11b}$$

$$P(2|1) = P_2^E - P_2^E \exp[-k_1/P_2^E t] \tag{11c}$$

$$P(2|2) = P_2^E + (1 - P_2^E)[-k_1/P_2^E t] \tag{11c}$$

The correlation function for internal motion that we seek, given previously in terms of a two-time joint probability, can now be cast in terms of conditional probabilities, so that

$$G(\tau) = \sum_{k=1}^{5} \sum_{i,j} f_k(i) \, P(i|j) \, P_j^E \, f_k(j), \qquad i,j = 1,2 \tag{12}$$

in which the $f$s are as previously defined. Executing the cyclic permutation of the indices $i, j$ and inserting the definition $\tau_{\text{int}} \equiv \tau_1 P_2^E$ produces four separate terms, the first one of which has the form

$$\sum_{k=1}^{5} f_k(1) \left[ (P_1^E)^2 + P_2^E(1 - P_1^E) \exp\left(\frac{-t}{\tau_{\text{int}}}\right) \right] f_k(1) + \cdots \tag{13}$$

The assumption of independent internal and overall motion allows the correlation functions for each to be averaged separately, then multiplied. Thus, because, for example, isotropic overall motion is represented by a single exponential correlation function, Eq. (13) becomes

$$\sum_{k=1}^{5} f_k(1) \left[ (P_1^E)^2 \exp\left(\frac{-t}{\tau_R}\right) + P_2^E(1 - P_1^E) \exp\left(\frac{-t}{\tau_{RC}}\right) \right] f_k(1) + \cdots \tag{14}$$

in which $\tau_R$ is the correlation time for overall motion, which, when Fourier transformed, gives the first term of the spectral density function

$$J(\omega) = \sum_{k=1}^{5} f_k(1) \left[ (P_1^E)^2 \frac{\tau_R}{1 + (\omega \tau_R)^2} + P_2^E(1 - P_1^E) \frac{\tau_{RC}}{1 + (\omega \tau_{RC})^2} \right] f_k(1) + \cdots \tag{15}$$

of which there are three corresponding to the combinations $\omega_I - \omega_S$, $\omega_I$, and $\omega_I + \omega_S$. The form of Eq. (2) is already apparent in Eq. (15), and for this two-state model the probability matrices are

$$P = \begin{bmatrix} (P_1^E)^2, & P_1^E P_2^E \\ P_1^E P_2^E, & (P_2^E)^2 \end{bmatrix}; \qquad P' = \begin{bmatrix} P_1^E P_2^E, & P_1^E P_2^E \\ P_1^E P_2^E, & P_1^E P_2^E \end{bmatrix}$$

I show how these matrices and the geometric factors can become expressions for a specific two-state model in Section IV,B.

### 3. The Three-State Jump, Joint Conformation Model

Tsutsumi (1979) has formulated the problem of a spin-pair jumping among three sites, two of which have equal potential energies. This formulation is far more complex than the two-site case, not only because nine terms appear in the autocorrelation function, but because internal motion can no longer be expressed in terms of a single exponential decay factor. Instead of trying to understand the Tsutsumi (1979) formulation of the three-site

## 11. Conformation of Nucleic Acids

model in light of Eq. (2), I move to a three-site model that forces one to grapple as well with the problem I wish to address in this section, the joint conformational change of the $\beta,\gamma$ bonds of the phosphodiester backbone of small, single-stranded oligonucleotides.

The joint three-site problem has been treated by P. A. Hart and C. F. Anderson (unpublished) using an approach for each P—O bond of the phosphodiester (or any other bond pairs of the phosphodiester) that is analogous to that of Tsutsumi (1979). To see how the analog of Eq. (2) takes shape, I focus on the phosphodiester model of Fig. 4 and the potential-energy profile of Fig. 5. The model of internal motion allows rotation to occur about each P—O bond randomly, independently, and discretely. For each bond the potential energy profile of Fig. 5 is required in which $P_1 \neq P_2 \neq P_3$ and $W_{21} = W_{31} = W_{32} = W$. Overall motion is required to be isotropic. As usual, direction cosines and P—H distances are computed using bond angles and distances, determined by X-ray analysis or estimated by other means, for the nine joint conformations.

In constructing the necessary correlation functions, we require expressions for the two-time *joint* conditional probability $P[\alpha(t)|\alpha'(0)]$ in which $\alpha$ and $\alpha'$ are different joint conformations of phosphodiester bonds. Because the A,B bond rotations are required to be statistically independent, $P[\alpha(t)|\alpha'(0)] = P_A(i|j) \cdot P_B(i|j)$; thus it is possible to develop a formulation for each bond separately and combine them in the final step.

In this more complicated case, the kinetic equations that pertain when the rate and equilibrium requirements just discussed are met, become

$$dP_1/dt = -(W_{12} + W_{13})P_1 + WP_2 + WP_3 \tag{16a}$$

$$dP_2/dt = W_{12}P_1 - (W + W_{23})P_2 + WP_3 \tag{16b}$$

$$dP_3/dt = W_{13}P_1 + W_{23}P_2 - 2WP_3 \tag{16c}$$

given standard equilibrium definitions as previously in the two-state example,

$$W_{12} = W\left(\frac{P_2^E}{P_1^E}\right); \quad W_{23} = W\left(\frac{P_3^E}{P_2^E}\right); \quad W_{13} = W\left(\frac{P_3^E}{P_1^E}\right)$$

and the solution of these coupled differential equations in terms of conditional probabilities has the general form for one bond:

$$P_A(i|j)P_A(j) = W_A^0 + W_A^1 P_1^A \exp\left(\frac{-t}{\tau_{A1}}\right)$$

$$+ W_A^2 \left(\frac{P_2^A P_3^A}{P_2^A + P_3^A}\right) \exp\left(\frac{-t}{\tau_{A2}}\right) \quad \text{(for } i,j = 1,3\text{)} \tag{17}$$

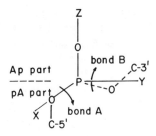

Fig. 4. The phosphodiester model and its relationship to the laboratory frame of reference.

in which $\tau_{A1} = P_1^A/W$ and $\tau_{A2} = P_2^A/W(2P_2^A + P_3^A)$ and, suppressing the bond designation,

$$W^0 \equiv \begin{bmatrix} P_1^2, & P_1P_2, & P_1P_3 \\ P_1P_2, & P_2^2, & P_2P_3 \\ P_1P_3, & P_2P_3, & P_3^2 \end{bmatrix}$$

$$W^1 \equiv \begin{bmatrix} P_2+P_3, & -P_2, & -P_3 \\ -P_2, & P_2/(P_2+P_3), & (P_2P_3)/(P_2+P_3) \\ -P_3, & (P_2P_3)/(P_2+P_3), & P_3/(P_2+P_3) \end{bmatrix}$$

$$W^2 \equiv \begin{bmatrix} 0 & 0 & 0 \\ 0 & 1 & -1 \\ 0 & -1 & 1 \end{bmatrix}$$

Exactly the same formulation is constructed for the other P—O bond (B); then the two expressions are combined to give the time dependence of the joint conditional probabilities. This combination results in a lengthy nine-term equation, the terms of which are expressed as direct products of the $W$ matrices, equilibrium populations, and exponential decay factors but are no more complex in principal than their analogs in the single two-state treatment. The first two terms are

$$\sum_{i,j=1}^{3} P_A(i|j)P_A(j) \cdot P_B(i|j)P_B(j) = W_A^0 \otimes W_B^0 + W_A^1 \otimes W_B^1 P_1^A P_2^B \exp\left[-t\left(\frac{1}{\tau_{A1}+\tau_{A2}}\right)\right] \quad (18)$$

The spectral densities are derived after overall motion is incorporated (exactly as was done in the two-state case) and after multiplication by the nine-element geometric factor vectors. The spectral densities are used to compute the separate contribution of each P—H pair to $1/T_1$ or $1/T_2$; the separate contributions are either examined separately or summed. The NOE

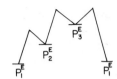

**Fig. 5.** Relative potential energy profile for the three-state model.

calculation requires the summation of all spectral densities prior to the computation because one requires $\langle r_{P-H}^{-6} \rangle$ to compute the NOE.

The specific application of the joint three-state model to phosphodiester conformation has been started and has been applied to the dinucleoside monophosphate ApA and its trideutero analog prepared by Yang (1980). It will become apparent why the deuterated compound was required. To apply the model at all, one needs to know the relevant number of significant P–H interactions. By analogy with the 5'- and 3'-AMP work cited previously, the phosphodiester phosphorus can interact with Ap (5', 5", 4', 3', 2', and 1') as well as pA (5', 5", 4', and 3'), a total of 10 potential interactions. Only 7 were used in the actual simulations: Ap (4', 3', 2', 1') and pA (5', 5", 4', and 3'). The NOE work of Hart (1978) and Yang (1980) clearly establishes that whereas Ap-3' and pA-5', 5" can be major contributors to the phosphodiester phosphorus relaxation, they are not by any means the only ones; indeed, certain of the other interactions listed can contribute on an equal footing in certain reasonable conformation states.

The use of the model to simulate the observed spin–lattice relaxation time (7.1 s) by summing the contribution of all P–H interactions revealed a few reasonable overall and internal correlation time combinations that gave values of $T_1$ in the range of the observed value. Within that correlation time range, there was a sensitivity to the choice of equilibrium probabilities. However, there was *not* an experimentally significant sensitivity to the choice of conformation distribution. That was a disturbing observation, because after all, that was the sensitivity required for the method to be a useful tool for conformational analysis. That outcome can be understood when one recognizes, by use of molecular models, that for many joint, $\beta,\gamma$ conformations, the phosphorus is situated pretty much symmetrically in a nest of hydrogens. Thus, the problem suffers from too much of a good thing as far as P–H interactions are concerned. When one follows the contribution of *each* P–H interaction rather than the summed contribution, an exquisite sensitivity to all geometric and dynamic parameters is seen. There are several experimental strategies available to measure single spin contributions, numbers that now become necessary if the simulations must be done on a per-spin interaction basis. However, they all require selective perturbation of particular protons or proton groups. The most promising method, providing the greatest flexibility in the variation of environmental conditions, is selective deuterium substitution. In the present case, that substitution(s) should be chosen which provides the maximum disruption of the symmetry of the intramolecular proton bath. One such choice is the pA-5′,5″,4′-trideutero analog mentioned previously. Indeed, simulation of the experimental $T_1$ (8.2 s) of the trideutero compound reduces the number of backbone conformation choices (within a reasonable range of overall and internal correlation times) to just a few. I think there is sufficient reason to continue the deuterium substitution studies even though they are difficult and even though the relaxation time method is fraught with other difficulties as well (see following discussion). These studies are potentially useful because they can provide an independent estimate of phosphodiester conformation that can serve as a basis for the phosphorus chemical-shift correlations with phosphodiester conformation pioneered and reviewed by Gorenstein (1981).

Despite its potential usefulness, it is entirely premature to claim success for the method. Although it is certainly possible to synthesize more than a sufficient number of deuterium-labeled ApA analogs to match the number of required parameters in the $T_1$ simulations, fundamental problems remain, and they appear beyond the specter of overparameterization. Perhaps the greatest difficulty with the relaxation method arises from the assumption of isotropic overall motion. Taking all molecular conformations into account, some (the extended forms) will definitely not execute isotropic motion. Further, these anisotropically reorientating forms are produced as a function of *internal motion,* thus the requirement of independent internal and overall motion is not met. These two coupled problems are generally significant, at least until they can be treated and the extent of their importance can be ascertained.

There are, however, experimental systems that do not manifest the coupling of internal and overall motion for certain carefully chosen but entirely reasonable modes of internal motion. One such system is double-stranded DNA. Moderately short segments of ds-DNA [20–200 base pairs (bp) long] behave in some experimental systems as though they were rigid rods (symmetric tops). Therefore, it is clear that their overall motion is anisotropic. The incorporation of anisotropic overall motion into these models while maintaining the requirement of independent internal and overall motion is discussed in the following section.

## B. Double-Stranded DNA

Many experiments based on a variety of techniques have disspelled the notion that double-stranded DNA in solution behaves as though it were, in effect, a sequence of rigid 500- to 600-Å-long cylindrical forms. Instead, it is clear that for each "rigid" segment, a number of modes of internal motion (bending, torsion, stretching, compression) on time scales much faster than overall motions must be considered. In general, these internal motions are viewed as small-amplitude motions (though that is not necessary as seen later); however, the B → Z transition involves substantial internal conformational changes that occur slowly, relative to overall motion. DNA dynamics is a subject that has been widely discussed, and it is not feasible to cite all the primary contributions that have been made. Keepers and James (1982), Allison et al. (1982), Hart et al. (1981), and Jardetzky (1981) have brought the relevant $^{31}$P-NMR literature into focus, and Hagerman (1981) and Diekmann et al. (1982) have clarified the experimental determination of overall DNA motion.

The observations that motivate this section are those made by Hart et al. (1981) of the temperature dependence of $^{31}$P $T_1$ and P{H} NOEs of DNA restriction fragments. Those authors reported that not only were measured $^{31}$P $T_1$ values fragment-length-dependent in the range 43–108 bp, but that each fragment showed a distinct temperature dependence of its $T_1$ in the range 30–50°C, well below the fragment $T_m$ values (70–80°C). The P{H} NOEs were not temperature-dependent. None of the relaxation observations could be modeled assuming purely rigid-rod behavior for the DNA fragments. Instead, it was necessary for us to include some internal motion correction, a common requirement as summarized in the references mentioned before. This section emphasizes the per-fragment relaxation behavior and shows in summary our attempts to rationalize the single-fragment, temperature-dependent, relaxation-time trends. The data are tabulated in Table III and plotted in Fig. 6. Lines are drawn through the points in the figure to guide the eye.

The rigid-rod calculation gave computed $T_1$ values shown in Table III, along with the experimental values. It is clear from this simulation that some other mode of motion must be included. Therefore, I show how a single two-state model for internal motion superimposed on independent, symmetric-top overall motion rationalizes the observed $T_1$ values at 30°C and how the model accounts in principle for the temperature-dependent $T_1$ changes. It is also shown that attempts to simulate observed $T_2$ values or linewidths using the same model fail, and a possible reason for the failure is given.

The choice of a specific molecular model for internal motion was guided by the need to maintain the independence of overall and internal motion, both because the mathematical formulation requires it and because Hagerman (1981) has shown temperature independence of the transient electric birefringence of specific DNA restriction fragments as large as 587 bp. In the Hagerman experiment, temperature did not exceed the $T_m$ of the fragment, so that this result means that diffusion of the long axis of the helix ($D_\perp$)

**TABLE III**

Experimental and Calculated $^{31}$P Spin–Lattice Relaxation Times for DNA Restriction Fragments

| Fragment length [a] | Observed $T_1$ [b] | Calculated $T_1$ [c] |
|---|---|---|
| 180 | 3.47 ± 0.2 | 15.8 |
| 84 | 3.05 ± 0.13 | 7.18 |
| 69 | 2.61 ± 0.14 | 5.86 |
| 43 | 2.12 ± 0.23 | 3.39 |

[a] Number of base pairs.
[b] Inversion–recovery sequence; 30°C.
[c] Rigid-rod approximation; no internal motion.

remains constant. Furthermore, as the apparent helix axis lengths measured by this method are entirely reasonable based on DNA double-helix geometry, there is no significant observed coupling between internal motion and $D_\perp$. Therefore, a molecular model of the B form of DNA was manipulated, with the objective being to find backbone conformational changes that could satisfy the independence criterion. A specific two-state fluctuation was found and it is symbolized in Fig. 7. In this two-state equilibrium, rotation about $\alpha$ is not allowed, whereas $\beta$, $\gamma$, and $\delta$ are allowed to assume the states shown in the figure in a correlated or concerted fashion. The geometric implications of this concerted conformational fluctuation are registered in Table IV. It is evident from the table that direction cosines are the geometric

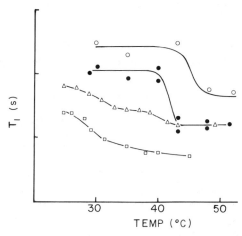

**Fig. 6.** Spin–lattice relaxation time ($T_1$) versus temperature for four DNA restriction fragments (length in base pairs): □, 43; △, 69; ●, 84; ○, 180. Adopted from Hart et al. (1981).

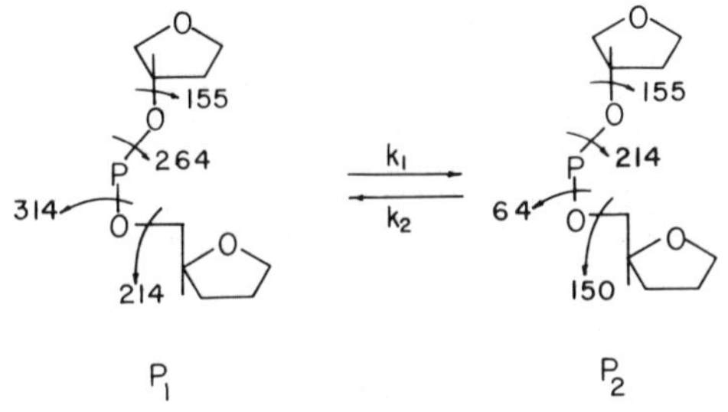

## 11. Conformation of Nucleic Acids

**Fig. 7.** The two DNA backbone states used in the $T_1$ simulations. Adapted from Hart et al. (1981).

parameters that vary widely and that P–H distances are only moderately altered. Olson (1981) has discovered a number of "crankshaft" rotations of the backbone that are analogous to the present case, and those rotations will eventually be used to simulate relaxation data using a more elaborate model.

This model of the restriction fragments can be used to simulate relaxation data via Eq. (15). The two-state concerted model is adopted for internal motion. Overall motion is assumed to be represented by diffusion of a rigid cylinder or symmetric top. In that case there are three time constants for motion that can be described as functions of two diffusion constants,

$$1/\tau_{RO} = 6D_\perp \tag{19a}$$

$$1/\tau_{R1} = D_\parallel + 5D_\perp \tag{19b}$$

$$1/\tau_{R2} = 4D_\parallel + 2D_\perp \tag{19c}$$

in which $D_\perp$ corresponds to diffusion about the short cylindrical axis (slow)

**TABLE IV**

Two-State DNA Geometric Parameters

| P–HX | X | State 1[a] | | | | State 2 | | | |
|---|---|---|---|---|---|---|---|---|---|
| | | $r$ (Å) | $l$ | $m$ | $n$ | $r$ (Å) | $l$ | $m$ | $n$ |
| Ap | 2' | 2.8 | .887 | .079 | .456 | 2.8 | .742 | .585 | .328 |
| | 2" | 3.9 | .796 | −.514 | .320 | 3.9 | .737 | .076 | .672 |
| | 3' | 2.7 | .599 | −.728 | .333 | 2.7 | .561 | −.103 | .821 |
| | 4' | 3.9 | .956 | .205 | −.209 | 3.9 | .958 | .210 | −.197 |
| pA | 5' | 3.0 | .309 | −.704 | −.639 | 2.6 | .619 | −.737 | .271 |
| | 5" | 2.5 | −.160 | −.492 | −.856 | 3.0 | .571 | −.767 | −.291 |
| | 4' | 4.5 | −.239 | −.736 | −.633 | 4.6 | .354 | −.935 | .020 |
| | 3' | 4.8 | −.587 | −.734 | −.340 | 5.2 | −.100 | −.982 | −.187 |

[a] B form of Arnott et al. (1969).

and $D_\parallel$ to diffusion about the long axis (fast). The values of these diffusion constants can be calculated using

$$D_\parallel = kT/8\pi\eta bL^2 \tag{20}$$

$$D_\perp = \frac{3kT}{\pi\eta L^3}\left\{\left[\ln\left(\frac{2L}{b}\right) - 1.57\right] + 7\left[\frac{1}{\ln(2L/b)} - 0.28\right]^2\right\} \tag{21}$$

in which $k$ is Boltzmann's constant, $T$ the absolute temperature, $\eta$ the solvent viscosity at $T$, $b$ the helix radius, and $L$ one-half the actual helical length ($3.4 \times n$ bp). This particular form of $D_\perp$ was used by Barkley and Zimm (1979) and comes originally from the work of Broersma (1969). More elaborate formulations of $D_\perp$ have been reported by Tirado and de la Torre (1980) that allow end-effect corrections; however, the level of the present analysis does not require them. The computed values of $D_\perp$ for various restriction fragments agree fairly well with those measured by transient electric birefringence; no experimental check on the $D_\parallel$ calculation is available.

The use of Eq. (2) to compute spectral densities to be converted to relaxation parameters of the symmetric top with internal motion requires an elaboration of $\tau_R$ in that equation. One simply expands the spectral densities into six terms rather than two, in which $\tau_R$ is successively $\tau_{RO}$, $\tau_{R1}$, and $\tau_{R2}$ of the symmetric top. All other features of the calculation are the same. Eight P–H interactions were used in the simulations: Ap (2′, 2″, 3′, 4″) and pA (5′, 5″, 4′, 3′).

## 1. Spin–Lattice Relaxation Time Simulations

Computed $T_1$ values corresponding to the four restriction fragments modeled as described in the preceding section responded to variation of *all* parameters of the model. They were least sensitive to the value chosen for helix diameter; however, they were distinctly sensitive to helix length, a sensitivity shown clearly in Fig. 8. Furthermore, Fig. 8 evidences sensitivity to internal conformation distribution when the time constant for internal motion is $10^{-9}$ s. Other time constants are associated with similar conformation-dependent variation, and in some cases the observed $T_1$ can be fitted using a fairly broad range of $\tau_1$. Total phosphorus–proton NOEs permit a narrowing of the allowed $\tau_1$ range to somewhere between $10^{-9}$ and $10^{-8}$ s. The $T_1$ simulations show clearly that quite subtle conformational changes can change the value of $^{31}$P $T_1$, and that longer fragment $T_1$ values are more sensitive to both internal conformational change and to the rate of conformational change.

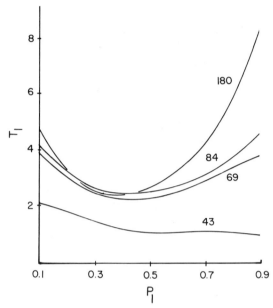

**Fig. 8.** Computed $T_1$ versus $P_1$ (the probability of state 1) for the four DNA restriction fragments. Time constant for internal motion $\tau_1$ is $10^{-9}$ s.

## 2. Spin–Spin Relaxation Time Simulations

In principle, for systems in which the time constant for overall motion is in the microsecond range and that for internal motion is one or two orders of magnitude shorter, $1/T_2$ is virtually dominated by the frequency-independent spectral density $J(0)$. The restriction fragments under discussion satisfy that condition; indeed, it is $\tau_{RO}$ of the symmetric top that is dominant. Consequently, time constants for internal motion are not reflected in simulations of $T_2$ unless they are in the microsecond range. However, because $J(0)$ retains the geometry and equilibrium-sensitive coefficient, $T_2$ simulations remain sensitive to them.

There are two overt $^{31}$P $T_2$ simulations in the DNA literature of which I am aware that are based on measured values of $T_2$, that is, simulations of experimental values of $T_2$ measured by spin–echo techniques rather than using linewidth measurements. Hogan and Jardetzky (1979, 1980) measured $^{31}$P $T_2$ values of various DNA fragments derived from sonicated DNA and simulated their data using the formulation of King and Jardetzky (1978), from which

$$1/T_2 = \tfrac{3}{5}(\alpha/\gamma_P^2\gamma_H^2\hbar^2 r^{-6}\tau_{RO}) \tag{22}$$

In the dynamic regime considered in this section, the $\alpha$ in Eq. (22) is a geometric factor that depends on the average orientaton of the (in this case) three P–H vectors and on the amplitude of internal motion. Lipari and Szabo (1982) have shown how their formulation is equivalent to that of King and Jardetzky when *internal* motion satisfies the extreme narrowing condition. By assuming that the three P–H vectors of their model are each 2.8 Å long and using values for $\tau_{RO}$ calculated by the Broersma formulation, Hogan and Jardetzky (1980) derived a value of $\alpha$ by fitting measured relaxation parameters for more than one fragment length. They made no effort to express their factor $\alpha$ in terms of a specific internal geometry. Therefore, what Hogan and Jardetzky have accomplished is a demonstration that their relaxation data vary in some consistent way with helix dimension and/or solvent viscosity. In contrast, if I use the two-state model discussed previously to fit the measured $T_2$ (0.032 s) of Hogan and Jardetzky's 140-bp (average) fragment, I require a value for $\tau_{RO}$ of $1 \times 10^{-6}$, corresponding to an 84-bp helix. In a similar vein, Allison *et al.* (1982) have used an elaborate model including long-axis reorientation of the rigid rod ($D_\perp$), internal torsion, and independent internal motion. Their model includes three P–H interactions of 2.6 Å, each at specified orientations relative to the helix axis. Allison *et al.* used their model to simulate a $T_2$ (0.012 s) measured by Hogan and Jardetzky for a 600-bp (average) fragment and found that a value of $\tau_{RO} = 4.9 \times 10^{-6}$ (corresponding to ~160 bp) was required.

It is clear that all models have adjustable elements that allow simulation of measured $T_2$ values. My two-state model, for example, could be adjusted by assuming different geometries that would change direction cosines or internuclear distances or both, and these changes would be acceptable within the context of DNA structure. It is also clear that $T_2$ simulations (or perhaps $T_{1\rho}$) in systems having the characteristics of the moderate-sized DNA fragments discussed here are the most severe test of one's choice of model. Because of that model sensitivity and because model geometry (rather than internal dynamics) is the critical collective parameter, $T_2$ simulations hold the greatest promise for studying internal geometry and environmental effects on it.

Experimental values of $T_2$ are not widely available because of the difficulty of measuring them accurately. Shindo (1980) used linewidths to estimate $T_2$ in 140-bp (average) DNA fragments. In these systems it is necessary to know the intrinsic chemical-shift dispersion before linewidths or linewidth changes can be estimated. Only in fairly short fragments (<43 bp) are groups of chemically shifted phosphorus lines observable. If the chemical-shift dispersion in these short fragments (~20 Hz at 36.44 MHz) can be extrapolated to the longer fragments where no resolution is seen, then an approximate value of the linewidth can be estimated. Such a procedure is frought with potential error.

Mention of the Shindo (1980) work exposes a further source of difficulty implicit in all of the foregoing, for Shindo was seeking an estimate of the contributions of chemical-shift anisotropy (CSA) to measured relaxation parameters. As mentioned at the outset, both internal and overall motion can contribute to relaxation via modulation of the anisotropic chemical shift. The principal axes of the chemical-shift tensor are related primarily to electronic distribution about the nucleus and secondarily to molecular geometry. Furthermore, this relaxation mechanism is proportional to $B_0^2$, where $B_0$ corresponds to the stationary magnetic field. The extent of the CSA contribution (at a given field) must be known before simulations of relaxation parameters based on models assuming pure dipolar relaxation can be taken seriously. It has been commonly assumed that at moderate fields (2.5 T or less) the CSA mechanism can be ignored; however, the works of Shindo (1980), Allison et al. (1982), and Keepers and James (1982) prompt caution. This topic is considered in greater detail in Section IV,C.

## C. Phosphoproteins

The phosphoproteins are ubiquitous and in many cases serve essential control mechanisms. A review by Greengard (1978), a monograph by Weller (1979), and a Cold Spring Harbor Conference proceedings edited by Rosen and Krebs (1981) serve nicely to display the functional scope of these proteins. It is a truism that the physical properties of these proteins can be profoundly altered by the presence or absence of a phosphorus-containing functional group. Of the very large number of known phosphorylated proteins, a few have been subjected to [31]P-NMR analysis (see Vogel, Chapter 4). Ho et al. (1969) and Ho and Kurland (1966) studied $\alpha_s$-casein of cow's milk and egg yolk phosvitin and concluded that most of the phosphorus appeared as serine monophosphate. A variety of enzymes have been studied and different types of covalent phosphate have been found. Thus phosphoserine residues (Bock and Sheard, 1975; Chlebowki et al., 1976; Coleman and Chelbowski, 1979; Hull et al., 1976; Ray et al., 1977), histidine N-phosphate (Gassner et al., 1977; Dooijewaard et al., 1979; Bridger, 1974; Vogel et al., 1982), and an aspartyl acyl phosphate (Fossel et al., 1981) have all been identified. Edmondson and James (1979) and James et al. (1981) have studied the covalently bound phosphate of some flavoproteins and have tentatively identified a phosphodiester linkage between two hydroxy amino acids.

Section IV,B,2 introduces the idea that chemical-shift anisotropy could become an important relaxation mechanism for phosphorus at high field. Vogel et al. (1982) have published the field dependence of the linewidth of the histidine 3-$N$-phosphate of succinyl-CoA synthetase and the serine phosphate of glycogen phosphorylase $a$. They found that the CSA mechanism did, indeed, dominate at 6.3- and 9.3-T fields, contributed substantially at 4.7 T, and contributed about 25% of the total relaxation at 2.1 T (36.4 MHz). Vogel et al. (1982) estimated for *Escherichia coli* succinyl-CoA synthetase (MW 140,000) a hydrated radius of 37 Å and an isotropic correlation time of 44 ns. These numbers are referenced to the phosphorus linewidth (sensitive only to overall motion) using an analysis very similar to that mentioned in Section IV,B on DNA phosphorus relaxation. In contrast, the *E. coli* phosphoryl carrier protein HPr (MW 9000) has a calculated hydrated radius of 17 Å and a correlation time of 4.2 ns, according to Vogel et al. (1982).

A group of phosphoproteins with no established catalytic function is the caseins of cow's milk. These proteins have molecular weights in the 20,000–30,000 range and have from one to nine covalently bound phosphates. P. A. Hart and R. C. Bookstaff (unpublished) have begun the study of $\alpha_s$-, $\beta$-, and $\kappa$-casein in order to establish basis values for the phosphorus relaxation parameters of these fairly well-characterized proteins under a variety of conditions. The caseins are known to aggregate and the conditions for aggregation have been thoroughly studied by a variety of techniques. Thus it seemed a good system for correlating phosphorus relaxation and state of aggregation was available. Typical $^{31}$P spectra of $\alpha_{s1}$- and $\beta$-caseins are shown in Fig. 9. The conditions are as given in the legend. $T_1$ values and P{H} NOEs are recorded in Table V. All relaxation times recorded for $\beta$-casein are the same for all resolved peaks within experimental error; however, the relaxation time recorded for $\alpha_{s1}$-casein is an average; there was a measurable differential recovery noted. An attempt to understand these data in terms of the proteins' molecular structure and other physical properties begins with the observation that $\alpha_{s1}$-casein may aggregate under the conditions employed by Waugh et al. (1970), whereas the more thoroughly studied $\beta$-casein should be monomeric at 4°C at the ionic strength used in the NMR experiments but should be aggregated (perhaps a 14-mer) at 20°C (Andrews et al., 1979). The radius of this disordered $\beta$-casein aggregate is 100 Å. Aggregation studies have been performed for selected buffers (not the imidazole buffer used by Hart and Bookstaff); the consensus is that aggregation varies with ionic strength, temperature, and concentration of calcium (and other divalent cations) and inversely with pH.

The relaxation data suggest that these systems have considerable mobility. The decrease in $T_1$ and NOE for $\beta$-casein as the pH is lowered (direction of

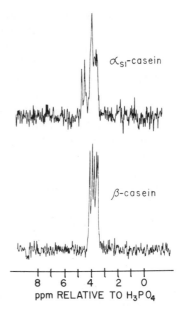

**Fig. 9.** $^{31}$P-NMR spectra (36.44 MHz) of $\alpha_{s1}$- and $\beta$-casein; 40 mg/ml, 0.1 $M$ imidazole buffer, pH 7. Collected 1000 transients. Broad-band proton decoupling, $T = 25°C$.

greater aggregation) show that the protein's effective correlation time is varying in the fast motion regime, that is, its correlation time is increasing and moving toward the $T_1$ minimum. This conclusion is supported by the narrow (3 Hz) lines observed (relatively long $T_2$ values). Similar conclusions can be reached for $\alpha_{s1}$-casein, though it has not been studied as thoroughly. The $\beta$-casein data were analyzed using a model that assumed isotropic overall motion and a single mode of internal motion, namely, rotation about the O—C bond of the serine monophosphate. The rotation was allowed access to three states (g$^+$, g$^-$, t), and the phosphorus was assumed to

TABLE V

Casein Spin–Lattice Relaxation Times and P{H} NOE Values

| Protein | pH | $T_1$ (s)[a] | NOE (%)[b] |
|---|---|---|---|
| $\beta$-Casein | 9.2 | 2.77 | 56 |
|  | 8.6 | — | 52 |
|  | 7.1 | 2.5 | 34 |
|  | 6.3 | 1.72 | 27 |
| $\alpha_{s1}$-Casein | 7 | 1.85 | 25 |

[a] Inversion recovery; 25°C.
[b] Standard method; see Hart (1978).

interact with the two immediately adjacent methylene protons. This model will yield only an approximation at best, for virtually nothing is known about internal geometry or overall shape of $\beta$-casein; thus more sophisticated models are unrealistic. The dynamic parameters suggested by this model are fairly rapid internal motion ($10^{-7}$–$10^{-9}$ s) superimposed on overall isotropic motion ($\tau_R$) in the $10^{-8}$- to $10^{-9}$-s range. If aggregation to the extent reported by Andrews *et al.* (1979) is assumed in the present case (a 14-mer with a 100-Å radius), one can estimate $\tau_R$ by standard means to be $5 \times 10^{-7}$. Thus either the NMR experiment is sensing only a limited and very mobile segment of the aggregate or $\beta$-casein does not aggregate under the conditions of these experiments. The dynamic parameters of $\beta$-casein characterize an even more mobile system than those reported by Vogel *et al.* (1982).

## V. Summary

I hope this chapter has given a fair assessment of what can be accomplished using phosphorus magnetic relaxation measurements as tools in the solution of conformation and dynamics problems pertinent to nucleic acids and phosphoproteins. I have shown that fairly detailed conformational analyses can be done of low-molecular-weight phosphates via the phosphorus–proton nuclear Overhauser effect (NOE) so long as relaxation mechanisms other than the dipolar mechanism do not intervene and so long as a sufficient number of interacting protons can be irradiated separately. I have shown how the method can be used to analyze more complex oligomers and have shown the torsion angles that are inaccessible in those systems, as well as the assumptions required. The possible use of phosphorus spin–lattice relaxation times in the analysis of phosphodiester conformation is given detailed coverage, and I hope the reader comes away from that section with some optimism but with the certainty that much work is needed both to prove and to improve the method. I then showed how the analysis of phosphorus relaxation data can be used to rationalize observed $^{31}$P $T_1$ values and NOEs of DNA restriction fragments and that it may be possible to use phosphorus $T_1$ values to detect conformational change in the phosphodiester backbone of double-stranded DNA. The observation that $T_2$ or $T_{1\rho}$ measurements are sensitive to helix length but that the observed values are not rationalized by existing reasonable models of DNA structure and dynamics needs to be pursued vigorously. Finally, I gave a brief summary of phosphorus NMR work on phosphoproteins and showed that not only are the phosphates of some phosphoproteins well resolved but also that they might serve as unique indicators of structural mobility.

# 11. Conformation of Nucleic Acids

## Acknowledgment

Part of the work summarized here was supported by the National Science Foundation (PCM 77 19927). The phosphoproteins were prepared by Hans Zoerb and Thomas Richardson, Dairy Science Department, University of Wisconsin.

## References

Abragam, A. (1961). "The Principles of Nuclear Magnetism." Oxford Univ. Press (Clarendon), London and New York.
Alderfer, J. L., and Ts'o, P. O. P. (1977). *Biochemistry* **16**, 2410–2416.
Allison, S. A., Shibata, J. H., Wilcoxon, J., and Schurr, J. M. (1982). *Biopolymers* **21**, 729–762.
Andrews, A. L., Atkinson, D., Evans, M. T. A., Finer, E. G., Green, J. P., Phillips, M. C., and Robertson, R. W. (1979). *Biopolymers* **18**, 1105–1121.
Arnott, S., Dover, S. D., and Wonecott, A. J. (1969). *Acta Crystallogr., Sect. B* **B25**, 2192–2206.
Barkley, M. D., and Zimm, B. H. (1979). *J. Chem. Phys.* **70**, 2991–3007.
Bock, J. L., and Sheard, B. (1975). *Biochem. Biophys. Res. Commun.* **66**, 24–30.
Broersma, S. (1969). *J. Chem. Phys.* **32**, 1626–1631.
Bridger, W. A. (1974). *In* "The Enzymes" (P. D. Boyer, ed.), 3rd ed., Vol. 10, pp. 581–606. Academic Press, New York.
Chlebowski, J. F., Armitage, I. M., Tusa, P. P., and Coleman, J. E. (1976). *J. Biol. Chem.* **251**, 1207–1216.
Coleman, J. E., and Chlebowski, J. F. (1979). *Adv. Inorg. Biochem.* **1**, 1–66.
Cozzone, P. J., and Jardetzky, O. (1976). *Biochemistry* **15**, 4860–4865.
Davies, D. B. (1978). *Prog. NMR Spectrosc.* **12**, 135–225.
Diekmann, S., Hillen, W., Morgeneyer, B., Wells, R. D., and Pörschke, D. (1982). *Biophys. Chem.* **15**, 263–270.
Dooijewaard, G., Roossien, F. F., and Robillard, G. T. (1979). *Biochemistry* **18**, 2996–3001.
Edmondson, D. E., and James, T. L. (1979). *Proc. Natl. Acad. Sci. U.S.A.* **76**, 3786–3789.
Evans, F. E., and Sarma, R. H. (1974). *J. Biol. Chem.* **249**, 4754–4759.
Farrar, T. C., and Becker, E. D. (1971). "Pulse and Fourier Transform NMR." Academic Press, New York.
Fossel, E. T., Post, R. L., O'Hara, D. S., and Smith, T. W. (1981). *Biochemistry* **20**, 7215–7219.
Gassner, M., Stehlik, D., Schrecker, O., Hengstenberg, W., Maurer, W., and Rüterjans, H. (1977). *Eur. J. Biochem.* **75**, 287–296.
Gorenstein, D. G. (1981). *Annu. Rev. Biophys. Bioeng.* **10**, 355–386.
Greengard, P. (1978). *Science* **199**, 146–152.
Hagerman, P. J. (1981). *Biopolymers* **20**, 1503–1535.
Hart, P. A. (1976). *J. Am. Chem. Soc.* **98**, 3735–3737.
Hart, P. A. (1978). *Biophys. J.* **24**, 833–848.
Hart, P. A., and Davis, J. P. (1969). *J. Am. Chem. Soc.* **91**, 512–513.
Hart, P. A., and Davis, J. P. (1972). *J. Am. Chem. Soc.* **94**, 2572–2577.
Hart, P. A., Anderson, C. F., Hillen, W., and Wells, R. D. (1981). *In* "Biomolecular Stereodynamics" (R. H. Sarma, ed.), Vol. 1, pp. 367–382. Adenine Press, New York.

Ho, C., and Kurland, R. J. (1966). *J. Biol. Chem.* **241,** 3002–3007.
Ho, C., Magnuson, J. A., Wilson, J. B., Magnuson, N. S., and Kurland, R. J. (1969). *Biochemistry* **8,** 2074–2082.
Hogan, M. E., and Jardetzky, O. (1979). *Proc. Natl. Acad. Sci. U.S.A.* **76,** 6341–6345.
Hogan, M. E., and Jardetzky, O. (1980). *Biochemistry* **19,** 3460–3468.
Hull, W. E., Halford, S. E., Gutfreund, H., and Sykes, B. D. (1976). *Biochemistry* **15,** 1547–1561.
James, T. L., Edmondson, D. E., and Husain, M. (1981). *Biochemistry* **20,** 617–621.
Jardetzky, O. (1981). *Acc. Chem. Res.* **14,** 291–298.
Keepers, J. W., and James, T. L. (1982). *J. Am. Chem. Soc.* **104,** 929–939.
King, R., and Jardetzky, O. (1978). *Chem. Phys. Lett.* **55,** 15–18.
King, R., Maas, R., Gassner, M., Nanda, R. K., Conover, W. W., and Jardetsky, O. (1978). *Biophys J.* **24,** 103–117.
Kuhlman, K. F., Grant, D. M., and Harris, R. K. (1970). *J. Chem. Phys.* **52,** 3439–3448.
Lapper, R. D., and Smith, I. C. P. (1973). *J. Am. Chem. Soc.* **95,** 2880–2884.
Lavalee, D. K., and Coulter, C. L. (1973). *J. Am. Chem. Soc.* **95,** 576–581.
Lee, C.-H., and Tinoco, I., Jr. (1977). *Biochemistry* **16,** 5403–5414.
Lee, C.-H., Ezra, F. S., Kondo, N. S., and Danyluk, S. S. (1976). *Biochemistry* **15,** 3627–3638.
Lenk, R. (1977). "Brownian Motion and Spin Relaxation." Am. Elsevier, New York.
Levine, Y. K., Partington, P., and Roberts, G. C. K. (1973). *Mol. Phys.* **25,** 497–514.
Lipari, G., and Szabo, A. (1982). *J. Am. Chem. Soc.* **104,** 4546–4559.
London, R. E., and Avitable, J. (1978). *J. Am. Chem. Soc.* **100,** 7159–7165.
Noggle, J. H., and Schirmer, R. E. (1971). "The Nuclear Overhauser Effect." Academic Press, New York.
Olson, W. K. (1981). *In* "Biomolecular Stereodynamics" (R. H. Sarma, ed.), pp. 327–344. Adenine Press, New York.
Ray, W. J., Mildvan, A. S., and Grutzner, J. B. (1977). *Arch. Biochem. Biophys.* **184,** 453–463.
Redfield, A. (1965). *Adv. Magn. Reson.* **1,** 1–30.
Rosen, O. M., and Krebs, E. G., eds. (1981). "Cold Spring Harbor Conferences on Cell Proliferation." Cold Spring Harbor Lab., Cold Spring Harbor, New York.
Sarma, R. H. (1980). *In* "Nucleic Acid Geometry and Dynamics" (R. H. Sarma, ed.), pp. 1–45. Pergamon, Oxford.
Sarma, R. H., Mynott, R. J., Wood, D. J., and Hruska, F. E. (1973). *J. Am. Chem. Soc.* **95,** 6457–6459.
Schirmer, R. E., Noggle, J. H., Davis, J. P., and Hart, P. A. (1970). *J. Am. Chem. Soc.* **92,** 3266–3273.
Schleich, R., Cross, B. P., and Smith, I. C. P. (1976). *Nucleic Acids Res.* **3,** 355–370.
Shindo, H. (1980). *Biopolymers* **19,** 509–522.
Slichter, C. P. (1963). "Principles of Magnetic Resonance." Harper, New York.
Son, T.-D., and Chachaty, C. (1973). *Biochim. Biophys. Acta* **335,** 1–13.
Son, T,-D., Guschlbauer, W., and Guéron, M. (1972). *J. Am. Chem. Soc.* **94,** 7903–7911.
Sundaralingam, M., Pullman, B., Saenger, W., Sasisekharan, V., and Wilson, H. R. (1973). *Jerusalem Symp. Quantum Chem. Biochem.* **5,** 815–820.
Tirado, M. M., and de la Torre, J. G. (1980). *J. Chem. Phys.* **73,** 1986–1993.
Tropp, J. (1980). *J. Chem. Phys.* **72,** 6035–6043.
Tsutsumi, A. (1979). *Mol. Phys.* **37,** 111–127.
Tsutsumi, A., Quaegebeur, J. P., and Chachaty, C. (1979). *Mol. Phys.* **38,** 1717–1735.
Vogel, H. J., Bridger, W. A., and Sykes, B. D. (1982). *Biochemistry* **21,** 1126–1132.
Vold, R. L., and Vold, R. R. (1978). *Prog. NMR Spectrosc.* **12,** 79–133.

Waugh, D. F., Creamer, L. K., Slattery, C. W., and Dresdner, G. W. (1970). *Biochemistry* **9,** 786–795.
Weller, M. (1979). "Protein Phosphorylation." Pion Ltd., London.
Werbelow, L. G., and Grant, D. M. (1977). *Adv. Magn. Reson.* **9,** 190–298.
Wittebort, R. J., and Szabo, A. (1978). *J. Chem. Phys.* **69,** 1722–1736.
Woessner, D. E. (1962). *J. Chem. Phys.* **36,** 1–4.
Yang, J. (1980). Ph.D. Thesis, University of Wisconsin, Madison.

CHAPTER 12

# Relaxation Behavior of Nucleic Acids: Dynamics and Structure

*Thomas L. James*
Department of Pharmaceutical Chemistry
School of Pharmacy
University of California
San Francisco, California

| | | |
|---|---|---|
| I. | Introduction | 349 |
| II. | Theory | 350 |
| | A. Relaxation Parameters | 350 |
| | B. Spectral Densities and Molecular Motions | 353 |
| III. | Structure and Possible Molecular Motions in Nucleic Acids | 368 |
| IV. | Considerations for Nucleic Acid Dynamics and Structure | 373 |
| | A. Dynamics | 373 |
| | B. Structure | 374 |
| V. | Relaxation Studies of Nucleic Acids | 377 |
| | A. DNA | 377 |
| | B. RNA | 386 |
| | C. Transfer RNA | 389 |
| | D. Nucleic Acid–Protein Systems | 390 |
| | E. Nucleic Acid–Drug Complexes | 396 |
| | References | 398 |

## I. Introduction

In recent years there has been a growing awareness in the scientific community that nucleic acids can exhibit a significant amount of conformational flexibility. The central cellular functions of replication, transcription, and translation involve alterations in the conformation of DNA and RNA. All the different steps in these processes involving nucleic acids, including control steps, indicate that the deformation of polynucleotide structure is intimately related to the biological role of nucleic acids. The conformational variants may exist for very long times ($> 1$ s) or may exist only transiently. Even those that exist only for a matter of nanoseconds may be functionally important structures. For example, a typical intercalating drug can diffuse 5–10 Å in a nanosecond.

NMR relaxation parameter measurements in general can yield information regarding the structure and dynamics of nucleic acids and nucleic acid complexes. $^{31}$P-NMR relaxation in particular will provide insight about the nucleic acid backbone. The reader should consult Chapter 8 (Chen and Cohen), 9 (Gorenstein), and 13 (Shindo) for additional discussion, principally regarding use of $^{31}$P-NMR parameters other than relaxation for study of nucleic acid structure. $^{31}$P-NMR relaxation has been used primarily to investigate dynamics in nucleic acids. Accordingly, the emphasis in this section is on dynamics; specifically, I address the questions of using $^{31}$P NMR to monitor the rates of conformational fluctuations, to estimate the amplitudes of the fluctuations, and to explore the nature of short-lived conformational changes. Second, I consider the use of $^{31}$P-NMR relaxation to examine the structure of nucleic acids. The discussion is limited to polynucleotides and their complexes, not oligomers or monomers. The smaller structures are discussed in Chapter 11 (Hart).

The following discourse deals primarily with the means by which one can extract information about motions (and secondarily about structure) in nucleic acids from NMR relaxation parameter measurements. A rigorous general discussion of NMR relaxation is provided by Abragam (1961) and Spiess (1978), whereas a general discussion oriented toward biochemical systems is given in the book by James (1975). London (1980) has also written an excellent review of the use of NMR to study protein dynamics. First, the relaxation parameters and several motional models are discussed. Second, various possible motions in nucleic acids are then considered. Finally, some of the $^{31}$P relaxation results regarding nucleic acids and their complexes with proteins and drugs are discussed.

## II. Theory

### A. Relaxation Parameters

Thermal motions in nucleic acids produce random, fluctuating magnetic fields that may influence the relaxation of nuclei on the nucleic acid which are interacting with the fluctuating fields. The strength of the interaction as well as the nature of and rate of the fluctuations will govern the nuclear spin relaxation. The strength of the interaction can be expressed in terms of a Hamiltonian. Molecular motions will cause the orientation of the Hamiltonian to change. How well the Hamiltonian at time zero ($t_0$) correlates with its orientation at some later time $t$ is expressed by a correlation function

$$G(t) = \langle D_{q0}^*(\Omega, 0) \, D_{q0}(\Omega, t) \rangle \tag{1}$$

where the orientation at time 0 and time $t$ can be expressed by the Wigner rotation matrix element $D_{q0}$ (Wallach, 1967) and the Euler angles $\Omega$ specify, with respect to the stationary magnetic field, either the orientation of the dipole–dipole internuclear vector in the case of dipolar (DD) relaxation or the orientation of the principal axis system of the chemical-shift anisotropy (CSA) tensor in the case of relaxation via the CSA mechanism. The angled brackets indicate an ensemble average over all possible configurations existing in the ensemble. Equation (1) becomes somewhat more complicated than indicated when the CSA tensor is not axially symmetric. Depending on the prevalent nuclear relaxation mechanism, one can write mathematical expressions relating NMR relaxation parameters to spectral densities $J(\omega)$ that contain the motional information. The spectral density is obtained by taking the Fourier transform of the correlation function:

$$J(\omega) = 2 \int_0^\infty G(t) \cos(\omega t) dt \tag{2}$$

Although other relaxation mechanisms may be important for some nuclei, the dipolar relaxation mechanism of $^{31}P$ that is coupled to protons and the chemical-shift anisotropy mechanism are most important for nucleic acids. In the case of dipolar relaxation, expressions for the spin–lattice relaxation time $T_1$, spin–spin relaxation time $T_2$, nuclear Overhauser effect (NOE),[1] rotating frame spin–lattice relaxation time in an off-resonance radiofrequency (rf) field $T_{1\rho}^{\text{off}}$, and off-resonance intensity ratio $R$ are given by (Doddrell et al., 1972; Kuhlmann et al., 1970; James et al., 1978; James, 1980)

$$\left(\frac{1}{T_1}\right)_{DD} = K \sum_1^N [J_0(\omega_H - \omega_P) + 3J_1(\omega_P) + 6J_2(\omega_H + \omega_P)] \tag{3}$$

$$\left(\frac{1}{T_2}\right)_{DD} = \frac{1}{2T_1} + K \sum_1^N [2J_0(0) + 3J_1(\omega_H)] = \pi(W_{1/2})_{DD} \tag{4}$$

$$\text{NOE}_{DD} = 1 + \frac{\gamma_H}{\gamma_P} \sum_1^N \frac{[6J_2(\omega_H + \omega_P) - J_0(\omega_H - \omega_P)]}{[J_0(\omega_H - \omega_P) + 3J_1(\omega_P) + 6J_2(\omega_H + \omega_P)]} \tag{5}$$

---

[1] Unless stated otherwise, any reference to NOE here means observation of the $^{31}P$-nucleus' response while applying broadband radiation to all protons; the usual nomenclature is $^{31}P\{^1H\}$ NOE.

$$\frac{1}{(T_{1\rho}^{\text{off}})_{\text{DD}}} = \frac{1}{T_1} + K \sin^2 \theta \sum_{1}^{N}$$

$$\times \left\{ 2J_0(\omega_e) + \frac{3}{2}[J_1(\omega_H + \omega_e)] + \frac{3}{2}[J_1(\omega_H - \omega_e)] \right\} \quad (6)$$

$$R_{\text{DD}} = \sum_{1}^{N}$$

$$\times \frac{J_0(\omega_H - \omega_P) + 3J_1(\omega_P) + 6J_2(\omega_H + \omega_P)}{2 \sin^2 \theta J_0(\omega_e) + J_0(\omega_H - \omega_P) + 3J_1(\omega_I) + 6J_2(\omega_H + \omega_P)} \quad (7)$$

where

$$K = \hbar^2 \gamma_H^2 \gamma_P^2 / (20 \, r^6) \quad (8)$$

and $\gamma_H$ and $\gamma_P$ are the respective gyromagnetic ratios of hydrogen and phosphorus, $\omega_H$ and $\omega_P$ the respective angular Larmor frequencies of hydrogen and phosphorus, $r$ is the distance between the phosphorus and the proton causing relaxation, $N$ the number of protons effecting relaxation, and $W_{\frac{1}{2}}$ the linewidth. The other terms are

$$\theta = \tan^{-1}(\gamma_P H_1 / 2\pi \nu_{\text{off}}) \quad (9)$$

and

$$\omega_e = 2\pi \nu_{\text{off}} / \cos \theta \quad (10)$$

where $\omega_e$ is the angular frequency about the effective field produced by application of the rf field $H_1$ at a frequency $\nu_{\text{off}}$ off-resonance. Equation (6) is valid only if $H_1$ is applied far off-resonance, that is, $\nu_{\text{off}} \gtrsim \gamma_P H_1$.

The summation indicates that more than one hydrogen spin could be interacting with the phosphorus spin and that the spectral densities may be different for each of these interactions. It will be noted that the dipolar relaxation equations given are in their conventional form. However, because the P—H internuclear distances may actually fluctuate with the nucleic acids' motions, $r$ may be incorporated explicitly into the spectral density calculations (see following discussion; Keepers and James, 1982).

The chemical-shift anisotropy (CSA) mechanism is also important for $^{31}$P relaxation. The pertinent equations for CSA relaxation are (Hull and Sykes, 1975; Bolton et al., 1981)

$$\left(\frac{1}{T_1}\right)_{\text{CSA}} = \frac{6}{40} (\gamma_P^2 H_0^2 \delta_z^2) \sum_{j=0}^{2} J_j(\omega) \quad (11)$$

$$\left(\frac{1}{T_2}\right)_{\text{CSA}} = \frac{1}{40} (\gamma_P^2 H_0^2 \delta_z^2) \sum_{j=0}^{2} [3J_j(\omega) + 4J_j(0)] = \pi(W_{1/2})_{\text{CSA}} \quad (12)$$

$$\left(\frac{1}{T_{1\rho}^{\text{off}}}\right)_{\text{CSA}} = \frac{1}{40}(\gamma_P^2 H_0^2 \delta_z^2) \sum_{j=0}^{2}[6J_j(\omega) + 4\sin^2\theta J_j(\omega_e)] \tag{13}$$

$$R_{\text{CSA}} = \sum_{j=0}^{2} J_j(\omega) \bigg/ \sum_{j=0}^{2}[6J_j(\omega) + 4\sin^2\theta J_j(\omega_e)] \tag{14}$$

where $H_0$ is the magnetic field strength, $\delta_z$ the anisotropy, and $\nu$ the asymmetry of the chemical-shift tensor ($\nu$ is a measure of deviation from axial symmetry).

If both relaxation mechanisms contribute, the combined relaxation rates are additive,

$$\frac{1}{T_{1\rho,1,2}^{\text{off}}} = \left(\frac{1}{T_{1\rho,1,2}^{\text{off}}}\right)_{\text{DD}} + \left(\frac{1}{T_{1\rho,1,2}^{\text{off}}}\right)_{\text{CSA}} \tag{15}$$

and

$$R = T_{1\rho}^{\text{off}}/T_1 \tag{16}$$

and

$$\text{NOE} = 1 + \frac{(\text{NOE}_{\text{DD}} - 1)T_1}{(T_1)_{\text{DD}}} \tag{17}$$

because only the dipolar relaxation mechanism contributes to a NOE.

Most of these relaxation parameters are familiar in the NMR literature. The off-resonance relaxation parameters $T_{1\rho}^{\text{off}}$ and $R$ are less familiar but have been discussed previously (James, 1980; James et al., 1978). The value of $R$ is obtained as the ratio of the intensity of a resonance in the presence to that in the absence of an off-resonance rf field. The $R$ value, as the NOE, is not explicitly dependent on the internuclear distance $r$ [namely, Eqs. (7) and (14)].

The possibility of more than one mechanism contributing to the relaxation of $^{31}$P presents complications to the interpretation of relaxation data. For the present purposes, an unwanted mechanism is that of relaxation caused by paramagnetic ions; consequently, we assume that efforts have been made to remove any paramagnetic metal ions that could contribute to relaxation of phosphorus in nucleic acids.

## B. Spectral Densities and Molecular Motions

Some $^{31}$P relaxation data for nucleic acids have been analyzed assuming the motion governing relaxation is simple isotropic motion, but usually it is necessary to consider the motion of nucleic acids as being more complicated. Generally, one calculates relaxation parameters using spectral densities based on reasonable (or at least simple) motional models. The calculated relaxation parameters are compared to experimental values to see if the model reproduces the data. This procedure will not necessarily yield a unique model. Measurement of several relaxation parameters at different spectrometer frequencies can nevertheless severely constrain the possible motional models that will fit the data. We now consider the form of the spectral densities for some of the possible motional models.

### 1. Random Isotropic Motion

The rate of the molecular motion effecting relaxation is often expressed in terms of a correlation time (James, 1975); so the spectral densities may also be expressed in terms of correlation times. In the simplest case, that is, that of random isotropic motion (which means there is no preference in the direction of motion), the spectral densities for dipolar relaxation are

$$J_n(\omega) = 2\tau_o/(1 + \omega^2\tau_o^2) \tag{18}$$

where $\tau_o$ is the rotational correlation time for the tumbling of the molecule. It is evident from Eqs. (3)–(7) that the different spectral densities vary in their influence on the different relaxation parameters. Figure 1 illustrates this for $^{31}P$ relaxed by three protons, each 2.6 Å away assuming random isotropic motion. The parameters obviously vary in their sensitivity to motions characterized by different correlation times.

For the CSA mechanism, the appropriate expression for the spectral density is

$$J_j(\omega) = \frac{2c_j\tau_j}{(1 + \omega^2\tau_j^2)} \tag{19}$$

where, for random isotropic motion, $\tau_j = \tau_o$ for $j = 0, 1, 2$; and $c_o = 1, c_1 = 0$, and $c_2 = \eta^2/3$.

It should be noted that the parameters are dependent on $\omega$ according to Eqs. (18) and (19), so spectrometer frequency is a variable that can be utilized. As shown in Fig. 1, the off-resonance parameters are also dependent on the off-resonance frequency, which can be another useful variable.

### 2. Free Internal Diffusion with Overall Isotropic Reorientation

Perhaps the next step up in complexity is a model in which the molecule as a whole tumbles isotropically, with additional free diffusion about some internal rotation axis. The case of random rotational diffusion of a spin pair about an axis of internal rotation which itself is reorienting isotropically has been developed by Woessner (1962a) and subsequently applied to motions in macromolecules (Kuhlmann *et al.*, 1970; Doddrell *et al.*, 1972; James,

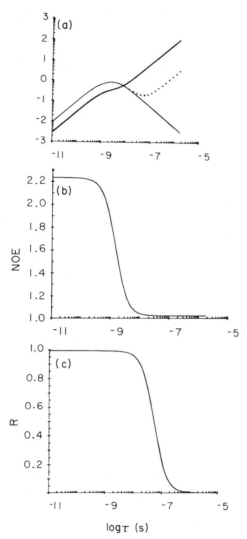

**Fig. 1.** Theoretical values of the spin–lattice relaxation rate (a; $1/T_1$; —), off-resonance nonselective nuclear Overhauser effect (NOE; b), rotating frame spin–lattice relaxation rate (a; $1/T_{1\rho}^{\text{off}}$; · · · ·), linewidth (a, —), and off-resonance intensity ratio ($R$; c) for $^{31}$P at 40.5 MHz. The random, isotropic motion model was utilized, as described in the text. The computed curves assume relaxation only from three protons that are 2.6 Å way from the phosphorus. The $T_{1\rho}^{\text{off}}$ and $R$ curves were calculated assuming an off-resonance field of 0.47 G applied 8 kHz off-resonance. From Bolton and James (1980a). Copyright 1980 American Chemical Society.

1980). The spectral densities appropriate for the dipolar relaxation mechanism for this model are

$$J_n(\omega) = A \frac{2\tau_o}{1 + \omega^2\tau_o^2} + B \frac{2\tau_1}{1 + \omega^2\tau_1^2} + C \frac{2\tau_2}{1 + \omega^2\tau_2^2} \quad (20)$$

with

$$A = \tfrac{1}{4}(3\cos^2\alpha - 1)^2 \qquad \tau_1 = \frac{1}{1/\tau_o + 1/6\tau_i}$$
$$B = \tfrac{3}{4}(\sin^2 2\alpha) \qquad\qquad\qquad\qquad\qquad\qquad (21)$$
$$C = \tfrac{3}{4}(\sin^4 \alpha) \qquad \tau_2 = \frac{1}{1/\tau_o + 2/3\tau_i}$$

where $\alpha$ is the angle between the P–H internuclear vector and the axis of internal rotation, $\tau_i$ the correlation time for the random reorientation of the P–H vector around the axis of internal rotation, and $\tau_o$ the correlation time for the isotropic reorientation of the vector. The inverses of the rotational diffusion coefficients for the isotropic reorientation and the internal rotation are $6\tau_o$ and $6\tau_i$, respectively.

For the CSA mechanism, the spectral densities are given by Eq. (19) with $\tau_o$, $\tau_1$, and $\tau_2$ as in Eq. (21) and

$$c_o = \tfrac{1}{4}[(3\cos^2\beta - 1) + \eta \sin^2\beta \cos 2\gamma]^2$$
$$c_1 = \tfrac{1}{3}\sin^2\beta[\cos^2\beta(3 - \eta \cos 2\gamma)^2 + \eta^2 \sin^2 2\gamma] \quad (22)$$
$$c_2 = [(\sqrt{3/4}\sin^2\beta) + (\eta/2\sqrt{3})(1 + \cos^2\beta)\cos 2\gamma]^2 + (\eta^2/3)\sin^2 2\gamma \cos^2\beta$$

The Euler angles $\beta$ and $\gamma$ result from rotation of the principal axes for the internal rotational diffusion into the chemical-shift tensor principal axes.

Figure 2 shows the effects of this correlation time model on the calculated relaxation parameters. The theoretical curves demonstrate the dependence of the relaxation parameters on the internal motion correlation time for a series of $\tau_o$ values ranging from $10^{-7}$ to $10^{-6}$ s. These curves are calculated for the case of a $^{31}$P nucleus relaxed by three protons, each 2.86 Å away, with rotation about an internal rotation axis such that $\alpha = 40°$ with an additional contribution from the CSA mechanism. The shape of the relaxation parameter curves for the single isotropic motion model and the two correlation time models can be compared in Figs. 1 and 2. Although use of only two measured relaxation parameters may or may not permit a distinction between the models, employment of additional parameters will permit a distinction between these two models. An illustration of the influence of the average internuclear distance on the relaxation when CSA contributions can

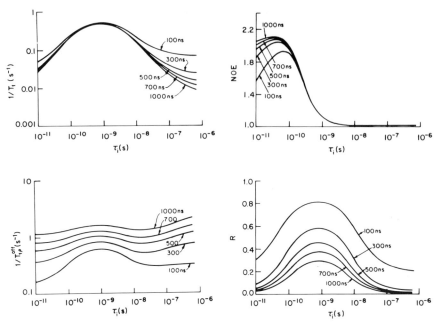

**Fig. 2.** Theoretical values of the NMR relaxation parameters for $^{31}$P at 40.5 MHz. The relaxation parameters were calculated assuming the free internal diffusion model presented in the text with an internal motion correlation time $\tau_i$ and an isotropic slower motion correlation time $\tau_o$. The calculations include contributions from chemical-shift anisotropy (CSA) as well as dipolar contributions from three protons at a distance of 2.86 Å from the phosphorus. The dipolar interaction is modulated by internal rotation with an angle of 40° between P–H vectors and the axis of internal motion. The CSA contributions utilized the following values: chemical shift anisotropy $\delta_z = 103$ ppm, asymmetry parameter $\eta = -0.63$ ppm, and Euler angles $\beta = 90°$, $\gamma = 0°$. The off-resonance NMR relaxation parameters $T_{1\rho}^{\text{off}}$ and $R$ were calculated assuming a radiofrequency (rf) field of strength 0.48 G applied 8.0 kHz off-resonance. The various curves were calculated assuming $\tau_o$ values listed next to the curves. From Bolton *et al.* (1982). Copyright 1982 American Chemical Society.

also be important is shown in Fig. 3. Both the CSA and DD contributions to the $^{31}$P $T_1$ relaxation time and NOE are shown for the case of free internal diffusion assuming that the average distance of three protons is either 2.70 or 2.86 Å from the phosphorus. The distance dependence might be anticipated for $T_1$, but Eq. (5) does not display an explicit dependence of the nuclear Overhauser effect on distance. Indeed, if the DD contribution solely determines the NOE (solid circles), then the curves would be identical for $r = 2.70$ and 2.86 Å. However, when CSA contributions are significant, the internuclear distance determines the relative contribution of DD and CSA

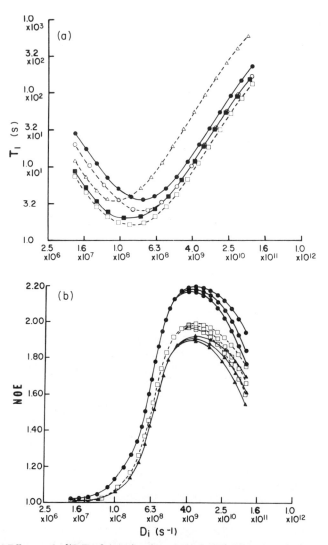

**Fig. 3.** (a) Effect on the $^{31}$P $T_1$ of changing the average P–H internuclear distance $r$ from 2.86 Å (●, ■) to 2.7 Å (○, □). The CSA contribution (△), which is the same in both cases, is calculated using $\delta = 99$ ppm and $\eta = -0.5825$ (see Shindo, 1980). The dipole contribution (●, ○) is due to three hydrogens. The total relaxation rate (■, □) is the sum of the CSA and DD contributions. Calculations were done at a field of 2.35 T, assuming the DD and CSA interactions were modulated by an isotropic motion with correlation time $\tau_o = 1000$ ns and free internal diffusion with rotational diffusion coefficient $D_i (= \frac{1}{6}\tau_i)$. (b) Effect on the calculated total NOE of using two different average distances. The solid line and circles represent the NOE assuming only dipolar contributions to $^{31}$P relaxation and are independent of the distance $r$. The NOE values taking into account the influence of the CSA contribution to relaxation are

contributions [see Eq. (17)], with a larger P–H distance decreasing the DD contribution to the total NOE with consequent diminution of the NOE values.

It should be noted that multiple free internal rotations, rather than just one, can also be considered (Wittebort and Szabo, 1978; London and Avitabile, 1978).

### 3. Rotational Tumbling of a Cylinder

Woessner (1962b) also treated the case of rotational reorientation of ellipsoidal molecules. For the case of a cylinder without internal motion, Eqs. (19)–(22) again express the spectral densities. However, for this case the definitions of $\alpha$ and the time constants $\tau_1$ and $\tau_2$ must be altered. Angle $\alpha$ is the angle between the P–H internuclear vector and the long axis of the cylinder. If the inverses of the diffusion coefficients for rotation about the short axis and about the long axis of the ellipsoid are $6\tau_\perp$ and $\tau_\parallel$, respectively, then the time constants to use with Eq. (20) are

$$\tau_o = \tau_\perp \qquad \begin{aligned} \tau_1 &= \left(\frac{5}{6\tau_\perp} + \frac{1}{6\tau_\parallel}\right)^{-1} \\ \tau_2 &= \left(\frac{1}{3\tau_\perp} + \frac{2}{3\tau_\parallel}\right)^{-1} \end{aligned} \qquad (23)$$

Curves similar to those in Fig. 2 could therefore be calculated for this model. With short segments of DNA, it may be possible to use this model, considering the double helix as a cylinder. In fact, it is possible to use this anisotropic overall motion with internal motions superimposed.

### 4. Distribution of Correlation Times

The motion can be described by a distribution of correlation times for random-coil polymers (Schaefer, 1973; Shindo, 1980). Although various types of distribution such as a log-normal distribution (Norwick and Berry, 1961) could be used for flexible polymers, Schaefer (1973) utilized the log $\chi^2$

---

dependent on the choice of 2.86 Å (▲) or 2.70 Å (□) for the P–H internuclear distance. Calculations were done at a field of 2.35 T, assuming the DD and CSA interactions were modulated by an isotropic motion and free internal diffusion with rotational diffusion coefficient $D_i$. At the larger values of the internal diffusion constant $D_i$, the curves for the three calculations break into three separate curves. In each type of calculation, the top curve is for $\tau_o = 1000$ ns, the middle for $\tau_o = 650$ ns, and the bottom for $\tau_o = 500$ ns. From Keepers and James (1982). Copyright 1982 American Chemical Society.

distribution function

$$F(s,p)ds = \frac{1}{\Gamma(p)} (ps)^{p-1} e^{-ps} ds \tag{24}$$

where $p$ is related to the width of the distribution, $\Gamma(p)$ is a gamma function, and $s = \log_b(1 + b\tau_c/\bar{\tau})$, where $\bar{\tau}$ is an average correlation time and $b$ reflects the symmetry and width of the distribution. The spectral density terms are then

$$J_n(\omega) = \int_0^\infty \frac{\bar{\tau} F(s,p)[(\exp_b(s) - 1)/(b - 1)]ds}{1 + \omega^2 \bar{\tau}^2 [(\exp_b(s) - 1)/(b - 1)]^2} \tag{25}$$

The spectral density represented by Eq. (25) can be used in place of any of the spectral densities discussed so far to provide a correlation time distribution; however, the appropriate geometric factors $A$, $B$, and $C$ in Eq. (21) or $c_0$, $c_1$, and $c_2$ in Eq. (22) would need to be coefficients to spectral density terms formed by Eq. (25). The effect of imposing a correlation time distribution is to (a) broaden the NMR resonance regardless of correlation time; (b) raise the $T_1$ minimum with a diminished slope ($T_1$ versus $\tau_c$ plot); and (c) decrease the NOE from its maximum in the extreme narrowing limit ($\tau \ll 1/\omega$) while maintaining a measurable NOE at longer correlation times.

## 5. Restricted Internal Diffusion

The models discussed treat molecular motions in nucleic acids in a fairly simplistic sense. For example, free rotation about bonds is certainly unreasonable in the case of a double-stranded helix, although it may happen for a single-stranded nucleic acid. More realistic models for motions in nucleic acids can limit the rotational freedom about particular rotation axes in order to account for excluded volume effects. We can merely impose boundary conditions on any rotation, permitting free diffusion within the boundaries. As discussed shortly, jumps among allowed conformations also serve to limit the motions to physically realistic situations.

First, we consider internal rotational diffusion with restrictions on the amplitude of the diffusion. The limits on the diffusion can be established using a square well potential or a harmonic function that approaches the limit asymptotically. Although relatively little work has employed restricted diffusion models, those that do generally utilize a square well potential (Wittebort and Szabo, 1978; London and Avitabile, 1978). Two general models can be considered (see Fig. 4), one which is most useful for local motions (Fig. 4a) and the other most useful for segmental motions (Fig. 4b).

The first model is appropriate for evaluating rotation about bonds with the rotation limited to an angular range of $\pm \gamma_0$ (Fig. 4a). For the case of

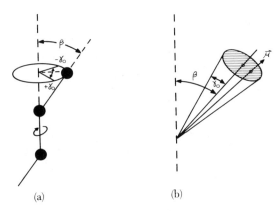

**Fig. 4.** Restricted diffusion models for internal motions. (a) Bond rotation limited in angular amplitude to $\pm\gamma_0$. $\beta$ is the angle between the rotational axis and the internuclear vector, just as in the free internal diffusion model. (b) With the "wobble-in-a-cone" model, the diffusion is limited to the conic angle $\gamma_0$, and the cone is tilted at angle $\beta$ from some reference axis.

overall isotropic motion and dipolar relaxation, the spectral density appropriate for this model of restricted diffusion is (Wittebort and Szabo, 1978)

$$J(\omega) = 2 \sum_{b=-2}^{2} \sum_{n=0}^{\infty} \frac{f_n}{f_n^2 + \omega^2} \Gamma_{bb'n}(\gamma_0) [d_{b0}(\beta)]^2 \qquad (26)$$

where the $d_{b0}(\beta)$ are reduced Wigner rotation matrices and $\beta$ is the angle between the P–H internuclear vector and the axis of internal rotation (Fig. 4a). The matrix elements $\Gamma_{bb'n}(\gamma_0)$ are set by the maximum amplitude of the rotation $\gamma_0$ and

$$f_n = 6D_o + \frac{D_i n^2 \pi^2}{4\gamma_0^2} \qquad (27)$$

where the diffusion constants for overall motion $D_o$ and internal motion $D_i$ are defined the same as in the case of free internal diffusion.[2] For the CSA relaxation mechanism, an additional transformation is required to align the laboratory coordinate system with the principal-axis system of the CSA tensor, and additional terms must be included if the tensor is not axially symmetric (Keepers, 1982).

The infinite series of Eq. (26) converges rapidly, so Wittebort and Szabo (1978) and London and Avitabile (1978) examined some of the conse-

---

[2] It should be noted that the conventional definition of the correlation time for overall tumbling is $\tau_o = \frac{1}{6}D_o$ and for free internal diffusion is $\tau_i = \frac{1}{6}D_i$, but $1/D_i$ is often used as an internal motion correlation time $\tau_i$ in jump models.

quences of restricted diffusion for the case of $^{13}$C relaxed by dipolar coupling to hydrogen. Two major conclusions stand out. First, restricting the amplitude of the motions results in a smaller $T_1$ when the internal motion is fast ($D_i > 1/\omega$) and a larger $T_1$ when the internal motion is slow ($D_i < 1/\omega$). Second, if the amplitude of diffusion is restricted to $\pm\gamma_0 = 25°$, the relaxation parameters are nearly those which obtain with no internal motion. The implication of this, of course, is that the NMR relaxation parameters will not be very sensitive to low amplitude fluctuations (unless the motions are concerted such that the net result of several small fluctuations results in a reorientation of $>25°$ in the DD vector or the CSA tensor relative to the stationary magnetic field.

The second type of restricted diffusion is illustrated in Fig. 4b and has sometimes been dubbed the "wobble-in-a-cone" model because the orientation of the interaction vector $\mu$ (either DD vector or CSA tensor principal axis) is permitted free diffusion within the cone of half angle $\gamma_0$. The cone is disposed toward a reference axis at a fixed angle $\beta$. This model can be adapted for internal motions but is especially useful for characterizing segmental motions, in which case the reference axis can be the helix axis of the nucleic acid. Librational motions, such as predicted by molecular dynamics calculations on proteins (McCammon et al., 1977), are readily accommodated by the wobble-in-a-cone model (Howarth, 1979; Richarz et al., 1980), but the model has also accommodated other studies of dynamics, for example, the situation of halide ions bound to proteins (Bull et al., 1978).

The wobble-in-a-cone model again entails an infinite series. Lipari and Szabo (1980, 1981a) have presented two closed-form, approximate expressions for the correlation function. These approximations were obtained by requiring the analytic expression to be exact at very short and at very long times and by requiring the area under the curve of $G(t)$ versus $D_w t$ to be exact, where $D_w$ is the diffusion constant for the wobbling. The correlation functions are expressed as an infinite sum of exponentials. The earlier version used a single exponential approximation, but the second version used a multiexponential approximation (illustrated by a triple exponential), which proved to give numerical results virtually indistinguishable from the exact correlation function.

The earlier paper of Lipari and Szabo (1980) gives a simple closed-form expression for the spectral density appropriate for cone diffusion superimposed on the isotropic tumbling of the macromolecule when the internuclear vector is coaxial with the wobble axis:

$$J(\omega) = \frac{2S^2\tau_o}{1 + \omega^2\tau_o^2} + (1 - S^2)\frac{2\tau_1^2}{1 + \omega^2\tau_1^2} \quad (28)$$

where the order parameter for cone diffusion, using $\chi_0 = \cos\gamma_0$, can be expressed by

$$S = \tfrac{1}{2}\chi_0(1 + \chi_0) \quad (29)$$

$$1/\tau_1 = 1/\tau_o + 1/\tau_{\text{eff}} \quad (30)$$

and

$$\tau_{\text{eff}} = \left( \frac{-\chi_0^2(1 + \chi_0)^2\{\ln[(1 + \chi_0)/2] + (1 - \chi_0)/2\}}{[2(1 - \chi_0)]} + \frac{(1 - \chi_0)(6 + 8\chi_0 - \chi_0^2 - 12\chi_0^3 - 7\chi_0^4)}{24} \right) \Big/ D_w(1 - S^2) \quad (31)$$

It will be seen that the spectral density simplifies considerably in the case that $\tau_{\text{eff}} \ll \tau_o$. It is apparent that the cone diffusion can be specified by merely two parameters, the diffusion constant $D_w$ and the semiangle of the cone $\gamma_0$.

## 6. Jump Models

In general, jump models have been developed that permit instantaneous jumps between two or three nonequivalent molecular configurations. A more general formulation permitting jumps between several configurations is discussed in terms of lattice models. Jump models are particularly useful, in comparison to diffusion models, in that fluctuations in the internuclear distance that may occur with the molecular motions can be accomodated easily. With the diffusion models, an average internuclear distance $r$ is used to calculate the dipolar contribution to relaxation parameters via Eq. (8). With jump models, the internuclear distance can be explicitly specified for each jump state; in this case the $r^{-6}$ is removed from Eq. (8), and the distance dependence is incorporated into the spectral density (Tropp, 1980; Keepers and James, 1982).

Two cases can be distinguished for jump models. In the first case, the dipolar-coupled atoms are bonded to each other. Further, the axis around which jumps occur must intersect or be rigidly attached to one of the atoms. This case is a propos of $^{13}$C relaxed by directly bonded protons and has been treated by London (1978). Several jump models have appeared (e.g., Woessner, 1962a; Marshall et al., 1972; Hubbard and Johnson, 1975). In this case, as in the models for free and restricted diffusion, the coordinate system can be defined such that only angle $\gamma_0$ is time-dependent. In effect, the drawing in Fig. 4a is also applicable for a two-state jump model; but the two states exist only with angles $+\gamma_0$ and $-\gamma_0$, not intermediate values. The spectral density for dipolar relaxation in this case is (London, 1978, 1980)

$$J(\omega) = (1 - C)\frac{2\tau_o}{1 + \omega^2\tau_o^2} + C\frac{2\tau_1}{1 + \omega^2\tau_1^2} \quad (32)$$

where

$$1/\tau_1 = 1/\tau_o + 1/\tau_A + 1/\tau_B \tag{33}$$

and

$$C = \frac{3\tau_A\tau_B}{(\tau_A + \tau_B)^2}[\sin^2\beta(1 - \cos 2\gamma_0)][2 - \sin^2\beta(1 - \cos 2\gamma_0)] \tag{34}$$

It will be noted that the form of the spectral density in Eq. (32) is similar to that for free internal diffusion [Eq. (20)] and for restricted diffusion [Eq. (28)]. This model permits the relative populations of the jump states $A$ and $B$ to be accounted for since the population in state $A$ is proportional to the lifetime $\tau_A$ in state $A$. London (1978) has shown that $\tau_A \sim \tau_B$ for relaxation to be substantially affected. In other words, the existence of a small amount of a minor configuration will have little influence on the measured relaxation parameters.

In the second class of jump models, the atoms defining the DD vector are not bonded to each other (Tropp, 1980; Keepers and James, 1982). The same treatment pertains to CSA contributions in which the jump axis does not intersect the relaxing nucleus (Keepers, 1982). This case is applicable to internal motions for $^{31}$P in nucleic acids because the P–H internuclear distances can vary with the motions and the jump axis does not necessarily intersect the phosphorus. In this case, the internuclear distance is specified for each jump state becoming incorporated into the spectral density, and the distance dependence in Eq. (8) is removed with the constant becoming $K = \hbar^2\gamma_H^2\gamma_P^2/20$. As well as $r$ changing with jump state, the Euler angles $\beta$ and $\gamma$ may also change.

In order to transform the interaction vector in the laboratory coordinate frame defined by the stationary magnetic field to the final coordinate frame $F$, defined by the jump axis, which is diagonal with the DD vector or aligned with the principal axis system of the CSA tensor, it may be useful to express the overall transformation as a series of transformations through intermediate coordinate frames $N$. This will enable the amplitude and frequency of each motion to be explicitly included in the correlation function. In the case of overall isotropic motion with diffusion constant $D_o$, the dipolar correlation function can be expressed as (Keepers and James, 1982)

$$G(\tau)_{DD} = \frac{1}{15}\exp(-6D_o t) \ldots \sum_n \exp(-\lambda_n t) \sum_i \sum_j c_0 c_n \xi_i^0 \xi_j^0 \xi_i^n \xi_j^n$$
$$\times \exp(ic\gamma_{Ni} - ic'\gamma_{Nj})\exp[i(b - b')\alpha_{N-1,N} + i(c - c')\alpha_{NF}] \tag{35}$$
$$\times d_{bc}(\beta_{N-1,N})d_{b'c'}(\beta_{N-1,N})\frac{d_{c0}(\beta_{iNF})d_{c'0}(\beta_{jNF})}{r_i^3 r_j^3}$$

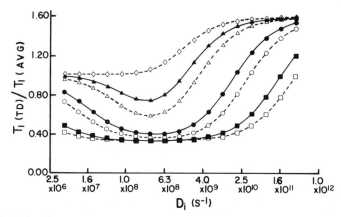

**Fig. 5.** Ratio of proton $T_1$ values calculated using explicit time-dependent internuclear distances $[T_1(\text{TD})]$ to the $T_1$ value calculated using an average value $[T_1(\text{AVG})]$ for the internuclear distances between the relaxing proton and the three protons on a methyl rotor that are undergoing a three-state jump. The ratio is obtained as a function of the elementary jump rate $D_i(=\frac{1}{6}\tau_i)$. Curves are computed for $\tau_o$ values of 1000 ns (□), 500 ns (■), 100 ns (○), 50 ns (●), 10 ns (△), 5 ns (▲), and 1 ns (◇). From Keepers and James (1982). Copyright 1982 American Chemical Society.

where the last transformations shown are representative of transformation through other intermediate coordinate frames for each of the configurations $i$ and $j$. The $d(\beta)$ are reduced Wigner rotation matrices (Spiess, 1978) and the term

$$\sum_n \exp(-\lambda_n t) \sum_i \sum_j c_0 c_n \, \xi_i^0 \xi_j^0 \xi_i^n \xi_j^n$$

is determined by the relative populations of the jump states. Note that the internuclear distances $r_i$ and $r_j$ in configurations $i$ and $j$, respectively, are now included in the correlation function. The spectral density may be obtained with this correlation function via Eq. (2). The appropriate correlation function for the CSA tensor is similar but excludes the distances and includes terms with the asymmetry parameter (assuming nonaxial symmetry of the CSA tensor), with an additional transformation to properly align the principal axis system of the CSA tensor.

Figure 5 illustrates the results of a test to see if an average internuclear distance can be chosen to simulate the relaxation time calculated by explicitly incorporating the internuclear distances in each jump state and for each DD interaction (Keepers and James, 1982). The example considers $T_1$ relaxation of a proton that is interacting with three other methyl protons in a three-state jump process in which the three states are equally probable. The ratio of the $T_1$ calculated using time-dependent internuclear distances to that calculated using an averaged $r^{-6}$ value is shown in Fig. 5 as a function of the jump rate $D_i$ for a series of isotropic tumbling correlation times $\tau_o$. It is evident that the average-value calculation would not mimic the explicit-distance calculation except in the case that the jump rate is much slower than the isotropic overall motion. The obvious implication is that explicit account of the internuclear distance fluctuations will be necessary for careful studies of molecular dynamics.

## 7. Lattice Models

This particular class of models can be similar to restricted diffusion models with multiple internal rotations or to jump models that permit several connected atoms to jump among a number of different configurations. Such models have been developed by London and Avitabile (1978) and Wittebort and Szabo (1978). We distinguish lattice models, however, by removing the requirement that a jump or rotation axis be defined. Instead, the orientation of the internuclear vector or the CSA tensor is referenced with respect to an axis system not connected to the atoms of the DD vector (Keepers and James, 1982; Keepers, 1982). For example, the reference axis might be the helix axis in a nucleic acid. The lattice model is therefore useful for analyzing concerted motions (in which rotation of a specified angle around one bond occurs simultaneously with rotations about other bonds) and for accounting for excluded volume in a natural manner. This kind of motion can permit local motions while maintaining approximately the same overall structure (see following discussion). In this case, the coordinates of each atom are known and the time dependence extends to Euler angles $\alpha$ and $\beta$ as well as the internuclear distance. With overall isotropic motion with diffusion constant $D_o$, the correlation function for the dipolar interaction can be written as (Keepers and James, 1982)

$$G(\tau)_{DD} = \frac{1}{15} \exp(-6D_o t) \sum_n \exp(-\lambda_n) \sum_i \sum_j c_0 c_n \zeta_i^{z0} \zeta_j^{z0} \zeta_i^{zn} \zeta_j^{zn}$$
$$\times \exp(i a \alpha_{Di} - i a \alpha_{Dj}) \frac{d_{a0}(\beta_{iDF}) d_{a0}(\beta_{jD})}{r_i^3 r_j^3}$$

(36)

where the reference axis (e.g., the DNA helix axis) defines the $D$ coordinate frame. The points noted for the jump model [Eq. (35)] regarding the populations of configurations i and j and modifications for the CSA correlation function also pertain. Correlation functions for lattice models with cylindrically anisotropic motion overall have also been developed (Keepers, 1982).

## 8. The Non-Model

It will be noted that the spectral densities expressed by Eqs. (20), (28), and (32) for free diffusion, restricted diffusion, and a two-state jump, respectively, are quite similar in form. For example, the spectral density given by Eq. (20) is a sum of three terms, each containing a coefficient based on geometry (e.g., $B$) and a time constant (e.g., $\tau_B$). The coefficients and time constants could be expressed for example by Eq. (21) or (23). If $\tau_i \ll \tau_o$ (or $\tau_\parallel \ll \tau_\perp$), Eq. (20) is reduced to two terms, each of which reflects only one of the motions. In general, as long as the motions are independent and have widely different correlation times, it is possible to express spectral densities as

$$J(\omega) = \sum_{j=1}^{M} c_j \frac{\tau_j}{1 + \omega^2 \tau_j^2} \tag{37}$$

where $\tau_j$ is the correlation time characteristic of a motion, $M$ the number of independent motions, and $c_j$ a coefficient related to the geometry of the interaction effecting relaxation and weights the contribution of any particular motion to relaxation.

It has been advocated that no specific motional models be assumed but that the spectral densities be obtained under the general assumption of the motions as Markovian processes; the general formulation is intractable for practical use, so an approximate expression was presented (Ribeiro *et al.*, 1980):

$$J(\omega) = \sum_{j=1}^{M} c_j \frac{\lambda_j}{\lambda_j^2 + \omega^2} \tag{38}$$

It can be seen that Eq. (38) is identical to Eq. (37) when $\lambda_j = 1/\tau_j$, (i.e., when the motions are independent and all occur at much different rates). In general, however, this is not the case, and $\lambda_j$ will be determined as a composite of different motions. Spectral densities of the form of Eq. (38) can be employed with $\lambda_j$ and $c_j$ values as unknown parameters to be determined from the measured NMR relaxation data. Although the $\lambda_j$ and $c_j$ values thus determined have no physical meaning, such a procedure should enable the minimum number of molecular motions contributing to relaxation to be determined. However, whenever a specific motional model is utilized, it is readily apparent when an additional motion is required to fit the data.

## 9. Molecular Dynamics–Derived Correlation Function Models

Correlation functions and spectral densities can be calculated using stochastic dynamics trajectories. So far, a single example of this potentially powerful technique has appeared in the literature (Levy et al., 1981). This method of deriving correlation functions is fundamentally different from the other motional models described. The other models postulate a particular type of motion and then entail taking an ensemble average to obtain the correlation function. With the molecular dynamics–derived model, a potential function (realistic, it is hoped) is used to predict the configuration of the molecule at very short time intervals (e.g., 1 ps). The correlation function is obtained by taking a time average over the configurations obtained in each time interval. Reorientation of a DD vector or a CSA tensor can be accomodated because the Cartesian coordinates of each atom at each time interval are known. The possible occurrence of excluded volume effects and concerted motions are naturally taken into account by proper construction of the potential energy function, which may include various repulsive and attractive energy terms. The calculations can become very computer-time consuming as the potential function becomes more sophisticated; however, this potential function must be as realistic as possible for the resulting trajectories to be believable.

The trajectories are generated by integrating simultaneously over all atom positions the diffusive Langevin equations of motion for each particle i:

$$m_i(d\mathbf{v}_i/dt) = -v_i\mathbf{v}_i + \mathbf{A}_i + \mathbf{F}_i \tag{39}$$

where $m_i$, $\mathbf{v}_i$, and $v_i$ are the mass, velocity, and friction constant and $\mathbf{A}_i$ is the force producing random Brownian motion. The systematic force $\mathbf{F}_i$ acting on particle i is determined by the potential energy function; in effect, the systematic force counters the perturbing Brownian force to minimize the molecular energy.

The primary limitation for wider application of this approach to spectral-density calculations is the tremendous amount of computer time required if good potential functions are utilized and long-time averages are used for studying the relatively slow motions of macromolecules.

## III. Structure and Possible Molecular Motions in Nucleic Acids

We consider here some of the possible motions of a double-stranded nucleic acid. If many of these are actually occurring simultaneously, a dynamic description of the nucleic acid becomes quite complicated. These motions could, however, permit the transient existence of conformations required for biological functions of the nucleic acid. Until recently, only two conformational forms of nucleic acids were generally considered as biologically important, the A form with 3'-endo sugar and the B form with 2'-endo sugar (see Chapters 8 and 11 by Chen and Cohen, and Hart, respectively, for a more complete description of the structures). Research into the details of nucleic acid structure via theoretical calculations (Levitt and Warshel, 1978; Olson, 1982; Keepers *et al.,* 1982), X-ray diffraction (Klug *et al.,* 1979; Wang *et al.,* 1979; Wing *et al.,* 1980), NMR [see Chapters 8 (Chen and Cohen) and 9 (Gorenstein)] and other techniques (Pohl and Jovin, 1972; Griffith, 1978) has revealed that variations in DNA structure from the usual B-DNA exist and probably are important. The various conformations are often separated by low-energy barriers. It is possible that these structures and perhaps other low-energy variants simply reflect either potential minima about which structural oscillations can occur or may be conformations between which jumps can occur.

Figure 6 shows some of the possible motions in double-stranded nucleic acids that may affect NMR relaxation parameters. The end-for-end tumbling and "speedometer-cable" rotation depicted in Fig. 6 become rotation around the short axis and around the long axis of the cylinder, if the nucleic acid is sufficiently short. Transient electric dichroism studies of rodlike DNA molecules have provided data about the end-for-end rotational diffusion (Hogan *et al.,* 1978). In contrast, not much experimental data are available for the speedometer-cable motion.

The possibility of bending DNA has been considered theoretically (Bloomfield *et al.,* 1974; Levitt, 1978; Barkley and Zimm, 1979). Barkley and Zimm (1979) derived dynamical equations describing the bending of DNA. Using a different approach, Levitt (1978) has employed empirical energy functions to calculate the energy for bending DNA. Relatively little energy is required to bend DNA. Therefore, it was proposed that bending motions may contribute significantly to NMR relaxation in nucleic acids (Bolton and James, 1980a). It should be pointed out that because bending can occur in any direction, it may be considered to a first approximation as an isotropic motion. Bending has also been seen in the crystal structure of a DNA dodecamer that is bent with a radius of curvature 112 Å (Wing *et al.,* 1980). Hydrodynamic studies indicate that double-stranded nucleic acids with chain lengths >500 Å [150 base pairs (bp), $10^5$ daltons] are neither freely flexible random coils nor completely, stiff rods (Bloomfield *et al.,* 1974). Rather, the semiflexible double-stranded nucleic acids are often described as wormlike chains.

The dynamics of helix twisting has also been considered with equations for torsional rotation of the DNA being developed (Barkley and Zimm, 1979). Assuming that intercalated dyes reflect the motional properties of nucleic acids, indications are that this motion can account for most of the fluorescence depolarization decay of ethidium intercalated into DNA,

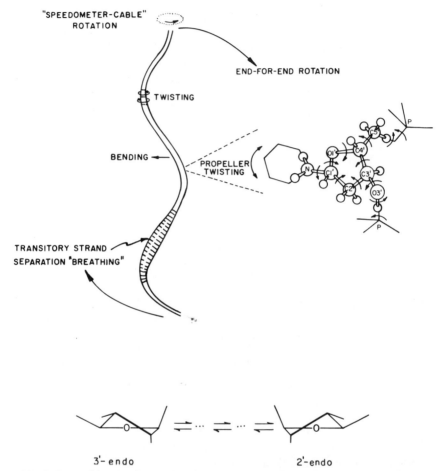

**Fig. 6.** Some of the plausible motions in a segment of double-standed nucleic acid. Refer to text for details.

which has a lifetime of the order of $10^{-8}$ s (Wahl et al., 1970; Millar et al., 1980). Motion in that same time range has been found from ESR studies of spin-labeled acridine dyes bound to DNA (Robinson et al., 1979).

The so-called breathing of double-stranded nucleic acids has been well documented (Englander and Englander, 1977). Motions that permit exchange of the hydrogen-bonded protons of polynucleotides have been studied by tritium exchange (Englander and Englander, 1977) and by NMR (Reid and Hurd, 1977; Johnston et al., 1979). These methods are useful for assessing the availability of amino and imino protons to the solvent water on the time scale of milliseconds to seconds for NMR and of many seconds to minutes for tritium exchange. Temperature-jump experiments monitoring various physical properties have yielded information regarding the rate of base pair formation, which is in the time scale of milliseconds (Bloomfield et al., 1974).

Another conceivable motion, that resulting from soliton excitation, has been proposed as a possible cause of the "breathing" (Englander et al., 1980). The thermally induced soliton can be thought of as a stable wave form, such as that pictured for breathing in Fig. 6, which possesses the dynamic character of moving along the polymer chain.

Two other possible segmental motions not depicted in Fig. 6 are in distinctly different time realms. At the slow end of the time scale (greater than milliseconds for reasonable amplitudes) are the Rouse–Zimm normal coordinate modes that result from the collective behavior of units of atoms (beads) along the chain pulling one another and acted on by Brownian forces, solvent frictional resistance, and other parts of the chain (Berne and Pecora, 1976). Librational motions, generally accorded to be of the order of $10^{11}$ s$^{-1}$, may also result from the thermally induced displacements of groups of atoms; such wobbling motions have been proposed as important factors in the NMR relaxation of proteins (Howarth, 1979).

The motions of bending, twisting, and breathing do involve some rotations about bonds in the nucleic acid. The concerted effect of these individual bond rotations wll be over a sizable segment of the polymer. Additionally, internal motions may occur that may change the conformation over a small part of the nucleic acid (e.g., a monomeric unit or less). Theoretical (Levitt, 1978) and experimental (Wing et al., 1980; Hogan et al., 1978) studies have indicated the existence of a propeller twist in contrast to the classical B-DNA structure, with each member of a hydrogen-bonded base pair in the same plane. The C-1'—N rotation in Fig. 6 could lead to twisting of the base-pair propeller, with motions altering the pitch of the propeller.

Rather large changes in local conformation can be obtained by concerted alterations in the torsion angles about some of the bonds (illustrated in Fig. 7) without significant effect on the overall conformational energy (Keepers et al., 1982; Olson, 1982). For example, it is possible that one member of a base pair could temporarily move away from the other or, even more moderate, that crankshaft motions could take place along the backbone of the nucleic acid (Keepers and James, 1982).

Most data indicate that RNA is in the A form (with the ribose ring in the 3'-endo conformation) and that DNA is usually in the B form (with the deoxyribose ring in the 2'-endo conformation) but sometimes in the A form (Altona and Sundaralingam, 1972); these conformations are shown in Fig. 6. Theoretical calculations, however, indicate that the energy barrier between the 3'-endo and the 2'-endo conformations is only 0.6 kcal/mol

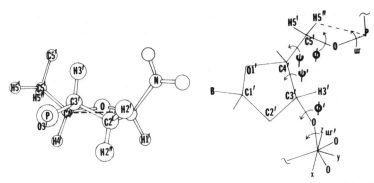

**Fig. 7.** Definition of the DNA backbone torsion angles. The torsion angles are defined as the clockwise rotation about B—C that aligns A—B with C—D in the sequence of atoms A—B—C—D: $\omega$ is for O-3'—P—O-5'—C-5'; $\phi$ is for P—O-5'—C-5'—C-4'; $\psi$ is for O-5'—C-5'—C-4'—C-3'; $\psi'$ is for C-5'—C-4'—C-3'—O-3'; $\phi'$ is for C-4'—C-3'—O-3'—P; $\omega'$ is for C-3'—O-3'—P—O-5'. The principal axis system is illustrated for rthe $^{31}$P chemical-shift tensor. From Keepers and James (1982). Copyright 1982 American Chemical Society.

(Levitt and Warshel, 1978) or 3–5 kcal/mol (Kollman *et al.*, 1981). This low barrier implies that there may be considerable flexibility in the sugar ring of nucleic acids, with the transient existence of many puckered conformations in the furanose or deoxyfuranose rings in the nucleic acid backbone.

In general, we can classify in three categories the motions that may alter the orientation of the DD vector or the CSA tensor of nucleic acid phosphorus: overall motions, segmental motions, and internal motions. The overall motions would include the end-for-end tumbling and speedometer-cable rotation. Although double-stranded nucleic acids are generally characterized as wormlike coils, as the polymer length decreases the nucleic acid behaves more as a rigid rod that can be treated using the anisotropic diffusion model. Possible segmental motions might include bending, twisting, librational motions, and Rouse–Zimm coil deformations. Although segmental motions can entail alterations in torsion angles about individual bonds, the collective nature of these alterations causes a segment of the polymer at least as large as a monomeric unit to reorient. Bending can be regarded as an isotropic motion to a first approximation; consideration of the double-helical structure implies that this is not strictly so at the local nucleotide level. Twisting and Rouse–Zimm deformations can be treated using jump models, lattice models, or restricted diffusion models. Librational motions are most easily handled with the wobble-in-a-cone model. Internal motions entail rotations about individual bonds and can be modeled using any of the formulations presented in the preceding section.

## IV. Considerations for Nucleic Acid Dynamics and Structure

Whether we are investigating dynamics per se or conformation, it will be necessary to establish reasonable forms of the spectral densities to be used in the equations for the relaxation parameters. This implies that some knowledge of the molecular motions must be derived from the experimental relaxation data, regardless of interest in mobility or structure.

### A. Dynamics

To describe the motion picture as well as possible for nucleic acids, several NMR relaxation parameters should be studied as a function of magnetic field strength for several nuclei. It is assumed that efforts to remove any paramagnetic metal ions, which have a strong influence on NMR relaxation (James, 1975), have been successful. After the relaxation parameters have been measured, plausible (or at least simple) motional models with their corresponding spectral densities are used in the mathematical expressions for the relaxation parameters. A dynamics model is tested by its ability to predict all of the parameters in agreement with the observed data. We emphasize that this procedure does not yield a unique description of the motions. Rather, the NMR data serve to eliminate possible motional models. Consequently, studies using as much relaxation data as possible hopefully will narrow the list of feasible models. Relaxation measurements on different nuclei should also be consistent, further limiting the acceptable motional models. Furthermore, the small amount of data from other techniques should be used to complement the NMR experiments.

We note that it is possible to use NMR data to obtain qualitative information about relative mobilities without a detailed knowledge of the dynamics. A semiquantitative description may be derived as well without knowledge of the type of motion actually occurring; for example, several different internal models may all fit the data with approximately the same correlation time, implying that *an* internal motion in that time range exists.

Phosphorus-31 NMR offers the observation of a single, readily assignable resonance with good sensitivity. In certain cases two resonances are observed (see Section III). Analysis of $^{31}$P relaxation must take into account the contribution from the chemical-shift anisotropy (CSA) mechanism as well as dipolar coupling to protons (see Section II,A). This additional complication can be dealt with by using chemical-shift anisotropy data provided by solid-state NMR (Terao *et al.*, 1977; Shindo, 1980; Opella *et al.*, 1981; Nall et al., 1981; Shindo, Chapter 13) and from the magnetic-field-strength dependence of relaxation parameters (Shindo, 1980; Bolton and James, 1979, 1980a). For DNA, values have been reported from 99 to 109 ppm for the CSA anisotropy $\delta_z$ and about $-0.6$ for the asymmetry $\eta$. As shown in Eqs. (11)–(13), contributions to relaxation via the CSA mechanism have an explicit dependence on the magnetic field strength as well as the field influencing the spectral densities as with all other mechanisms. Actually, because the dipolar and CSA mechanisms have different orientation dependences on the motions, satisfactory separation of CSA and dipolar contributions should provide complementary information.

Bendel and James (1982) have studied the possible contribution of another relaxation mechanism, that of scalar relaxation of the second kind, to $^{31}$P relaxation in phosphoryl-containing compounds using 5'-AMP as a model. Indeed, it was found that in the absence of proton decoupling, the dominant contribution to spin–spin relaxation of the phosphorus comes from scalar coupling to the two 5' protons. The potential problem of spin–echo attenuation owing to magnetization transfer in the presence of proton decoupling was found to be negligible. Consequently, broadband proton decoupling will be sufficient to remove any contribution from this potential complicating mechanism. Contributions from chemical-shift heterogeneity to the measured linewidth can exist because the observed $^{31}$P resonance is due to many individual nuclei on the nucleic acid. (An obvious exception to this is the spectrum of an aminoacyl-specific tRNA containing individual nuclear resonances.) The chemical-shift inequivalence of the different $^{31}$P nuclei in the double-stranded nucleic acids has been estimated to be at least 0.4 ppm by Bolton and James (1980a) and 0.5 ppm by Shindo (1980). Therefore, linewidth measurements are considered to be qualitatively valuable but less useful for quantitative studies, except when the linewidth is much larger than the chemical-shift inequivalence.

It was mentioned that spin diffusion (cross-relaxation) can be an important factor for $^1$H relaxation; $^{31}$P experiments with nucleic acids generally will not suffer the complication of cross-relaxation as long as proton decoupling is employed. Cross-relaxation can be important, indeed utilized, when proton coupling is permitted (Prestegard and Grant, 1978).

## B. Structure

NMR relaxation parameters, in particular the NOE, have been used to obtain structural information on small molecules including nucleotides (Noggle and Schirmer, 1971). (See Hart, Chapter 11, for a more complete description of this work, including experiments with $^{31}$P.) Owing to the small size of the molecules, extreme narrowing conditions ($\omega\tau_c \ll 1$) generally apply, so the spectral densities are particularly simple.

Even if a molecule is large, the situation can still be simple under certain very stringent conditions. If we are dealing with a two-spin system, say a $^{31}$P and a $^1$H nucleus, and dipolar relaxation via interaction between the two spins is the only source of relaxation, then the $^{31}$P NOE is

$$\text{NOE} = 1 + \frac{\gamma_H^2 \gamma_P^2 \hbar^2}{20r^6} [6J(\omega_H + \omega_P) - J(\omega_H - \omega_P)]T_1 \quad (40)$$

This expression allows the P–H distance $r$ to be determined readily from the $^{31}$P NOE and $T_1$ measurement, especially in the case of isotropic motion where the spectral densities are especially simple. The problem is that this case of a single dipole–dipole interaction being the only source of relaxation is extremely rare in interesting systems. In fact, as Kalk and Berendsen (1976) demonstrated for protons, extensive self-diffusion among the nuclear spins can make the situation quite complicated.

To overcome some of the problems, efforts have been made to measure transient NOE values rather than steady-state NOE values. Bothner-By and Noggle (1979) computed the time development of the NOE in a system of many spins, and Olejniczak et al. (1981) used that approach to study proton–proton interactions in lysozyme. So far, such experiments have not been conducted in heteronuclear systems at all or even for protons in nucleic acids.

It is also possible to use $T_1$ measurements to derive distance data. The procedure for this entails a combination of nonselective, semiselective, and selective relaxation time measurements. The nonselective $T_1$ measurement is one in which the 180° pulse in the usual inversion–recovery experiment (James, 1975) is applied to all of the interacting nuclei. Only one nucleus is subjected to the inverting 180° pulse in the selective $T_1$ measurement and only two nuclei in the biselective experiment. The measured $T_1$ values may vary in these three experiments and may be shown to give (Solomon, 1955; Nicollai and Tiezzi, 1979)

$$\left(\frac{1}{T_1}\right)_j^{ns} = \left(\frac{1}{T_1}\right)_j^s + \sum_i \frac{\gamma_i^2 \gamma_j^2 \hbar^2}{20r_{ij}^6} [6J(\omega_H + \omega_P) - J(\omega_H - \omega_P)] \quad (41)$$

$$\left(\frac{1}{T_1}\right)_j^{bs} = \left(\frac{1}{T_1}\right)_j^s + \frac{\gamma_i^2 \gamma_j^2 \hbar^2}{20r_{ij}^6} [6J(\omega_H + \omega_P) - J(\omega_H - \omega_P)] \quad (42)$$

where the initial relaxation of nucleus j is measured in the selective (superscript s), nonselective (ns), and biselective (bs) experiment. The sum in the $(1/T_1)_j^{ns}$ expression is over all nuclei i dipolar-coupled to nucleus j, and subscript i in the $(1/T_1)_j^{bs}$ expression refers to the particular nucleus whose spin is also inverted in the biselective experiment. The Larmor frequencies $\omega_H$ and $\omega_P$ have been written in the spectral densities for $^{31}P$ (nucleus j) relaxed by protons (nuclei i) at distances $r_{ij}$. It should be noted that this approach should even be valid when other relaxation mechanisms (such as chemical-shift anisotropy) apply since other contributions will appear in $(1/T_1)^s$.

So far, Eqs. (41) and (42) have not been used to determine P–H distances in any system. In fact, only one case has been reported for a nucleic acid using similar equations with proton–proton interactions. In a preliminary account, Broido and Kearns (1980) used the procedure to study poly(C) in neutral solution, with the results providing the bulk of the evidence for a left-handed helical structure for poly(C) (Broido and Kearns, 1982).

Another relaxation technique that is likely to see much utilization for structural studies in the future is the two-dimensional (2D) NOE experiment. The basic experiment has been largely developed in Ernst's lab with applications to proteins illustrated in Wüthrich's lab. The pulse sequence $[(\pi/2)_x - t_1 - (\pi/2)_x - \tau_M - (\pi/2)_x - t_2 -]$ generates a 2D spectrum following the two Fourier transforms. The intensity of a peak in the 2D-NOE spectrum arising from interaction of nucleus k with nucleus l is (Jeener *et al.*, 1979; Macura and Ernst, 1980)

$$[\text{Intensity}]_{kl} \propto C[\exp(-R\tau_M)]_{kl} \tag{43}$$

where $\tau_M$ is the experimental mixing time and the relaxation matrix $R$ composed of transition probability terms that are proportional to $J(\omega)/r_{kl}^6$. cfAlthough there are difficulties with phasing 2D spectra, the appearance of cross-peaks owing to scalar coupling, and the relatively long time required for an experiment, the technique holds tremendous promise for the future, with the possibility of determining solution structures at least to the resolution obtainable for crystals by X-ray diffraction. This potential should be considerably augmented by use of the distance geometry algorithm (Kuntz *et al.*, 1979).

So far, only proton homonuclear 2D-NOE experiments have been used in a qualitative (or perhaps semiquantitative) sense to obtain structural information from proteins (Kumar *et al.*, 1981; Wüthrich *et al.*, 1982) and, in one case, from poly(C) (Broido and Kearns, 1982). The possibility exists that 2D "accordion" NMR may also prove useful if the resolution problem is not too severe (Bodenhausen and Ernst, 1982).

## V. Relaxation Studies of Nucleic Acids

Although other studies have examined certain aspects of nucleic acid structure predominantly using NMR chemical shifts, the following discussion is only concerned with dynamics information in nucleic acids that has been derived from NMR relaxation experiments.

### A. DNA

Several investigations have utilized $^{31}$P NMR to study molecular motions in double-stranded DNA. Table I summarizes most of the relaxation results. In all of the studies, the DNA was maintained in a double-stranded helix with no single strands or fraying of the ends of the polynucleotide.

Hanlon et al. (1976) showed that the $^{31}$P-NMR linewidth decreased on sonication of high-molecular-weight DNA. The linewidth decreased by a factor of only 2–3 while the length of DNA was decreased more than 60-fold. Klevan et al. (1979) made $T_1$ and $^{31}$P{$^1$H} NOE measurements on DNA of uniform chain length (140 bp) obtained by nuclease digestion of chromatin. Their relaxation results could not be fit with a single isotropic motion, implying that internal motions might exist. Therefore, the free internal diffusion model described by Eqs. (20) and (21) was used assuming $^{31}$P relaxation was effected only by dipolar coupling to three protons, each at a distance of 2.8 Å from the phosphorus, and that angle $\alpha$ was 60°. It was postulated that $\tau_o$ is $\sim 10^{-7}$ s. It can be seen in Fig. 2 that the NOE is rather insensitive to the precise $\tau_o$ value in that range. An internal motion correlation time $\tau_i$ of 0.4 ns predicted the measured NOE and was also in agreement with the measured $T_1$. To a first approximation, the NOE is relatively insensitive to the P—H internuclear distance [see Eq. (5)] and, for that particular combination of $\tau_o$ and $\tau_i$ values, is not very sensitive to the exact value of the angle $\alpha$ either.

The off-resonance parameters $T_{1\rho}^{\text{off}}$ and $R$ in addition to $T_1$ and NOE values were determined by Bolton and James (1979, 1980a) for $^{31}$P relaxation in DNA with a higher molecular weight but with a molecular-weight dispersion (Table I). Again, it was concluded that there are internal motions in the DNA backbone. The two-correlation time model of Eqs. (20) and (21) was also utilized, but with the P—H distances all 2.6 Å and the angle of internal rotation 40°. The angle $\alpha$ was varied and the best fit of all the relaxation data was obtained with $\alpha = 40°$, although the fit was not too sensitive to the value of $\alpha$ within 10°. A value of 40° is in agreement with a molecular model constructed from X-ray data. Klevan et al. (1979) and

## TABLE I
### $^{31}$P-NMR Relaxation Parameters for Linear DNA

| Sample (base-pair concentration, m$M$) | Length (bp) | Frequency (MHz) | Temperature (°C) | $Te_1$ (s) | NOE | $T_2$ (s) | $T_{1\rho}^{\text{off } a}$ (s) | $R^a$ | $W_{1/2}$ (Hz) | Reference[b] |
|---|---|---|---|---|---|---|---|---|---|---|
| Chicken erythrocyte DNA (1.5) | 145 ± 3 | 109.3 | 18.5 | 3.4 | 1.3 | — | — | — | 103 | 1 |
| Calf thymus DNA (10) | 190 ± 20 (67% of sample) | 101.2 | 30 | 3.7 | — | — | — | — | 53 | 2 |
| Calf thymus DNA (15) | 700 ± 300 | 81 | 35 | 2.2 | 1.3 | — | — | — | — | 3 |
|  |  | 40.5 | 20 | 2.5 | 1.56 | — | 0.9 | 0.37 | 95 | 3 |
|  |  |  | 40 | 2.3 | 1.6 | — | 1.2 | 0.53 | 45 |  |
| Calf thymus DNA (4–12) | 600 ± 150 | 40.5 | 23 | 3.9 | — | 0.012 | — | — | 56 | 4 |
|  | 300 ± 75 |  |  | 3.5 | — | 0.014 | — | — | 35 |  |
|  | 260 ± 25 |  |  | 3.6 | 1.37 | 0.013 | — | — | 39 |  |
|  | 140 ± 20 |  |  | 3.2 | 1.35 | 0.032 | — | — | 22 |  |
|  | 140 ± 20 |  | 39 | 3.0 | 1.44 | — | — | — | 19 |  |
| pIns 36 plasmid DNA, linear (2) | 7,200 | 40.5 | 25 | 2.4 | 1.3 | 0.019 | — | — | 45 | 5 |
| Chicken erythrocyte DNA (1.5) | 145 ± 3 | 40.3 | 18.5 | — | 1.28 | — | — | — | 28 | 1 |
| Calf thymus nucleosome DNA (8) | 140 | 36.4 | 8 | 2.8 | 1.6 | — | — | — | 37 | 6 |
| Calf thymus DNA (0.9) | 12,800 | 36.4 | 27 | — | — | — | — | — | 41 | 7 |
|  | 210 |  |  |  |  |  |  |  | 17 |  |
| pVH51 restriction fragments (30) | 180 | 36.4 | 30 | 3.5 | 1.20 | — | — | — | — | 8 |
|  | 84 |  |  | 3.0 | 1.21 | — | — | — | — |  |
|  | 69 |  |  | 2.6 | — | — | — | — | — |  |
|  | 43 |  |  | 2.1 | 1.23 | — | — | — | — |  |
| Calf thymus DNA (10) | 1400 ± 300 | 24.2 | 30 | 2.2 | — | — | — | — | 40 | 2, 9 |
|  | 190 ± 20 (67% of sample) |  | 30 | 2.2 | 1.3 | — | — | — | 16 |  |
|  |  |  | 50 | 2.0 | — | — | — | — | — |  |

[a] $R$ and $T_{1\rho}^{\text{off}}$ were determined with an rf field of 0.47 G applied 8 kHz off-resonance.
[b] References: 1, Shindo (1980); 2, Wilson et al. (1981); 3, Bolton and James (1979, 1980a); 4, Hogan and Jardetzky (1979, 1980a); 5, Bendel et al. (1982); 6, Klevan et al. (1979); 7, Hanlon et al. (1976); 8, Hart et al. (1981); 9, Wilson and Jones (1982).

Bolton and James (1979, 1980a) found that protons from bound water were not responsible for the relaxation.

In accord with Klevan et al. (1979), $\tau_i$ was determined to be 0.3 ns by fitting all five $^{31}$P relaxation parameters to the motional model (Bolton and James, 1979, 1980a). The $\tau_i$ value was roughly independent of temperature from 20 to 40°C. Although rotations about the O-3'—P and P—O-5' bonds were specifically considered, C—O bond rotation and possibly others could also be contributing. It should be apparent from the structural constraints of DNA (see Fig. 7) that the free internal diffusion model employed by Klevan et al. (1979) and Bolton and James (1979, 1980a) is physically unrealistic but may provide some information regarding the rate of internal motion without addressing questions of the exact nature of the motion.

The $^{31}$P relaxation parameters were all fit with a $\tau_o$ value of 1.0 µs at 20° for DNA, which decreased to 0.5 µs at 40° (Bolton and James, 1979, 1980a). It was suggested that $\tau_o$ may be the correlation time for bending of DNA that would be consistent with theoretical considerations of the energy required for bending DNA (Bloomfield et al., 1974; Barkley and Zimm, 1979; Levitt, 1978). As mentioned before, the NMR results do not explicitly determine the type of motion, only that it is characterized by a correlation time of ~1 µs. The motion characterized by $\tau_o$ is inconsistent with end-for-end tumbling of DNA that would have a correlation time >20 µs for the DNA used in this study (Hogan et al., 1978). If the speedometer-cable motion illustrated in Fig. 6 exists, it could also contribute to $\tau_o$. I consider that possibility after the results of other studies are discussed.

The $^{31}$P $T_1$, NOE, and linewidth of 145-bp DNA were studied at three spectrometer frequencies by Shindo (1980). Measurements of relaxation parameters at different field strengths permit the chemical-shift anisotropy and dipolar contributions to be partitioned once a particular motional model is assumed. The CSA contribution, unlike the dipolar, is explicitly dependent on the magnetic field strength [see Eqs. (11)–(13)]. Assuming that 145-bp DNA can be represented by a rigid rod with no internal motion [Eqs. (20) and (23)], Shindo (1980) attributed 95% of the $^{31}$P relaxation to CSA at 109.3 MHz, 70% at 40.3 MHz, and 49% at 24.3 MHz. In comparison, Bolton and James (1980a) ascribed 35% to CSA at 81 MHz and 10% at 40.5 MHz, when an internal motion model was used. It should be noted that the cylinder model used by Shindo (1980) did not yield a very good fit to his relaxation, so he suggested that significant torsional and bending motions may also be occurring with DNA.

We can consider how much of the $^{31}$P relaxation should be attributed to the dipolar mechanism and how much to the chemical-shift anisotropy mechanism. Paramagnetic contributions to the relaxation from transition-metal ions are assumed negligible; paramagnetic metal ions could influence any of the relaxation parameters but would be especially notable in reducing the NOE toward 1 because the NOE is manifest only with the dipolar mechanism. Unfortunately, distinction between the dipolar and CSA contributions is not entirely possible unless all molecular motions are so fast that extreme narrowing conditions ($\omega^2\tau^2 \ll 1$) apply; otherwise, partitoning the relaxation between dipolar and CSA depends on the motional model selected.

Opella *et al.* (1981) and DiVerdi and Opella (1981a) examined the $^{31}$P-NMR spectrum of very large (90,000 bp) calf thymus DNA in solution and in the solid state using the solid-state NMR techniques of magic-angle sample spinning, proton dipolar decoupling, and cross-polarization. (These techniques are described by Shindo, Chapter 13.) The spectra are shown in Fig. 8. The anisotropic line shape is retained but with decreasing width as the temperature is increased until an isotropically averaged line results at 50°C. DiVerdi and Opella (1981a) compared their $^{31}$P results with the $^2$H-NMR spectrum obtained from DNA with the 8 position of the purine bases labeled with deuterium; the deuterium spectrum showed no change in the quadrupolar splitting over the same temperature range, exhibiting a coupling constant indicative of no base motion. They concluded that reorientation of DNA about the long axis, as proposed by Shindo (1980), was not occurring because the $^2$H-NMR spectrum should have undergone changes as the $^{31}$P-NMR spectrum did with temperature (Fig. 8). Instead, they concluded that the backbone was reorienting on a time scale of microseconds, but the bases were more rigid.

Hart *et al.* (1981) have obtained $^{31}$P $T_1$ values of several DNA restriction fragments as a function of temperature (some values are listed in Table I). As these results are discussed in more detail in Chapter 11 by Hart, they are not described further here.

Hogan and Jardetzky (1979, 1980a) reported $^{31}$P relaxation data for DNA that had been fractionated to varying average chain lengths (see Table I). They assumed only dipolar relaxation and used a motional model with the anisotropic overall motion of a cylinder plus a two-site jump as an internal motion. The correlation time used for the end-to-end tumbling was that determined by transient electric dichroism measurements (Hogan *et al.*, 1978), and the correlation time used for rotation about the long axis of the rod was that calculated from the theoretical expression for a cylinder of length $2L$ and radius $b$ according to Lamb (1945; Barkley and Zimm, 1979):

$$\tau_\parallel = 4\pi\eta b^2 L/3kT \tag{44}$$

where $k$ is Boltzmann's constant, $T$ the absolute temperature, and $\eta$ the solvent viscosity. The rotational correlation time $\tau_\parallel$ is defined in Eq. (23). A correlation time of 1 ns (Hogan and Jardetzky, 1979) and later revised to 2.2

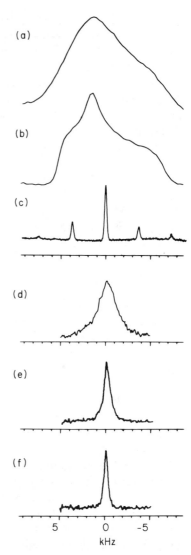

**Fig. 8.** $^{31}$P-NMR spectra of calf thymus DNA. (a) Solid DNA, cross-polarized but undecoupled spectrum at 61 MHz. (b) Solid DNA, proton-decoupled at 61 MHz. (c) Solid DNA, magic-angle spinning and proton-decoupled at 61 MHz. (d) DNA in solution at 30°C, 145 MHz (8.5 T). (e) DNA in solution at 30°C, 61 MHz (3.5 T). (f) DNA in solution at 30°C, proton-decoupled at 61 MHz (3.5 T). From Opella *et al.* (1981). Copyright 1981 American Chemical Society.

ns for a jump of ±27° (Hogan and Jardetzky, 1980a) was calculated for the internal motion. Because the definitions of correlation times for jump models and diffusion models differ by a factor of 6 (see footnote 2), the value of the internal motion correlation time obtained by Hogan and Jardetzky (1980a) is surprisingly close to that of Klevan *et al.* (1979) and Bolton and James (1979, 1980a). Most importantly, it is consistent with the idea that there is motion in the phosphodiester region of the DNA backbone on the time scale of a nanosecond. It should be noted that both Lipari and Szabo (1981b) and Allison *et al.* (1982) have presented corrections in the mathematical formula for the motional model used by Hogan and Jardetzky (1979, 1980a).

Lipari and Szabo (1981b) compared the fit of experimental relaxation data from the literature with various models having no internal motion, twisting motion, wobbling-in-a-cone, and a two-state jump. It was assumed that only dipolar contributions from the two H-5' and the H-3' protons contribute to the $^{31}$P relaxation. It was concluded that none of the models was substantially superior to the others. For each model, however, the experimental data could be fit only if a large amplitude motion in the nanosecond time range existed.

The importance of torsional deformation models (in effect, the twisting shown in Fig. 6) for $^{31}$P relaxation in DNA was promulgated by Allison *et al.* (1982). This conclusion was based on a comparison of experimental relaxation data in the literature with calculations assuming only dipolar relaxation. The model is effectively that of "beads" connected by springs, where the size of the beads can range from one to hundreds of base pairs. Earlier, fluorescence depolarization of intercalated ethidium, assuming the intercalation does not modify the motional properties of DNA, lead Allison and Schurr (1979) to the conclusion that the bead was rodlike with a length of 86 bp, although a length of 1 bp was also consistent with the data. Such a model assumes the bead is rigid (i.e., there are no motions within the nucleotides). A very important aspect of this model is that the beads are connected by springs, so that the twisting motion of one bead will be propagated to others (i.e., the torsional deformations represent a collective behavior). Allison *et al.* (1982) concluded that these torsional modes occurring on the nanosecond time scale provide the dominant source of relaxation for $T_1$ and NOE; because the calculated $T_1$ and NOE values could not entirely account for the experimental values, it was suggested that internal motions also of the order of a nanosecond made further contributions. Although the root mean square (RMS) amplitude of the twist between adjacent beads (base pairs) is only 6°, the collective behavior predicts an rms azimuthal displacement of 18.2° in 1 ns, which together with a 7° amplitude of local motion was sufficient to reproduce the experimental data. Further considerations of possible CSA contributions to $T_1$ relaxation indicated that the CSA contribution amounts to 31% using this model.

Not noted by Allison *et al.* (1982) in their conception of DNA as a collection of rigid rods connected by springs, is that when one considers DNA on a molecular level, any torsional deformation (twisting) must be a result of alterations in the torsion angles about backbone bonds. In other words, twisting is a result of internal motions that act in a concerted fashion.

**TABLE II**

Plausible Motional Models for $^{13}$C and $^{31}$P Relaxation in DNA Backbone[a]

| Internal motion model | Ribose $^{13}$C, $^{31}$P data fit | 1/jump rate (ns) |
|---|---|---|
| 1. Twisting (jump between 8 and 11.3 bp/turn); note: B-DNA has 10 bp/turn | No | |
| 2. Base tilting (jump between $-10$ and $+17°$) | No | |
| 3. Base propeller twisting (jump between 0 and 6°) | No | |
| 4. Combination base twist and tilt | No | |
| 5. Sugar puckering (jump between 2'-endo and 3'-endo; $\pm 35°$); note: smaller amplitude insufficient | All $^{13}$C except C-5' | 5 |
| 6. Sugar puckering (2'-endo $\rightleftarrows$ 3'-endo plus rotation about C-4'—C-5' ($\pm 15$–$25°$) | All $^{13}$C | 5 (0.4–3.0) |
| 7. Sugar puckering (2'-endo $\rightleftarrows$ 3'-endo) plus rotation about C-3'—O-3' ($\pm 25°$) | All $^{13}$C (except C-5') and $^{31}$P | 5 (1.5–2.5) |

[a] Based on calculation of Keepers and James (1982).

Starting with the viewpoint that some structural information is already available for DNA, Keepers and James (1982) considered some of the plausible motions for DNA shown in Fig. 6 to see which were in accord with $^{31}$P and $^{13}$C relaxation data for backbone nuclei and which were not. The models employed were more complete than any previously used. Both DD and CSA mechanisms were included. In addition, fluctuations in the P–H internuclear distances with the molecular motions were explicitly included in the calculations that employed two different types of jump models (see Section II,B,6). One type of jump (lattice) model could be used for concerted motions such as twisting (winding and unwinding motions), base-tilting motions, and base propeller-twisting motions, using the helix axis as a reference axis. The second class of models entails motions of repuckering in the sugar ring and bond rotations in the phosphodiester moiety of the backbone. The essential conclusions regarding which motions would fit the data are listed in Table II.

As shown in Fig. 9, even a large deviation in the amount of winding or unwinding (from 11.3 to 8 bp/turn) changes the orientation of the $^{31}$P CSA tensor (principal axes shown in Fig. 9) or P–H vectors very little. It should be noted that the calculations did not include the possibility that the small reorientations resulting from twisting could be additive, progressing along the helix from base pair to base pair [see discussion regarding Allison et al. (1982)]. However, theoretical calculations indicate that any conformational changes in DNA structure are compensated by nearby torsion-angle adjustments to minimize the energy of DNA (Olson, 1982; Keepers et al., 1982).

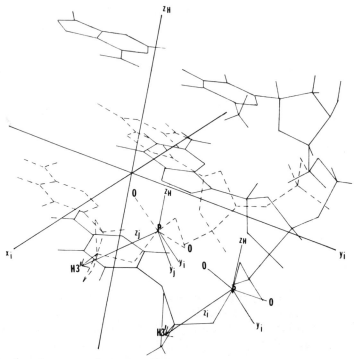

**Fig. 9.** Reorientation of the DNA backbone torsion angles between structures wound to 11.3 base pairs per turn (solid line) and unwound to 8 base pairs per turn (dashed line). The $z_H$ is defined by the helix axis of DNA.

In fact, the adjustment often entails a 1–3 crankshaft type of motion. The conclusion from the study of Keepers and James (1982) is that sugar-repuckering motions and rotational wobbling in the phosphodiester moiety are most probably sources for the earlier conclusions of nanosecond motions. Furthermore, it would appear that any such motions are highly localized, being compensated by adjustments within a few bonds. It is noted that the best-fit internal motion model contains a contribution of 37% from the CSA mechanism at 40.5 MHz.

Hogan and Jardetzky (1979) performed $^{31}$P-NMR relaxation measurements on DNA fractions of different lengths (Table I). Their $T_2$ measurements may help to elucidate the nature of the slower motions that dominate $T_2$, $R$, and $T_{1\rho}^{\text{off}}$ relaxation. They attributed most of the $T_2$ relaxation for 140-bp DNA to the slow end-for-end tumbling ($\tau_\perp$), with a smaller contribution from rotation about the long axis ($\tau_\parallel$). In contrast, Bolton and James (1979, 1980a,b) suggested that the slower motion influencing $^{31}$P relaxation values in larger DNA was due to bending, and Allison *et al.* (1982) suggested it was due to Rouse–Zimm coil-deformation modes. According to theoretical considerations (Bloomfield *et al.,* 1974) and transient electric dichroism studies (Hogan *et al.,* 1978), $\tau_\perp$ is strongly dependent on DNA chain length. Likewise, according to the Lamb equation [Eq. (44)], $\tau_\parallel$ is linearly dependent on chain length, and the collective Rouse–Zimm modes depend on length as $L^{1.6}$ (Schurr, 1977). Although the development of Barkley and Zimm (1979) may not be strictly applicable, it would appear from their analysis that a bending motion correlation time does not depend on chain length if a DNA longer than one persistence length (variously determined from 270 to 680 Å) is used.

For motions that are in the range of $10^{-7}-10^{-5}$ s, the spectral density $J_0(0)$ will dominate $T_2$ relaxation, and $1/T_2$ should be linearly dependent on the correlation time. Although there is only a limited amount of data, the data of Hogan and Jardetzky (1979, 1980b) and Bendel and James (1982) listed in Table I indicate that the value of $T_2$ has a dependence on chain length that is at least linear between 140- and 260-bp DNA. However, within experimental error, the $1/T_2$ values of the 260-, 300-, and 600-bp DNA samples are the same. Furthermore, the linear DNA sample of Bendel and James (1982) that is more than an order of magnitude longer has nearly the same $T_2$ value (the somewhat larger $T_2$ may be due to the lower concentration of DNA used). These $T_2$ data are therefore consistent with the notion that the 140-bp DNA behaves approximately as a rigid rod. But this lack of dependence of $T_2$ on chain length is consistent only with bending as the dominant motion for $T_2$ relaxation for longer DNA. Further NMR experiments need to be performed to substantiate that conclusion, but it is in accordance with sedimentation studies (Record *et al.,* 1975) and calculations (Schellman, 1974) that indicate 700-bp DNA, for example, extends on average to only 50% of its contour length.

The first NMR experiments on intact plasmid DNA have been reported (Bendel *et al.,* 1982). The native plasmid (in closed duplex, supercoiled form) was pIns36 (7200 bp long). Measurements of the $^{31}$P $T_1$, $T_2$, and NOE were conducted on the supercoiled, nicked circular, and linear forms of pIns36. The results are given in Table III. The $T_1$ and NOE values in all three forms conform to prior descriptions of DNA having internal motions on the nanosecond time scale. The observed $T_2$ values are substantially different in the three forms, increasing to a surprising 1.17 s for the supercoiled pIns36. The two-order of magnitude increase in $T_2$ is much too large to be accounted for by considering the overall tumbling of the molecules (although the trend of these results is predicted by the decrease in the radius of gyration from the linear through the circular to the supercoiled form). This is further evidence for the notion that the isotropic reorientation effective in the $^{31}$P-NMR relaxation is caused by bending motions of the chain for molecules longer

**TABLE III**

Experimental $^{31}$P Relaxation Time and NOE for pIns36 DNA[a,b]

| Form | $T_1$ (s) | $T_2$ (s) | NOE |
|---|---|---|---|
| Linear | 2.4 ± 0.2 | 0.019 ± 0.003 | 1.3 ± 0.1 |
| Circular | 2.5 ± 0.2 | 0.25 ± 0.07 | 1.4 ± 0.15 |
| Supercoiled | 1.64 ± 0.15 | 1.17 ± 0.10 | 1.45 ± 0.18 |

[a] From Bendel and James (1982).
[b] At 40.5 MHz, 25°C, in 0.1 $M$ NaCl, 10 m$M$ Na-cacodylate, pH 7.0. Most of the values listed represent averages of two or three measurements on each of two separate preparations of pIns36. Experiments were carried out on different days, and the uncertainties reflect both variations in reproducibility as well as error margins of the individual results.

than one persistence length. It was suggested that the increased effective frequency of these bending motions observed for the closed DNA forms results mainly from an excess of conformational free energy and coupling to higher-frequency torsional motions (Bendel *et al.*, 1982).

Mariam and Wilson (1979) obtained $^{31}$P spectra of DNA in a study of the helix-to-coil transition. At temperatures near the melting temperature, they observed four different signals. The results indicated different $T_1$ values for the various peaks, suggesting the possibility of different mobilities. Wilson and Jones (1982) reported the $T_1$ value of DNA first decreased and then increased slightly as the temperature was raised from 30 to 70°C. This is consistent with a change in the internal motion correlation time for $^{31}$P in DNA, $\tau_i$ being near the $T_1$ minimum (see Fig. 2).

In the investigations performed to date on double-stranded DNA, it has been presumed that the relaxation behavior of each $^{31}$P nucleus is the same as all others. However, that presumption is probably not precisely correct. Work has indicated that the preferred conformation of the nucleotide units in the helix varies somewhat with base composition and sequence (e.g., Wang *et al.*, 1979; Wing *et al.*, 1980; Levitt and Warshel, 1978; Keepers *et al.*, 1982). The relaxation of $^{31}$P depends on the magnitude of the P–H internuclear distances that may vary with conformation of the monomeric unit. It should be anticipated that variations, perhaps small, in $^{31}$P relaxation values from DNAs with different composition will be reported in the future.

## B. RNA

### 1. Double-Stranded Polynucleotides

Tritton and Armitage (1978) measured the $^{31}$P $T_1$ of 16-S rRNA and the NOE for a mixture of 16- and 23-S rRNA obtained from *Escherichia coli* ribosomes. Bolton and James (1979, 1980a) measured the $^{31}$P $T_1$, NOE, $R$, $T_{1\rho}^{off}$, and linewidth of poly(I) · poly(C) as a function of temperature. The $T_1$ and NOE values were in excellent agreement with those measured by Tritton and Armitage (1978) for rRNA. The data were analyzed using the free internal diffusion model described by Eqs. (20) and (21), which provided a very good fit of all the relaxation data (Bolton and James, 1979, 1980a). The analysis yielded a value of 0.5 ns for the internal motion correlation time $\tau_i$ that was independent of temperature, but the slower motion correlation time $\tau_o$ was 1 $\mu$s at 20°C and had an activation energy of $\sim$ 3 kcal/mol over the range 20–40°C.

The values of the internal motion correlation time are nearly the same for DNA and poly(I) · poly(C), showing that this motion is not substantially coupled to the overall conformation of the polynucleotide because DNA exhibits B-form structure and poly(I) · poly(C) exhibits A-form structure (Bolton and James, 1980a). In addition, the slower motion correlation time found for poly(I) · poly(C) had the same value and temperature dependence of the larger DNA, providing additional support for the suggestion that $\tau_o$ reflects a bending motion that is independent of chain length.

Addition of 20 m$M$ Mg(II) to poly(I) · poly(C) modified some of the $^{31}$P-NMR relaxation parameters (Bolton and James, 1980a). It was found that the internal motion was not affected, but the $\tau_o$ value increased from 1 to 3 $\mu$s. The precise cause of this is not known, but we note that Mg(II) does inhibit thermal denaturation of double-stranded polynucleotides (Record, 1975).

## 2. Single-Stranded Polynucleotides

$^{31}$P-NMR spectra of poly(I) at 36.4 and 111.6 MHz were examined by Neumann and Tran-Dinh (1981). Assuming that the relaxation could be effected by a single random isotropic motion, it was concluded from $^{31}$P-NMR $T_1$ and NOE measurements that the CSA contribution to relaxation amounted to 12% at 36.4 MHz and 72% at 111.6 MHz. From the $^{31}$P $T_1$ and NOE data, a temperature-dependent correlation time of $\sim$ 1 ns was calculated. $^{31}$P-NMR $T_1$ and NOE measurements have also been carried out with poly(U), poly(A), and poly(C) as a function of temperature (Akasaka et al., 1977). Again, assuming random, isotropic motion, it was estimated that the CSA mechanism may account for 20% of the $^{31}$P relaxation for these three polynucleotides at 40.5 MHz and 72°C.

Shindo (1980) measured the linewidth, $T_1$, and NOE of poly(U) at 109.3 and 24.3 MHz. Because poly(U) in solution has a random-coil conformation, a motional model with a log $\chi^2$ correlation time distribution about a single average correlation time $\bar{\tau}$ value was utilized [see Eqs. (24) and (25)].

The $T_1$ and linewidth data were fitted with an average correlation time of 3.3 ns and a distribution parameter $p$ of 14 for a $b$ value of 1000. A poorer fit of the model to the experimental NOE value at 109.3 MHz was attributed to the CSA relaxation contribution.

Unlike poly(U), poly(A) exists in a single-stranded structure with the bases mostly stacked, the extent of stacking decreasing with temperature noncooperatively (Stannard and Felsenfield, 1975). Akasaka *et al.* (1977) measured the $^{31}$P $T_1$ and NOE of poly(A). Evidence that the rotamer populations in ApA and poly(A) did not vary with temperature was implied by the negligible changes in coupling-constant values of ApA over a wide temperature range. Assuming the motion governing $^{31}$P relaxation was isotropic, a correlation time of 6 ns at 20°C and an activation energy of 5.3–6.0 kcal/mol for the motion fit the data.

Although their $^{31}$P $T_1$ and NOE measurements were in good agreement with those of Akasaka *et al.* (1977), Bolton and James (1979, 1980a) discovered that the additional information supplied by the off-resonance relaxation parameters $R$ and $T_{1\rho}^{\text{off}}$ required a two-correlation time model to fit adequately all the relaxation parameters for poly(A). Using the free internal diffusion with spectral densities given by Eqs. (20) and (21), it was found that an internal motion characterized by a 0.5-ns correlation time that was independent of temperature and a much slower motion that depends strongly on temperature over the 6–40°C range, provided an excellent fit for all of the relaxation parameters. The strong temperature dependence of the slower motion $\tau_o$ was attributed to unstacking of the poly(A) bases, resulting in a polynucelotide with a smaller radius of gyration. It is also interesting that the internal motion correlation time obtained for $^{31}$P in single-stranded poly(A) is the same as that obtained in double-stranded poly(I) · poly(C) and DNA.

Yamada *et al.* (1978) examined $^{31}$P-NMR spectra of poly(G) as a function of temperature. The $^{31}$P $T_2$ values determined from the spin–echo method and from the linewidth were considerably different, implying much chemical-shift heterogeneity among the phosphorus nuclei. Assuming a single isotropic motion to be applicable, Yamada *et al.* (1978) used the $T_1/T_2$ ratio to calculate a correlation time at 27°C of 70 ns. It was pointed out that this value indicates much slower motion than previously concluded for poly(A), poly(C), and poly(U).

Akasaka *et al.* (1977) investigated the temperature dependence of the $^{31}$P NOE and $T_1$ for poly(C). In comparison with results from poly(A) and poly(U), it was suggested that the flexibility of these single-stranded polynucleotides increased in the order poly(U) < poly(C) < poly(A), which is inversely correlated with the tendency for base stacking.

## C. Transfer RNA

Transfer RNAs (tRNAs) are composed of double-stranded and single-stranded regions with the additional complications of tertiary structure. A few $^{31}$P-NMR relaxation studies of amino acyl-specific tRNAs have been performed. Guéron and Shulman (1975) observed several $^{31}$P resonances in the spectra of yeast tRNA$^{Phe}$ and *E. coli* tRNA$_2^{Glu}$, which could be monitored as the tRNA was melted or Mg(II) was added. On the basis of the increased $^{31}$P linewidths observed in the 109-MHz spectra compared to 40.5-MHz spectra, it was concluded that the CSA contribution dominated the relaxation at 109 MHz.

Two of the partially resolved $^{31}$P signals of yeast tRNA$^{Phe}$ at 40.5 MHz were observed by Hayashi *et al.* (1977) to have quite different $T_1$ values. Assuming a single isotropic motion model, it was concluded from $T_1$ and NOE measurements that both the CSA and dipolar mechanisms contribute to relaxation at 40.5 MHz, with the dipolar contribution being the greatest. Applying the isotropic motion model to the experimental $T_1$ and NOE values, it was estimated that the $^{31}$P correlation time was 3 ns in yeast tRNA$^{Phe}$.

The tRNA$^{Phe}$ was investigated in more detail by Gorenstein and Luxon (1979). The $^{31}$P-NMR linewidths and $T_1$ values were obtained at 146 and 32 MHz. Values of $T_1$ at 146 MHz ranged from 2.1 to 3.5 s for 13 different resonances observed. With the assumption that the single isotropic motion model was applicable, it was estimated that >90% of the relaxation is due to the CSA mechanism at 146 MHz. Utilizing the expressions appropriate for the CSA mechanism [Eqs. (11) and (12)] with the isotropic motion model, Gorenstein and Luxon (1979) used their $T_1$ and $T_2$ values to compute a $^{31}$P correlation time of 11 ns and a chemical-shift anisotropy of 180 ppm. This value for the CSA is somewhat higher than that of 140 ppm estimated by Guéron and Shulman (1975) or 120–130 ppm estimated by Hayashi *et al.* (1977).

The discrepancy between the $^{31}$P correlation time estimated by Gorenstein and Luxon (1979) and that by Hayashi *et al.* (1977) for tRNA$^{Phe}$ may result from using different relaxation parameters, with the simple assumption of a single correlation time in the two studies. The phosphorus nuclei may experience two or more motions with different correlation times (e.g., anisotropic tumbling of tRNA or probably internal motion superimposed on the overall tumbling of tRNA). The NOE and $T_1$ are more sensitive to faster motions and will lead to a smaller estimation of the average correlation time. But $T_2$ is more sensitive to slower motions, so the average correlation time estimated from $T_1$ and $T_2$ measurements would be larger.

## D. Nucleic Acid – Protein Systems

For nearly all of their biological processes, nucleic acids are intimately involved with proteins. Consequently, a number of investigations of natural and model systems exploring such possible facets of interaction as ionic attractions, stacking of aromatic amino acid residues with nucleic acid bases, and topology (the $\beta$ structure of proteins fits the major groove of DNA quite well). However, relatively few NMR studies have been carried out with nucleic acid – protein systems owing to the complexity of the systems. The following discussion is limited to a consideration of the $^{31}$P-NMR relaxation behavior of nucleic acids in nucleoprotein complexes. Table IV summarizes some of the relaxation parameters measured for an array of nucleic acid – protein systems. Perhaps the most obvious generalization is that the relaxation results for the various nucleic acid – protein systems cannot be generalized.

### 1. DNA in Chromatin and Nucleosomes

Most cellular DNA is in the huge protein – nucleic acid complex chromatin that contains nucleosomes as one of the two repeating structural units in chromatin. Even a nucleosome is a complicated system, being composed of a DNA helix with several proteins forming a roughly disk-shaped object with dimensions 57 Å high by 110 Å across. A few studies on this important nucleoprotein have utilized $^{31}$P NMR (Hanlon et al., 1976; Cotter and Lilley, 1977; Kallenbach et al., 1978). It should be pointed out that only a single $^{31}$P signal is observed in studies of nucleoproteins, so any conclusions must take that into account.

Klevan et al. (1979) measured the $^{31}$P $T_1$ and NOE for calf thymus nucleosomes (Table IV) and compared the results with the DNA extracted from the nucleosomes (Table I). Using the internal diffusion model [Eqs. (20) and (21)] with an isotropic tumbling time estimated from electric dichroism measurements for the overall reorientation and assuming only dipolar relaxation, the internal motion effecting $^{31}$P relaxation was determined to have a correlation time of 0.4 ns. That value is the same as calculated for DNA, leading to the conclusion that protein binding to DNA in nucleosomes has little effect on the internal motion. Substitution of $H_2O$ for $D_2O$ does have an influence on $T_1$ (Table IV but not much on NOE, unlike $^{31}$P relaxation in pure DNA which is not affected by $H_2O$ substitution.

The large NOE value obtained by Klevan et al. (1979) led them to conclude dipolar relaxation was strongly dominant. On the other hand, Shindo et al. (1980) maintained that CSA contributions were quite important in $^{31}$P relaxation of nucleosomes. Chemical-shift inequivalence was shown to be an important contributor to the $^{31}$P-NMR linewidth, being 0.77 ppm in nucleosomes compared to 0.50 ppm in pure DNA; the values were determined by treating the inequivalence as a parameter to be fit with the experimental linewidth and relaxation data. The experimental relaxation data (Table IV) were fitted using a motional model with a distribution of correlation times [Eqs. (24) and (25)]. The data were best-fit with an average correlation time of 30 ns and a distribution covering about two orders of magnitude. The average correlation time is about an order of magnitude smaller than the rotational tumbling time of a nucleosome, implying that the DNA $^{31}$P nuclei possess motional freedom in addition to the tumbling of the nucleosomes. Because a different motional model was employed in the accompanying paper on pure DNA (Shindo, 1980; and see preceding discussion), a direct comparison is not possible.

Larger, more complicated chromosomal complexes were investigated by DiVerdi et al. (1981), using solid-state NMR techniques. A soluble chromatin fibril constituted of ~10 nucleosomes connected by linear DNA was examined using cross-polarization, an NMR technique that detects only solid-like DNA. The reorientation rate of the fibril is so slow that for NMR purposes it behaves as a solid. The spectrum is shown in Fig. 10b. The isotropic spectrum shown in Fig. 10d using ordinary 90° pulses was reported to be due to a small amount of degraded DNA in the sample. The anisotropic pattern of the solid-like DNA in the soluble chromatin, as well as that from bull sperm heads (Fig. 10c), is diminished about 30% in width relative to that of rigid DNA (Fig. 10a). It was suggested that the reduction may arise from limited-amplitude fluctuations in the DNA or motions of the particles. Furthermore, it was noted that the DNA motions must be restricted in these chromosomal complexes relative to free DNA, because free DNA in solution exhibits an isotropic line shape, indicating that motions faster than the chemical-shift anisotropy (11.8 kHz in this case) are occurring (see Shindo, Chapter 13 for additional discussion of solid-state $^{31}$P NMR).

## 2. RNA in Ribosomes and Messenger Ribonucleoprotein

Cellular protein synthesis occurs on ribosomes which are complicated complexes of several proteins associated with long, single-stranded RNA segments. Few NMR studies have been performed with ribosomes, probably because of the complicated structure and large size, ranging from 2.7 megadaltons (prokaryotes) to 4.0 megadaltons (eukaryotes).

Tritton and Armitage (1978) obtained the $^{31}$P $T_1$ and NOE for the 50- and 30-S subunits as well as the intact 70-S ribosomes from *E. coli* (Table IV). They found that substitution of $D_2O$ for $H_2O$ influences the $^{31}$P relaxation as

## TABLE IV
### $^{31}$P-NMR Relaxation Parameters for Nucleoproteins

| Sample | Solvent | Frequency (MHz) | Temperature (°C) | $T_1$ (s) | NOE | $T_{1\rho}^{off\,a}$ (s) | $R^a$ | $W_{1/2}$ (Hz) | Reference[b] |
|---|---|---|---|---|---|---|---|---|---|
| Calf thymus nucleosome (10 m$M$ Tris, 5 m$M$ EDTA, pH 7.6) | H$_2$O<br>D$_2$O | 36.4 | 8 | 2.8<br>3.6 | 1.7<br>1.6 | —<br>— | —<br>— | 33<br>37 | 1 |
| Chicken erythrocyte nucleosome core (5 m$M$ Tris, 0.1 m$M$ EDTA, pH 8) + 6 $M$ urea | | 109.3<br>40.3<br>24.3<br>109.3<br>24.3 | 19<br><br><br>19 | 3.34<br>3.00<br>2.70<br>2.3<br>2.2 | 1.22<br>—<br>1.52<br>1.20<br>— | —<br>—<br>—<br>—<br>— | —<br>—<br>—<br>—<br>— | 130<br>45<br>25<br>205<br>35 | 2 |
| *Escherichia coli* 70-S Ribosome 50 m$M$ Tris, 10 m$M$ MgOAc, 0.1 $M$ KCl, pH 7.5 | H$_2$O<br>D$_2$O | 36.4 | 23 | 2.0 (2.3)$^c$<br>2.3 | 1.32<br>1.08 | —<br>— | —<br>— | 208<br>— | 3 |
| 50-S Ribosomal subunit (20 m$M$ Tris, 0.1 $M$ NH$_4$Cl, 10 m$M$ MgOAc, pH 7.5) | H$_2$O | 36.4 | 23 | 1.3 | 1.18 | — | — | — | |
| 30-S Ribosomal subunit (10 m$M$ Tris, 30 m$M$ NH$_4$Cl, 10 m$M$ MgCl$_2$, 6 m$M$ mercaptoethanol, pH | H$_2$O | 36.4 | 23 | 1.7 | 1.21 | — | — | — | |

| Sample | Solvent | | | | | | | | Reference |
|---|---|---|---|---|---|---|---|---|---|
| Rat liver ribosomes intact (50 mM Tris, 10 mM MgOAc, 0.1 M KCl, pH 7.5) | $D_2O$ | 40.5<br>81 | 20 | 2.35<br>2.45 | 1.1<br>1.1 | 0.77<br>— | 0.33<br>— | 80<br>90 | 4 |
| limited-digest "core" following RNase treatment | $D_2O$ | 40.5 | 20 | 2.5 | 1.4 | 1.0 | 0.4 | 60 | |
| Rat liver messenger ribonucleoprotein (0.3% SDS, 0.1 M NaCl, 50 mM NaOAc, 10 mM EDTA, pH 5.2) | $D_2O$ | 40.5 | 5 | 0.46 | 1.5 | 0.05 (4 kHz)<br>0.13<br>0.28 (16 kHz) | 0.11 (4 kHz)<br>0.29<br>0.60 (16 kHz) | 140 | 4 |
| MS2 virus $D_2O$ | $H_2O$ | 81<br>20 | 22–45<br>4.2 | 3.1<br>1.15 | 1.4<br>— | —<br>— | —<br>— | 80 | 4 |
| QB virus | $H_2O$ | 81 | 20 | 3.1 | 1.4 | — | — | 80 | 4 |

[a] Experiment performed with an rf field strength 0.48 G applied 8.0 kHz off-resonance unless noted otherwise.
[b] References: 1, Klevan et al. (1979); 2, Shindo et al. (1980); 3, Tritton and Armitage (1978); 4, Bolton et al. (1982).
[c] The value in parentheses was obtained when 0.1 M EDTA was added.

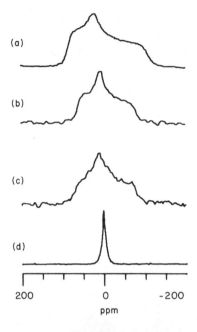

**Fig. 10.** ³¹P-NMR spectra of DNA and nucleoprotein complexes. (a) Solid fibrous calf thymus DNA. (b) Chicken erythrocyte chromatin in solution. (c) Bull sperm heads in solution. (d) High-molecular-weight calf thymus DNA in solution. Spectra (a), (b), and (c) resulted from cross-polarization with 1-ms mix time and 1-s recycle delay. Data were acquired for 10 ms with a 2.3-mT ¹H decoupling field. Spectrum (d) was from $\pi/2$ pulses rather than cross-polarization. From DiVerdi et al. (1981).

it did with nucleosomes but which it did not with pure DNA. With the free internal diffusion model, assuming only dipolar relaxation and independent values for the rotational tumbling of intact ribosomes and subunits, it was concluded from the experimental NOE values that the correlation time for internal motion is 3–5 ns. The ³¹P spectra published by Tritton and Armitage (1978) exhibited peaks characteristic of degradation products from the action of ribonuclease.

³¹P-NMR experiments with undegraded eukaryotic ribosomes from rat liver have also been carried out with the off-resonance relaxation parameters $T_{1\rho}^{off}$ and $R$ being measured in addition to $T_1$, NOE, and $W_{1/2}$ (Bolton et al., 1982; and see Table IV). Using the motional model of Eqs. (20) and (21) and including both dipolar and CSA contributions to relaxation, all of the relaxation parameters at both frequencies were consistent with an internal motion with $\tau_i = 2$ ns, in reasonable agreement with Tritton and Armitage (1978). This is a factor of 4–5 slower than the internal motion in uncomplexed double-stranded RNA; the relaxation data of Bolton and James (1979) were reanalyzed, with the resulting $\tau_i$ being decreased to 0.4 ns from 0.5 ns when the CSA contribution is included. The limited-digest ribosome was obtained by removing 30% of RNA susceptible to ribonuclease cleavage. Presumably, the remaining 70% of the RNA would be in intimate contact with protein because the limited-digest ribosome would retain

almost all of the protein and have essentially the same sedimentation properties as the intact ribosome.

As seen in Table IV, the $^{31}$P NOE of the limited-digest ribosome is increased relative to the value from the intact ribosome, thus indicating a greater degree of motional freedom. Analysis of all the relaxation data quantified this observation; the RNA in the limited-digest ribosome exhibited the same internal motion correlation time as free RNA. Apparently, the local motion of the RNA backbone is restricted in the intact ribosome, but this strain is relieved by enzyme cleavage of the long RNA strands, yielding shorter polynucleotides in the limited-digest ribosome.

Bolton *et al.* (1982) have also performed $^{31}$P-NMR experiments with messenger ribonucleoprotein (mRNP), which is about the same size as ribosomes. The $T_1$, NOE, $T_{1\rho}^{\text{off}}$, $R$, and $W_{1/2}$ values are listed in Table IV. The $^{31}$P $T_1$ value of mRNP is much smaller than that of any other nucleic acid–protein complex in Table IV; in fact, it is smaller than one could expect for a $^{31}$P nucleus relaxed solely by protons on the RNA. Consequently, selective $^{31}$P{$^1$H} NOE experiments were carried out in which the $^{31}$P resonance intensity was monitored, as 50-Hz "windows" in the $^1$H-NMR spectrum were strongly irradiated rather than irradiating all proton frequencies, as is usually done with heteronuclear NOE experiments. The selective NOE experiments showed that the phosphorus in the mRNP was dipolar-coupled to protein protons as well as ribose protons. Apparently, the nature of the protein–nucleic acid interactions differ in comparison with the other nucleic acid–protein complexes of Table IV, in that protons from the protein are much closer to the phosphorus of RNA.

## 3. Viruses

There is a great diversity in the structure and properties of viruses. In certain instances, their structures can be relatively simple, being composed of a nucleic acid (either RNA or DNA) and several copies of a single protein. In other cases, structures can become more complicated, taking on the characteristics of primitive cells with lipids and carbohydrates added. Experience to date has shown that liquid-state NMR methods may be used with some viruses, but solid-state NMR methods are necessary for others.

Bolton *et al.* (1982) investigated the $^{31}$P-NMR relaxation properties of the two nearly spherical viruses QB and MS2 (see Table IV), which have particle weights of 4.2 and 3.6 megadaltons, respectively, and contain single-stranded RNA. Relaxation measurements were made for MS2 virus in both $D_2O$ and $H_2O$ solutions. As seen in Table IV, the solvent isotope influenced the measurements. Selective $^{31}$P{$^1$H} NOE experiments were carried out for the MS2 virus in $H_2O$ solution. In addition to the H-5' and H-3' protons on the ribose, protons resonating at ~ 5 ppm (attributed to $H_2O$ or exchangeable protons) were found to be dipolar-coupled to the $^{31}P$. As mentioned previously, $H_2O$ influences $^{31}P$ relaxation in ribosomes (Tritton and Armitage, 1978) and nucleosomes (Klevan et al., 1979) but not pure RNA or DNA. It was suggested by Bolton and James (1980a) that the $H_2O$ did not influence $^{31}P$ relaxation in pure nucleic acids because of rapid exchange of bound water with bulk water, together with rapid reorientation of any bound water molecules. It was proposed by Bolton et al. (1982) that this situation may be altered with nucleic acid–protein complexes. Although it is possible that the rate of $H_2O$ exchange itself may be sufficient, it is also likely that positively charged protein groups, such as $NH_3^+$ of lysine, bring protons sufficiently close to the nucleic acid $^{31}P$ to affect its relaxation. The exchange of these particular protons with $H_2O$ results in an averaged chemical shift in agreement with the selective $^{31}P\{^1H\}$ NOE experiments on MS2 virus. Further, $D_2O$ exchange for $H_2O$ will give $ND_3^+$, which is only 2% as effective as $NH_3^+$ in promoting relaxation.

Analysis of the relaxation results for MS2 virus in $D_2O$ indicated that the internal motion was restricted by about an order of magnitude compared with free RNA (Bolton et al., 1982). It is not clear, however, if this result is due to a change in the rate of internal motion or if the structure of the RNA packaged into the virus is modified, resulting in altered relaxation parameters; note that a modified structure could change P—H internuclear distances.

The larger (16.4 megadaltons), morphologically more complex (filamentous rod), DNA-containing filamentous virus fd required solid-state NMR methods for its investigation (DiVerdi and Opella, 1981b). Dipolar proton decoupling, cross-polarization, and magic-angle sample spinning were employed in this study of the viral DNA that was effectively immobilized when packaged in the virus by interaction with the viral coat proteins. The linewidth was shown to be dominated by chemical-shift anisotropy rather than dipolar coupling to protons. The chemical-shift anisotropy is of the order of 10 KHz, leading to the observation that there are few motions of the DNA in the virus faster than $10^4$ Hz.

### E. Nucleic Acid–Drug Complexes

The mode of action of several classes of drugs apparently entails binding of the drugs to nucleic acids. Although NMR has been a valuable tool for studies of drugs in complexes with oligonucleotides, few investigations have

## 12. Relaxation Behavior of Nucleic Acids

used polynucleotides (see, however, Gorenstein and Goldfield, Chapter 10), and fewer still have employed relaxation-time measurements. However, conformational fluctuations in nucleic acids can play a major role in drug action. Nucleic acid conformational changes are apparently required for drug binding and drug release. Little is known about the effect of drugs on such conformational changes, and even less is known about their influence on the rate and amplitude of conformational fluctuations.

Of the few relaxation studies that have been made, binding of the intercalator ethidium bromide to nucleic acids has been of interest in most

investigations. Hogan and Jardetzky (1980b) titrated DNA with ethidium bromide and observed the concomitant intensity decrease for the $^{31}$P resonance, suggesting that the drug immobilizes the DNA at the binding site resulting in resonances so broad that they cannot be detected. It was concluded that four phosphates were immobilized for every ethidium bound. Following an earlier report of Jones and Wilson (1980) in which the $^{31}$P resonance intensity of DNA had not been observed to decrease with ethidium addition, Wilson et al. (1981) carefully investigated the $^{31}$P NMR of DNA complexed with ethidium as a function of molecular weight (190- versus 1400-bp DNA), temperature, and magnetic field strength. No signal intensity loss was observed in any case in contrast to the report of Hogan and Jardetzky (1980b). Allison et al. (1982) also addressed this question and reported that the conclusion of Hogan and Jardetzky (1980b) was erroneous owing to the existence of a phase separation in their DNA–ethidium bromide sample, the more concentrated phase having a $T_2$ value so short as to preclude observation.

Wilson and Jones (1982) monitored the DNA $^{31}$P-NMR signal with addition of actinomycin D, ethidium, quinacrine, daunorubicin, and tetralysine. In each case it was found that the $T_1$ and NOE change negligibly with the addition, but the linewidth increased. Thus it was concluded that the internal motions in the backbone are unaffected by intercalation but that overall motion is diminished.

## Acknowledgments

Foremost, I wish to gratefully acknowledge the original research contributions of my colleagues who are joint authors with me of any publications listed here. Financial support for our work has been received as research grants GM25018 and CA27343 from the National Institutes of Health. I also want to acknowledge receipt of Research Career Development Award AM00291 from NIH.

## References

Abragam, A. (1961). "The Principles of Nuclear Magnetism." Oxford Univ. Press (Clarendon), London and New York.
Akasaka, K., Yamada, A., and Hatano, H. (1977). *Bull. Chem. Soc. Jpn.* **50**, 2858–2862.
Allison, S. A., and Schurr, J. M. (1979). *Chem. Phys.* **41**, 35–59.
Allison, S. A., Shibata, J. H., Wilcoxon, J., and Schurr, J. M. (1982). *Biopolymers* **21**, 729–762.
Altona, C., and Sundaralingam, M. (1972). *J. Am. Chem. Soc.* **94**, 8205–8212.
Barkley, M. D., and Zimm, B. H. (1979). *J. Chem. Phys.* **70**, 2991–3007.
Bendel, P., and James, T. L. (1982). *J. Magn. Reson.* **48**, 76–85.
Bendel, P., Laub, O., and James, T. L. (1982). *J. Am. Chem. Soc.* **104**, in press.
Berne, B. J., and Pecora, R. (1976). "Dynamic Light Scattering." Wiley, New York.
Bloomfield, V. A., Crothers, D. M., and Tinoco, I. (1974). "Physical Chemistry of Nucleic Acids." Harper, New York.
Bodenhausen, G., and Ernst, R. R. (1982). *J. Am. Chem. Soc.* **104**, 1304–1309.
Bolton, P. H., and James, T. L. (1979). *J. Phys. Chem.* **83**, 3359–3366.
Bolton, P. H., and James, T. L. (1980a). *J. Am. Chem. Soc.* **102**, 25–31.
Bolton, P. H., and James, T. L. (1980b). *Biochemistry* **19**, 1388–1392.
Bolton, P. H., and Mirau, P. A., Shafer, R. H., and James, T. L. (1981). *Biopolymers* **20**, 435–449.
Bolton, P. H., Clawson, G., Basus, V. J., and James, T. L. (1982). *Biochemistry* **24**, 6073–6081.
Bothner-By, A. A., and Noggle, J. H. (1979). *J. Am. Chem. Soc.* **101**, 5152–5155.
Broido, M. S., and Kearns, D. R. (1980). *J. Magn. Reson.* **41**, 496–501.
Broido, M. S., and Kearns, D. R. (1982). *J. Am. Chem. Soc.* **104**, 5207–5216.
Bull, T. E., Norne, J. E., Reimarsson, P., and Lindman, B. (1978). *J. Am. Chem. Soc.* **100**, 4643–4649.
Cotter, R. I., and Lilley, D. M. J. (1977). *FEBS Lett.* **82**, 63–68.
DiVerdi, J. A., and Opella, S. J. (1981a). *J. Mol. Biol.* **149**, 307–311.
DiVerdi, J. A., and Opella, S. J. (1981b). *Biochemistry* **20**, 280–284.
DiVerdi, J. A., Opella, S. J., Ma, R.-I., Kallenbach, N. R., and Seeman, N. C. (1981). *Biochem. Biophys. Res. Commun.* **102**, 885–890.
Doddrell, D., Glushko, V., and Allerhand, A. (1972). *J. Chem. Phys.* **56**, 3683–3689.
Englander, S. W., and Englander, J. J. *In* "Methods in Enzymology" (C. H. W. Hirs, ed.), Vol. 49, Part G, pp. 24–39. Academic Press, New York.
Englander, S. W., Kallenbach, N. R., Heeger, A. J., Krumhansl, J. A., and Litwin, S. (1980). *Proc. Natl. Acad. Sci. U.S.A.* **77**, 7222–7226.
Gorenstein, D. G., and Luxon, B. A. (1979). *Biochemistry* **18**, 3796–3804.
Griffith, J. D. (1978). *Science* **201**, 525–527.
Guéron, M., and Shulman, R. G. (1975). *Proc. Natl. Acad. Sci. U.S.A.* **72**, 3482–3485.

## 12. Relaxation Behavior of Nucleic Acids

Hanlon, S., Glonek, T., and Chan, A. (1976). *Biochemistry* **15**, 3869–3875.
Hart, P. A., Anderson, C. F., Hillen, W., and Wells, R. D. (1981). *In* "Biomolecular Stereodynamics" (R. H. Sarma, ed.), Vol. 1, pp. 367–382. Adenine Press, New York.
Hayashi, F., Akasaka, K., and Hatano, H. (1977). *Biopolymers* **16**, 655–667.
Hogan, M. E., and Jardetzky, O. (1979). *Proc. Natl. Acad. Sci. U.S.A.* **76**, 6341–6345.
Hogan, M. E., Jardetzky, O. (1980a). *Biochemistry* **19**, 3460–3468.
Hogan, M. E., and Jardetzky, O. (1980b). *Biochemistry* **19**, 2079–2085.
Hogan, M., Dattagupta, N., and Crothers, D. M. (1978). *Proc. Natl. Acad. Sci. U.S.A.* **75**, 195–199.
Howarth, O. W. (1979). *J. Chem. Soc., Faraday Trans. 2* **275**, 863–873.
Hubbard, T. S., and Johnson, C. S., Jr. (1975). *J. Chem. Phys.* **63**, 4933–4940.
Hull, W. E., and Sykes, B. D. (1975). *J. Mol. Biol.* **98**, 121–153.
James, T. L. (1975). "Nuclear Magnetic Resonance in Biochemistry." Academic Press, New York.
James, T. L. (1980). *J. Magn. Reson.* **39**, 141–153.
James, T. L., Matson, G. B., and Kuntz, I. D. (1978). *J. Am. Chem. Soc.* **100**, 3590–3594.
Jeener, J., Meier, B. H., Bachmann, P., and Ernst, R. R. (1979). *J. Chem. Phys.* **71**, 4546–4553.
Johnston, P. D., Figueroa, N., and Redfield, A. G. (1979). *Proc. Natl. Acad. Sci. U.S.A.* **76**, 3130–3134.
Jones, R. L., and Wilson, W. D. (1980). *J. Am. Chem. Soc.* **102**, 7776–7778.
Kalk, A., and Berendsen, H. J. C. (1976). *J. Magn. Reson.* **24**, 343–366.
Kallenbach, N. R., Appleby, D. W., and Bradley, C. H. (1978). *Nature (London)* **272**, 134–138.
Keepers, J. W. (1982). Doctoral Dissertation, University of California, San Francisco.
Keepers, J. W., and James, T. L. (1982). *J. Am. Chem. Soc.* **104**, 929–939.
Keepers, J. W., Kollman, P. A., Weiner, P. K., and James, T. L. (1982). *Proc. Natl. Acad. Sci. U.S.A.* **79**, 5537–5541.
Klevan, L., Armitage, I. M., and Crothers, D. M. (1979). *Nucleic Acids Res.* **6**, 1607–1616.
Klug, A., Jack, A., Viswamitra, M. A., Kennard, O., Shakkad, Z., and Steitz, T. A. (1979). *J. Mol. Biol.* **131**, 669–680.
Kollman, P. A., Weiner, P. K., and Dearing, A. (1981). *Biopolymers* **20**, 2583–2681.
Kuhlmann, K. F., Grant, D. M., and Harris, R. K. (1970). *J. Chem. Phys.* **52**, 3439–3448.
Kumar, A., Wagner, G., Ernst, R. R., and Wüthrich, K. (1981). *J. Am. Chem. Soc.* **103**, 3654–3658.
Kuntz, I. D., Crippen, G. M., and Kollman, P. A. (1979). *Biopolymers* **18**, 939–957.
Lamb, H. (1945). "Hydrodynamics." Dover, New York.
Levitt, M. (1978). *Proc. Natl. Acad. Sci. U.S.A.* **75**, 640–644.
Levitt, M., and Warshel, A. (1978). *J. Am. Chem. Soc.* **100**, 2607–2613.
Levy, R. M., Karplus, M., and Wolynes, P. G. (1981). *J. Am. Chem. Soc.* **103**, 5998–6011.
Lipari, G., and Szabo, A. (1980). *Biophys. J.* **30**, 489–506.
Lipari, G., and Szabo, A. (1981a). *J. Chem. Phys.* **75**, 2971–2976.
Lipari, G., and Szabo, A. (1981b). *Biochemistry* **20**, 6250–6256.
London, R. E. (1978). *J. Am. Chem. Soc.* **100**, 2678–2685.
London, R. E. (1980). *In* "Magnetic Resonance in Biology" (J. S. Cohen, ed.), Vol. 1, pp. 1–69. Wiley (Interscience), New York.
London, R. E., and Avitabile, J. (1978). *J. Am. Chem. Soc.* **100**, 7159–7165.
McCammon, J. A., Gelin, B. R., and Karplus, M. (1977). *Nature (London)* **267**, 585–590.
Macura, S., and Ernst, R. R. (1980). *Mol. Phys.* **41**, 95–117.
Mariam, Y. H., and Wilson, W. D. (1979). *Biochem. Biophys. Res. Commun.* **88**, 861–866.

Marshall, A. G., Schmidt, P. G., and Sykes, B. D. (1972). *Biochemistry* **11**, 3875–3879.
Millar, D. P., Robbins, R. J., and Zewail, A. H. (1980). *Proc. Natl. Acad. Sci. U.S.A.* **77**, 5593–5597.
Nall, B. T., Rothwell, W. P., Waugh, J. S., and Rupprecht, A. (1981). *Biochemistry* **20**, 1881–1887.
Neumann, J. M., and Tran-Dinh, S. (1981). *Biopolymers* **20**, 89–109.
Niccolai, N., and Tiezzi, J. (1979). *J. Phys. Chem.* **83**, 3249–3256.
Noggle, J. H., and Schirmer, R. E. (1971). "The Nuclear Overhauser Effect." Academic Press, New York.
Norwick, A. S., and Berry, B. S. (1961). *IBM J. Res. Dev.* **5**, 297–305.
Olejniczak, E. T., Poulsen, F. M., and Dobson, C. M. (1981). *J. Am. Chem. Soc.* **103**, 6574–6580.
Olson, W. K. (1982). *Nucleic Acids Res.* **10**, 777–787.
Opella, S. J., Wise, W. B., and DiVerdi, J. A. (1981). *Biochemistry* **20**, 284–290.
Pohl, F. M., and Jovin, T. M. (1972). *J. Mol. Biol.* **67**, 375–396.
Prestegard, J. H., and Grant, D. M. (1978). *J. Am. Chem. Soc.* **100**, 4664–4669.
Record, M. T., Jr. (1975). *Biopolymers* **14**, 2137–2155.
Record, M. T., Jr., Woodbury, C. P., and Inman, R. B. (1975). *Biopolymers* **14**, 393–408.
Reid, B. R., and Hurd, R. E. (1977). *Acc. Chem. Res.* **10**, 396–402.
Ribeiro, A. A., King, R., Restivo, C., and Jardetzky, O. (1980). *J. Am. Chem. Soc.* **102**, 4040–4047.
Richarz, R., Nagayama, K., and Wüthrich, K. (1980). *Biochemistry* **19**, 5189–5198.
Robinson, B. H., Hurley, I., Scholes, C. P., and Lerman, L. S. (1979). *In* "Stereodynamics of Molecular Systems" (R. H. Sarma, ed.), p. 283. Pergamon, Oxford.
Scheafer, J. (1973). *Macromolecules* **6**, 881–888.
Schellman, J. A. (1974). *Biopolymers* **13**, 217–226.
Schurr, J. M. (1977). *CRC Crit. Rev. Biochem.* **4**, 371–431.
Shindo, H. (1980). *Biopolymers* **19**, 509–522.
Shindo, H., McGhee, J. D., and Cohen, J. S. (1980). *Biopolymers* **19**, 523–537.
Solomon, I. (1955). *Phys. Rev.* **99**, 559–565.
Spiess, H. W. (1978). *NMR: Basic Princ. Prog.* **15**, 55–214.
Stannard, B. S., and Felsenfeld, G. (1975). *Biopolymers* **14**, 229–313.
Terao, T., Matsui, S., and Akasaka, K. (1977). *J. Am. Chem. Soc.* **99**, 6136–6138.
Tritton, T. R., and Armitage, I. (1978). *Nucleic Acids Res.* **5**, 3855–3869.
Tropp, J. (1980). *J. Chem. Phys.* **72**, 6035–6043.
Wahl, P., Paoletti, J., and LePecq, J.-B. (1970). *Proc. Natl. Acad. Sci. U.S.A.* **65**, 417–421.
Wallach, D. (1967). *J. Chem. Phys.* **47**, 5258–5268.
Wang, A. H.-J., Quigley, E. J., Kolpak, F. J., Crawford, J. L., van Boom, J. H., van der Marel, G., and Rich, A. (1979). *Nature (London)* **282**, 680–686.
Wilson, W. D., and Jones, R. L. (1982). *Nucleic Acids Res.* **10**, 1399–1410.
Wilson, W. D., Keel, R. A., and Mariam, Y. H. (1981). *J. Am. Chem. Soc.* **103**, 6267–6269.
Wing, R., Drew, H., Takano, T., Broka, C., Tanaka, S., Itakura, K., and Dickerson, R. E. (1980). *Nature (London)* **287**, 755–758.
Wittebort, R. J., and Szabo, A. (1978). *J. Chem. Phys.* **69**, 1722–1736.
Woessner, D. E. (1962a). *J. Chem. Phys.* **36**, 1–4.
Woessner, D. E. (1962b). *J. Chem. Phys.* **37**, 647–654.
Wüthrich, K., Wider, G., Wagner, G., and Braun, W. (1982). *J. Mol. Biol.* **155**, 311–319.
Yamada, A., Akasaka, K., and Hatano, H. (1978). *Biopolymers* **17**, 749–757.

CHAPTER 13

# Solid-State Phosphorus-31 NMR: Theory and Applications to Nucleic Acids

*Heisaburo Shindo*
Tokyo College of Pharmacy
Tokyo, Japan

I. Introduction 401
II. Basic Concepts of Solid-State $^{31}$P NMR 402
   A. Cross-Polarization–Dipolar Decoupling 402
   B. Chemical-Shielding Tensor 403
III. Studies of Oriented DNA Fibers 408
   A. Effects of Hydration on DNA Conformation 409
   B. The A Form of DNA 410
   C. The B Form of DNA 414
IV. Studies of DNA Complexes 418
V. Concluding Remarks 421
   References 421

## I. Introduction

A wide variety of structures of DNA has been proposed on the basis of X-ray fiber diffraction patterns. Such fiber diffractions provide two major types of information — helical parameters and intensity distribution patterns — which are used as constraints for building detailed structured "models." New structures were occasionally proposed but there often appeared to be other possible structures infered from very similar X-ray diffraction patterns. Thus the detailed structure so determined is not necessarily a unique solution of X-ray diffraction data. It is, therefore, important to test the accuracy of these proposed models by methods other than X-ray diffraction; NMR can be a particularly powerful means for such a test.

Whereas the gross structure of DNA in highly hydrated DNA fibers and in solution has been designated the B conformation, the detailed structural features in aqueous solution and the effects of binding to proteins and membranes are not well characterized. We believe that NMR can fill a gap between the knowledge of DNA structure in solution and in the solid state.

We describe here how the conformation of the phosphodiester backbone of DNA can be deduced from analysis of the chemical-shielding anisotropy of the $^{31}$P nuclei and also show how the $^{31}$P-NMR spectral pattern is sensitive to the conformational state of DNA in the solid state.

## II. Basic Concepts of Solid-State $^{31}$P NMR

Nuclear magnetic resonances of a given sample in the solid state are usually very broad owing to the dominance of dipole–dipole interactions between nuclear spins. When these interactions can be removed by, for example, spin-decoupling techniques, the resultant spectrum exhibits chemical-shielding anisotropy that is directly related to the nature of the electronic state of the chemical bonds and hence to the structure of particular atomic groups. Because of the behavior of chemical-shielding anisotropy, this technique has been employed to investigate the structure and dynamics of individual atomic groups and the molecule as a whole (Mehring, 1976; Haeberlen, 1976).

### A. Cross-Polarization–Dipolar Decoupling

Although elimination of spin-dipolar interactions (especially with abundant protons in the sample) can be achieved by various means (Pines et al., 1973; also see Mehring, 1976), we briefly describe a double-resonance technique that is currently most used [i.e., cross-polarization–dipolar decoupling (Pines et al., 1973)]. This technique differs in two respects from ordinary proton decoupling (so-called scalar decoupling) in solution. First, the applied decoupling power is much higher by one order or more than that of scalar decoupling and, second, signal enhancement can be obtained by polarization through more abundant protons.

Consider two unlike spin systems: one is a diluted spin S system and the other is an abundant spin I system. As shown schematically in Fig. 1, a short pulse at the resonance frequency of spin I is applied to rotate the magnetization $M_I$ onto the $y$ axis, followed by a long, high-power pulse $H_{1I}$, which differs by 90° in phase from the first pulse. This procedure is called "spin locking" of spin I (Solomon, 1950). At the same time of the second pulse $H_{1I}$, a long radio-frequency pulse $H_{1S}$ is applied to the spin S system. Only if $\gamma_I H_{1I} = \gamma_S H_{1S}$ is satisfied (the requirement called the "Hartmann–Hahn condition" Hartmann and Hahn, 1962), does the polarization occur most efficiently in the S system through the I system. When polarization or the magnetization of the spin S system $M_S$ reaches a maximum for an appropri-

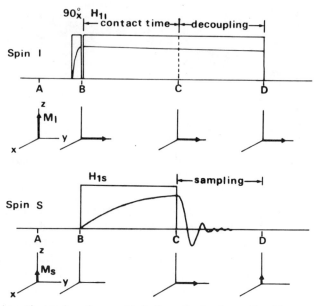

**Fig. 1.** Schematic drawing of cross-polarization–dipolar decoupling. The magnitudes and orientations of the magnetization are shown by the curves inside rectangular pulses and arrows at instant times A, B, C, and D in the rotating frame.

ate contact time, the $H_{1S}$ is turned off to observe a free induction decay (FID). During observation of the FID, $H_{1I}$ continues to be applied to remove dipolar coupling between spin I and spin S. In the case of $^{31}$P nuclei as spin S, a signal enhancement factor (about $\gamma_I/\gamma_S = 2.5$) may be achieved. Another great advantage of this method is that the observation of the FID can be repeated as fast as the proton relaxation time $T_1(H)$ but not the phosphorus relaxation time $T_1(P)$; here $T_1(H)$ is usually much shorter than $T_1(P)$ in the solid state. Thus, with use of cross-polarization–dipolar decoupling, signal enhancement will be several times as great as that by a simple double-resonance method (Pines *et al.*, 1973). The dipolar-decoupled spectrum is comprised mainly by chemical-shielding anisotropy and in part by homonuclear spin interactions. The latter term tends to be small if the spin S is dilute in the sample; this is usually the case for $^{31}$P nuclei in biological systems.

## B. Chemical-Shielding Tensor

The observed chemical shift is interpretable only if sufficient information is known about the magnitude and orientation of the chemical-shielding tensor in the molecular frame. The origin of the chemical-shielding anisot-

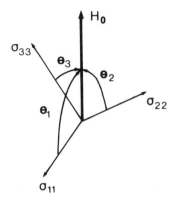

**Fig. 2.** Direction cosines of the principal axes of the shielding tensor with respect to the magnetic field direction.

ropy (CSA) is obvious: the nuclear spin interacts with the static magnetic field via surrounding electrons. In general, the chemical shieldings have contributions owing to the electron distribution in both ground and excited states (Pople et al., 1959). Quantitative predictions of chemical shielding, therefore, must involve molecular orbital calculations and are beyond the scope of this chapter. We outline here the treatment for oriented samples such as DNA fibers, general cases of which were thoroughly discussed for single crystals and randomly oriented samples by Haeberlen (1976) and Mehring (1976).

Suppose $A'$ is a second-rank tensor of chemical-shielding anisotropy in the laboratory frame $(x',y',z')$

$$A' = \begin{bmatrix} A'_{11} & A'_{12} & A'_{13} \\ A'_{21} & A'_{22} & A'_{23} \\ A'_{31} & A'_{32} & A'_{33} \end{bmatrix} \tag{1}$$

An appropriate choice of the axis system orthogonalizes the shielding tensor as

$$A = \begin{bmatrix} \sigma_{11} & 0 & 0 \\ 0 & \sigma_{22} & 0 \\ 0 & 0 & \sigma_{33} \end{bmatrix} \tag{2}$$

This coordinate system $(x,y,z)$ is called the principal-axis system (PAS) of the tensor. The chemical-shielding spectrum is given in the manner shown in Fig. 2, where the observed chemical shift $\sigma_{obs}$ is

$$\sigma_{obs} = \sigma_{11} \cos^2 \theta_1 + \sigma_{22} \cos^2 \theta_2 + \sigma_{33} \cos^2 \theta_3 \tag{3}$$

and

$$\cos^2 \theta_1 + \cos^2 \theta_2 + \cos^2 \theta_3 = 1$$

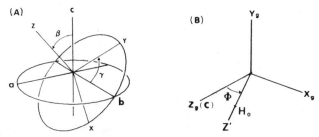

**Fig. 3.** Relationships between coordinate systems: (A) the PAS $(x,y,z)$, the crystal frame $(a,b,c)$; (B) the goniometer frame $(x_g,y_g,z_g)$ and the laboratory frame $(x',y',z')$.

In order to simulate the spectrum, we must know the orientation or the probability of orientation of the PAS relative to the laboratory frame (the magnetic field direction). If the transformation to the PAS of the tensor from the laboratory frame is known, all the elements of the tensor **A'** (i.e., all the direction cosines) can be calculated. In general, the transformation may be performed by three steps:

$$\text{Lab } (x',y',z') \xrightarrow{R(0,\Phi,0)} \text{Goniometer } (x_g,y_g,z_g) \xrightarrow{R(\alpha',\beta',\gamma')} \text{Crystal } (a,b,c) \xrightarrow{R(\alpha,\beta,\gamma)} \text{PAS } (x,y,z)$$

where $R(a,b,c)$ represents the Wigner rotation matrix. The Euler angles $\alpha, \beta,$ and $\gamma$ specify the orientation of the PAS with respect to the crystal axis system (Fig. 3A) and the angle $\Phi$ is the rotation angle about the goniometer $y_g$ axis. In case of oriented fibers, because identical goniometer and the crystal axis systems can be chosen (Fig. 3B), the transformation is effected in the manner (Shindo et al., 1980)

$$\underset{A'}{\text{Lab}} \xrightarrow{R(0,\Phi,0)} \underset{A}{\text{Goniometer}} \xrightarrow{R(\alpha,\beta,\gamma)} \underset{A}{\text{PAS}}$$

Thus the tensor **A** or **A'** is given by

$$A = R(\alpha,\beta,\gamma)R(0,\Phi,0)A'R(0,\Phi,0)^{-1}R(\alpha,\beta,\gamma)^{-1} \qquad (4)$$

or

$$A' = R(0,\Phi,0)^{-1}R(\alpha,\beta,\gamma)^{-1}AR(\alpha,\beta,\gamma)R(0,\Phi,0) \qquad (5)$$

Because the magnetic field is applied along the $z'$ axis (Fig. 3B), the observed chemical shift $\sigma_{\text{obsd}}$ is given by the $z'$ component of the tensor **A'**,

$$\sigma_{\text{obs}} = \tfrac{1}{2}(A_{11} + A_{22}) + \tfrac{1}{2}(A_{11} - A_{22})\cos(2\Phi) + \tfrac{1}{2}(A_{12} + A_{21})\sin(2\Phi) \qquad (6)$$

where the $A_{ij}$ are the elements of the tensor **A** in the crystal (or goniometer) axis system. Formulas equivalent to Eq. (6) have been derived in a more elegant way (Nall et al., 1981). Clearly, $\sigma_{\text{obs}}$ is a function of angles $\alpha, \beta, \gamma,$ and $\Phi$.

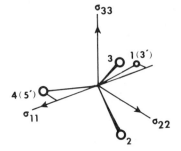

**Fig. 4.** Orientation of the PAS of the shielding tensor relative to the atomic coordinates of phosphodiester. Phosphorus atom is at the origin. The principal axis of $\sigma_{22}$ is a bisector of the plane O(2) PO(3).

Consider the phosphorus atoms in highly oriented DNA fiber. It is reasonable to assume that the helical axis of DNA may be identical to the fiber axis. Assuming that the orientation of all the phophodiesters are equivalent with respect to the helical axis and hence the fiber axis, the Euler angles $\beta$ and $\gamma$ are fixed, whereas the angle $\alpha$ is randomly distributed. Thus the spectral intensity at a rotation angle $\Phi$ of the goniometer is given by the derivative

$$I(\sigma) = \frac{d\alpha}{d|\sigma_{obs}(\alpha)|} \qquad (7)$$

where $\alpha$ varies from 0 to 360°. The observed chemical-shielding spectrum may be broadened owing to residual dipolar interactions and field inhomogeneity. Therefore, the calculated spectrum may be convoluted by a Gaussian distribution function

$$H(\sigma) = \int_{-\infty}^{\infty} I(\sigma)G(\sigma - \sigma')d\sigma' \qquad (8)$$

If the orientation of the PAS relative to the crystal frame is known, that is, Euler angles $\beta$ and $\gamma$ are known, we can easily calculate the spectrum at any rotation angle of $\Phi$.

The detailed orientation of the PAS of the $^{31}$P shielding tensor may be different for the phosphodiester with different ester groups, but the phosphodiester in the polymer with nearly identical repeating units, such as DNA, appears to have the same orientation relative to the atomic coordinates. Kohler and Klein (1976) evaluated its orientation from $^{31}$P-NMR studied, depicted in Fig. 4. This result was nearly in agreement with that found for a single crystal of barium diethyl phosphate by Herzfeld et al. (1978).

Figure 5 shows typical examples of the spectral patterns calculated for the fiber parallel and perpendicular to the magnetic field, that is, $\Phi = 0°$ and $\Phi = 90°$, respectively. As predicted, the parallel spectrum exhibits a singlet whereas the perpendicular spectrum displays a bimodal pattern.

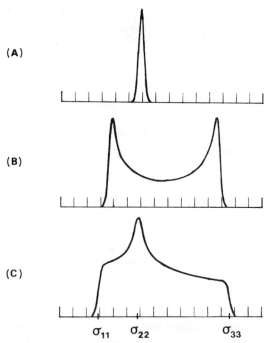

**Fig. 5.** Representative spectral patterns of chemical-shielding anisotropy. Spectra A (parallel) and B (perpendicular) were calculated for oriented DNA fibers, assuming Euler angles of 77 and 79° for $\beta$ and $\gamma$, respectively, that correspond to the A-DNA model of Arnott and Hukins (1972). Spectrum C is a typical powder pattern of asymmetric shielding anisotropy.

This procedure was carried out by assuming a hypothetical perfect orientation of DNA molecules along the fiber axis. In order to take into account dispersion in their orientation about the fiber axis, we assume a Gaussian distribution such that DNA molecules most probably align along the fiber axis with a standard deviation $\langle\theta\rangle$ from the axis. Each spectrum at a given $\theta$ are weighted by the Gaussian function

$$G(\theta) = \frac{1}{(2\pi)^{1/2}\langle\theta\rangle} \exp\left[\frac{-\theta^2}{2}\langle\theta\rangle^2\right] \qquad (9)$$

and all spectra must be added. Such a deviation in alignment of DNA molecules may be one of the main factors causing line broadening.

When $\theta$ is randomly distributed, that is, the sample is polycrystalline or powder, a so-called powder pattern will be observed as shown in Fig. 5C. The magnitudes of the principal values of the tensor $\sigma_{11}$, $\sigma_{22}$, and $\sigma_{33}$ can be obtained directly from the powder pattern. For the phosphodiester of various DNA and RNA the principal value of the $^{31}$P chemical-shielding tensor is almost identical and independent of base difference (Terao *et al.,*

1977). The principal value was reported to be in the range of 80 to 85, 18 to 30, and $-100$ to $-110$ ppm for $\sigma_{11}$, $\sigma_{22}$, and $\sigma_{33}$, respectively (Terao et al., 1977; Opella et al., 1981; Nall et al., 1981).

## III. Studies of Oriented DNA Fibers

Three well-defined types of natural DNA can be prepared relatively easily by adjusting the salt content and relative humidity in oriented fibers. As shown in Fig. 6, for example, the fibers give distinct X-ray diffraction patterns that are designated as the A, B, and so on. Using X-ray diffraction data as constraints, the atomic coordinates of each structure were determined by model building (Langridge et al., 1960; Marvin et al., 1961; Fuller et al., 1965), and from time to time they were refined further by other structural information such as bond angles and lengths and by bond-energy calculations (Arnott and Hukins, 1972; Levitt, 1978).

Only a few methods other than X-ray fiber diffraction were attempted to study the structure of DNA in the solid state; infrared was one (for example, see Pilet and Brahms, 1972). However, thin films of DNA were used for these studies instead of the fiber samples. On the other hand, $^{31}$P NMR can be measured for the fiber sample, and thus a parallel study of NMR and X-ray fiber diffraction methods can be carried out (Shindo et al., 1980, 1981; Shindo and Zimmerman, 1980; Nall et al., 1981). However, $^{31}$P NMR can only probe phosphodiester groups and requires larger samples than the

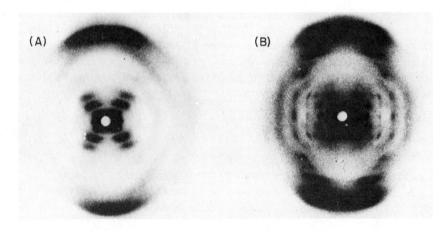

**Fig. 6.** Typical X-ray fiber diffraction images of DNA: (A) for the B form of sodium DNA at 98% R.H.; (B) for the A form of DNA at 79% R.H.

## 13. Solid-State ³¹P NMR and Nucleic Acids

X-ray method. A detailed experimental method on how to mount DNA fibers in the NMR probe and how to control the relative humidity may be obtained from the original papers (Shindo *et al.*, 1980; Nall *et al.*, 1981).

Two representative structures of DNA, namely, the A and B forms of DNA, are now discussed.

### A. Effects of Hydration on DNA Conformation

The conformational changes from the A to B form is known to occur with increasing relative humidity (Cooper and Hamilton, 1966; Arnott and Selsing, 1974). The transition occurs in the range 80–90% R.H.: the A form is stable below and the B form is stable above the range. Figure 7 shows the solid-state ³¹P-NMR spectra of DNA fibers at various humidities (H. Shindo and H. Akutsu, unpublished), where the parallel spectra refer to the spectrum that is obtained from the fibers oriented parallel to the magnetic field and the perpendicular spectra for the fibers oriented normal to the field. At low relative humidities, the parallel spectrum exhibits a singlet whereas the

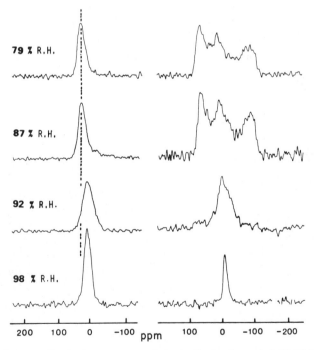

**Fig. 7.** Influence of hydration on cross-polarized–dipolar-decoupled ³¹P-NMR spectra of DNA fibers. The parallel spectra are shown on the left and the perpendicular on the right.

perpendicular displays a trimodal pattern unlike the bimodal as expected theoretically. Nonetheless, it is clear from the perpendicular spectra that the transition occurs in a very narrow humidity range, between 87 and 92% where the trimodal pattern collapses into a singlet with increasing relative humidity. On the other hand, the parallel spectrum exhibits a singlet in the entire range of relative humidity but shifts slightly upfield during the transition. Strikingly, the linewidth broadens above the transition and then becomes narrower at higher humidities. As discussed in the previous sections, chemical-shift data provide structural information, and linewidth data also provide information about the distribution of phosphodiester orientations, including molecular dynamics (see Hart, Chapter 11, and James, Chapter 12).

## B. The A Form of DNA

The A form of sodium DNA (NaDNA) gives a typical crystalline fiber pattern in the X-ray diffraction image (Fig. 6B) that is characterized by a pitch of 28 Å and 11 residues per helical turn (Fuller *et al.*, 1965). If the backbone conformation is regular as is expected from the double-helical structure of Watson–Crick base pairs (Watson and Crick, 1953), the spectra as shown in Fig. 5 should be seen, depending on the fiber orientation relative to the magnetic field direction. The observed spectra are as simple as expected, but many factors contribute to the line shape of the $^{31}$P-NMR spectra from DNA fibers.

### 1. Spectral Simulation

Shindo *et al.* (1980) attempted to investigate the $^{31}$P NMR of oriented DNA fibers but failed to interpret the spectra of the A form of DNA. Subsequently, Shindo *et al.* (1981) and Nall *et al.* (1981) determined the phosphodiester orientation in the A form of DNA relative to the helical axis. In Fig. 8 are the spectra obtained by Nall *et al.* (1981), showing the strong dependence of the $^{31}$P-NMR spectrum of the A form of NaDNA on the rotation angle between the fiber axis and the magnetic field. Calculation of the line shapes can be carried out easily by using Eq. (6), but the following assumptions must be made: (1) there is no motional averaging of $^{31}$P chemical shift; (2) the orientation of the PAS of the shielding tensor is known relative to the crystal axis system; (3) the backbone conformation of DNA is uniform irrespective of base sequence; (4) imperfect orientation of DNA molecules exists about the fiber axis; (5) amorphous regions cannot be neglected in connection with the crystallinity of the DNA sample; and (6)

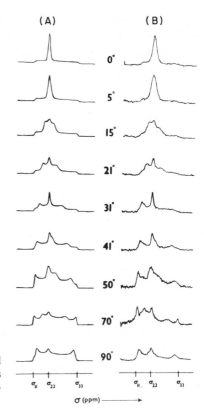

**Fig. 8.** Comparison of the calculated (A) and the observed (B) spectra at various rotation angles of the goniometer. From Nall et al. (1981). Copyright 1981 American Chemical Society.

the intrinsic linewidth for fibrous A-form DNA is ~ 600 Hz. For assumption (1), the abnormal spectra of A-form NaDNA measured at 24.3 MHz were interpreted in terms of dispersion of phosphodiester orientation and reorientational motion about the helical axis (Shindo et al., 1980). However, the latter effect was excluded because the phosphodiester backbone was found to be virtually rigid (Shindo et al., 1981; Nall et al., 1981). Assumption (2) was already discussed in Section II,B and we assumed the coincidence of the molecular symmetry axes and the chemical-shielding tensor (Kohler and Klein, 1976). Assumption (4) is taken into account by deviation of DNA alignment in the crystalline parts in the sample. A Gaussian distribution was assumed. Assumption (5) is responsible for the intervening area between the crystalline regions; that is, the region contains amorphous material that is randomly oriented. The amorphous part gives a powder pattern that must be superimposed on the spectrum from the crystalline parts in the sample. Assumption (6) concerns line broadening due to dipole–dipole interactions. It is granted that $^{31}P-^{1}H$ interaction almost vanishes by application of dipolar decoupling. The magnitude of $^{31}P-^{31}P$ dipolar interaction was estimated to be ~250 Hz, corresponding to a distance of 5.6 Å between nearest-neighbor phosphorus atoms (Nall et al., 1981). Opella et al. (1981) experimentally determined it to be ~400 Hz for DNA in solution. The distance between $^{31}P$ and $^{23}Na$ counterions was estimated to be 3.5 Å, corresponding to the linewidth of ~890 Hz owing to $^{31}P-^{23}Na$ dipolar interaction (Nall et al., 1981). Thus the total dipolar contribution to linewidth could be 1300 Hz. Because dipolar interaction is independent of resonance frequency, the higher the magnetic field used, the more resolved the spectrum should become. Considering these factors, the calculation of simulated spectra can be performed by using Eqs. (6)–(9).

## 2. Orientation of the $^{31}P$ Shielding Tensor Relative to the Helical Axis

From the best fit of the simulated spectrum to the observed, Euler angles $\beta$ and $\gamma$ for the PAS of the tensor were obtained, and these are readily converted to the direction cosines $\theta_1$, $\theta_2$, and $\theta_3$. Table I lists the direction cosine angles of the PAS relative to the helical axis determined from the $^{31}P$-NMR spectra of the A form of DNA (Nall et al., 1981), together with the A form of poly(dAdT)·poly(dAdT) (Shindo et al., 1981). The results obtained for different specimens are in good agreement within ±5° of each other, suggesting that the A form of DNA exhibits no base-sequence dependence of the backbone structure, or little if any. If the detailed structure of DNA is known, the orientation of the tensor can easily be evaluated with respect to the helical axis, assuming the relative orientation of the PAS and

**TABLE I**

Orientation of the Shielding Tensor of the A Form of DNA[a]

| DNA | $\beta$ | $\gamma$ | $\theta_1$ | $\theta_2$ | $\theta_3$ | Reference |
|---|---|---|---|---|---|---|
| Experimental | | | | | | |
| Poly(dAdT)·poly(dAdT)[b] | 70 | 52 | 55 | 138 | 110 | Shindo et al. (1981) |
| Salmon sperm NaDNA[b] | 71 | 59 | 61 | 144 | 109 | H. Shindo and H. Akutsu (unpublished) |
| Calf thymus NaDNA[c] | 75 | 63 | 64 | 149 | 105 | Nall et al. (1981) |
| Calculated[d] | | | | | | |
| A-DNA model | 77 | 79 | 79 | 163 | 103 | Arnott and Hukins (1972) |

[a] Angles $\theta_1$, $\theta_2$, and $\theta_3$ and Euler angles $\beta$ and $\gamma$ in degrees describe the tensor orientation relative to the helical axis (see Figs. 2 and 3).
[b] Obtained by spectral simulation by using the principal values $\sigma_{11} = 82$, $\sigma_{22} = 22$, and $\sigma_{33} = -113$ ppm.
[c] Obtained by spectral simulation by using $\sigma_{11} = 83$, $\sigma_{22} = 18$, and $\sigma_{33} = -104$ ppm.
[d] Assumed the orientation of the tensor having a symmetry relative to the atomic coordinates of the phosphodiester.

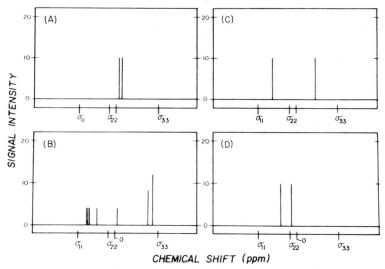

**Fig. 9.** Predicted $^{31}$P-NMR spectra of oriented (A) Z-DNA (Wang *et al.*, 1979); (B) "side-by-side model" (Rodley *et al.*, 1976); (C) S-DNA (Arnott *et al.*, 1980); and (D) "alternating B-DNA" (Klug *et al.*, 1979) for the case where the fiber is oriented parallel to the magnetic field. From Nall *et al.* (1981). Copyright 1981 American Chemical Society.

the phosphodiester backbone (see Fig. 4). This was determined from the structure of A-form DNA most commonly used at present (Arnott and Hukins, 1972), and its Euler angles and direction consines are presented in Table I. The calculated values are not in good agreement with the experimental, but the later refined structure of the A-DNA model (S. Arnott and R. Chandrasekaran, unpublished) gave better agreement (Nall *et al.*, 1981).

It is best to end this section by presenting Fig. 9, showing the great ability of solid-state NMR to distinguish between structural models of DNA. Here, the DNA fibers are aligned along the direction of the magnetic field, and all line-broadening effects are neglected. It is clear that these structures may be distinguished at a glance on the basis of one such NMR spectrum.

### 3. Regularity of the Phosphodiester Backbone

As shown in Fig. 8, the excellent agreement between the simulated and observed spectra demonstrates that the A form of DNA has a single, regular conformation of the phosphodiester backbone. To be certain of this point, the linewidth should be interpreted. Consider the frequency dependence of the linewidth of the parallel spectra of A-DNA. The linewidths were 30 ppm (1208 Hz) at 40.3 MHz (see Fig. 7) and 21 ppm (2041 Hz) at 97.2 MHz (Fig. 8; Nall *et al.*, 1981). Assuming the properties of the fibers are the same

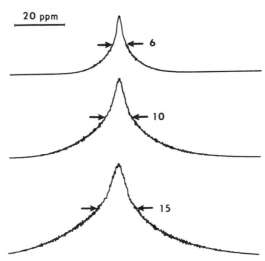

**Fig. 10.** Line broadening caused by a dispersion in DNA alignments along the fiber axis. Standard deviation from the fiber axis $\langle \theta \rangle$, is 4, 8, and 12° from the top spectrum. The numbers are half-height widths. From Shindo *et al.* (1980). Copyright 1980 American Chemical Society.

for the two experiments, a component of the linewidth in hertz that is proportional to the magnetic field strength is considered to be the same. Assuming 600 Hz as the total contribution of the dipolar interactions in Hertz that are independent of the applied field, subtraction of this contribution from the observed linewidth results in 608 Hz (15 ppm) at 40.3 MHz and 1441 Hz (14.8 ppm) at 97.2 MHz. Note that the resultant linewidth is nearly the same at two different fields when measured in parts per million. These values are attributed to the contribution of a dispersion of the phosphodiester orientation, inhomogeneity of the magnetic field, and application of an exponential window to the FID. Thus the contribution associated with the phosphodiester orientation would be ~10 ppm in width, assuming other miscellaneous factors of ~5 ppm. The magnitude of 10 ppm can be predicted in terms of a standard deviation of $\langle \theta \rangle = 8°$ for imperfect alignment of DNA molecules along the fiber axis, as indicated in Fig. 10. These arguments make it clear that the phosphodiester backbone in the A form of DNA has regularity with respect to the helical axis.

## C. The B Form of DNA

The B form of DNA usually occurs in NaDNA fibers at high relative humidity and also in lithium salt at medium and high relative humidities (Langridge *et al.*, 1960). Investigation of the structure and dynamics of highly hydrated DNA is especially important because the characteristics

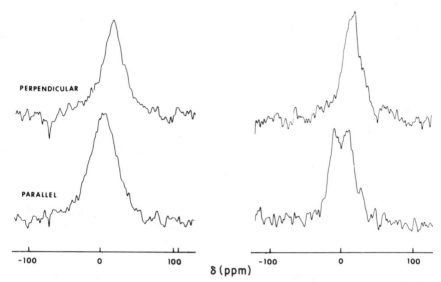

**Fig. 11.** Cross-polarization–dipolar-decoupled ³¹P-NMR spectra (24.7 MHz) at 25°C from poly(dAdT)·poly(dAdT) fibers (right) and salmon sperm DNA fiber (left) at 98% relative humidity. From Shindo and Zimmerman (1980), by permission from *Nature (London)* **283**, 691. Copyright 1980 Macmillan Journals Ltd.

observed for such DNA fibers may be extended to DNA in solution. There are many lines of evidence that synthetic DNA with a particular base sequence does not form some of the familiar conformations occurring in DNA but rather forms a unique conformation (Leslie *et al.,* 1980). Such structural versatility of DNA may be important in conjunction with DNA recognition by a particular protein.

In Fig. 11, the ³¹P NMR spectra at 24.3 MHz from natural DNA and poly(dAdT)·poly(dAdT) fibers at two fiber orientations relative to the magnetic field are compared. Strikingly, poly(dAdT)·poly(dAdT) fibers when oriented parallel to the magnetic field exhibit two well-defined resonances, whereas natural DNA exhibits a single, broad resonance under the same conditions. For the singlet of the perpendicular spectra, the bimodal pattern observed at low relative humidity is collapsed into a singlet at high humidity because the increased reorientational motion about the helical axis is brought about by hydration of the DNA fibers. Surprisingly, the perpendicular spectrum linewidth is narrower than the parallel.

*1. Analysis of the Linewidth*

We encounter two questions about the linewidth of the ³¹P-NMR spectra for the B form of DNA: (1) What causes the parallel spectrum broadening (40 ppm at 92% R.H. and 28 ppm at 98% R.H.), and (2) why is the

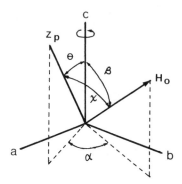

Fig. 12. Illustration of the principal axis $z_p$ of the shielding tensor and the direction of the magnetic field $H_0$ relative to the molecular frame $(a,b,c)$. The axis $c$ is coincident with the helical axis of DNA, and a rapid reorientational motion takes place about this axis.

perpendicular spectrum narrower than the parallel by ~ 10 ppm? For the first question, one might imagine that the orientation of DNA decreases, because of either lack of truly parallel alignment among the fibers or a dispersion of orientation of DNA molecules or both. Both types of disorder could artifactually broaden the linewidth. The first effect may be trivial; the second one was discussed in the previous section and it was estimated to be ~ 10 ppm, which corresponds to a 8° standard deviation. The transverse relaxation time $T_2$ was measured to be ~ 1.7 ms, corresponding to ~ 200 Hz (~ 5 ppm at 40.3 MHz) for the parallel spectrum (H. Shindo and H. Akutsu, unpublished result). This result indicates that the contribution of static dipolar interaction to the linewidth is very small in the B form of DNA; this situation contrasts with the A form of DNA. This is probably due to molecular motions averaging out such static interactions. By subtracting those contributions attributable to the imperfect orientation of DNA molecules (10 ppm as in A-DNA) and miscellaneous contributions (~ 5 ppm) from the total linewidth, we have an extra contribution of 10-25 ppm in width for the parallel spectrum.

These arguments lead to the conclusion that the variation in the orientation of the phosphodiester groups of DNA in the fiber is the predominant factor in the line broadening. In fact, it is easy to show that rotation of a phosphodiester group about an arbitrary axis by 20° can cause a change in chemical shift by as much as 30 ppm. This working hypothesis may help to explain the second question.

We briefly discuss the effect of variation of the phosphodiester orientation on spectral linewidth. Figure 12 shows the crystal axis system $(a,b,c)$ with the principal axis $z_p$ of the tensor and the direction of the magnetic field, all of which are related by angles, $\alpha$, $\beta$, $\theta$, and $\chi$. According to Eq. (3), the observable chemical shift can be given by

$$\sigma_{obs} = \sigma_{11} \cos^2 \chi_1 + \sigma_{22} \cos^2 \chi_2 + \sigma_{33} \cos^2 \chi_3$$

where $\cos \chi$ represents the direction cones of the tensor axes relative to the laboratory frame $H_0$. For simplicity, assuming axial symmetry of the tensor (i.e., $\sigma_\perp = \sigma_{11} = \sigma_{22}$ and $\sigma_\parallel = \sigma_{33}$), then we have

$$\sigma_{obs} = (\sigma_\parallel - \sigma_\perp)\cos \chi + \sigma_\perp \qquad (10)$$

where a subscript of $\chi$ is dropped out to match notation in Fig. 12. The angle $\chi$ can easily be related with $\theta$, $\beta$, and $\alpha$:

$$\cos \chi = \cos \beta \cos \theta + \sin \beta \sin \theta \cos \alpha \qquad (11)$$

When the DNA molecule rapidly rotates about the helical axis $c$, an average of $\sigma_{obs}$ in Eq. (11) must be taken over angle $\alpha$. After rearrangement, we have

$$\bar{\sigma}_{obs} = \frac{\sigma_\parallel - \sigma_\perp}{6}(3\cos^2 \beta - 1)(3\cos^2 \theta - 1) + \sigma_i \qquad (12)$$

Considering a distribution of the tensor orientation over $\theta \pm \Delta\theta$ relative to the helical axis, line broadening because of this distribution becomes

$$\Delta\bar{\sigma}_c = \frac{\sigma_\parallel - \sigma_\perp}{2}|\sin 2\theta \sin 2\Delta\theta| \qquad (13)$$

for the DNA fibers parallel to the magnetic field (i.e., $\beta = 0°$), and

$$\Delta\bar{\sigma}_a = \frac{\sigma_\parallel - \sigma_\perp}{4}|\sin 2\theta \sin 2\Delta\theta| \qquad (14)$$

for the perpendicular (i.e., $\beta = 90°$). Equations (13) and (14) indicate that the line broadening for the parallel spectrum is twice as wide as that for the perpendicular. For a detailed example, the angle $\theta = 119°$ was calculated from the B-DNA model of Langridge et al. (1960), and using the value $\sigma_\parallel - \sigma_\perp = 150$ ppm for the phosphodiester and assuming $\Delta\theta = 10°$, the line broadening is calculated to be 26 and 13 ppm for the parallel and the perpendicular spectra, respectively. This argument explains why the linewidth of the parallel spectrum is broader than that of the perpendicular. It can be concluded that the $^{31}$P-NMR spectra from B-DNA fibers are interpreted consistently in terms of the presence of significant variations in the backbone conformation.

## 2. Base-Sequence Dependence of the Backbone Conformation

As was seen in Section III,C,1 the broad $^{31}$P-NMR lines of the B form of natural DNA suggested that the phosphodiesters have a considerable dispersion in their orientation relative to the helical axis. If such a variation of the backbone conformation was induced by difference in base sequence, synthetic polynucleotide with a known sequence would emphasize a unique structure reflected in the spectral pattern. In fact, poly(dAdT)·poly(dAdT) fibers at 98% R.H. exhibited two well-defined peaks for the parallel spectrum as shown in Fig. 11. This doublet pattern clearly indicates that two distinct orientations of the phosphodiester are present, to which two sequences, ApT and TpA, can be assigned on the basis of the alternating sequence of this polymer. Klug *et al.* (1979) previously proposed "an alternating B form of DNA" model on the basis of X-ray diffraction and DNase digestion patterns of poly(dAdT)·poly(dAdT). Using atomic coordinates of their model (Klug *et al.*, 1979), the chemical-shift values for the fibers parallel to the magnetic field were calculated to be 28.8 and $-1.2$ ppm (cf. Fig. 9D). These values are in fairly good agreement with the observed values (20 and $-1$ ppm, respectively). It is likely that the B form of DNA has structural characteristics induced by base sequence. This contrasts with the observation that the A form of DNA has a single conformation, irrespective of base sequence. Furthermore, such characteristics are also observed for various synthetic polynucleotides in solution (see Chen and Cohen, Chapter 8). Therefore, we can state that irregularity of the backbone conformation is ubiquitious for natural DNA, both in highly hydrated fibers and in solution.

X-Ray diffraction studies of a single crystal of the dodecamer d(CGCGAATTCGCG)$_2$ have offered unambiguous knowledge of the nature of DNA structure (Wing *et al.*, 1980); its overall structure is more like the canonical model for the B form of DNA, but there are significant variations in the local structure. For example, the 11 local helical pitches corresponds to 9.4, 9.6, 9.1, 10.8, 9.6, 11.2, 10.0, 8.7, 11.1, 8.0, and 9.7 residues per turn (Drew *et al.*, 1981), although their average of 9.75 residues per turn and a rise of 3.33 Å per residue are within a few percent of those inferred from the fiber diffraction of B-DNA (Langridge *et al.*, 1960). Another local variation is seen in the conformations that span almost the whole range possible for the sugar moiety. These results certainly reveal the striking static heterogeniety of B-DNA.

## IV. Studies of DNA Complexes

The state of DNA in living things is very complicated indeed: tremendous condensation of DNA in the cell, packaging, and superhelicity of DNA within viruses may be induced by interactions of DNA with proteins and certain binding substances. Such a kaleidoscopic transformation of DNA is clearly important for its function and may be explained by the conformational flexibility of DNA. For such complex systems *in vivo* and *in vitro*, conventional high-resolution NMR studies are problematical because of the

# 13. Solid-State $^{31}$P NMR and Nucleic Acids

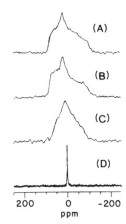

**Fig. 13.** $^{31}$P-NMR spectra of fd and fd DNA. (A) Stationary powder of fd at 0% R.H. (B) Stationary powder of fd at 92% R.H. (C) Solution of fd. (D) Isolated fd DNA in solution. (See also James, Chapter 12, Fig. 10.) From DiVerdi and Opella (1981a). Copyright 1981 American Chemical Society.

very broad linewidth resulting from the slow motion of DNA. On the other hand, solid-state NMR techniques may be appropriate for obtaining reliable and interpretable data from, for example, viruses (Akutsu *et al.*, 1980; DiVerdi and Opella, 1981a), chromatins (DiVerdi *et al.*, 1981), and high-molecular-weight DNA (Opella *et al.*, 1981; DiVerdi and Opella, 1981b).

Using solid-state $^{31}$P NMR at 61 MHz, DiVerdi and Opella (1981a) studied a filamentous virus (fd) known to be a protein–DNA complex with no associated membrane components. The particle weight is $16.4 \times 10^6$ daltons, and circular, single-stranded DNA is packed inside the coat-protein shell. Figure 13 shows the powder spectral patterns of the $^{31}$P chemical shielding for fd DNA under various situations. Changes in shape of the powder pattern can be described in terms of molecular motions. The magnitude of the $^{31}$P chemical-shielding interaction is $\sim 10^4$ Hz at 61 MHz (i.e., $\Delta\sigma = \sigma_{11} - \sigma_{33} = \sim 200$ ppm). The motions that occur comparable to $10^4$ s$^{-1}$ will strongly influence the line shape of the powder pattern and motion much faster than $10^4$ s$^{-1}$ should average out the shielding interaction to a single line.

The $^{31}$P NMR spectrum of solid fd in Fig. 13 exhibits the full chemical-shielding anisotropy of the phosphodiester. Therefore, no significant motions are present in the phosphodiester linkages of lyophilized and frozen solutions of fd. Figure 13B displays the spectrum from fd equilibrated in an atmosphere of 92% R.H. although B-DNA fibers at 92% R.H. were previously shown to exhibit significant motion (Fig. 7), the magnitudes of the principal elements of the shielding tensor of the DNA are almost identical between hydrated and completely dehydrated fd. This result rules out greater molecular motion of DNA in hydrated fd than previously suggested. Figure 13C,D shows the $^{31}$P spectra of fd in solution and the solution

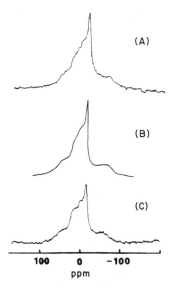

Fig. 14. $^{31}$P-NMR spectra of PM2 virus and nucleocapsid. (A) Bacteriophage PM2. (B) Calculated spectrum. (C) Nucleocapsid. From Akutsu et al. (1980). Copyright 1980 American Chemical Society.

spectrum of isolated, protein-free DNA. The narrow linewidth of protein-free DNA indicates a rapid isotropic motion of the phosphodiester groups. Furthermore, as was expected, its linewidth is much narrower than that of calf thymus DNA in solution, suggesting that fd DNA is single-stranded DNA. The linewidth of fd in solution, however, retains static chemical-shielding anisotropy that is not averaged by motion. These and other CPMAS experiments show no evidence of structural change in the phosphodiester backbone induced by the coat-protein shell and also demonstrate that the DNA in the virion has a limited amount of motion faster than $\sim 10^4$ $s^{-1}$, indicating that the DNA is immobilized by being packaged inside the virus particles.

Akutsu et al. (1980) have reported $^{31}$P-NMR studies of lipid-containing viruses (e.g., bacteriophage PM2) that are spherical in shape with a hydrated diameter of 600 Å and possessing a lipid bilayer. The virus contains only four proteins, namely, proteins I, II, III, and IV. The representative $^{31}$P-NMR spectra are shown in Fig. 14, where a 60% sucrose solvent was used to eliminate the influence of the overall rotational motion of the virus on the spectrum. It is clear from the spectra A and C that there are two major components: an axial symmetric powder pattern superimposed on the broad component. Akutsu and co-workers assigned these two components to a liquid-crystalline bilayer and the DNA inside the virus. Simulated spectrum B based on the spectra of extracted lipids and T4 phage, which is known to have no lipid membrane, is in good agreement with the observed spectrum A or C. In the presence of 4–6 $M$ urea, the nucleocapsid was isolated and exhibited spectrum C, which reveals no significant difference from that of intact virus (spectrum A). This observation means that the nucleocapid contains the lipid bilayer; this result argues against the structure with no lipid bilayer that was previously proposed.

## V. Concluding Remarks

A distinct advantage of solid-state NMR over solution NMR is the fact that solid material is more or less anisotropic in nature and hence changes in structure and molecular motion are directly visualized as changes in the spectral pattern. On the other hand, in solution NMR such an anisotropy is averaged out in the spectrum, and it can only be deduced through an exhaustive analysis of NMR relaxation mechanisms. In that sense it can be said that solid-state NMR is, in principle, more informative than solution NMR. Yet a combination of both, of course, leads to a much better understanding of the structure and dynamics of biological substances because the majority of these naturally occur in solution or in solution-like states.

We saw heretofore that solid-state $^{31}$P-NMR provides us with a powerful means to investigate the properties of nucleic acids, both in solution and in the solid state. Nucleic acids have essentially the same kind of phosphorus atoms, namely, phosphodiester groups; this makes it simple to analyze $^{31}$P-NMR spectra of DNA. However, in general cases, exact analyses of the spectral line shape and relaxation parameters are still problematical because of strong anisotropic properties and the complexity of biological materials. Nevertheless, this is becoming of increasing interest for further investigation of such anisotropy because it must contain fruitful information on molecular motions and its structure.

## Acknowlegement

The author thanks Dr. H. Akutsu for critical reading of this manuscript and also Mr. T. Fujiwara for calculating the spectral patterns in Fig. 5.

## References

Akutsu, H., Satake, H., and Franklin, R. M. (1980). *Biochemistry* **19**, 5264–5270.
Arnott, S., and Hukins, D. W. L. (1972). *Biochem. Biophys. Res. Commun.* **47**, 1504–1509.
Arnott, S., and Selsing, E. (1974). *J. Mol. Biol.* **88**, 551–552.

Arnott, S., Chandrasekaran, R. Birdsall, D. L. Leslie, A. G. W., and Ratliff, R. L. (1980). *Nature (London)* **283**, 743–745.
Cooper, P. J., and Hamilton, L. D. (1966). *J. Mol. Biol.* **16**, 562–563.
DiVerdi, J. A., and Opella, S. J. (1981a). *Biochemistry* **20**, 280–284.
DiVerdi, J. A., and Opella, S. J. (1981b). *J. Mol. Biol.* **149**, 307–311.
DiVerdi, J. A., Opella, S. J., Ma, R.-I., Kallenback, N. R., and Seeman, N. C. (1981). *Biochem. Biophys. Res. Commun.* **102**, 885–890.
Drew, H. R., Wing, R. M., Takano, T., Brake, C., Tanaka, S., Itakura, I., and Dickerson, R. E. (1981). *Proc. Natl. Acad. Sci. U.S.A.* **78**, 2179–2183.
Fuller, W., Wilkins, M. H. F., Wilson, H. R., and Hamilton, L. D. (1965). *J. Mol. Biol.* **12**, 60–76.
Haeberlen, U. (1976). *Adv. Magn. Reson.*, Suppl. **1**, 17–35.
Hartmann, S. R., and Hahn, E. L. (1962). *Phys. Rev.* **128**, 2042–2045.
Herzfeld, J., Griffin, R. G., and Harberkorn, R. A. (1978). *Biochemistry* **17**, 2711–2718.
Klug, A., Jack, A., Viswamitra, M. A., Kennard, O., Shakked, Z., and Steitz, T. A. (1979). *J. Mol. Biol.* **131**, 669–680.
Kohler, S. J., and Klein, M. P. (1976). *Biochemistry* **15**, 967–973.
Langridge, R., Wilson, H. R., Hooper, C. W., Wilkins, M. H. F., and Hamilton, L. D. (1960). *J. Mol. Biol.* **2**, 19–37.
Leslie, A. G. W., Arnott, S., Chandrasekaran, R., and Ratliff, R. L. (1980). *J. Am. Chem. Soc.* **143**, 49–72.
Levitt, M. (1978). *Proc. Natl. Acad. Sci. U.S.A.* **75**, 640–644.
Marvin, D. A., Spencer, M., Wilkins, M. H. F., and Hamilton, L. D. (1961). *J. Mol. Biol.* **3**, 547–565.
Mehring, M. (1976). *NMR: Basic Princ. Prog.* **11**, 7–39, 167–191.
Nall, B. T., Rothwell, W. P., Waugh, J. S., and Rupprecht, A. (1981). *Biochemistry* **20**, 1881–1887.
Opella, S. J., Wise, W. B., and DiVerdi, J. A. (1981). *Biochemistry* **20**, 284–290.
Pilet, J., and Brahms, J. (1972). *Nature (London), New Biol.* **236**, 99–100.
Pines, A., Gibby, M. G., and Waugh, J. S. (1973). *J. Chem. Phys.* **56**, 569–590.
Pople, J. A., Schneider, W. G., and Bernstein, H. J. (1959). "High-Resolution Nuclear Magnetic Resonance." McGraw-Hill, New York.
Rodley, G. A., Scobie, R. S., Bates, R. H. T., and Lewitt, R. (1976). *Proc. Natl. Acad. Sci. U.S.A.* **73**, 2959–2963.
Shindo, H., and Zimmerman, S. B. (1980). *Nature (London)* **283**, 690–691.
Shindo, H., Wooten, J. B., Pheiffer, B. H., and Zimmerman, S. B. (1980). *Biochemistry* **19**, 518–526.
Shindo, H., Wooten, J. B., and Zimmerman, S. B. (1981). *Biochemistry* **20**, 745–750.
Solomon, I. (1950). *C. R. Helid. Seances Acad. Sci.* **248**, 92–95.
Terao, T., Matsui, S., and Akasaka, K. (1977). *J. Am. Chem. Soc.* **99**, 6136–6138.
Wang, A. H.-J., Quigley, G. J., Kolpak, F. J., Crawford, J. L., van Boom, J. H., van der Marel, G., and Rich, A. (1979). *Nature (London)* **282**, 680–686.
Watson, J. D., and Crick, F. H. C. (1953). *Nature (London)* **171**, 737–738.
Wing, R., Drew, H. R., Takano, T., Broka, C., Tanaka, S., Itakura, K., and Dikerson, R. E. (1980). *Nature (London)* **287**, 755–758.

# CHAPTER 14

# Phosphorus-31 NMR of Phospholipids in Micelles

### Edward A. Dennis
### Andreas Plückthun

Department of Chemistry
University of California at San Diego
La Jolla, California

| | | |
|---|---|---|
| I. | Introduction and Perspective | 423 |
| II. | Spectral Characteristics of Phospholipids in Mixed Micelles with Detergents | 425 |
| III. | $T_1$, Nuclear Overhauser Effect, and Quantitative Analysis | 431 |
| IV. | Solubilization of Phospholipids by Detergents | 433 |
| V. | Critical Micelle Concentration Determinations and Micellization of Monomeric Phospholipids by Detergents | 435 |
| VI. | Lysophospholipids: Acyl and Phosphoryl Migration | 438 |
| VII. | Phospholipases: Specificity and Kinetics | 442 |
| | References | 444 |

## I. Introduction and Perspective

NMR spectra of phospholipids in natural membranes and of synthetic phospholipids in model bilayer membranes are generally characterized by broad lines (see Smith and Ekiel, Chapter 15). However, when the phospholipids are dispersed in mixed micelles with detergents, significant improvement in the resolution of the phospholipid resonances is observed in $^{31}$P (London and Feigenson, 1979; Roberts et al., 1979) as well as in $^{1}$H NMR (Dennis and Owens, 1973; Ribeiro and Dennis, 1974) and in $^{13}$C NMR (Ribeiro and Dennis, 1976). Such spectra can be characterized as high-resolution spectra. Although they are not bilayer membranes, mixed micelles of phospholipids and detergents do provide useful membrane models for examining lipid–protein interactions and the mechanism of action of lipolytic enzymes (Dennis et al., 1981; Plückthun and Dennis, 1982b). $^{31}$P-NMR spectra (with $^{1}$H decoupling) of mixed micelles have the advantage of allowing resolution and quantitation of individual classes of phospholipids in single peaks in the presence of a large variety of detergents (London and Feigenson, 1979; Roberts et al., 1979). For lysophospholipids (Plückthun and Dennis, 1982a) and synthetic phospholipids containing short chain fatty acids (Plückthun and Dennis, 1981), which form micelles in aqueous solution without detergent added, $^{31}$P NMR can be used to identify individual isomers, as well as to determine critical micelle concentrations (CMCs)[1] and aggregation states.

This chapter limits itself to $^{31}$P-NMR studies on phospholipids and lysophospholipids in monomeric and micellar states and focuses on the identification of species and aggregation states, as well as on the dynamic processes of migration and reaction kinetics. For this purpose, micelles and mixed micelles are defined as dilute isotropic solutions of phospholipids, either with or without detergents that form spontaneously and are at thermodynamic equilibrium; this definition specifically excludes sonicated or small unilamellar vesicles as well as membranes. These are covered by Smith and Ekiel (Chapter 15). This limitation in scope necessarily requires heavy reliance on work from the laboratory of the authors of this chapter.

The $^{31}$P-NMR characteristics of phospholipids in micelles and mixed micelles with detergents have not been reviewed previously, but many reviews have appeared that concentrate in full (Seelig, 1978; Yeagle, 1978; Cullis and de Kruyff, 1979) or in part (Bocian and Chan, 1978) on $^{31}$P NMR of phospholipids in crystalline or powder form, multibilayers, natural membranes, and sonicated vesicles. These studies have emphasized headgroup orientation, phospholipid conformation, and packing in membranes, as well as the asymmetric distribution of phospholipids in sonicated vesicles (Michaelson et al., 1973; Berden et al., 1975; Nolden and Ackermann, 1976). Interest has also focused on the hexagonal phase (Cullis and de Kruyff, 1976, 1979) of phospholipids such as PE which, like the multibilayers, is not isotropic. In mixtures of certain phospholipids, an isotropic "lipidic particle" phase (presumably consisting of inverted micelles sand-

---

[1] Abbreviations: PC, 1,2-diacyl-*sn*-glycero-3-phosphorylcholine, phosphatidylcholine, or lecithin; β-PC, 1,3-diacyl-*sn*-glycero-2-phosphorylcholine; PE, 1,2-diacyl-*sn*-glycero-3-phosphorylethanolamine or phosphatidylethanolamine; β-PE, 1,3-diacyl-*sn*-glycero-2-phosphorylethanolamine; PS, 1,2-diacyl-*sn*-glycero-3-phosphorylserine or phosphatidylserine; PI, 1,2-diacyl-*sn*-glycero-3-phosphorylinositol or phosphatidylinositol; PG, 1,2-diacyl-*sn*-glycero-3-phosphorylglycerol or phosphatidylglycerol; PA, 1,2-diacyl-*sn*-glycero-3-phosphate or phosphatidic acid; CL, diphosphatidylglycerol or cardiolipin; SM, sphingomyelin; lyso-PC, 1-acyl-*sn*-glycero-3-phosphorylcholine; 2-lyso-PC, 2-acyl-*sn*-glycero-3-phosphorylcholine; lyso-PE, 1-acyl-*sn*-glycero-3-phosphorylethanolamine; lyso-β-PC, 3-acyl-*sn*-glycero-2-phosphorylcholine; lyso-β-PE, 3-acyl-*sn*-glycero-2-phosphorylethanolamine; $P_i$, inorganic phosphate; NMR, nuclear magnetic resonance; NOE, nuclear Overhauser effect; $T_1$, spin–lattice relaxation time. For phospholipids in which the identity of the acyl group or its natural source is known, this will be specified.

wiched between the two monolayers of the bilayer) has been suggested from $^{31}$P-NMR studies (de Kruyff et al., 1979). Different phospholipids have been resolved in organic solvents and sonicated vesicles (Henderson et al., 1974; Berden et al., 1975), and phospholipids in lipoprotein structures have also been studied and found to give very sharp resonances (Assmann et al., 1974; Glonek et al., 1974; Henderson et al., 1975).

Most of the data summarized herein were originally reported using the $^{31}$P-NMR convention of positive chemical shifts designating increasing field strength. All tables and figures reproduced here have been altered to reflect the newer convention that positive chemical shifts are in the direction of decreasing field strength, which has always been used for NMR studies with most nuclei other than phosphorus. When referring to the original literature cited herein, the interconversion of (+) and (−) must be made. Chemical shifts are reported relative to external 85% phosphoric acid but have usually been determined with other secondary standards. Samples usually contain some $D_2O$ as a lock signal. With mixed micelles, data have generally been obtained with broadband $^1H$ decoupling to achieve narrow linewidths.

## II. Spectral Characteristics of Phospholipids in Mixed Micelles with Detergents

In 1979 London and Feigenson showed that individual classes of phospholipids in the presence of sodium cholate could be resolved readily by $^{31}$P NMR with $^1H$ broadband decoupling as illustrated in Fig. 1. In their studies they included an excess of the detergent, and mixed micelles were presumably formed. The peaks appeared better resolved and the linewidths narrower than in earlier spectra of similar mixtures of phospholipids in organic solvents, such as 2 : 1 chloroform : methanol (which were, however, obtained without $^1H$ broadband decoupling) as reported by Henderson et al. (1974). With $^1H$ broadband decoupling, other investigators (Berden et al., 1975) found the linewidths to be as small as 1.5 Hz in organic solvents. London and Feigenson (1979) report linewidths of <1 Hz in mixed micelles. The chemical shifts for the various phospholipids in mixed micelles with cholate as shown in Table I differ from those measured in organic solvents as shown in Table II.

In the same year Roberts et al. (1979) also showed that various phospholipids in mixed micelles, in this case with the nonionic surfactant Triton X-100, could be resolved, as well as the corresponding lysophospholipids as illustrated in Fig. 2. $^{31}$P-NMR chemical shifts of various phospholipids,

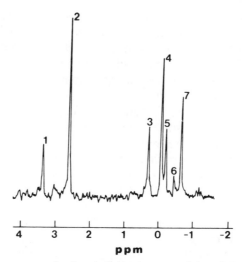

**Fig. 1.** $^{31}$P-NMR spectrum of a phospholipid mixture consisting of 5% w:v cholate, 50 m$M$ EDTA, and the following: peak 1, 10 mg of PA; peak 2, 6.6 mg of K$_2$HPO$_4$; peak 3, 8 mg CL; peak 4, 13 mg of PE; peak 5, 12.5 mg of PS; peak 6, PI present as an impurity in PS; peak 7, 12.5 mg of dipalmitoyl-PC. Total volume 1 ml, pH ~ 8. A Varian CFT-20 NMR spectrometer operating at 32.19 MHz was employed and spectra were recorded at 40°C with $^1$H broadband decoupling; 100 transients were collected with an acquisition time of 2 s per transient, no delay between transients, and a filtering time constant of 1 s. From London and Feigenson (1979), but note that the original figure has been altered to now show positive chemical shifts in the direction of decreasing field strength.

### TABLE I

$^{31}$P-NMR Chemical Shifts of Various Phospholipids in Potassium Cholate at pH ≃ 8[a]

| Phospholipid | Chemical shift[b] (ppm) |
|---|---|
| PC | −0.65 |
| PI | −0.40 |
| lyso-PC | −0.15 |
| PS | −0.12 |
| PE | 0.00 |
| SM | 0.00 |
| CL | 0.31 |
| PG | 0.43 |
| PA | 3.80 |

[a] Adapted from London and Feigenson (1979).

[b] Note that the signs of the chemical shifts have been switched from the original report so that positive chemical shifts are now shown in the direction of decreasing field strength.

**TABLE II**

$^{31}$P-NMR Chemical Shifts of Phospholipids in Chloroform–Methanol and Analogs in Aqueous Solution[a]

| Compound (20 mM) | Solvent[b] | Chemical Shift[c] (ppm) |
|---|---|---|
| Egg PC | C–M | −0.9 |
| N,N-Dimethyl-PE | C–M | 0.3 |
| N-Methyl-PE | C–M | 0.0 |
| PE | C–M | 0.2 |
| PS | C–M | 0.0 |
| Dipalmitoyl-PC | C–M | −0.8 |
| Distearyl-PC | C–M | −0.8 |
| Dioleoyl-PC | C–M | −0.8 |
| Distearyl-PE | C–M | −0.1 |
| Lyso-PC | C–M | −0.2 |
| Lyso-PE[d] | C–M | 0.2 |
| Lyso-PS[e] | C–M | 0.2 |
| Plasmalogen PE | C–M | 0.2 |
| SM | C–M | 0.0 |
| PI | C–M | 0.2 |
| PA | C–M | 2.8 |
| PG | C–M | 1.2 |
| Cl | | |
|   Low-field band | C–M | 1.0 |
|   High-field band | | 0.8 |
| Phosphorylcholine | Water | 3.2 |
| Phosphorylethanolamine | Water | 3.7 |
| Phosphorylserine | Water | 3.7 |
| Glycerophosphorylcholine | Water | −0.1 |
| Glycerophosphorylethanolamine | Water | 0.4 |
| Glycerophosphorylserine | Water | 0.1 |

[a] Adapted from Henderson *et al.* (1974). Copyright 1974 American Chemical Society.

[b] C–M is chloroform–methanol 2:1 (v:v); water contained 0.2 M EDTA, Na$^+$ ion, pH 7.0.

[c] Note that the signs of the chemical shifts have been switched from the original report so that positive chemical shifts are now shown in the direction of decreasing field strength.

[d] Concentration 7 mM.

[e] Concentration 5 mM.

lysophospholipids, and related analogs in Triton X-100 or aqueous solution are given in Table III. Chemical shifts for some water-soluble analogs also included in the studies of Henderson *et al.* (1974) are summarized in Table II. The chemical shifts of additional water-soluble analogs are reported by Assmann *et al.* (1974). Relative chemical shifts of PC, PE, monomethyl-PE,

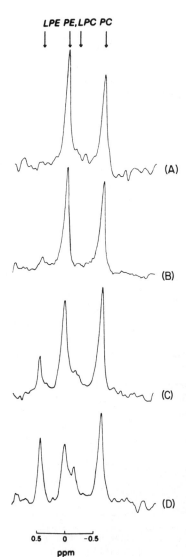

**Fig. 2.** Cobra venom phospholipase $A_2$ (0.05 μg) was added to 1.4 ml of solution containing egg PC (2.6 m$M$) and egg PE (3.0 m$M$) in 48 m$M$ Triton X-100, 50 m$M$ Tris-HCl, and 10 m$M$ $CaCl_2$; pH 8.0, 40°C. $^{31}$P-NMR spectra are shown at the following times (min) after the initiation of reaction: A, 0; B, 7; C, 27; D, 87. A JEOL PFT-100 NMR spectrometer operating at 40.3 MHz was employed and spectra were recorded with $^1$H broadband decoupling. Peaks corresponding to lyso-PE (LPE), lyso-PC (LPC), PE, and PC are indicated. From Roberts *et al.* (1979). Copyright 1979 American Chemical Society. Note that the original figure has been altered to now show positive chemical shifts in the direction of decreasing field strength.

and dimethyl-PE and their lyso derivatives with the fatty acid composition of egg PC have also been reported (Roberts *et al.,* 1979), as well as chemical shifts for PC and PE in Triton X-100 (London and Feigenson, 1979) and, within experimental error, do not differ significantly from those given in Table III for the same phospholipids but of defined fatty acid composition. In both organic solvents and mixed micelles with detergents, it appears that

## TABLE III

$^{31}$P-NMR Shifts of Phospholipids and Analogs in Aqueous Solution with Triton X-100[a,b]

| Compound | Triton X-100 (48 m$M$) | Chemical shift (ppm) | Reference[c] |
|---|---|---|---|
| Dipalmitoyl-PC | + | −0.86 | 1,2 |
| 1-Palmitoyl-lyso-PC | − | −0.34 | 2 |
| 1-Palmitoyl-lyso-PC | + | −0.38 | 2 |
| 2-Palmitoyl-lyso-PC | − | −0.52 | 2 |
| 2-Palmitoyl-lyso-PC | + | −0.55 | 2 |
| Dipalmitoyl-$\beta$-PC | + | −1.45 | 2 |
| Palmitoyl-lyso-$\beta$-PC | − | −1.13 | 2 |
| Dibutyryl-PC | − | −0.60 | 2 |
| Dibutyryl-PC | + | −0.61 | 1 |
| 1-Butyryl-lyso-PC | − | −0.24 | 2 |
| 1-Butyryl-lyso-PC | + | −0.25 | 1 |
| 2-Butyryl-lyso-PC | − | −0.44 | 2 |
| Dipalmitoyl-PE | + | −0.15 | 1 |
| Palmitoyl-lyso-PE | + | 0.26 | 1 |
| Dipalmitoyl-$\beta$-PE | + | −0.71 | 3 |
| Palmitoyl-lyso $\beta$-PE | + | −0.46 | 3 |
| Dipalmitoyl-$N$-methyl-PE | + | −0.30 | 3 |
| Palmitoyl-lyso-$N$-methyl-PE | + | 0.10 | 3 |
| Dipalmitoyl-$N$-methyl-$\beta$-PE | + | −0.86 | 3 |
| Palmitoyl-lyso-$N$-methyl-$\beta$-PE | + | −0.60 | 3 |
| Dipalmitoyl $N,N$-dimethyl-PE | + | −0.42 | 3 |
| Palmitoyl-lyso-$N,N$-dimethyl-PE | + | −0.02 | 3 |
| Dipalmitoyl-$N,N$-dimethyl-$\beta$-PE | + | −1.02 | 3 |
| Palmitoyl-lyso-$N,N$-dimethyl-$\beta$-PE | + | −0.75 | 3 |
| Glycero-3-phosphorylcholine | − | −0.08 | 2 |
| Glycero-3-phosphorylcholine | + | −0.09 | 1 |
| Glycero-2-phosphorylcholine | − | −0.79 | 4 |
| Glycero-3-phosphate[d] | − | 4.3 | 4 |
| Glycero-2-phosphate[d] | − | 3.9 | 4 |
| Dodecylphosphorylcholine | + | −0.39 | 1 |
| Bis(monoacylglyceryl)phosphate | + | −0.70 | 1 |

[a] Note that the signs of the chemical shifts have been switched from the original report so that positive chemical shifts are now shown in the direction of decreasing field strength.

[b] CaCl$_2$ was generally present. When the compound readily dissolves in water in the absence of detergent, its chemical shift is also indicated. All compounds were analyzed at pH 8.0.

[c] References: 1, Plückthun and Dennis (1981); 2, Plückthun and Dennis (1982a); 3, A. Plückthun, J. de Bony, and E. A. Dennis (manuscript in preparation); 4, A. Plückthun and E. A. Dennis (unpublished).

[d] Contained 10 m$M$ EDTA.

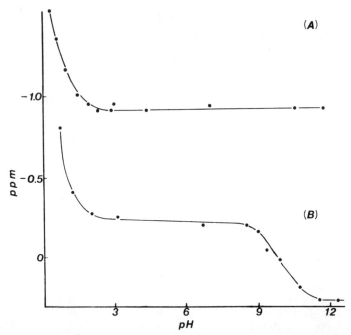

**Fig. 3.** The pH dependence of the chemical shift of phospholipids dissolved in Triton X-100. (A) 4 mg ml$^{-1}$ dimyristoyl-PC dissolved in 2.5% (w:v) Triton X-100; (B) 2.1 mg ml$^{-1}$ dilauryl-PE dissolved in 2.5% (w:v) Triton X-100. Samples contained an external D$_2$O field-frequency lock. From London and Feigenson (1979), but note that the original figure has been altered to now show positive chemical shifts in the direction of decreasing field strength.

the specific fatty acid composition of normal long-chain phospholipids does not affect chemical shifts within experimental error (Henderson et al., 1974; London and Feigenson, 1979). Absolute chemical shifts in these tables are probably only reproducible to ±0.1 ppm, although relative chemical shifts within a given series of measurements have much less error (Plückthun and Dennis, 1981).

Another advantage of examining $^{31}$P-NMR chemical shifts in aqueous solution with detergents is that p$K_a$ values can be obtained readily from chemical-shift changes. As illustrated in Fig. 3, the ionization of the phosphate in PC and PE can be followed readily in Triton X-100 mixed micelles where the apparent p$K_a$ is < 1, as can the ionization of the amino group on PE, where the p$K_a \simeq 9.75$ (London and Feigenson, 1979). Thus $^{31}$P NMR is an excellent method for determining p$K_a$ values of phospholipids in mixed micelles. In addition to being pH-dependent, chemical shifts of phospholipids in detergents are temperature-dependent, as illustrated in Fig. 4. Chemical shifts may also depend on the presence of metal ions, especially multivalent cations, if the lipids and/or detergents are ionic. In organic

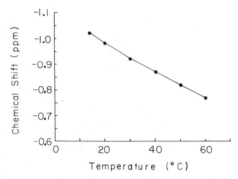

**Fig. 4.** ³¹P-NMR chemical shift of 10 m$M$ egg PC in mixed micelles with 80 m$M$ Triton X-100 as a function of temperature. From Plückthun and Dennis (1981). Copyright 1981 American Chemical Society. Note that the original figure has been altered to now show positive chemical shifts in the direction of decreasing field strength.

solvents the addition of EDTA gives narrower lines and better resolution (Henderson et al., 1974), but a comprehensive study of metal-ion dependence in mixed micelles has not been reported. London and Feigenson (1979) routinely included EDTA in their studies, whereas Dennis and co-workers (Roberts et al., 1979; Plückthun and Dennis, 1981) usually included $CaCl_2$ to obtain data under conditions used in enzymatic studies; this may in certain cases broaden the lines but has a negligible effect on the chemical shift of zwitterionic phospholipids such as PC in Triton X-100. Detergent, phospholipid, and $D_2O$ concentration probably have only small effects on chemical shifts (Plückthun and Dennis, 1981).

## III. $T_1$, Nuclear Overhauser Effect, and Quantitative Analysis

The spin–lattice relaxation time $T_1$ (determined at 32.2 MHz) for PC, PE, and PA in potassium cholate in the presence of EDTA at pH $\simeq$ 8 was found to be 3.3, 3.2, and 3.3 s, respectively, and shorter than 4.3 s for $P_i$ (London and Feigenson, 1979). In mixed micelles with Triton X-100 in the presence of EDTA, Roberts et al. (1979) found the $T_1$ (determined at 40.3 MHz) for PC and PE to be 2.5 and 2.2 s, respectively. Plückthun and Dennis (1981) found a $T_1$ of 2.5 s for PE and 12.4 s for the monomeric phospholipid dibutyryl-PC in the presence of Triton X-100 under slightly different experimental conditions; the dibutyryl-PC was not incorporated into the micelles. $T_1$ for 1-palmitoyl-lyso-PC and 2-palmitoyl-lyso-PC is 2.3 s (Plückthun and Dennis, 1982a), and for egg lyso-PC, $T_1$ is 2.3 s (Yeagle, 1979).

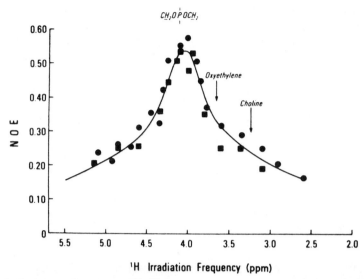

**Fig. 5.** Frequency dependence of $^{31}$P{$^1$H}NOE of 10 m$M$ egg PC (●) and 10 m$M$ egg PE (■) in 40 m$M$ Triton X-100 at 40°C plotted as a function of the continuous-wave proton-decoupler frequency. The chemical shifts for choline methyl groups, methylene groups adjacent to the phosphate of the phospholipid, and Triton oxyethylene groups are indicated in parts per million from tetramethylsilane. From Roberts et al. (1979). Copyright 1979 American Chemical Society.

The nuclear Overhauser effect (NOE) for PC, PE, and PA in potassium cholate was found to be 60, 60, and 50% compared to 5% for $P_i$. Roberts et al. (1979) found the NOEs for PC, PE, lyso-PC, and lyso-PE in mixed micelles to be indistinguishable (60–70%) within experimental error. Yeagle (1979) reported an NOE for lyso-PC of 70%. In mixed micelles with Triton X-100, the NOE arises predominantly from the intramolecular methylene protons adjacent to the phosphate moiety (Roberts et al., 1979) as illustrated in Fig. 5, in contrast to sonicated vesicles where Yeagle et al. (1977) found that the choline methyl groups dominate intermolecularly the NOE of both PC and PE in mixed vesicles. Yeagle (1979) found that the NOE of micelles of lyso-PC derived from egg PC is also dominated by intramolecular methylene groups rather than by intermolecular interactions, supporting the contention that in micelles the headgroups are further apart than in vesicles, thereby defeating the intermolecular interactions found in the bilayer. In the case of mixed micelles with detergents, the intermolecular interactions are, in addition, presumably weakened by the separation of phospholipid molecules by detergent (Roberts et al., 1979). On the other hand, it has been suggested that in egg PC–sodium taurocholate mixed micelles, which are thought to be structurally different (Mazer et al.,

1980), the NOE is due to intermolecular choline methyl interactions in patches of PC in the mixed micelles (Castellino and Violand, 1979). However, the frequency dependence of the NOE was not determined in the latter case.

With broadband $^1$H decoupling and pulse intervals of the order of $T_1$, the similar $T_1$ and NOE values for various phospholipids in mixed micelles suggest that peak areas or intensities (Plückthun and Dennis, 1982a) can be used reliably for quantitative analysis of the phospholipids present in a mixture. With a significant loss of signal to noise for greater reliability, gated broadband decoupling and longer pulse intervals of the order of $5-10 \times T_1$ can be used; indeed, where comparison of $^{31}$P-NMR intensities with weighed mixtures or traditional TLC separation and $P_i$ analysis has been carried out, good agreement is obtained (London and Feigenson, 1979).

## IV. Solubilization of Phospholipids by Detergents

Detergents have been widely used to solubilize and purify membrane-bound proteins (Helenius and Simons, 1975; Tanford and Reynolds, 1976; Helenius et al., 1979; Lichtenberg et al., 1983). Central to the effect of detergents is the solubilization of the phospholipid components into mixed micelles. This can be followed by $^{31}$P NMR, as illustrated in Fig. 6, where cholate was used to solubilize sarcoplasmic reticulum membranes. In the spectrum, the phospholipids are well resolved and the spectrum is similar to that obtained when the phospholipids are extracted with chloroform–methanol first and then solubilized with cholate (London and Feigenson, 1979). This shows that in the presence of the protein, all of the phospholipids are solubilized as in its absence. Without cholate, the spectrum of the membranes would be quite broad and the individual phospholipid classes would not be resolved.

In principle, $^{31}$P NMR can be used to follow the conversion of membranes to mixed micelles by the appearance of sharp peaks for the phospholipids. This would also be the case for model membranes of pure phospholipids. For multibilayers or hexagonal phases (Cullis and de Kruyff, 1979), high-resolution sharp peaks would appear out of the broad baseline; for sonicated vesicles, peaks would become sharper and inside–outside signals would disappear. This is shown in Table IV, where Castellino and Violand (1979) followed the decrease in the linewidth as well as the $T_1$ and NOE for egg PC vesicles on the addition of sodium taurocholate. The values of the $T_1$ and NOE in mixed micelles (low phospholipid/detergent ratio) are consistent with those reported in Section III. When only a small amount of detergent is

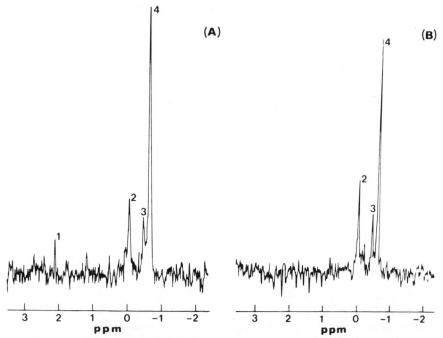

**Fig. 6.** $^{31}$P-NMR spectra of sarcoplasmic reticulum phospholipids using the spectrometer described in Fig. 1. (A) Sarcoplasmic reticulum (12 mg of protein per milliliter) dissolved in cholate, pH ~ 8. Peak assignments: 1, $P_i$; 2, PE; 3, PI; 4, PC. 6000 transients were collected with an acquisition time of 2 s per transient, no delay between transients, and a filtering time constant of 1 s; (B) lipid extract of sarcoplasmic reticulum (10 mg of lipid per milliliter) dissolved in cholate. Peak assignments as in (A); 600 transients were collected with other spectrometer settings as in (A). From London and Feigenson (1979), but note that the original figure has been altered to now show positive chemical shifts in the direction of decreasing field strength.

added to the phospholipid, the basic vesicular bilayer structure is retained (Castellino and Violand, 1979), whereas with PE, the addition of small amounts of a variety of detergents can result in the transformation of hexagonal structures into bilayer structures (Madden and Cullis, 1982) but only at detergent/phospholipid ratios below mixed-micelle formation. Jackson et al. (1982) have utilized $^{31}$P NMR to follow the solubilization of large, unilamellar PC vesicles (which have a very broad spectrum) by octylglucoside to form mixed micelles (with quite narrow lines) and to analyze the solubilization process. $^{31}$P-NMR studies of mixtures of PC multibilayers and phosphorus-containing detergents from decyl to hexadecyl phosphonate, in which the signals from both the phospholipid and detergent can be followed, have been carried out by Klose and Hollerbuhl (1981). By examining mixtures at molar ratios when both micellar and lamellar structures are

**TABLE IV**

$^{31}$P-NMR Parameters for Egg PC–Sodium Taurocholate Mixtures[a,b]

| Mole ratio (egg PC/sodium taurocholate) | Linewidth (Hz) | $T_1$ (s) | NOE (%) |
|---|---|---|---|
| ∞ | 13 | 1.7 | 52 |
| 35 | 13 | 1.7 | 50 |
| 10 | 12 | 1.7 | 41 |
| 4 | 13 | 1.8 | 39 |
| 1.3 | 5.4 | 2.0 | 42 |
| 0.9 | 3.8 | 2.2 | 43 |
| 0.6 | 2.5 | 2.2 | 40 |
| 0.1 | 2.0 | 2.6 | 49 |

[a] Adapted from Castellino and Violand (1979).
[b] The egg PC was in the form of sonicated vesicles to which the detergent was added. All values are ±10%.

present as a function of time after mixing, the process of equilibration and the equilibrium state can be determined.

## V. Critical Micelle Concentration Determinations and Micellization of Monomeric Phospholipids by Detergents

Phospholipids can be synthesized with short fatty acid chains to make them water soluble, and the physical properties of such phospholipids have been studied extensively by Tausk et al. (1974a,b,c). Their critical micelle concentration (CMC) depends on chain length and has been determined by a variety of methods for a number of PC derivatives. It was found that the $^{31}$P-NMR signal is chemically shifted in going from the monomeric form to the micellar form of dihexanoyl PC (Roberts et al., 1979). This difference can be used (Plückthun and Dennis, 1981) to determine the CMC according to

$$\nu = x_{mono}\nu_{mono} + x_{mic}\nu_{mic} \tag{1}$$

where $\nu$ is the observed chemical shift, $\nu_{mono}$ the chemical shift below the CMC, $\nu_{mic}$ the chemical shift of dihexanoyl PC micelles, $x_{mono}$ the molar fraction of phospholipid in the monomeric state, and $x_{mic}$ the molar fraction of phospholipid in the micellar state. The values for $\nu_{mono}$ and $\nu_{mic}$ can be estimated from the experimental data, along with the CMC, and all three

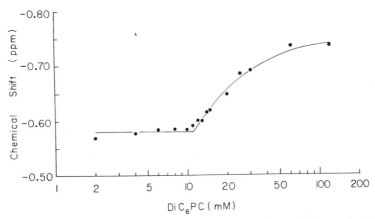

**Fig. 7.** $^{31}$P-NMR chemical shift of dihexanoyl-PC (DiC$_6$PC) in D$_2$O as a function of concentration. The solution contained no further additives. The best-fit curve is calculated by using Eq. (1) and assigning $v_{mono} = 0.58$ ppm, $v_{mic} = 0.76$ ppm, and CMC = 11 m$M$. From Plückthun and Dennis (1981). Copyright 1981 American Chemical Society. Note that the original figure has been altered to now show positive chemical shifts in the direction of decreasing field strength.

varied to obtain the best fit of the theoretical curve to experimental data. Experimental results are shown in Fig. 7. The resulting CMC is in agreement with that determined by other methods, including $^1$H and $^{13}$C NMR. The $^{31}$P-NMR method can be used for other phospholipids or to determine if a given phospholipid is monomeric or micellar under certain experimental conditions. $^{31}$P-NMR chemical-shift changes similarly have been used to follow micelle formation by decyldimethyl phosphine oxide by Kresheck and Jones (1980). These workers plotted the observed chemical shift $v$ against the inverse of the concentration of surfactant, which has the advantage of graphically giving $v_{mic}$ and the CMC.

Similarly, incorporation of monomeric phospholipids by detergents into mixed micelles results in a chemical-shift change in $^{31}$P NMR, and this can be used to follow the micellization process. This is illustrated in Fig. 8, where the micellization of dihexanoyl-PC by titration with Triton X-100 was followed by changes in the $^{31}$P-NMR chemical shift. In contrast, with the same amount of Triton X-100, dibutyryl-PC is hardly micellized at all. A control titration of the completely water-soluble analog glycerophosphorylcholine is also included. For the dihexanoyl-PC, a partition coefficient between Triton X-100 micelles and free solution was calculated as a function of the Triton X-100 concentration, using the phase-separation approximation for mixed-micelle formation and considering the mixed micelles to be a pseudo-phase. With the data in Fig. 8, the fraction of dihexanoyl-PC

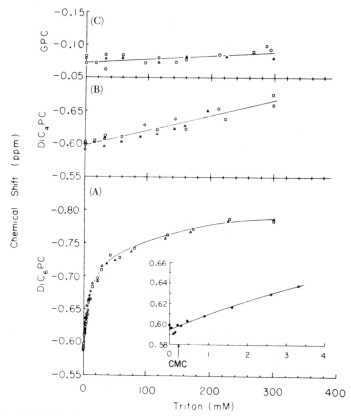

**Fig. 8.** ³¹P-NMR chemical shift of (A) dihexanoyl-PC (DiC$_6$PC), (B) dibutyryl-PC (DiC$_4$PC), and (C) glycerophosphorylcholine (GPC) as a function of the concentration of Triton X-100. The initial concentration of phospholipid or GPC was 7.5 m$M$. The titration was carried out with 500 m$M$ detergent except for the insert to (A) in which 7.5 m$M$ Triton X-100 was employed. In each panel, the results of three experiments are plotted (○, □, △). In (B), the titration was also carried out with mixed micelles (▲) consisting of Triton X-100 (500 m$M$) and PE (125 m$M$). From Plückthun and Dennis (1981). Copyright 1981 American Chemical Society. Note that the original figure has been altered to now show positive chemical shifts in the direction of decreasing field strength.

present in Triton X-100 was calculated by Plückthun and Dennis (1981) from Eq. (1) by using the relationship $x_{\text{mono}} = 1 - x_{\text{mic}}$, which results in

$$x_{\text{mic}} = (\nu - \nu_{\text{mono}})/(\nu_{\text{mic}} - \nu_{\text{mono}}) \qquad (2)$$

A partition coefficient $K$ can be defined as shown in

$$K = X_{\text{mic}} c_{\text{Pmono}} \qquad (3)$$

Here $c_{Pmono}$ is the concentration of phospholipid in the monomeric state and $X_{mic}$ the molar fraction of phospholipid in the micellar phase consisting of Triton and phospholipid as defined in

$$X_{mic} = c_{Pmic}/(c_{Pmic} + c_{Tmic}) \qquad (4)$$

where $c_{Pmic}$ and $c_{Tmic}$ are the concentrations of phospholipid and Triton, respectively, that are in micelles. The value for $c_{Tmic}$ was calculated from the relation $c_{Tmic} = c_{Ttot} - cmc_T$, where $c_{Ttot}$ is the total Triton concentration employed and $cmc_T$ the critical micelle concentration of pure Triton. Using the relations $c_{Pmic} = x_{mic}c_{Ptot}$ and $c_{Pmono} = (1 - x_{mic})c_{Ptot}$, where $c_{Ptot}$ is the total phospholipid concentration, and by combining Eqs. (3) and (4), one can then determine $K$ from the chemical-shift data as shown in

$$K = \frac{1}{x_{mic}c_{Ptot} + (c_{Ttot} - cmc_T)} \frac{x_{mic}}{1 - x_{mic}} \qquad (5)$$

The partition coefficient $K$ was found to be $\sim 40\ M^{-1}$ at 50 m$M$ Triton X-100 and varied somewhat with Triton concentration (Plückthun and Dennis, 1981). The chemical shift of the short-chain phospholipids incorporated into mixed micelles approaches that of long-chain "normal" phospholipids in the micelles, suggesting similar structures for the mixed micelles.

Interestingly, Burns *et al.* (1983) have reported that micelles formed by short-chain phospholipids can themselves be used as detergents to solubilize triglycerides containing short fatty chains to form microemulsion particles in which the $^{31}$P-NMR linewidths are narrower than for the pure phospholipid micelles. These particles serve as models for lipoproteins.

## VI. Lysophospholipids: Acyl and Phosphoryl Migration

Lysophospholipids lack one of the acyl groups on phospholipids and form micelles by themselves without added detergent. The spectral characteristics of lyso-PC, both alone and in the presence of detergents, have been considered in preceding sections along with normal phospholipids. Lysophospholipids possess the ability to rearrange via migration of either the acyl or phosphoryl group and the sensitivity of the $^{31}$P-NMR chemical shift to the resulting isomers has been particularly useful in following these migration reactions (Plückthun and Dennis, 1982a). Phosphoryl and acyl migration of lysophospholipids must be taken into account when evaluating phospholipase specificity, the chemical synthesis of mixed acyl phospholipids, and the biosynthesis of phospholipids, so that the determination of the kinetics of

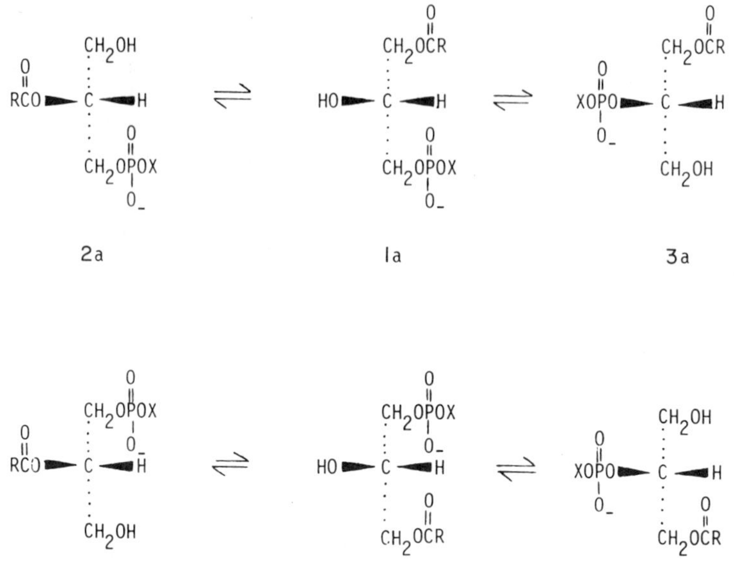

**Fig. 9.** Possible interconversions via acyl ($1 \rightleftharpoons 2$) and phosphoryl ($1 \rightleftharpoons 3$) migration of lysophospholipids derived from natural phospholipids (a series) and their enantiomorphs (b series). When RCOO is palmitic acid and X is choline, the structures correspond to the following compounds: 1-palmitoyl-*sn*-glycero-3-phosphorylcholine (**1a**), 2-palmitoyl-*sn*-glycero-3-phosphorylcholine (**2a**), 1-palmitoyl-*sn*-glycero-2-phosphorylcholine (**3a**), 3-palmitoyl-*sn*-glycero-1-phosphorylcholine (**1b**), 2-palmitoyl-*sn*-glycero-1-phosphorylcholine (**2b**), and 3-palmitoyl-*sn*-glycero-2-phosphorylcholine (**3b**). From Plückthun and Dennis (1982a). Copyright 1982 American Chemical Society.

such processes has been important. The lack of a suitable analytical tool prior to the use of $^{31}$P NMR probably prevented an earlier detailed evaluation of these processes. There are six different possible lyso-PCs consisting of three enantiomeric pairs of positional isomers as shown in Fig. 9.

Migration of the acyl group in the 1 position of lyso-PC (**1a**) to the 2 position (**2a**) as well as from **2a** to **1a** is shown in Fig. 10 as a function of time. From data of this type, the rate constant for acyl migration can be determined. The migration was found to be first order in both lysophospholipid and acid or base with a base-catalyzed, second-order rate constant of ~160 $M^{-1}$ $s^{-1}$. The pH dependence of the migration reaction is shown in Fig. 11. At alkaline pH values, the equilibrium mixture contains about 90% of the 1-acyl and about 10% of the 2-acyl isomer. A slow acyl migration also occurs in organic solvents, most notably in the presence of basic catalysts used in common acylation procedures for the synthesis of phospholipids from

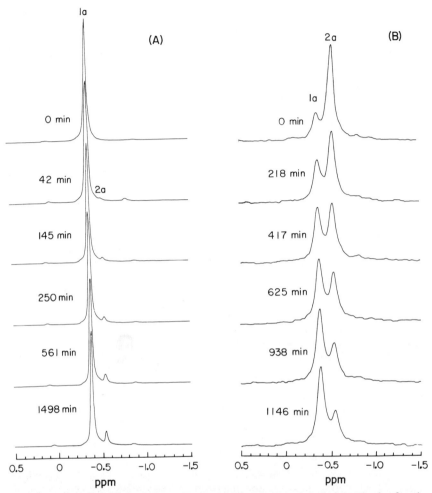

**Fig. 10.** ³¹P-NMR spectra obtained using the spectrometer described in Fig. 2, after the times indicated, of (A) 30 m$M$ 1-palmitoyl-lyso-PC (**1a**) at pH 7.0 and (B) 20 m$M$ 2-palmitoyl-lyso-PC (**2a**), of which some had migrated to **1a** during its preparation, at pH 7.0. From Plückthun and Dennis (1982a). Copyright 1982 American Chemical Society. Note that the original figure has been altered to now show positive chemical shifts in the direction of decreasing field strength.

lysophospholipids. At alkaline pH, no phosphoryl migration was detected in the time scale of acyl migration and hydrolysis, although in acid, phosphoryl migration does occur. Because of competing hydrolysis reactions in acid, the determination of the precise rate constants for phosphoryl migration is quite

14. ³¹P NMR of Phospholipids in Micelles 441

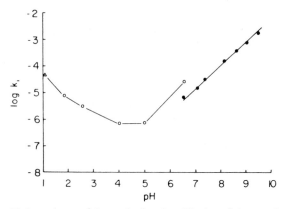

**Fig. 11.** The pH dependence of the acyl migration. The log of the pseudo-first-order rate constant $k_1$ for the rearrangement of 2-palmitoyl-lyso-PC (**2a**) into 1-palmitoyl-lyso-PC (**1a**) is plotted against the pH buffered with (●) 50 m$M$ Tris-HCl, (○) 50 m$M$ citrate, or (△) 0.1 $M$ HCl alone. At pH 1.0, Triton X-100 was included. From Plückthun and Dennis (1982a). Copyright 1982 American Chemical Society.

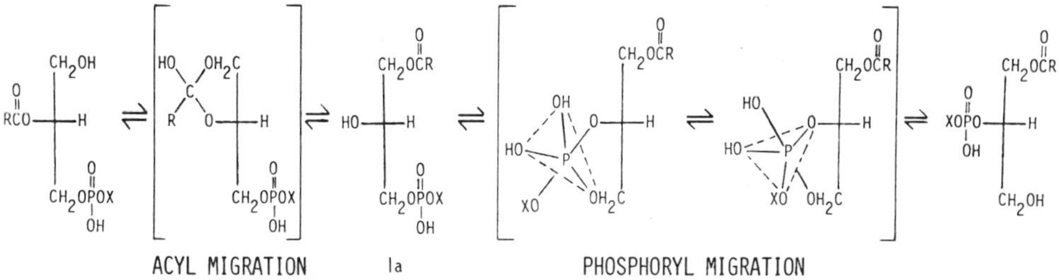

**Fig. 12.** Mechanism for the acyl migration and phosphoryl migration in lysophospholipids. For simplicity, the acid-catalyzed reactions with fully protonated intermediates are shown. The phosphoryl migration has to involve pseudorotation of the trigonal bipyramidal intermediate as indicated. Although the OX group is shown in an apical position after initial attack of the hydroyxl group of **1a**, the hydroxyl group of the phosphate could instead initially occupy the apical position. Pseudorotation cannot occur in base because it would bring an oxyanion from an equatorial to an apical position. Pathways leading to hydrolysis of the OX group and formation of a tetracoordinated cyclic phosphate diester followed by hydrolysis and migration of the phosphate group are not included in this figure. The acyl migration probably goes through a cyclic ortho ester intermediate. The basic catalyst probably partially removes the proton of the glycerol hydroxyl group and facilitates the attack at the carbonyl group. Under these conditions, the phosphate group bears a negative charge, and the attack of the glycerol oxyanion on the phosphorus atom would be expected to be slow. From Plückthun and Dennis (1982a). Copyright 1982 American Chemical Society.

complex (A. Plückthun and E. A. Dennis, unpublished), and the rate constants have not been determined. The mechanisms of acyl and phosphoryl migration are summarized in Fig. 12.

## VII. Phospholipases: Specificity and Kinetics

Phospholipases are enzymes that hydrolyze phospholipids (Dennis, 1983). Among the products of hydrolysis are lysophospholipids, phosphatidic acid (PA), and glycerophosphorylcholine and glycerophosphorylethanolamine. Because these can be resolved readily in the presence of the phospholipid substrates by $^{31}$P NMR (Section II), this method provides a ready assay that in many cases has advantages over more traditional methods. Of the phospholipases, the most well-studied and characterized is phospholipase $A_2$. The substrate phospholipid is most advantageously studied when it is solubilized in mixed micelles with detergents such as Triton X-100 or deoxycholate; for the enzyme from pancreas, its physiological substrates are mixed micelles of bile salts and phospholipids. The usefulness of $^{31}$P NMR in following the hydrolysis of PC and PE in Triton X-100 mixed micelles was shown in Fig. 2, where both PC, PE, and their lyso products can be resolved. In this case, following the hydrolysis of phospholipid mixtures would be particularly laborious by traditional methods, and mechanistic studies required carrying out kinetic studies in the presence of mixtures of phospholipids (Roberts et al., 1979). Phosphorus-31 NMR has been employed to follow the kinetics of phospholipase $A_2$ from a variety of sources in mixed micelles with a large number of types of detergents and with a range of different phospholipids and analogs (Adamich et al., 1979; Roberts et al., 1979; Plückthun and Dennis, 1982b). Phosphorus-31 NMR is particularly advantageous for kinetic studies on mixtures of normal phospholipids in mixed micelles with detergents and synthetic short-chain phospholipids such as dibutyryl-PC, where $^{31}$P NMR can at the same time show this compound to be in a monomeric state (Section V; Plückthun and Dennis, 1982b). Implications for the results of these studies are beyond the scope of this chapter (Dennis et al., 1981; Dennis, 1983).

Because 1-lyso- and 2-lyso-PC can be resolved by $^{31}$P NMR (Section VI), $^{31}$P-NMR can also be used to demonstrate directly the positional specificity of phospholipase $A_2$ and lipase, which acts as a phospholipase $A_1$, by the direct observation of the lyso products formed under conditions where migration is slow, as shown in Fig. 13. Historically, the specificities of phospholipases have been determined by utilizing specifically radiolabeled phospholipids, which were synthesized either chemically or by using

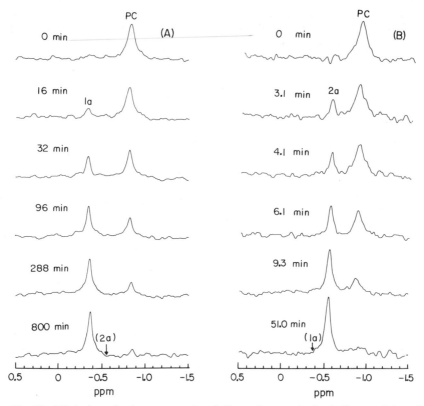

**Fig. 13.** (A) Action of cobra venom phospholipase $A_2$ on mixed micelles consisting of 15 m$M$ dipalmitoyl-PC (PC) and 120 m$M$ Triton X-100. The enzyme concentration was 12.5 μg ml$^{-1}$. The solution contained 10 m$M$ CaCl$_2$, 20% D$_2$O, and 50 m$M$ Tris-HCl buffer, pH 6.0. (B) Action of *Rhizopus arrhizus* lipase (375 μg ml$^{-1}$) under identical conditions. Spectra were obtained using the spectrometer described in Fig. 2. From Plückthun and Dennis (1982a). Copyright 1982 American Chemical Society. Note that the original figure has been altered to now show positive chemical shifts in the direction of decreasing field strength.

known phospholipases. The $^{31}$P-NMR procedure allows one now to make a direct structural determination of phospholipase products that results in a truly independent specificity determination. Whereas it is well-known that phospholipase $A_2$ is specific for the *sn*-2 position of phospholipids in micelles and bilayer membranes, it was demonstrated by this technique that this specificity also holds for the monomeric phospholipid dibutyryl-PC (Plückthun and Dennis, 1982a).

Phosphorus-31 NMR has also been used to follow phospholipase $A_2$ hydrolysis in substrate forms other than micelles and mixed micelles, such as lipoproteins (Brasure *et al.*, 1978) where the major phospholipid is PC. In these studies on lipoproteins, the linewidths of the PC and lyso-PC peaks were ~7.5 Hz, and the small amount of sphingomyelin (SM) present was not resolved from the lyso-PC product, although it could be easily taken into account in the kinetic analysis. In this case, as well as when red blood cells were treated with various phospholipases (Van Meer et al., 1980), $^{31}$P NMR was used to identify the physical state of the resulting lysophospholipid product, for if it is dissociated from the lipoprotein particles or the membranes, it should give rise to a sharp resonance indicative of its monomeric or micellar state. For both cases (Brasure et al., 1978; Van Meer et al., 1980), the lysophospholipid product after phospholipase $A_2$ digestion stays associated with the particle because such a peak does not appear.

An interesting use of $^{31}$P NMR in the study of phospholipases $A_2$ and C is in the resolution of the two diastereomers of the phosphorothioate analog of PE (Orr et al., 1982). The racemic phospholipid gives rise to two peaks (separated by 0.14 ppm) in its $^{31}$P-NMR spectrum in chloroform; hydrolysis by phospholipase $A_2$ gave rise to the loss of one of the peaks, and hydrolysis by phospholipase C gave rise to the loss of the other. This allowed the preparation and identification by $^{31}$P NMR of the pure diastereomers, although the absolute configuration of each could not be determined. $^{31}$P NMR has also been used with the diastereomers of PE containing one $^{18}$O in the phosphate moiety to establish the stereochemical course of the transphosphatidylation reaction catalyzed by phospholipase D under certain experimental conditions (Bruzik and Tsai, 1982; Chapter 6, Tsai). The use of $^{31}$P NMR in studying enzymes that can differentiate phosphate-containing compounds that are chiral at phosphorus is receiving attention as discussed elsewhere in this volume, and the identification of diastereomers of phospholipids by $^{31}$P NMR should lead to increased study of the phospholipases (Dennis, 1983); perhaps resolution of the diastereomers will be aided by incorporating them into mixed micelles.

## Acknowledgment

Support for this work was provided by National Science Foundation grant PCM 82-16963.

## References

Adamich, M., Roberts, M. F., and Dennis, E. A. (1979). *Biochemistry* **15**, 3308–3314.
Assmann, G., Sokoloski, E. A., and Brewer, H. B., Jr. (1974). *Proc. Natl. Acad. Sci. U.S.A.* **71**, 549–553.

Berden, J. A., Barker, R. W., and Radda, G. K. (1975). *Biochim. Biophys. Acta* **375**, 186–208.
Bocian, D. F., and Chan, S. I. (1978). *Annu. Rev. Phys. Chem.* **29**, 307–335.
Brasure, E. B., Henderson, T. O., Glonek, T., Pattnaik, N. M., and Scanu, A. M. (1978). *Biochemistry* **17**, 3934–3938.
Bruzik, K., and Tsai, M-D. (1982). *J. Am. Chem. Soc.* **104**, 863–865.
Burns, R. A., Donovan, J. M., and Roberts, M. F. (1983). *Biochemistry* **22**, 964–973.
Castellino, F. J., and Violand, B. N. (1979). *Arch. Biochem. Biophys.* **193**, 543–550.
Cullis, P. R., and de Kruyff, B. (1976). *Biochim. Biophys. Acta* **436**, 523–540.
Cullis, P. R., and de Kruyff, B. (1979). *Biochim. Biophys. Acta* **559**, 399–420.
de Kruyff, B., Verkley, A. J., Van Echteld, C. J. A., Gerritsen, W. J., Mombers, C., Noordam, P. C., and de Gier, J. (1979). *Biochim. Biophys. Acta* **555**, 200–209.
Dennis, E. A. (1983). "The Enzymes" (P. Boyer, ed.), 3rd ed., Vol. XVI, pp. 307–353. Academic Press, New York.
Dennis, E. A., and Owens, J. M. (1973). *J. Supramol. Struct.* **1**, 165–176.
Dennis, E. A., Darke, P. L., Deems, R. A., Kensil, C. R., and Plückthun, A. (1981). *Mol. Cell. Biochem.* **36**, 37–45.
Glonek, T., Henderson, T. O., Kruski, A. W., and Scanu, A. M. (1974). *Biochim. Biophys. Acta* **348**, 155–161.
Helenius, A., and Simons, K. (1975). *Biochim. Biophys. Acta* **415**, 29–79.
Helenius, A., McCaslin, D. R., Fries, E., and Tanford, C. (1979). *In* "Methods in Enzymology" (S. Fleischer and L. Packer, eds.), Vol. 56, pp. 734–749. Academic Press, New York.
Henderson, T. O., Glonek, T., and Myers, T. C. (1974). *Biochemistry* **13**, 623–628.
Henderson, T. O., Kruski, A. W., Davis, L. G., Glonek, T., and Scanu, A. M. (1975). *Biochemistry* **14**, 1915–1920.
Jackson, M. L., Schmidt, C. F., Lichtenberg, D., Litman, B. J., and Albert, A. D. (1982). *Biochemistry* **21**, 4576–4582.
Klose, G., and Hollerbuhl, T. (1981). *Stud. Biophys.* **83**, 35–40.
Kresheck, G. C., and Jones, C. (1980). *J. Colloid Interface Sci.* **77**, 278–279.
Lichtenberg, D., Robson, R. J., and Dennis, E. A. (1983). *Biochim. Biophys. Acta* **737**, 285–304.
London, E., and Feigenson, G. W. (1979). *J. Lipid Res.* **20**, 408–412.
Madden, T. D., and Cullis, P. R. (1982). *Biochim. Biophys. Acta* **684**, 149–153.
Mazer, N. A., Benedek, G. B., and Carey, M. C. (1980). *Biochemistry* **19**, 601–615.
Michaelson, D. M., Horwitz, A. F., and Klein, M. P. (1973). *Biochemistry* **12**, 2637–2645.
Nolden, P. W., and Ackermann, T. (1976). *Biophys. Chem.* **4**, 297–304.
Orr, G. A., Brewer, C. F., and Heney, G. (1982). *Biochemistry* **21**, 3202–3206.
Plückthun, A., and Dennis, E. A. (1981). *J. Phys. Chem.* **85**, 678–683.
Plückthun, A., and Dennis, E. A. (1982a). *Biochemistry* **21**, 1743–1750.
Plückthun, A., and Dennis, E. A. (1982b). *Biochemistry* **21**, 1750–1756.
Ribeiro, A. A., and Dennis, E. A. (1974). *Biochim. Biophys. Acta* **332**, 26–35.
Ribeiro, A. A., and Dennis, E. A. (1976). *J. Colloid Interface Sci.* **55**, 94–101.
Roberts, M. F., Adamich, M., Robson, R. J., and Dennis, E. A. (1979). *Biochemistry* **15**, 3301–3308.
Seelig, J. (1978). *Biochim. Biophys. Acta* **515**, 105–140.
Tanford, C., and Reynolds, J. A. (1976). *Biochim. Biophys. Acta* **457**, 133–170.
Tausk, R. J. M., Karmiggelt, J., Oudshoorn, C., and Overbeek, J. T. G. (1974a). *Biophys. Chem.* **1**, 175–183.
Tausk, R. J. M., van Esch, J., Karmiggelt, J., Voordouw, G., and Overbeek, J. T. G. (1974b). *Biophys. Chem.* **1**, 184–203.

Tausk, R. J. M., Oudshoorn, C., and Overbeek, J. T. G. (1974c). *Biophys. Chem.* **2**, 53-63.
Van Meer, G., de Kruyff, B., Op Den Kamp, J. A. F., and Van Deenen, L. L. M. (1980). *Biochim. Biophys. Acta* **596**, 1-9.
Yeagle, P. L. (1978). *Acc. Chem. Res.* **11**, 321-327.
Yeagle, P. L. (1979). *Arch. Biochem. Biophys.* **198**, 501-505.
Yeagle, P. L., Hutton, W. C., Huang, C.-H., and Martin, R. B. (1977). *Biochemistry* **16**, 4344-4349.

CHAPTER 15

# Phosphorus-31 NMR of Phospholipids in Membranes

*Ian C. P. Smith*
*Irena H. Ekiel*

Division of Biological Sciences
National Research Council of Canada
Ottawa, Canada

| | |
|---|---:|
| I. Properties of Membranes | 447 |
| II. $^{31}$P-NMR Spectra of Ordered Systems | 449 |
|    A. Chemical-Shift Anisotropy | 449 |
|    B. Influence of Motional Rates | 452 |
|    C. Types of Phases | 454 |
| III. Obtaining $^{31}$P-NMR Spectra | 457 |
| IV. Applications to Biomembranes | 461 |
|    A. Lipid Polymorphism | 461 |
|    B. *Tetrahymena* Phosphonolipids | 464 |
|    C. The Purple Membrane | 466 |
|    D. Effects of Membrane-Active Agents | 468 |
|    E. Relaxation Times | 470 |
| V. Conclusion | 472 |
|    References | 472 |

## I. Properties of Membranes

Biological membranes contain a wide variety of chemical species, the major ones being lipids, proteins, and carbohydrates (Harrison and Lunt, 1980). The lipid components are thought to interact as shown in Fig. 1 to form phases with a variety of superstructures. The most common structure is the bilayer, where two layers of lipid lie tail-to-tail to form a membrane with hydrophobic center and hydrophilic surfaces, suitable for enclosing the hydrophilic biochemical machinery of a cell within a surrounding hydro-

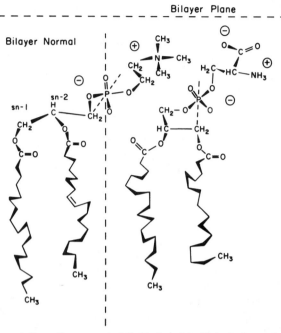

Fig. 1. Representation of some typical lipids found in biological membranes; the headgroups shown are phosphatidylcholine and phosphatidylserine.

philic medium. Figure 1 shows two principal phospholipids, phosphatidylcholine and phosphatidylserine, with fatty acyl chains that may be either saturated or unsaturated. Note the many conformations available to the fatty acyl chains and to the hydrophilic headgroups. The former have been extensively studied by deuterium NMR (Seelig, 1977; Seelig and Seelig, 1980; Smith, 1981, 1983; Davis, 1983) and the latter by both deuterium and phosphorus NMR (Seelig, 1978; Cullis and de Kruijff, 1979; Griffin, 1981; Smith and Jarrell, 1983). The nature of these conformations, and the rates of conversion between them, are thought to influence strongly the biological properties of membranes. This chapter deals with the use of $^{31}$P NMR to study headgroup conformations in membranes as well as the nature of the phases formed by the lipids. The influence on the $^{31}$P-NMR parameters of the rates of the various motions available to a lipid molecule is also discussed. We say very little about high-resolution spectra of lipid systems because this is dealt with by Dennis and Plückthun in Chapter 14.

## II. ³¹P-NMR Spectra of Ordered Systems

### A. Chemical-Shift Anisotropy

The $^{31}$P chemical shift of a rigid phosphodiester, such as is found in membrane lipids (Fig. 1), depends on the orientation of the group with respect to the magnetic field of the spectrometer. This is represented in Fig. 2A, where the chemical shifts expected along the three principal directions are labeled $\sigma_{11}$, $\sigma_{22}$, and $\sigma_{33}$. For orientations other than along the three principal axes, the chemical shifts will lie somewhere between $\sigma_{11}$, $\sigma_{22}$, and $\sigma_{33}$. These $\sigma$ values are the components of the chemical-shift (or shielding) tensor, which describes in general the Zeeman interaction between the $^{31}$P nucleus and the applied magnetic field (see also Gorenstein, Chapter 1, and Dennis and Pluckthün, Chapter 14). When one talks about the principal components of the tensor, this means that the axis system has been chosen in the phosphate moiety to cause cross-terms in the tensor, such as $\sigma_{12}$, to be zero. Knowledge of the orientation of the axis system (and of $\sigma_{11}$, $\sigma_{22}$, and $\sigma_{33}$) is all that is required to calculate the chemical shift expected for any orientation. For example, in the plane containing the 2 and 3 axes, for an angle $\theta$ between the applied magnetic field and the 2 axis,

$$\sigma(\theta) = \sigma_{22} \cos^2 \theta + \sigma_{33} \sin^2 \theta \tag{1}$$

If one has a randomly oriented sample, all angles between the applied field and the principal axes are populated. The $^{31}$P-NMR spectrum is thus the sum of spectra for all possible orientations. Rather than a featureless lump, the so-called powder spectrum has several distinctive features: $\sigma_{11}$ and $\sigma_{33}$ define the outermost edges, and $\sigma_{22}$ leads to a peak. The principal values of the chemical-shift tensor can be estimated directly from the spectrum, as shown in Fig. 2; the directions of the principal axes in the molecular system can only be found by studying a single crystal of known structure. Note the values given in Fig. 2A for $\sigma_{11}$ and $\sigma_{33}$—the powder spectrum spans some 190 ppm. Principal tensor components for the phosphate moiety in a variety of compounds have been compiled (Seelig, 1978).

From the case of the rigid solid, consider now an example closer to the situation found in membranes. We shall allow rapid anisotropic motion that can average some of the components of the chemical-shift tensor. Suppose, for simplicity, that this motion is about the 1 axis, thus averaging $\sigma_{22}$ and $\sigma_{33}$:

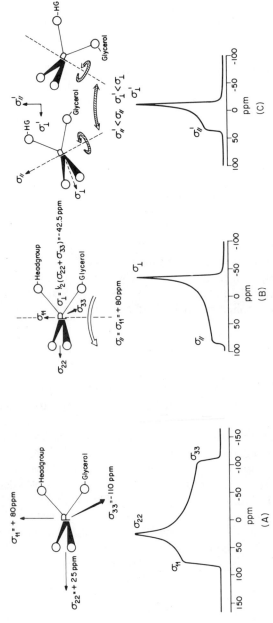

**Fig. 2.** Various possible motional states of the phosphodiester moiety of a membrane lipid and the expected $^{31}$P-NMR spectra: (A) static phosphodiester; (B) ordered phosphodiester, rapid axial rotation; (C) disordered phosphodiester, rapid axial rotation. In the case of motional averaging of the chemical-shift tensor to axial symmetry, $\sigma_\parallel$ refers to the chemical shift for the external magnetic field parallel to the unique axis, and $\sigma_\perp$ to that for the field in the equatorial plane.

## 15. $^{31}$P NMR of Phospholipids in Membranes

$$\sigma_\perp = \int_0^{2\pi} (\sigma_{22} \cos^2\theta + \sigma_{33} \sin^2\theta) d\theta \bigg/ \int_0^{2\pi} d\theta$$

$$= \frac{\sigma_{22} + \sigma_{33}}{2} \qquad (2)$$

Thus the chemical shift will have the same value for the field anywhere in the 2-3 plane but a different value when the field is perpendicular to the plane; the chemical-shift tensor is axially symmetric. The chemical shift expected when the field is parallel to the unique axis is labeled $\sigma_\parallel$, whereas that expected for the field perpendicular to this axis is $\sigma_\perp$. Figure 2B demonstrates this averaging and the resultant powder spectrum. Note the characteristic shape with the buildup of intensity at $\sigma_\perp$. Both $\sigma_\parallel$ and $\sigma_\perp$ can be measured from the spectrum as shown, but this will involve some error, the magnitude of which depends on the linewidths of the individual components of the powder line shape (Seelig, 1978).

Proceeding to the situation most relevant to membranes, let us now allow rapid motion of limited amplitude of the 1 axis of the phosphodiester moiety, retaining the rapid motion *about* the 1 axis (Fig. 2C). The 1 axis now moves in a cone, and there is partial averaging of the former $\sigma_\perp$ and $\sigma_\parallel$ to yield new, smaller effective values $\sigma'_\perp$ and $\sigma'_\parallel$. The effective tensor still has axial symmetry, but the total chemical-shift anisotropy (CSA) represented by the spectrum, $\Delta\sigma = \sigma'_\parallel - \sigma'_\perp$, is reduced. The amount by which it is reduced is related to the allowed amplitude of the motion. Should the amplitude of this rapid motion become totally unrestricted, the motion is effectively isotropic, and the pattern collapses to a single line with a chemical shift identical to that which would be observed in nonviscous solution:

$$\sigma_{\text{iso}} = \tfrac{1}{3}(\sigma_{11} + \sigma_{22} + \sigma_{33})$$
$$= \tfrac{1}{3}(\sigma_\parallel + 2\sigma_\perp)$$
$$= \tfrac{1}{3}(\sigma'_\parallel + 2\sigma'_\perp) \qquad (3)$$

This value varies slightly from lipid to lipid and with conditions such as pH, presence of ions, etc., but is in the region of $\pm 5$ ppm.

The spectrum shown in Fig. 2C is typical of many seen for membrane systems. It is sometimes said that the observed $\Delta\sigma$ is a measure of the degree of order of the phosphate group—the amplitude of angular excursion during motional averaging. This is unfortunately only partially true, for the shape and width of the pattern depend critically on the orientation of the axis of motional averaging with respect to the principal components of the chemical-shift tensor $\sigma_{11}, \sigma_{22}$, and $\sigma_{33}$, as has been pointed out forcefully by Thayer and Kohler (1981). Furthermore, it is often considered that the type of motional averaging, and the resultant powder spectrum just described, is characteristic of a bilayer arrangement of lipids. This is also not necessarily true, although most systems known to contain a bilayer have indeed yielded spectra similar to that shown in Fig. 2C. The rate of motion within a partially ordered environment also influences the shape of the $^{31}$P powder spectrum. This aspect, which has been largely neglected in $^{31}$P-NMR studies of membranes, is treated in detail in the next section.

## B. Influence of Motional Rates

Section II,A deals with the spectra of systems with very slow or very fast rotational rates. In the intermediate region (for diffusion coefficients $10 < R < 10^7$ s$^{-1}$ and the spectral widths usually encountered with membranes), the $^{31}$P-NMR spectra are sensitive to both type and rate of the motion. Although the lipids of liquid crystalline membranes usually have motional rates in the fast-limit region, this is often not so at lower temperatures (Fig. 9). In such cases, the spectra can be used as a source of information about the nature of the motion and orientation of the headgroup.

Campbell et al. (1979) have developed computational methods for simulating spectra for nuclei of spin $\frac{1}{2}$. They assumed, following Mason et al. (1974), that motion is very anisotropic and that rotation about one axis is much faster than about the others. In such a case, an asymmetric rotation can be treated as if it were axially symmetric and can be described by two diffusion coefficients $R_{\parallel}$ (fast motion) and $R_{\perp}$ (net slow motion).

Figure 3 shows how the shapes of the spectra respond to changes in the rate of the fast motion ($R_{\parallel}$). This is similar to Fig. 5 in the original paper by Campbell et al. (1979) but was generated using an improved version of the program (courtesy of Dr. C. F. Polnaszek), which yields correct simulations for very slow motions. Campbell et al. (1979) showed how such simulations could provide information about the orientation of the rotational axes and the rates of motion about these axes. One should be cautious in view of the various assumptions made in such calculations. Simulated spectra will depend on the model of motion (Brownian motion was used by Campbell et al., but others such as free diffusion or discrete-jump diffusion are possible). The motion of the phosphate group in lipids is a superposition of internal motion and axial diffusion about the long molecular axis; more studies on the general problem of motion of lipid molecules in membranes are necessary before a complete interpretation of $^{31}$P-NMR spectra can be performed with confidence. Furthermore, for proper simulations, good quality, undistorted experimental spectra are necessary (Section III). Campbell et al. (1979) assumed in their calculations that the component linewidth, and hence the relaxation time $T_2$, was constant across the pattern. This is good

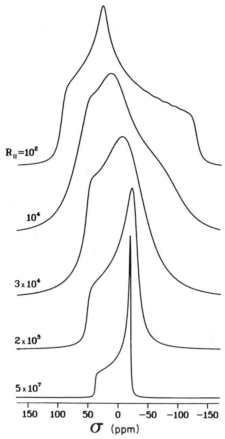

**Fig. 3.** Simulated $^{31}$P-NMR spectra for anisotropic Brownian diffusion. The rates of fast diffusion $R_{\parallel}$ are given for each spectrum, $R_{\perp} = 15$ rad s$^{-1}$. Spectrometer frequency 109 MHz, chemical-shielding tensor: $\sigma_{11} = 100$, $\sigma_{22} = 30$, $\sigma_{33} = -130$ ppm. Assumptions about the orientation of the diffusion axis and about linewidths were taken from the Campbell *et al.* (1979) example of the simulation of DPPC spectra (Fig. 5 in their paper).

only as a rather crude approximation (Section IV,E). Considerable insight can be gained by studies of proximal nuclei ($^{13}$C, $^2$H, $^1$H) and by measuring relaxation times of the various nuclei at different magnetic field values. The present simulations give a good idea of the behavior expected for various models, but analysis in terms of these models is ambiguous at present.

Thus far we have discussed only the effect of varying the rate of motion about one axis. For relatively small membrane fragments (<10,000 Å), the rate of overall isotropic diffusion is also important, leading to further averaging of spectral components. This problem was studied by Burnell *et al.* (1980b). For relatively small, spherical vesicles, two diffusion processes contribute to the averaging of the spectra: rapid Brownian tumbling of the entire vesicle (characterized by the diffusion coefficient $D_t$) and lateral diffusion of lipids around the vesicle (with diffusion coefficient $D_{diff}$). The correlation time $\tau_c$ is given by the equation

$$\frac{1}{\tau_c} = \frac{6}{R^2}(D_t + D_{diff}) \tag{4}$$

where $R$ is the radius of a vesicle and $D_t = KT/8\pi R\eta$, where $\eta$ is the viscosity (Bloom *et al.*, 1975). Using such a motional model and Freed's theory for the motional dependence of line shapes (Freed *et al.*, 1971), Burnell *et al.* (1980b) were able to simulate experimental spectra of dioleoylphosphatidylcholine vesicles for different temperatures and viscosities of the medium (Fig. 4). As seen from Eq. (4), the effectiveness of the averaging process will increase with decreasing size of the vesicle. Figure 5 shows the dependence of the spectral shape on the size of the vesicle (note that the assumed value of $D_{diff}$ is important and that it does depend on the type of lipid). In the limiting case of very fast motion (very small vesicles), averaging of anisotropy is complete, and a narrow line is observed at the isotropic chemical shift. Of course, even for large vesicles, partial averaging similar to that shown in Fig. 4 is possible by an increase of $D_{diff}$. For a proper interpretation of spectra such as that for 2500 Å in Fig. 5, it is essential to estimate the size of the membrane fragments. Furthermore, measuring chemical-shift anisotropy from such a spectrum can lead to quite a large error, which can be reduced by applying simulation including all motions.

## C. Types of Phases

In Section II,A, we consider the types of spectra expected for rapid anisotropic motion within a bilayer superstructure (Fig. 6A). It has been known for some time that certain lipids will form a completely different type of structure in which the molecules project radially from the center of a cylinder (Fig. 6B; Luzzati and Husson, 1962). As mentioned in Section II,B, rapid lateral diffusion within a bilayer phase only affects $^{31}$P-NMR line shapes if the lamellae enclose a particle of small radius. The cylinders in a hexagonal phase have a very small radius, and therefore lateral diffusion about the cylinder axis can cause further averaging of tensor components. The unique axis of the system now becomes the axis of the cylinder, and thus we label its chemical-shift component $\sigma_\parallel^H$. However, noticing that along this axis the field would be roughly normal to the fatty acyl chains, we would expect the value of this chemical shift to be similar to $\sigma_\perp^L$. On the other hand,

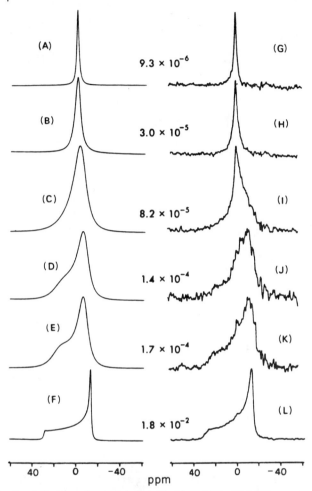

**Fig. 4.** Simulated (A–F) and experimental (G–L) 80-MHz $^{31}$P-NMR spectra of dioleoylphosphatidylcholine vesicles ($R = 980$ Å) for different values of $\tau_c$ (in seconds) as indicated. The spectra were simulated using $(\sigma_\| - \sigma_\perp) = 3550$ Hz (43.8 ppm). The experimental spectra were measured under the following conditions: (G) 60°C; (H) 25°C; (I) 0°C; (J) −10°C, 30% glycerol; (K) −15°C, 30% glycerol; (L) 30°C, unsonicated liposomes. From Burnell et al. (1980b).

$\sigma_\perp^H$ will be an average of $\sigma_\perp^L$ and $\sigma_\|^L$, owing to rapid motion around the cylinder axis. Hence

$$\sigma_\|^H = \sigma_\perp^L \tag{5}$$

$$\sigma_\perp^H = \frac{\sigma_\perp^L + \sigma_\|^L}{2} \tag{6}$$

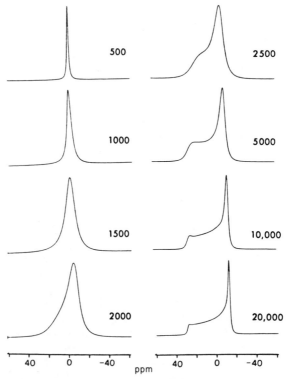

**Fig. 5.** Simulated $^{31}$P NMR spectra of dioleoylphosphatidylcholine vesicles of different sizes at 30°C. The spectra were simulated using $(\sigma_\| - \sigma_\perp) = 3550$ Hz (43.8 ppm), $\eta = 0.008$ P, and radius (in angstroms) as indicated. From Burnell et al. (1980b).

and

$$\Delta\sigma^H = \frac{-\Delta\sigma^L}{2} \quad (7)$$

The net result of this is that the $^{31}$P powder pattern for the hexagonal phase has a $\Delta\sigma$ that is roughly half that of a corresponding lamellar phase and a shape that has the buildup of intensity owing to the axial component on the other side of the chemical-shift zero. One might be tempted to use the breadth and sidedness of the $^{31}$P-NMR patterns as diagnostic for the presence of either phase. This should always be done in the presence of other corroborating data because a change in the location of the axis of motional averaging relative to the phosphodiester moiety, or a change in the headgroup conformation, with no change in the overall symmetry of the superstructure can cause a transition from the type of spectrum in Fig. 6A to that in Fig. 6B (Thayer and Kohler, 1981).

Another type of lipid phase that has been invoked on the basis of X-ray

## 15. 31P NMR of Phospholipids in Membranes

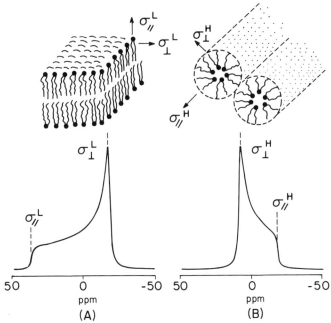

**Fig. 6.** Representation of the (A) bilayer and (B) hexagonal ($H_{II}$) phases formed by membrane lipids and their expected $^{31}$P-NMR spectra.

diffraction, freeze-fracture electron microscopy, and NMR data (Wieslander *et al.*, 1978, 1981; Burnell *et al.*, 1980a; de Kruijff *et al.*, 1980) contains an effectively isotropic distribution of lipids. This could come about in various ways, such as a fish-net structure comprised of bilayers or small micellar structures similar to those found with detergents and sonicated phospholipids. In either case, rapid motion over an effectively isotropic path will lead to a $^{31}$P-NMR spectrum consisting of a single resonance centered at $\sigma_{iso}$. It is even possible to obtain such a spectrum from a single conformation of the lipid headgroup making particular angles with the axis for overall motion of the entire lipid molecule (Thayer and Kohler, 1981). As with the spectra indicative of hexagonal phases, independent confirmation of the presence and nature of the isotropic phase must be sought, such as $^2$H or $^{14}$N NMR of the same system (Deslauriers *et al.*, 1982).

## III. Obtaining $^{31}$P-NMR Spectra

Up to this point we have discussed the basis for and the interpretation of $^{31}$P-NMR spectra of membrane systems, but we have not questioned the fidelity of the spectra themselves. For a rigid phosphodiester, the $^{31}$P spectra

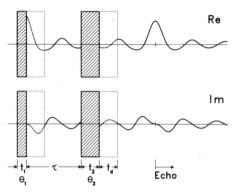

**Fig. 7.** Schematic diagram of the events during a two-pulse Hahn echo sequence. Re and Im refer to the so-called real and imaginary NMR signal components, that is, the two channels of the quadrature phase-sensitive detector. Ideally, $\theta_1 = 90°$ and $\theta_2 = 180°$, with phase cycling of $\theta_1$ and $\theta_2$. The dashed regions of the NMR signals following the pulses represent the "deadtime" $t_d$ of the receiver. From Rance and Byrd (1983).

have widths of ~190 ppm (23.1 kHz for a spectrometer operating at 7.05 T); the $T_2$ values can be quite short and the $T_1$ values quite long. This combination implies that the powder spectra could be badly distorted by insufficient attention to the methods of spectral acquisition. Most of the membrane spectra reported have, in fact, suffered from greater or lesser degrees of distortion, in some cases leading to erroneous conclusions.

The first problem arises from the spectral width. The effective strength of the radio-frequency pulse must be such to excite equally the nuclei contributing to all frequencies within the powder spectrum; in practical terms this means that the 90° pulses should be of the order of 5–10 µs in duration. A second problem arises from the relatively short $T_2$ values expected in these spectra. Most NMR spectrometers are the high-resolution type, which implies probes of high $Q$ factor and therefore long ringdown time after application of the radio-frequency pulse. To avoid spectral distortion because of probe and receiver ringdown, it is common high-resolution practice to insert a delay between excitation of the system and the beginning of the acquisition of the free induction decay $t_d$ (Fig. 7). This procedure necessarily yields a loss of signal intensity and the need for a first-order phase correction of the data. In the limit of short $T_2$ and long $t_d$, one ends up with a wobbly baseline and no signal!

The problems referred to here were experienced earlier with $^2$H NMR of membranes and to a large part were resolved by means of a quadrupole echo (Davis *et al.*, 1976). Considerable improvement in $^{31}$P-NMR spectra of membranes has been reported by use of the analogous Hahn echo based on residual chemical-shift anisotropy (Rance and Byrd, 1983). A schematic of

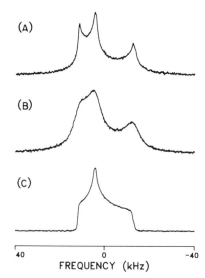

Fig. 8. Comparison of the $^{31}$P-NMR spectra (121.47 MHz) of a dry powder of dipalmitoylphosphatidylcholine (DPPC) acquired with (A) a single 90° pulse and $^1$H decoupling; (B) a single 90° pulse and no $^1$H decoupling; and (C) a Hahn echo and $^1$H decoupling, where $\theta_1 = 90°$, $\theta_2 = 180°$, and the pulse spacing was 60 µs. The 90° pulse width was 7 µs and the acquisition delay for the single pulse spectra was 12 µs. From Rance and Byrd (1983).

the behavior of the magnetization in the real (Re) and imaginary (Im) channels of a quadrature spectrometer is shown in Fig. 7. A 90° pulse ($\theta_1$) excites magnetization in the Re channel that decays in amplitude according to $T_2$; a 180° pulse ($\theta_2$) after a delay $\tau$ results in a refocusing of this magnetization to reach an echo maximum in the Re channel after a further time $\tau$. If $\tau$ is longer than the ringdown time of the system, no spurious signal from this latter source will remain. Note that when the echo reaches a maximum in the Re channel, the magnetization in the Im channel passes through zero; this condition can be used to adjust the phase relationship between the quadrature channels using the time-domain signals. If the exact maximum of the echo can be determined and a Fourier transform done on half the echo starting from the time of maximum amplitude, no first-order phase correction should be required on the transformed signal. Extensive phase cycling of the two pulses serves to minimize other aberrations owing to miss-set pulses, channel imbalance, and residual baseline instability. To avoid saturation effects, some estimate of the $T_1$ values of the system must be made and an appropriate cycle time employed.

Figure 8 shows the experimental $^{31}$P-NMR spectra of a dry powder of dipalmitoylphosphatidylcholine (DPPC) under a variety of conditions. Spectrum (A) was obtained under conditions similar to the best one might hope for from a high-resolution spectrometer, using single 90° pulses and a preacquisition delay of 12 µs. This is to be compared with spectrum (C), obtained with the Hahn echo, and with that simulated for a rigid powder in Fig. 2A. The single-pulse method clearly causes extreme spectral distortion;

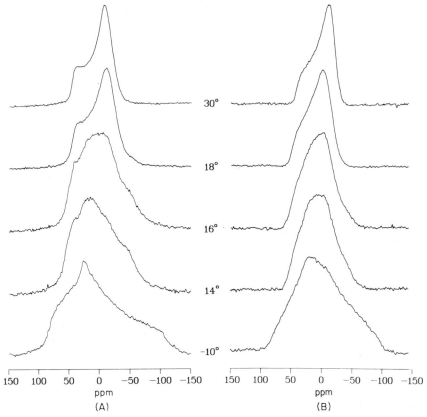

**Fig. 9.** Comparison of the $^{31}$P-NMR spectra (121.47 MHz) of hydrated DPPC at various temperatures taken with single 90° pulses (A) or with Hahn echoes (B). Unpublished data of D. Marsh, I. Ekiel, and I. C. P. Smith.

attempts to measure the principal components of the chemical-shift tensor would lead to overestimates by 20%. The proton-coupled spectrum (B) in Fig. 8 is shown to demonstrate that incomplete proton decoupling can also cause serious distortions; in this case it actually masks some of the distortions owing to the use of single 90° pulses.

Figure 9 shows the 121-MHz $^{31}$P-NMR spectra of hydrated DPPC taken over a range of temperatures. DPPC undergoes three separate phase transitions (liquid crystal-to-gel, 42°; pretransition, 36°; subtransition, ~14°). Those in Fig. 9A were taken with the single-pulse method and those in Fig. 9B using the Hahn echo. The aberrations in the single-pulse spectra over the range 5–16°C misleadingly suggest the presence of two components of different $\Delta\sigma$. This structure is completely absent from the spectra obtained via the echo method.

## IV. Applications to Biomembranes

### A. Lipid Polymorphism

Biological membranes contain a variety of different lipids, and many studies have been performed to investigate the properties of simple model membranes composed of one or two lipidic species. Although biological membranes usually give the bilayer type of $^{31}$P-NMR spectra, this is not always true for lipid dispersions (Cullis and de Kruijff, 1979). As already mentioned in Section II,C, $^{31}$P NMR alone cannot be used to determine the type of lipid organization (bilayer or hexagonal phase, small vesicles). However, large amounts of information about lipid polymorphism are available from X-ray (Tardieu *et al.*, 1973; Rand and Sengupta, 1972) and freeze-fracture electron microscopy studies (Papahadjopoulos *et al.*, 1976), allowing detailed $^{31}$P-NMR studies of polymorphic phase transitions and of the factors influencing phase preferences in simple and mixed lipid systems.

Some lipids that are present in significant concentrations in biological membranes form bilayer structures, for example, phosphatidylcholines (PC), sphingomyelin, and diglucosyl and digalactosyl diglycerides (Cullis *et al.*, 1982). Also, the sodium salts of acidic lipids—cardiolipin, egg phosphatidylserine, egg phosphatidylglycerol, and soya phosphatidylinositol—adopt bilayer structures at neutral pH and ambient temperatures (Cullis *et al.*, 1982). Other lipids, especially unsaturated phosphatidylethanolamines, exhibit preference for the hexagonal $H_{II}$ phase (Cullis and de Kruijff, 1978a,b). This preference is dependent on the nature of the fatty acids, and for unsaturated chains, a relatively sharp phase transition from a bilayer structure at low temperature to a hexagonal ($H_{II}$) structure at high temperature can be observed. Figure 10 shows such a transition in phosphatidylethanolamine (PE) derived from soya phosphatidylcholine (Cullis and Hope, 1980). In this case the fatty acids are highly unsaturated and the transition temperature low. The hexagonal phase is strongly preferred above 0°C. With increasing degree of saturation of the fatty acyl chains, the transition temperature increases, for example, for PE from *Escherichia coli* the transition to the hexagonal $H_{II}$ phase occurs around 54°C (Cullis and de Kruijff, 1978b). The behavior of acidic lipids depends very much on pH and on interactions with cations, especially $Ca^{2+}$ (Cullis *et al.*, 1982). For example, $Ca^{2+}$ triggers the transition from a bilayer to a hexagonal $H_{II}$ phase in beef heart cardiolipin and phosphatidic acid (Cullis *et al.*, 1982). In the case of egg phosphatidylserine, decreasing pH causes a transition from bilayer to hexagonal phase at approximately the p$K$ of the headgroup (Hope and Cullis, 1980).

In addition to the study of the phase behavior of individual lipids, the question arises as to how or whether they exhibit their preferences in

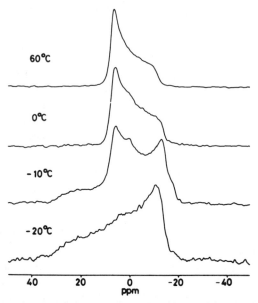

**Fig. 10.** Temperature dependence of the 81-MHz $^{31}$P-NMR spectra (proton decoupled) of phosphatidylethanolamine dispersed in buffer. From Cullis and Hope (1980).

complex biological membranes. A first step toward the answer comes from studies of the properties of model bi- or tricomponent lipid systems (Cullis *et al.*, 1982; Tilcock and Cullis, 1982; Cullis and Hope, 1980).

Very often for lipid mixtures undergoing transition from a bilayer to an $H_{II}$ hexagonal phase, a narrow spectral component appears at the isotropic chemical shift (Cullis and de Kruijff, 1979; de Kruijff *et al.*, 1979). Similar narrow resonances were observed in lipids extracted from biological membranes (de Grip *et al.*, 1979; Burnell *et al.*, 1980a; Cullis *et al.*, 1980b; Deslauriers *et al.*, 1982).

The same type of narrow resonance was observed during fusion of membrane vesicles (Verkleij *et al.*, 1979, 1980). This was interpreted in terms of a structure allowing rapid isotropic averaging (Cullis *et al.*, 1978). A variety of possibilities can lead to such averaging—small vesicles, micelles, inverted micelles, even cubic or rhombic phases. In some cases the narrow resonances were correlated with the appearance of "lipidic particles" in freeze-fracture electron micrographs (Verkleij *et al.*, 1979; de Kruijff *et al.*, 1979) and interpreted as intramembrane inverted micelles.

The possible biological involvement of lipid polymorphism in many events such as membrane fusion, exocytosis, transport across bilayers, and intermembrane connections was extensively discussed by Cullis and de Kruijff (1979) and Cullis *et al.* (1982).

## 15. ³¹P NMR of Phospholipids in Membranes

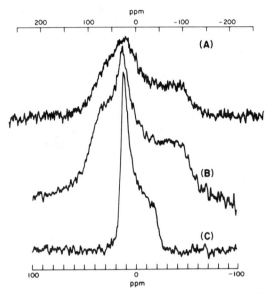

**Fig. 11.** ³¹P-NMR spectra of the sodium salt of didodecanoylphosphatidyldidodecanoylglycerol (A) in powder form; (B) as aqueous dispersions below the phase transition temperature; and (C) as aqueous dispersions above the phase-transition temperature. From Noggle et al. (1982).

Once again it should be stressed that any conclusions about the type of phase adopted by lipids should be supported by other techniques (X-ray diffraction, freeze-fracture). Figure 11 shows that aqueous dispersions of a synthetic phospholipid, phosphatidyldiacylglycerol, give "hexagonal-type" spectra, although freeze-fracture electron microscopy demonstrates a bilayer type of organization of the lipids (Noggle et al., 1982). The observed shape of the ³¹P-NMR spectrum can be explained by a phosphodiester conformation different from that usually found in phospholipids (as a consequence of the unusual structure of the molecule). The difference in conformation leads to a change in the orientation of the axis of rapid motional averaging with respect to the principal axes of the chemical-shift tensor (Noggle et al., 1982). Below the temperature of the liquid crystal-to-gel transition of the acyl chains didodecanoylphosphatidyldecanoylglycerol also does not give a typical bilayer spectrum but one very similar to that of the dry sample. This means that below the transition temperature, all motions that could average the ³¹P shielding tensor are slow on the NMR scale. This situation is not very common for membranes, but has been observed in several other cases. One such case is dipalmitoylphosphatidylcholine below the subtransition temperature (Füldner, 1981) (see Fig. 9, $-10°$C). Similar behavior was reported for some acidic phospholipids as an effect of interaction with $Ca^{2+}$ (Hope and Cullis, 1980; Farren and Cullis, 1980).

## B. *Tetrahymena* Phosphonolipids

*Tetrahymena* is one of the organisms that contains, in addition to phospholipids, phosphonolipids (derivatives of 2-aminoethylphosphonic acid). The phase behavior of these unusual lipids was studied by Jarrell *et al.* (1981) and Deslauriers *et al.* (1982). Because of the presence of C—P linkages, one cannot assume that the $^{31}P$ shielding tensor is oriented with respect to the molecular reference plane in the same manner as in the usual reference compounds, phosphoethanolamine and barium diethyl phosphate. A difference in the orientation of the shielding tensor can completely change the line shape of the $^{31}P$-NMR spectra (Thayer and Kohler, 1981) so that one cannot draw conclusions about the type of lipid phase based only on $^{31}P$-NMR spectra (Jarrell *et al.*, 1981). This difficulty was overcome by mixing phosphonolipid with phospholipids of well-known phase behavior —dipalmitoylphosphatidylcholine and egg phosphatidylethanolamine (PE). Figure 12A shows the spectra for a phosphonolipid–DPPC mixture. There is a difference of ~28 ppm between the isotropic chemical shifts of these two lipids, so that the two powder patterns are partially separated. To effect a better resolution of the two patterns, that of DPPC was saturated using the DANTE pulse sequence (Morris and Freeman, 1978), leaving only the pattern of the phosphonolipid (Fig. 12B). The powder pattern of DPPC was obtained by subtraction (Fig. 12C). One can see that both lipid species give $^{31}P$ powder patterns of the same shape, with that of the phosphonolipid having a smaller chemical-shift anisotropy. For the phosphonolipid–PE mixture, a temperature-induced phase transition to a hexagonal structure was observed (Jarrell *et al.*, 1981). Again, in the hexagonal phase the shapes of both patterns were similar. Based on these results obtained for the simple lipid mixtures, the phase behavior of the lipids extracted from *Tetrahymena* was studied (Jarrell *et al.*, 1981; Deslauriers *et al.*, 1982). The $^{31}P$-NMR spectra suggested a bilayer organization for the lipids between $-20$ and $20°C$. At higher temperatures, relatively narrow lines appeared in both phospho- and phosphonolipid region, and at $35°C$ only non-bilayer components remained in the spectra. Parallel to the changes monitored by $^{31}P$ NMR, a drastic narrowing of the $^{14}N$- and $^{2}H$-NMR patterns was observed [10% w/w of 2-(14-$^{2}H_2$)palmitoylphosphatidylcholine was added to the lipid extract for the deuterium spectra]. At $40°C$, both $^{2}H$ and $^{14}N$ spectra collapsed to a single narrow resonance, which demonstrated that the changes seen by $^{31}P$ NMR were not the result of local changes in the

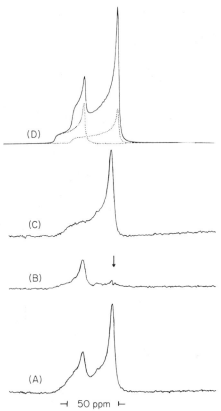

**Fig. 12.** ³¹P-NMR spectra (121.47 MHz) of a phosphonolipid–DPPC mixture (1:4, w/w) in water at pH 7.0 at 50°C. (A) 90° pulses, 2400 accumulations at 1 scan s⁻¹. (B) As in (A) but using a DANTE pulse sequence at the indicated frequency (arrow) with 4° pulses separated by 50 μs and a total saturation time of 800 ms. (C) Difference spectrum of (A)–(B). (D) Computer simulation of spectrum (A) in terms of two axially symmetric powder patterns characterized by a difference in isotropic chemical shift of 28 ppm, angular-independent and angular-dependent linewidths of 70 Hz, chemical-shift anisotropies of +46 and +30 ppm for phospho- and phosphonolipid, respectively. From Jarrell *et al.* (1981).

headgroup conformation. It is very probable that the observed narrow features represent a motionally averaged phase similar to that reported for other membrane systems (Cullis and de Kruijff, 1978a,b; Burnell *et al.*, 1980a; Cullis *et al.*, 1980b). The observed polymorphism of the lipids extracted from *Tetrahymena* was attributed to the high content of ethanolamine headgroups and highly unsaturated fatty acids (Deslauriers *et al.*, 1982).

## C. The Purple Membrane

The purple membrane of *Halobacterium cutirubrum* is one of the most simple natural membranes, containing only a single protein — bacteriorhodopsin. The lipids of the purple membrane are very unusual; rather than the usual fatty acids, they contain phytanyl chains with ether linkages (Kates and Kushwaha, 1978). The major phospholipid, phosphatidylglycerophosphate (PGP), constituting 85% of the phospholipids, contains two phosphate groups, one mono- and one diesterified. As a consequence, the $^{31}$P-NMR spectra of the purple membrane are a superposition of two patterns (Fig. 13), which can be separated by computer simulation (dashed spectra in Fig. 13). There are contradictory reports about the presence of a lipid phase transition between 5 and 35°C in the purple membrane (Degani *et al.*, 1978; Jackson and Sturtevant, 1978) and in lipids extracted from halobacteria (Plachy *et al.*, 1974; Chen *et al.*, 1974). $^{31}$P-NMR studies of the temperature dependence of the chemical-shift anisotropy (Fig. 14) showed no discontinuity for either the purple membrane or the major phospholipid PGP (Ekiel *et al.*, 1981). This is consistent with calorimetric results arguing against the presence of a phase transition (Chen *et al.*, 1974; Jackson and

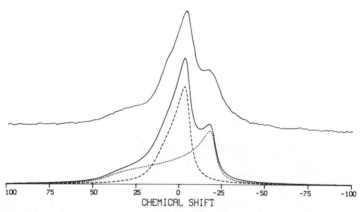

**Fig. 13.** (Upper) $^{31}$P-NMR spectrum (121.5 MHz) of the purple membrane from *Halobacterium cutirubrum* obtained with high-power $^{1}$H decoupling at 15°C; spectral width 125 kHz; acquisition time 6 ms; recycle time 1 s; pulse width 45°; decoupler on 4 µs before acquisition and during acquisition but off during remainder of cycle; Fourier transform of 16,384 data points after zero filling. (Lower) Computer simulation of the above in terms of two axially symmetric powder patterns characterized by effective chemical-shift anisotropies of +61 and +18.5 ppm; the angular-independent and -dependent linewidths were 300 and 200 Hz, and 350 and 250 Hz, for the phosphomonoester and phosphodiester, respectively. The broken curves are the powder patterns for the two types of phosphate ester present. From Ekiel *et al.* (1981).

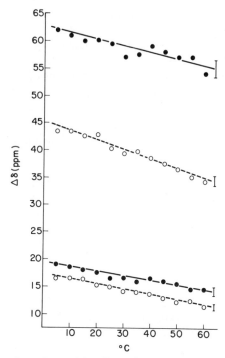

**Fig. 14.** Temperature dependence of the effective $^{31}$P chemical-shift anisotropies ($\Delta\sigma$) for the phosphodiester (upper data) and phosphomonoester (lower data) moieties in the purple membrane (●) and phosphatidylglycerophosphate (○) in excess water. The anisotropies were taken from computer simulations of the spectra. The bars to the right of the curves are estimates of the precision of the measurements. From Ekiel et al. (1981).

Sturtevant, 1978). It is possible that abrupt changes detected by $^{13}$C- and $^1$H-NMR spectroscopy (Degani et al., 1980) and EPR (Chignell and Chignell, 1975); Plachy et al., 1974) represent changes in the rate of motion of only the phytanyl chains.

Figure 14 shows that there is a large difference in chemical-shift anisotropy of the disubstituted phosphate group of PGP in purple membrane and in aqueous dispersions. This difference, attributable to interaction with bacteriorhodopsin, can be interpreted as either a decrease of the rate or amplitude of lipid motion or as a change in the orientation of the phosphate groups relative to the axis of motional averaging in the membrane. It is one of the largest spectral effects of lipid–protein interaction reported so far (see following discussion), but could be due to relatively small changes in parameters.

## D. Effects of Membrane-Active Agents

The $^{31}$P nucleus is a natural probe of events ocurring at the headgroup region of membrane lipid. As $^{31}$P NMR of membranes is relatively easy to do, many studies have been reported on the effects of agents thought to act by perturbing membranes—anesthetics (Cullis *et al.*, 1980a; Boulanger *et al.*, (1981; Hornby and Cullis, 1981), peptide hormones (Van Echteld *et al.*, 1981; Rajan *et al.*, 1981), fusion-inducing compounds (Hope and Cullis, 1981), ions (Hope and Cullis, 1980; Akutsu and Seelig, 1981), psychotropic drugs (Frenzel *et al.*, 1978), proteins (Seelig and Seelig, 1978; Rajan *et al.*, 1981; McLaughlin *et al.*, 1981; Seelig *et al.*, 1981; Deese *et al.*, 1981; Yeagle, 1982), and sterols (Brown and Seelig, 1978; Rajan *et al.*, 1981). In view of the large number of studies, we confine ourselves to only a few in order to demonstrate the methods.

Figure 15 shows the $^{31}$P-NMR spectra of dimyristoylphosphatidylcholine (DMPC) alone (Fig. 15A) and in the presence of several proteins (Fig. 15B–D) and cholesterol (Fig. 15E). In all cases the spectra are typical of that found for bilayers, but the proteins cause a broadening of the line shape and very little change in $\Delta\sigma$ (slight decreases). In contrast, both the spin–spin ($T_2$) and spin–lattice ($T_1$) relaxation times of $^{31}$P were very much shorter in the presence of protein. No extra spectra are seen because of boundary lipid—that constrained to remain in the immediate proximity of the protein. Earlier spin-label ESR studies on the cytochrome oxidase system had demonstrated clearly the presence of two spectra, one from lipid bound to protein and one from normal bilayer lipid (Jost *et al.*, 1977). $^2$H-NMR spectra of chain-labeled lipids had also failed to detect a separate protein-bound lipid phase (Paddy *et al.*, 1981). The apparent discrepancy was rationalized in terms of a model where lipid in proximity of protein is disorganized slightly and immobilized strongly but remains in equilibrium with the remaining boundary lipid. The rate of interconversion between the bound and unbound states is assumed to be rapid on the time scale of $^2$H and $^{31}$P NMR but slow on the scale of ESR, thus yielding an average spectrum in the former case but two spectra in the latter. An exchange rate of $10^6 - 10^7$ s$^{-1}$ has been estimated (Paddy *et al.*, 1981). The strong effects on the relaxation times $T_1$ and $T_2$ are due to the immobilization of the lipid near the protein.

On the question of boundary lipid, another point of view is that the absence in the high-resolution spectrum of a separate component attributable to immobilized lipid could be due to the difficulty of observing a very wide pattern of long $T_1$ (Yeagle, 1982). Systematic attempts to quantitate the amount of "missing" lipid have yielded reproducible results (Yeagle and Romans, 1981). This missing lipid could in principle be observed by spectrometers with faster data acquisition and the echo technique (to over-

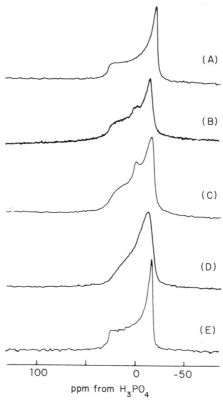

**Fig. 15.** Proton-decoupled $^{31}$P-NMR spectra (at 60.7 MHz) of pure DMPC and of protein- or cholesterol-containing complexes at $32 \pm 2°C$ in excess water. (A) Pure DMPC. (B) DMPC sample containing $\sim 80$ wt % cytochrome $c$ oxidase. (C) DMPC sample containing $\sim 70$ wt % sarcoplasmic reticulum ATPase. (D) DMPC sample containing $\sim 70$ wt % human lipophilin (N2 protein). (E) DMPC system containing $\sim 25$ wt % cholesterol. Spectral conditions were typically a 50-kHz spectral width, 1-s recycle time, 4-$\mu$s 90° pulse, 50-$\mu$s data-acquisition delay time, 8192 data points, and 50-Hz line broadening. The number of scans varied between 4000 and 16,000. Sample volume was $\sim 250 \mu$l. Gated proton decoupling with pulses of 50–100 ms, $\sim 40$ W, were used. From Rajan *et al.* (1981).

come short $T_2$), cross-polarization (to enhance intensity), and long cycle times (to counter long $T_1$). In our laboratory we have tried all these for the system DMPC (or egg lecithin):glycophorin (M. Rance, H. Jarrell, K. W. Butler, R. A. Byrd, I. C. P. Smith, and P. L. Yeagle, unpublished). Thus far we have been unable to demonstrate the presence of the bound lipid implied in the report of Yeagle and Romans (1981). The presence of protein does cause a significant decrease in the $T_2$ value of lipid $^{31}$P. It is conceivable that the missing lipid is due to the loss of several data points on the free induction decay because of the need for an acquisition delay. Time will tell, but this aspect does underline the care that must be taken in the acquisition of powder spectra.

The effects of cholesterol on lipid acyl chains have been studied for some time. With liquid-crystalline lipids, cholesterol usually leads to an increase in molecular ordering (tighter packing). The impact this rearrangement would have on the conformations in the head group region is not obvious. Early $^2$H-NMR studies showed little effect of cholesterol addition on the ordering of the choline methyl groups of phosphatidylcholine (Stockton *et al.*, 1974). Brown and Seelig (1978) found a very slight decrease in the $^{31}$P $\Delta\sigma$ on addition of as much as 50 mol % cholesterol to phosphatidylcholine and phosphatidylethanolamine. A similar result is seen in Fig. 15E. Thus cholesterol packs between the lipid acyl chains and restricts their conformational choices but in doing so separates phospholipid molecules, allowing perhaps more conformation freedom for the phosphodiester moiety (*perhaps* is used deliberately here in view of our earlier caveats regarding the difficulty of extracting information about order parameters from $^{31}$P $\Delta\sigma$ values).

### E. Relaxation Times

Relaxation times are much used in high-resolution NMR spectroscopy to gain information about the rates of molecular and intramolecular motion (Wehrli and Wirthlin, 1976; Levy *et al.*, 1980). They have been measured in membrane systems for $^{13}$C (Lee *et al.*, 1974; Smith, 1979; Brainard and Cordes, 1981; Coddington *et al.*, 1981), $^2$H (Stockton *et al.*, 1974, 1976; Brown and Davis, 1981; Brown, 1982), and $^{31}$P (Yeagle *et al.*, 1975; Rajan *et al.*, 1981; Seelig *et al.*, 1981; Banaszak and Seelig, 1982). However, even with nuclei such as $^2$H and $^{13}$C, where the relaxation mechanisms are fairly straightforward, it has been difficult to formulate the dependence of the relaxation times on the various motional correlation times and order parameters expected to occur. With $^{31}$P the problem is even greater because the mechanism is a mixture of chemical-shift and dipolar components, and the latter can arise from various proton sites. A further complication encountered with powder spectra is that the relaxation times may vary across the powder pattern. For example, in the DMPC referred to in Section IV,D, $T_2$ $(\sigma_\perp)$ = 7.4, ms, $T_2$ $(\sigma_\parallel)$ = 2.4 ms. Finally, although the $T_1$ values usually increase with increasing temperature, indicating the motional narrowing limit, $T_1 \gg T_2$; in DMPC at 30°, $T_1$ = 660 ms. This suggests the presence of several motions with different correlation times. Therefore, at this point it is not possible to calculate motional correlation times from $^{31}$P relaxation

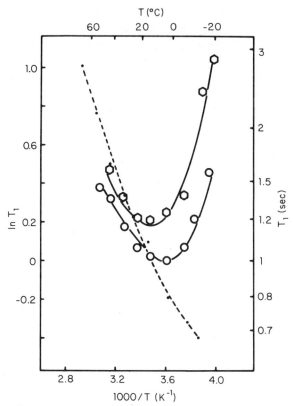

**Fig. 16.** Variation of the phosphorus relaxation time $T_1$ with temperature: (○) DOPC liposomes measured at 121.4 MHz resonance frequency; (○) sarcoplasmic reticulum membrane exchanged with DOPC measured at 121.4 MHz; (---) DOPC liposomes at 36.4 MHz without proton decoupling. From Seelig et al. (1981). Copyright 1981 American Chemical Society.

times with confidence. One can say that longer $T_1$ and $T_2$ values imply more rapid motion.

A convenient way around this inadequacy can occasionally be found by measuring the temperature dependence of relaxation. Figure 16 shows this for dioleoylphosphatidylcholine alone and in reconstituted sarcoplasmic reticulum membranes (Seelig et al., 1981). The observed minima allow a direct and accurate calculation of the correlation time for motions occurring at rates near the resonance frequency: for a minimum to occur $\omega_0^2 \tau_c^2 \simeq 1$, where $\omega_0$ is the Larmor frequency and $\tau_c$ the correlation time for the motion. Thus, for DOPC bilayers at the minimum in $T_1$, $\tau_c \approx 1 \times 10^{-9}$ s. Note that in the reconstituted sarcoplasmic reticulum membranes, the $T_1$ minimum is shifted toward higher temperatures by 10°C, indicating that the presence of protein reduces the rate of rapid motions undergone by the phosphate moiety. A similar conclusion for the acyl chains was reached via $^2$H-NMR $T_1$ values (Seelig et al., 1981). Interestingly, the estimated correlation time for the C-9 and C-10 positions of the chains in DOPC was $0.17 \times 10^{-9}$ s, a factor of 5 shorter than that for the phosphate group.

It should be pointed out that in Fig. 16, the $^{31}$P $T_1$ data for DOPC at 36.4 MHz are also given (dashed curve). There is no minimum, and at only one temperature are the $T_1$ values equal for measurements at 36.4 and 121.4 MHz. This behavior is due to the dominance of $T_1$ relaxation at 36.4 MHz by $^{31}$P–$^1$H dipolar interaction, whereas at 121.4 MHz this mechanism and chemical-shift anisotropy contribute about equally. This study points out very clearly that great care must be taken in the use of $^{31}$P $T_1$ values. Temperature and frequency dependences are necessary for unequivocal interpretation of the data.

## V. Conclusion

There is no doubt that $^{31}$P NMR has already led to valuable insight into the behavior of lipids in membranes. On the other hand, owing to the relative ease of obtaining spectra, insufficient attention has been paid to the exigencies of acquiring powder spectra with correct shapes. Some misleading conclusions are bound to have resulted. Contemporary spectrometers are better equipped for this task, so the problem should be less severe in the future. More detailed spectral analysis is required, and whenever possible spectra and relaxation times should be measured at several magnetic field strengths. Resorting to other techniques or NMR of other nuclei is often necessary for unequivocal interpretation.

Much work has been done on $^{31}$P NMR of membranes in microorganisms. The future holds the possibility of similar detailed studies on multicellular systems and animals. We shall probably see more use of $^1$H–$^{31}$P cross-polarization, two-dimensional, and other techniques for better deconvolution of complex spectra, as well as a better understanding of the relaxation behavior. The spectra still contain more information than we can conveniently extract from them.

## References

Akutsu, H., and Seelig, J. (1981). *Biochemistry* **20**, 7366–7373.
Banaszak, L. J., and Seelig, J. (1982). *Biochemistry* **21**, 2436–2443.

Bloom, M., Burnell, E. E., Valic, M. I. and Weeks, G. (1975). *Chem. Phys. Lipids* **14,** 107–112.
Boulanger, Y., Schreier, S., and Smith, I. C. P. (1981). *Biochemistry* **20,** 6824–6830.
Brainard, J., and Cordes, E. H. (1981). *Biochemistry* **20,** 4607–4617.
Brown, M. F. (1982). *J. Chem. Phys.* **77,** 1576–1599.
Brown, M. F., and Davis, J. H. (1981). *Chem. Phys. Lett.* **79,** 431–435.
Brown, M. F., and Seelig, J. (1978). *Biochemistry* **17,** 381–384.
Burnell, E. E., Van Alphen, L., Verkleij, A. J., and de Kruijff, B. (1980a). *Biochim. Biophys. Acta* **597,** 492–501.
Burnell, E. E., Cullis, P. R., and de Kruijff, B. (1980b). *Biochim. Biophys. Acta* **603,** 63–69.
Campbell, R. F., Meirovitch, E., and Freed, J. H. (1979). *J. Phys. Chem.* **83,** 525–533.
Chen, J. S., Barton, P. G., Brown, D., and Kates, M. (1974). *Biochim. Biophys. Acta* **352,** 202–217.
Chignell, C. F., and Chignell, D. A. (1975). *Biochem. Biophys. Res. Commun.* **62,** 136–143.
Coddington, J. M., Johns, S. R., Leslie, D. R., Willing, R. I., and Bishop, D. O. (1981). *Biochim. Biophys. Acta* **663,** 653–660.
Cullis, P. R., and de Kruijff, B. (1978a). *Biochim. Biophys. Acta* **507,** 207–218.
Cullis, P. R., and de Kruijff, B. (1978b). *Biochim. Biophys. Acta* **513,** 31–42.
Cullis, P. R., and de Kruijff, B. (1979). *Biochim. Biophys. Acta* **559,** 399–420.
Cullis, P. R., and Hope, M. J. (1980). *Biochim. Biophys. Acta* **597,** 533–542.
Cullis, P. R., Hornby, A. P., and Hope, M. J. (1980a). *Prog. Anesthesiol.* **2,** 367–403.
Cullis, P. R., de Kruijff, B., Hope, M. J., Nayar, R., Rietveld, A., and Verkleij, A. J. (1980b). *Biochim. Biophys. Acta* **600,** 625–635.
Cullis, P. R., de Kruijff, B., Hope, M. J., Verkleij, A. J., Nayar, R., Farren, S. B., Tilcock, C., Madden, T. D., and Bally, M. B. (1982). In "Membrane Fluidity in Biology" (R. C. Aloia, ed.), Vol. 2, pp. 39–81. Academic Press, New York.
Davis, J. H. (1983). *Biochim. Biophys. Acta* **737,** 117–171.
Davis, J. H., Jeffrey, K. R., Bloom, M., Valic, M. I., and Higgs, T. P. (1976). *Chem. Phys. Lett.* **42,** 390–394.
Deese, A. J., Dratz, E. A., and Brown, M. F. (1981). *FEBS Lett.* **124,** 93–99.
Degani, H., Bach, D., Danon, A., Garty, H., Eisenbach, M., and Caplan, S. R. (1978). In "Energetics and Structure of Halophilic Organisms" (S. R. Caplan and M. Ginzburg, eds.), pp. 225–232. Elsevier/North-Holland, Amsterdam.
Degani, H., Danon, A., and Caplan, S. R. (1980). *Biochemistry* **19,** 1626–1631.
de Grip, W. J., Drenth, E. H. S., van Echteld, C. J. A., de Kruijff, B., and Verkleij, A. J. (1979). *Biochim. Biophys. Acta* **558,** 330–337.
de Kruijff, B., Verkleij, A. J., van Echteld, C. J. A., Gerritsen, W. J., Mombers, C., Noordam, P. C., and de Gier, J. (1979). *Biochim. Biophys. Acta* **555,** 200–209.
de Kruijff, B., Rietveld, A., and Cullis, P. R. (1980). *Biochim. Biophys. Acta* **600,** 343–357.
Deslauriers, R., Ekiel, I., Byrd, R. A., Jarrell, H. C., and Smith, I. C. P. (1982). *Biochim. Biophys. Acta* **720,** 329–337.
Ekiel, I., Marsh, D., Smallbone, B. W., Kates, M., and Smith, I. C. P. (1981). *Biochem. Biophys. Res. Commun.* **100,** 105–110.
Farren, S. B., and Cullis, P. R. (1980). *Biochem. Biophys. Res. Commun.* **97,** 182–191.
Freed, G. H., Bruno, G. V., and Polnaszek, C. F. (1971). *J. Phys. Chem.* **73,** 3385–3399.
Frenzel, J., Arnold, K., and Nuhn, P. (1978). *Biochim. Biophys. Acta* **507,** 185–197.
Füldner, H. H. (1981). *Biochemistry* **20,** 5707–5710.
Griffin, R. G. (1981). *In* "Methods in Enzymology" (J. M. Lowenstein, ed.), Vol. 72, Part D, pp. 108–174. Academic Press, New York.
Harrison, R., and Lunt, G. G. (1980). "Biological Membranes: Their Structure and Function." Blackie, Glasgow.

Hope, M. J., and Cullis, P. R. (1980). *Biochem. Biophys. Res. Commun.* **92**, 846–852.
Hope, M. J., and Cullis, P. R. (1981). *Biochim. Biophys. Acta* **640**, 82–90.
Hornby, A. P., and Cullis, P. R. (1981). *Biochim. Biophys. Acta* **647**, 285–292.
Jackson, M. B., and Sturtevant, J. M. (1978). *Biochemistry* **17**, 911–915.
Jarrell, H. C., Byrd, R. A., Deslauriers, R., Ekiel, I., and Smith, I. C. P. (1981). *Biochim. Biophys. Acta* **648**, 80–86.
Jost, P. C., Nadakavukaren, K. K., and Griffith, O. H. (1977). *Biochemistry* **16**, 3110–3114.
Kates, M., and Kushwaha, S. C. (1978). *In* "Energetics and Structure of Halophilic Microorganisms" (S. R. Caplan and M. Ginzburg, eds.), pp. 461–480. Elsevier/North-Holland, Amsterdam.
Lee, A. G., Birdsall, N. J. M., and Metcalfe, J. C. (1974). *Methods Membr. Biol.* **2**, 1–156.
Levy, G. C., Lichter, R. L., and Nelson, G. C. (1980). "Carbon-13 Nuclear Magnetic Resonance Spectroscopy." Wiley, New York.
Luzzati, V., and Husson, F. (1962). *J. Cell Biol.* **12**, 207–219.
McLaughlin, A. C., Herbette, L., Blasie, J. K., Wang, C. T., Hymel, L., and Fleischer, S. (1981). *Biochim. Biophys. Acta* **643**, 1–16.
Mason, R. P., Polnaszek, C. F., and Freed, J. H. (1974). *J. Phys. Chem.* **78**, 1324–1329.
Morris, G. A., and Freeman, R. (1978). *J. Magn. Reson.* **29**, 433–453.
Noggle, J. H., Maracek, J. F., Mandal, S. B., Van Venetie, R., Rogers, J., Jain, M. K., and Ramirez, F. (1982). *Biochim. Biophys. Acta* **691**, 240–248.
Paddy, M., Dahlquist, F. W., Davis, J. H., and Bloom, M. (1981). *Biochemistry* **20**, 3152–3162.
Papahadjopolous, D., Vail, W. J., Pangborn, W. A., and Poste, G. (1976). *Biochim. Biophys. Acta* **448**, 265–283.
Plachy, W. Z., Lanyi, J. K., and Kates, M. (1974). *Biochemistry* **13**, 4906–4913.
Rajan, S., Kang, S.-Y., Gutowsky, H. S., and Oldfield, E. (1981). *J. Biol. Chem.* **256**, 1160–1166.
Rance, M., and Byrd, R. A. (1983). *J. Magn. Reson.* **52**, 221–240.
Rand, R. P., and Sengupta, S. (1972). *Biochim. Biophys. Acta* **255**, 484–492.
Seelig, A., and Seelig, J. (1978). *Hoppe-Seyler's Z. Physiol. Chem.* **359**, 1747–1756.
Seelig, J. (1977). *Q. Rev. Biophys.* **10**, 353–418.
Seelig, J. (1978). *Biochim. Biophys. Acta* **515**, 105–140.
Seelig, J., and Seelig, A. (1980). *Q. Rev. Biophys.* **13**, 19–61.
Seelig, J., Tamm, L., Hymel, L., and Fleischer, S. (1981). *Biochemistry* **20**, 3922–3932.
Smith, I. C. P. (1979). *Can. J. Biochem.* **57**, 1–14.
Smith, I. C. P. (1981). *Bull. Magn. Reson.* **3**, 120–133.
Smith, I. C. P. (1983). *Biomembranes* **12** (in press).
Smith, I. C. P., and Jarrell, H. C. (1983). *Acc. Chem. Res.* **16**, 266–272.
Stockton, G. W., Polnaszek, C. F., Leitch, L. C., Tulloch, A. P., and Smith, I. C. P. (1974). *Biochem. Biophys. Res. Commun.* **60**, 844–850.
Stockton, G. W., Polnaszek, C. F., Tulloch, A. P., Hasan, F., and Smith, I. C. P. (1976). *Biochemistry* **15**, 954–966.
Tardieu, A., Luzzati, V., and Reman, F. C. (1973). *J. Mol. Biol.* **75**, 711–733.
Thayer, A. M., and Kohler, S. J. (1981). *Biochemistry* **20**, 6831–6834.
Tilcock, C. P. S., and Cullis, P. R. (1982). *Biochim. Biophys. Acta* **684**, 212–218.
Van Echteld, C. J. A., Van Stigt, R., de Kruijff, B., Leunissen-Bijvelt, J., Verkleij, A. J., and de Gier, J. (1981). *Biochim. Biophys. Acta* **648**, 287–291.
Verkleij, A. J., Mombers, C., Gerritsen, W. J., Leunissen-Bijvelt, L., and Cullis, P. R. (1979). *Biochim. Biophys. Acta* **555**, 358–361.

Verkleij, A. J., Van Echteld, C. J. A., Gerritsen, W. J., Cullis, P. R., and de Kruijff, B. (1980). *Biochim. Biophys. Acta* **600,** 620–624.

Wehrli, F. W., and Wirthlin, T. (1976). "Interpretation of Carbon-13 NMR Spectra." Heyden, London.

Wieslander, Å., Ulmius, J., Lindblom, G., and Fontell, K. (1978). *Biochim. Biophys. Acta* **512,** 241–253.

Wieslander, Å., Rilfors, L., Johansson, L. B.-Å., and Lindblom, G. (1981). *Biochemistry* **20,** 730–735.

Yeagle, P. L. (1982). *Biophys. J.* **37,** 227–239.

Yeagle, P. L., and Romans, A. Y. (1981). *Biophys. J.* **33,** 243–252.

Yeagle, P. L., Hutton, W. C., Huang, C., and Martin, R. B. (1975). *Proc. Natl. Acad. Sci. U.S.A.* **72,** 3477–3481.

# PART 3

*Application to Biology and Medicine:
Future Directions*

CHAPTER 16

# Two-Dimensional Phosphorus-31 NMR

*William C. Hutton*
Department of Chemistry
University of Virginia
Charlottesville, Virginia

| | |
|---|---:|
| I. Introduction | 479 |
|   A. Scope | 480 |
|   B. Historical Summary | 480 |
|   C. Definition of Two-Dimensional NMR | 480 |
|   D. Instrumental Requirements | 483 |
| II. General Principles of Two-Dimensional NMR: The Homonuclear, J-Correlated Experiment | 484 |
|   A. Jeener Pulse Sequence | 485 |
|   B. Collection of the Two-Dimensional Data Set | 489 |
|   C. Two-Dimensional Data-Set Processing | 489 |
|   D. Graphic Display of the Two-Dimensional Data Set | 491 |
| III. Applications of Two-Dimensional Spectroscopy to $^{31}$P NMR | 492 |
|   A. Homonuclear Chemical-Shift Correlated Spectra | 492 |
|   B. Heteronuclear $^{31}$P–$^{1}$H Chemical-Shift Correlated Spectra | 497 |
|   C. Chemical-Exchange Correlated Spectra | 506 |
| IV. Concluding Remarks | 509 |
| References | 510 |

## I. Introduction

Two-dimensional spectroscopy provides an experimentalist with the means for clarifying assignments in a conventional NMR spectrum. It is often difficult to analyze $^{31}$P spectra of mixtures without ambiguity. Phosphorus nuclei in similar chemical environments often yield overlapping resonances. The resonances for a mixture of di-, tri-, and higher phosphorylated compounds will fall into three regions, each with a range of < 1.5 ppm. The increased field dispersion of high magnetic fields (> 4 T) mitigates the overlap to some extent. However, for biological samples in particular, the advantage of a high field is compromised by an increase in linewidth attributable to more efficient chemical-shift anisotropy relaxation and inhomogeneous samples. Additional loss of resolution can occur in some cases by the presence of paramagnetic metal ions or from line broadening induced from exchange processes with diamagnetic ions. A homonuclear, two-dimensional, chemical-shift correlation map will, in a single experiment, define all of the homonuclear spin–spin coupling networks in the spectrum. With these data in hand, analysis of the spectrum is straightforward.

## A. Scope

This chapter addresses applications of two-dimensional techniques for isotropic solutions though applications to solid-state $^{31}$P NMR are practical. A detailed, theoretical description is not appropriate here. Instead, the general principles of two-dimensional NMR are discussed in Section II.

The subject has been reviewed (Freeman and Morris, 1979) and a monograph on two-dimensional NMR has been published (Bax, 1982). Accordingly, a complete treatment of the literature is not attempted. The examples presented are drawn mostly from spectra relating to cellular, bioenergetic processes. However, application of the two-dimensional technique will be of general use when complicated $^{31}$P-NMR spectra are encountered.

## B. Historical Summary

Jeener first described a two-dimensional NMR experiment in 1971,[1] but it was not until 1975 that the first two-dimensional spectra were published by Ernst (1975). The pioneering efforts of Ernst and co-workers in developing the technique should be appreciated by those who glean information from two-dimensional experiments. The NMR community also owes a debt of gratitude to the laboratory of Ray Freeman. His contributions and those of his associates to two-dimensional NMR have expanded the technique and promoted its growth. The NMR spectrometer manufacturers have also been instrumental in the development and application of two-dimensional NMR. By 1980 all of the commercially available Fourier-transform spectrometer systems provided two-dimensional data processing as part of their standard software package.

## C. Definition of Two-Dimensional NMR

A two-dimensional data set is one that is collected as a function of two independent time periods:

[1] J. Jeener, in an unpublished lecture to the Ampere International Summer School, Basko Polje, Yugoslavia, 1971.

$$D(t_1,t_2) \xrightarrow{FT_{t_1}, FT_{t_2}} D(v_1,v_2) \qquad (1)$$

Only data collected using pulse Fourier-transform (FT) techniques are considered. Fourier transformation carried out twice with respect to the time periods $t_1$ and $t_2$ will yield a data matrix defined by two frequency domains $v_1$ and $v_2$. More rigorously one may write

$$D(v_1,v_2) = \int_0^\infty dt_1 \exp(-i2\pi v_1 t_1) \int_0^\infty dt_2 \exp(-i2\pi v_2 t_2) v(t_1,t_2) \qquad (2)$$

which indicates that the principles of Fourier-transform NMR apply.

Ernst has shown that all two-dimensional experiments can be described by the general scheme shown in Fig. 1a (Aue *et al.*, 1976). During the preparation period, the spin system is allowed to reach an equilibrium state appropriate for the desired experiment. The preparation period ends with the preparation pulse. The preparation pulse serves to create transverse magnetization as in a conventional-pulse NMR sequence. During $t_1$, the evolution period, the magnetization precesses in a fashion determined by events during the preparation period. At this point it is important to avoid confusion between $t_1$, the evolution period, and $T_1$, the spin–lattice relaxation time. Similarly, $t_2$, the detection period, may be confused with $T_2$, the spin–spin relaxation time. This problem often arises when using the spoken word and it is recommended that $t_1$ and $t_2$ be referred to as time-one and time-two in verbal discussions.

The evolution period is followed by the mixing period. The mixing

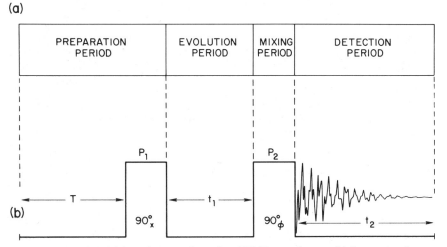

**Fig. 1.** (a) Time division of all two-dimensional NMR experiments. (b) Jeener two-dimensional NMR pulse sequence.

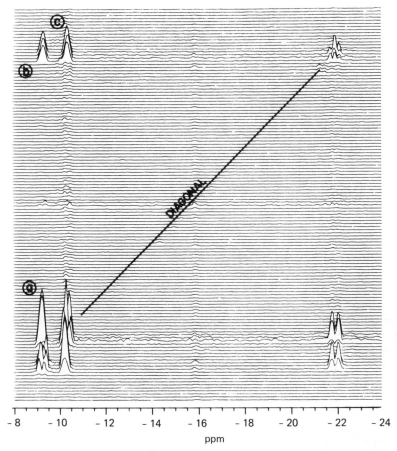

**Fig. 2.** A 146-MHz, $^{31}$P homonuclear chemical-shift correlated two-dimensional spectrum of an equimolar (50 m$M$) mixture of ADP and ATP. The data set (128 × 256) is displayed in the stacked-plot mode. Each trace corresponds to an increment in the evolution time. The data were acquired in 45 min. The stacked plot took 15 min to complete. The diagonal, which makes a 45° angle with the $F_1$ (ordinate) and $F_2$ (abscissa) axes, reproduces the one-dimensional spectrum. The off-diagonal peaks correlate peaks on the diagonal that are spin–spin coupled to each other. Peak identifications can be found in Fig. 3.

period, which in the case of Fig. 1b is the second radiofrequency (rf) pulse, allows the transfer of information between related spins. The preparation and mixing periods determine the type of information contained in the two-dimensional data set.

The detection period corresponds to the data-acquisition time in a pulse Fourier-transform experiment, with the exception that the data being digitized are a function of two variables, $t_1$ and $t_2$.

Figure 2 depicts a twice-Fourier-transformed, two-dimensional data set resulting from the pulse sequence in Fig. 1b. The stacked-plot presentation serves to demonstrate how the second dimension is created. The $F_1$ axis consists of 128 individually collected and stored one-dimensional spectra, or data blocks. These data blocks were made by incrementing the evolution period $(t_1)$ by the inverse of the spectrum width as each block was obtained. Thus the second dimension is obtained by incrementing the evolution period. The $F_2$ axis, corresponding to the detection period $(t_2)$, was determined as it would be in a regular Fourier-transform spectrum.

### D. Instrumental Requirements

In order for two-dimensional experiments to be conducted on a routine basis, the NMR instrumentation used must fulfill certain criteria. Whereas the requirements discussed next are essential for producing quality $^{31}$P two-dimensional spectra, they will also benefit the variety of one-dimensional techniques available to the spectroscopist.

Two-dimensional spectra are too often compromised by instrumental artifacts. Suggestions for experimental techniques that avoid or suppress artifacts have been summarized (Bax, 1982). Instrumental instabilities cause an undesirable phenomenon called $t_1$ noise. The $t_1$ noise will, at least, reduce the sensitivity of a two-dimensional spectrum. The entire spectrometer system should be designed to facilitate the execution, collection, and processing of the two-dimensional data set.

The heart of any NMR spectrometer is the magnet. The inherent stability, increased sensitivity, and advantageous field dispersion of superconducting magnets make them ideal for two-dimensional experiments. In general, a two-dimensional data set is best collected with the highest possible magnetic field. This statement is particularly relevant for $^{31}$P measurements. Wide-bore, superconducting magnets are especially well suited to $^{31}$P two-dimensional studies, owing to their ability to accept large-volume samples and to their flexibility for probe construction.

Because the number of spectra, or data blocks, necessary to produce a two-dimensional spectrum is one to two orders of magnitude larger than that needed for a conventional NMR spectrum, the sensitivity of the probe is important. The probe design should provide good pulse homogeneity so that artifacts are minimized. For $^{31}$P applications on superconducting systems, a probe utilizing transverse coil geometry (Oldfield and Meadows, 1978) is desirable.

Quality two-dimensional spectra demand high accuracy from the pulse programmer. Accurately timed, well-shaped pulses with minimal phase errors must be generated for both the observe transmitter and for the spin decoupler. The programming and hardware control capabilities of the pulse programmer must be versatile. It will be necessary to produce independently the phase shifts needed in both the observe transmitter and decoupler required for quadrature detection and channel cycling.

Of all the NMR techniques, two-dimensional spectroscopy places the greatest demands on the data system. The computer and peripherals are responsible for collecting, monitoring, processing, storing, and displaying a two-dimensional data set of up to 1024 × 1024 data points. Accordingly, the central processing unit and interface must be fast and flexible. In order to minimize the number of disk accesses, it is advantageous to have the computer equipped with a large amount of random-access memory. A large memory will allow the NMR program to reside in memory and will also minimize the time required to transpose (see Section II,C) a two-dimensional data set. An array processor provides additional speed.

Because the data processing will require that the original $D(t_1,t_2)$ data be manipulated and stored many times, a quick, reliable mass-storage disk system is essential. Two-dimensional data fill disk cartridges quickly, making long-term storage of data on magnetic tape attractive. The NMR software package should be facile with regard to two-dimensional experiments. Efficient use of the spectrometer will be enhanced if two-dimensional data sets can be collected, processed, and displayed as a foreground–background operation.

Long-term, two-dimensional experiments will benefit from interleaving the data-block signal averaging. The $t_1$ noise resulting from instrumental and environmental instabilities will be reduced by signal averaging into the 256 blocks four or more times, rather than passing through the blocks only a single time. The NMR software should allow the operator to automate fully the two-dimensional data-set collection, processing, and plotting.

The final step in processing a two-dimensional data set is displaying the data. Rapid, temporary data display is best achieved using a raster-scan CRT system. It is possible to obtain a hard copy, two-dimensional spectrum in less than 10 min using a high-speed, dot-matrix printer. High-quality data sets are best obtained with a digital plotter. Finally, the most efficient usage of magnet time will be realized if an independent, remote data station is available for processing two-dimensional data sets.

## II. General Principles of Two-Dimensional NMR: The Homonuclear, J-Correlated Experiment

In order to demonstrate the methodology for generating a two-dimensional NMR spectrum, the homonuclear, *J*-correlated experiment first proposed by Jeener will be described. This experiment has been named COSY for

# 16. Two-Dimensional $^{31}$P NMR

correlated spectroscopy (Aue et al., 1976; Wagner et al., 1981; Nagayama et al., 1980). What follows is a description of how the experiment works, the procedures necessary to obtain and process the two-dimensional data matrix, and ways the spectrum can be displayed.

The purpose of the pulse sequence is to map out simultaneously the homonuclear spin–spin coupling networks in a NMR spectrum. In $^{31}$P magnetic resonance this information will be useful for unambiguously identifying the components in a mixture of polyphosphorylated molecules. Figure 3 contains a contour plot of the ATP–ADP mixture shown in Fig. 2. The signal intensity that falls on a line 45° with respect to the chemical-shift scale portrays the regular $^{31}$P-NMR spectrum. The signal intensity located off the diagonal, or $J$ cross-peaks, arise from spin–spin coupling between resonances on the diagonal. The off-diagonal peaks correlate nuclei that share $J$-coupling energy levels.

## A. Jeener Pulse Sequence

The pulse sequence used to generate the two-dimensional data set is shown in Fig. 1b. The preparation period begins with an interval $T$ when the spin systems are allowed to reach thermal equilibrium with the external magnetic field. $P_1$, the 90° preparatory pulse, generates transverse magnetization. The magnetization precesses during the evolution period $t_1$ in the rotating frame of reference at a frequency equal to $v_0 - v_n$; where $v_0$ equals the frequency of the preparation pulse and $v_n$ equals the Larmor frequency of nucleus $n$.

The mixing pulse $P_2$ produces three types of signals in the two-dimensional spectrum. For the time being, we consider the case where $P_2$ is also a 90° pulse. The second 90° pulse mutually exchanges the transverse magnetization established during $t_1$ among the transitions of all spins that are connected via homonuclear, spin–spin coupling networks. The change in the precessional frequency of the transitions after the mixing pulse may be zero or only $\pm J$, in which case signal intensity will be observed on or close to the diagonal. For some transitions the frequency change will correspond to the chemical-shift difference of the coupled nuclei. In this case, off-diagonal or $J$ cross-peaks will appear. These are the signals that provide information unique to the Jeener technique. They correlate nuclei that are $J$-coupled to each other. The mixing pulse also can produce undesirable axial peaks in the two-dimensional map. The axial peaks arise from any $Z$ magnetization that builds up during the evolution period. Axial peaks are routinely suppressed using appropriate phase shifts for the mixing pulse relative to the receiver (Bax et al., 1981).

Immediately after the mixing pulse, the free induction decay of the

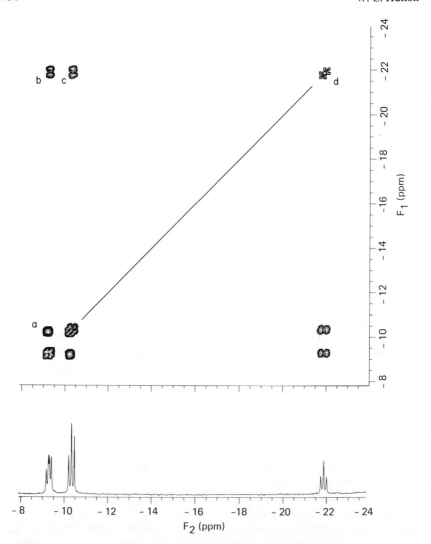

**Fig. 3.** One-dimensional (bottom) and contour (top) plots of the data set in Fig. 2. The off-diagonal peak **a** correlates the $\alpha$ (−10.3 ppm) and $\beta$ (−9.2 ppm) resonances of ADP; **b** correlates the $\gamma$ (−9.4 ppm) and $\beta$ (−21.9 ppm, peak **d**) resonances of ATP, and **c** correlates $\alpha$ (−10.4 ppm) and $\beta$ resonances of ATP. The data were plotted in 3 min. All of the homonuclear chemical-shift correlated spectra were obtained at 8.5 T with a Nicolet Magnetics Corporation NT-360 equipped with the 1280/293B data system. The spectra were measured at 25 °C using a 10-mm broadband probe with an internal $^2$H lock. Uninterrupted incoherent $^1$H decoupling at low levels (~1 W) was employed. Standard NMC software was used to collect and process the two-dimensional data sets.

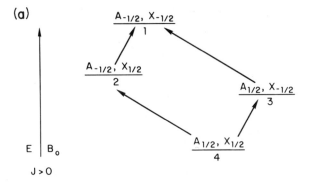

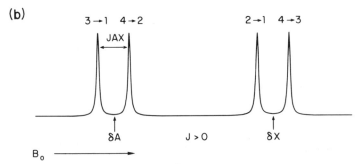

**Fig. 4.** Energy-level diagram (a) of a scalar-coupled AX spin system where $J > 0$, and the resulting NMR spectrum (b).

magnetization is acquired as in a conventional Fourier-transform experiment. During this detection period $t_2$ the properties of the magnetization—its precessional frequency, phase, and amplitude—depend on both $t_1$ and $t_2$.

By incrementing the duration of the evolution period systematically and collecting a free induction decay during the detection period for each value of $t_1$, the two-dimensional data matrix is created. In the two-dimensional data set of the Jeener experiment, the amplitude of a coupled resonance is modulated as a function of $t_1$. More specifically (Bax, 1982), the modulation of a transition $P_{3\rightarrow 1}$ in an AX spin system (refer to Fig. 4) by transition $P_{2\rightarrow 1}$ is given by

$$P_{3\rightarrow 1}(t_1,t_2) = \sum_{1,2} K_{1,2-3,4} \sin(v_{1\rightarrow 2}t_1 + \theta_{1,2-3,4}) \exp(iv_{3\rightarrow 1}t_2 + i\Delta_{1,2-3,4}) \quad (3)$$

where $K$ is a constant that is the same for all the transitions in the case under discussion, $\theta$ and $\Delta$ are phase constants equal to $\pi/2$, $-\pi/2$, or 0 depending

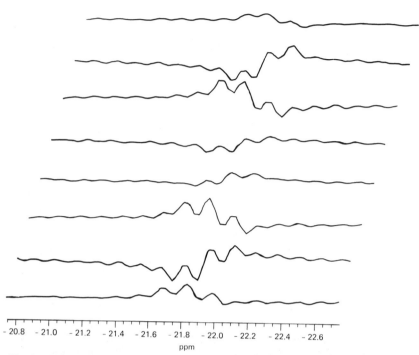

**Fig. 5.** Section of the $D(t_1,F_2)$ data matrix demonstrating the phase modulation of the $\beta$ resonance of ATP. The spectra were extracted from the data set used to generate Figs. 2 and 3. The $t_1$ data blocks were incremented by 429 $\mu$s.

on the connectivity of P (Aue et al., 1976), and $\nu$ is the precessional frequency in the rotating frame of reference. Equation (3) considers single quantum transitions only and neglects relaxation effects. The modulation described by Eq. (3) is amplitude modulation, which implies that it is not possible to distinguish resonances on either side of the carrier frequency $\nu_0$. In order to implement quadrature detection in both the $F_1$ and $F_2$ dimensions, it is necessary to collect a minimum of four transients per $t_1$ increment with the phase of the second pulse $\phi$ shifted by 90° for each transient (Nagayama et al., 1980; Bax et al., 1981). These 0°, 90°, 180°, 270° phase shifts convert the amplitude modulation to phase modulation and suppress axial peaks as well.

Figure 5 demonstrates phase modulation of the $\beta$ $^{31}$P resonance of ATP. This figure is taken from the two-dimensional data set used to generate Figs. 2 and 3. The $D(t_1,t_2)$ data was Fourier transformed once to create a $D(t_1,F_2)$ data set. The modulation of a resonance by another nucleus with which it shares some interaction is fundamental to the two-dimensional technique. It is this modulation that creates the second dimension.

## B. Collection of the Two-Dimensional Data Set

A two-dimensional data set is composed of a matrix of $[LX(R + i)]$ data points. Here $L$ is the number of increments in $t_1$, and $R$ and $i$ are the number of real and imaginary data points digitized into memory locations; thus $R = i$. The two-dimensional data set is collected by storing a signal-averaged, free induction decay for each value of $t_1$. Typically, the free induction decays are sequentially stored on a disk drive as individual data blocks. For the first block of data, the duration of $t_1$ is the inverse of the desired spectral width that is equivalent to the dwell time of the one-dimensional, free induction decay digitization. The evolution time is then incremented by the dwell time $L$ times to create $L$ data blocks. $L$ should be chosen to be a power of 2 and is usually 128, 256, or 512. For the homonuclear, chemical-shift correlation experiment, when $L*t_1 = t_2$ and $L = R$, the two-dimensional data set will be symmetrical after double Fourier transformation.

## C. Two-Dimensional Data-Set Processing

When the two-dimensional data-block collection has been completed, each free induction decay is Fourier transformed after being treated with a suitable apodization function (Bax, 1982):

$$D(t_1 t_2) \xrightarrow{\text{apodize}} D_a(t_1,t_2) \xrightarrow[\text{scale}]{\text{FT}} D(t_1,F_2) \qquad (4)$$

The free induction decay may be zero-filled to improve the digital resolution in the $t_2$ domain. After Fourier transformation, the data are scaled and $L$ blocks of frequency domain data $D(t_1,F_2)$ are then stored. Figure 5 depicts part of a $D(t_1,F_2)$ data set. As the resonance lines in the $D(t_1,F_2)$ data set are phase modulated as a function of $t_1$, no phase corrections need be applied to the transformed data set.

At this point, the modulation information in the $D(t_1,F_2)$ data set must be transposed so that the second Fourier transformation can be accomplished. The transpose is affected by reversing the coordinates of the two-dimensional data set:

$$D(t_1 F_2) \xrightarrow{\text{transpose}} D(F_2,t_1) \qquad (5)$$

This process creates $L$ blocks of new free induction decays. These free induction decays contain the time-domain information from the evolution period. Figure 6 shows the transposed data blocks corresponding to the frequency-domain data in Fig. 5.

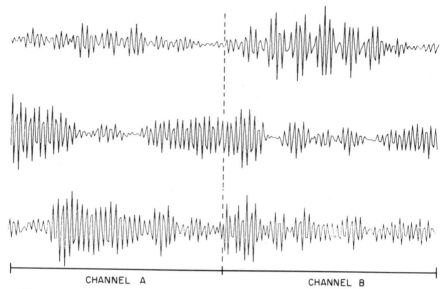

**Fig. 6.** Free induction decays from the transposed $D(F_2,t_1)$ data set. The quadrature data was created by transposing the complete data set partially depicted in Fig. 5. The data have not been apodized.

The $D(F_2,t_1)$ data set is now apodized and the second Fourier transform is performed on each data block:

$$D(F_2,t_1) \xrightarrow{\text{apodize}} D_a(F_2,t_1) \xrightarrow[\text{scale}]{\text{FT}} D(F_2,F_1) \qquad (6)$$

The results are scaled and stored. As was the case with the first Fourier transformation, the $D(F_2,t_1)$ free induction decays may be zero-filled to increase the digital resolution in the $t_1$ dimension.

Equation 3 indicates that the peaks in the two-dimensional spectrum may be positive or negative, depending on their relative connectivities in the spin–spin coupling energy levels. Normally this information is lost as it is most convenient to display the data set in the absolute-value mode (Freeman and Morris, 1979). The phase correction for a two-dimensional data set is considerably more complicated than with one-dimensional data. Even when the cumbersome phase correction is carried out, the resulting line shape is twisted, making distinction of overlapping peaks difficult (Bodenhausen et al., 1977). Therefore, the $D(F_2,F_1)$ data set is stored as absolute-value spectra. In order to obtain the most satisfactory line shapes, the free induction decays are often apodized with a sine–bell (DeMarco and Wüthrich, 1976) function, although other apodization functions may be used (Bax, 1982).

As the data matrix for a homonuclear chemical-shift correlation experiment is symmetrical, the $D(F_2,F_1)$ data set is identical to a $D(F_1,F_2)$ data set. However, because homonuclear, J-resolved and heteronuclear, two-dimensional data matrices are not symmetrical, a second transpose operation is required in these cases:

$$D(F_2 F_1) \xrightarrow{\text{transpose}} D(F_1,F_2) \qquad (7)$$

The symmetry of the Jeener data matrix may be exploited to improve the two-dimensional spectrum. The off-diagonal peaks are always found in pairs that are symmetrical about the diagonal. By comparing the intensities of the peaks on either side of the diagonal and replacing the higher point of each pair by the lower point, artifacts present in the spectrum are suppressed and the signal-to-noise ratio of the off-diagonal peaks is enhanced (Bauman et al., 1981). The symmetrization process provides a tangible improvement of the spectrum:

$$D(F_2 F_1) \xrightarrow{\text{symmetrize}} D_{sy}(F_2,F_1) \qquad (8)$$

## D. Graphic Display of the Two-Dimensional Data Set

Figures 2 and 3 contain the two methods available for plotting the two-dimensional data set. A stacked-plot display can be obtained with a conventional plotter and requires a minimum of additional software. A contour plot is an intensity map of the peak positions of a stacked-plot, two-dimensional data set. Although it requires additional software and hardware to generate a contour plot, the spectrum is obtained much faster than with a stacked plot. Because the contour plot gives the viewer the ability to look down on the spectrum, small peaks cannot be hidden behind large peaks. The contour plot can be displayed with various slice heights through the peaks. It is efficient to display the contour plot on a raster-scan display. The raster spectra are generated quickly and allow the user to select the best slice levels for the hard-copy plot.

Depending on the type of two-dimensional experiment selected, useful information can be obtained from a projection of the data set. A projection is obtained by summing or integrating the intensities for each data block of the data set perpendicular to the dimension ($F_1$ or $F_2$) of interest. The projection produces a low-resolution, one-dimensional spectrum.

As most of the two-dimensional spectrum contains no information, it is often useful to plot individual cross-sections from the $D(F_1,F_2)$ data set. For the chemical-shift correlated spectra, a cross-section of the two-dimensional map taken at the chemical shift of a resonance on the $F_1$ axis contains only

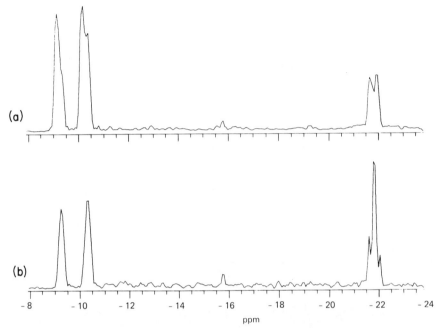

**Fig. 7.** A 0° projection (a) of the two-dimensional data matrix of Fig. 2. The cross-section (b) for the same data matrix selected from the $F_1$ dimension at $-21.9$ ppm (the $\beta$ phosphate of ATP). This cross-section contains the other two resonances spin coupled to the $\beta$ resonance.

peaks that are spin–spin coupled to the selected resonance. Examples of a projection and of a cross-section for the ATP–ADP Jeener spectrum appear in Fig. 7.

## III. Applications of Two-Dimensional Spectroscopy to $^{31}$P NMR

### A. Homonuclear Chemical-Shift Correlated Spectra

In Section II the methodology of the homonuclear chemical-shift correlated technique is discussed. Applications of the Jeener experiment are now described, showing how two-dimensional, chemical-shift correlated spectra provide insight into the assignment of complicated $^{31}$P spectra.

Figure 8 contains the conventional and correlated 146-MHz $^{31}$P spectra for Ni ($\alpha$-[18]ane P$_4$O$_2$)·2BF$_4^-$ in acetonitrile. Inspection of the homonuclear chemical-shift correlation map unambiguously shows there are four

16. Two-Dimensional $^{31}$P NMR 493

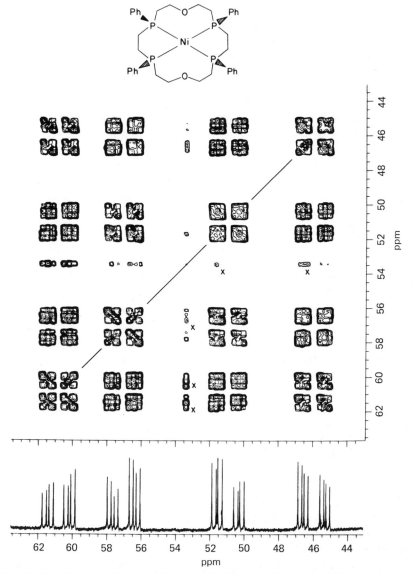

**Fig. 8.** One-dimensional and two-dimensional chemical-shift correlated $^{31}$P-NMR spectra of Ni($\alpha$-[18]ane P$_4$O$_2$)·2BF$_4^-$ in acetonitrile. X, Designates artifact peaks at the rf carrier position. $P_2$ (see Fig. 1b) was a 90° pulse. The nickel complex was kindly provided by Professor Mario Ciampolini.

nonequivalent phosphorus nuclei in the complex mutually coupled via the nickel atom. This spectrum is consistent with the proposed structure for the nickel-α-isomer (Ciampolini *et al.,* 1982). In Fig. 9 the mixing pulse in the Jeener sequence (Fig. 1b) was shortened to provide a 45° flip angle (Bax, 1982; Bax and Freeman, 1981). This modification of the pulse sequence suppresses any indirectly connected, spin–spin coupled transitions in the two-dimensional spectrum. The diagonal peaks are simplified (a useful result in general for crowded spectral regions), and only the directly connected transitions appear causing the cross-peaks to become asymmetric. The slant or tilt of a cross-peak with respect to the diagonal indicates the relative sign of the coupling constant $J$ for the two resonances correlated by the cross-peak. Analysis of the spectrum in Fig. 9 leads to the conclusion that one of the three couplings for each phosphorus is of the opposite sign. Because the structure of the complex dictates that each phosphorus has two couplings via ~90° bond angles and one coupling via an ~180° bond angle, the data infer that the sign of the two-bond coupling is a function of the bond angle. Indeed, coupling involving trans phosphorus donor atoms are expected to be positive, whereas the coupling with cis donors is negative (Hyde *et al.,* 1977). The analysis of the cross-peaks also shows the resonances at 60.8 and 46.0 ppm, and those at 57.1 and 50.8 ppm must be from $^{31}$P nuclei that are directly across from one another.

Phosphorus-31 NMR has been invaluable for probing cellular energetics on a molecular level. In many cases, mixtures of polyphosphorylated metabolites appear in the $^{31}$P spectra of *in vivo* and *in vitro* systems. Previously, ambiguities in the assignment of $^{31}$P spectra of such mixtures have been overcome by using time-consuming metal and/or pH titrations of cell extracts. The Jeener technique is a powerful alternative for unraveling mixtures of molecules containing a polyphosphate chain with at least two phosphoryl groups (Van Divender and Hutton, 1982). Figure 10 shows the application of the homonuclear correlated experiment to a mixture of cell metabolites. Phosphorus-31 NMR has been successfully applied to studying pH gradients in intracellular compartments. In Fig. 10, separate cellular pH environments were simulated by placing a 5-mm tube concentrically inside the 10-mm NMR tube. The pH gradient used is similar to one found from $^{31}$P-NMR measurements during the life cycle of aerobic *Escherichia coli* cells (Navon *et al.,* 1977). The cross-peaks for $^{31}$P nuclei in the same phosphorylated chain lie along a horizontal line above the diagonal parallel to the $F_2$ axis. The homonuclear chemical-shift correlation map resolves peak overlap and identifies the nucleotide and to which pH environment the resonances belong. Homonuclear correlated spectra may prove to be the method of choice for *in vivo* cellular-compartment analysis of metabolites with a phosphate chain.

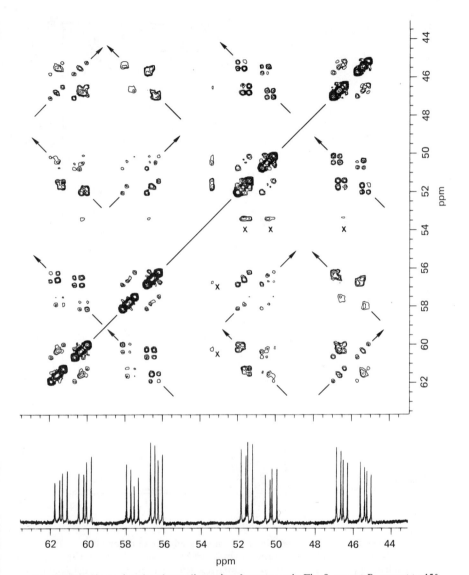

**Fig. 9.** One-dimensional and two-dimensional spectra as in Fig. 8, except $P_2$ was set to 45°. Indirectly connected transitions are suppressed, and the resultant slope of the $J$ cross-peaks indicates the relative spin for the coupling constants.

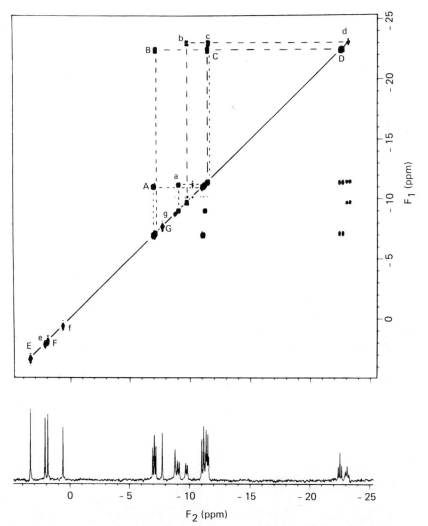

Fig. 10. Two-dimensional correlated spectrum for a mixture of phosphate metabolites in separate pH compartments. The experiment employed concentric NMR tubes with the inside (capital letters) at pH 7.5 and the outside (lower-case letters) at pH 6.5. The diagonal peaks labeled **D–d**, **E–e**, **F–f**, and **G–g**, represent the phosphate resonances of $\beta$-ATP$\beta$, AMP, inorganic phosphate, and pyrophosphate, respectively. Cross-peak **A–a** correlates the $\alpha$- and $\beta$-phosphorus resonances of ADP, **B–b** correlates the $\gamma$ and $\beta$ resonances of ATP, and **C–c** correlates the $\alpha$ and $\beta$ resonances of ATP. The resonances were broadened by the addition GdCl$_3$. From Van Divender and Hutton (1982).

The ability of the homonuclear correlated experiment to resolve severely overlapped resonances is demonstrated in Fig. 11. This mixture of four polyphosphorylated nucleotides cannot be deciphered, even at 146 MHz, as resonances from the nucleotides all lie within three 1-ppm-wide regions. The two-dimensional spectrum contains four sets of parallel cross-peaks above the diagonal, indicating that there are four separate phosphorylated chains in the mixture. Analysis of this mixture would be difficult with one-dimensional techniques. The relatively large $^{31}P-^{31}P$ coupling constants (18–25 Hz) in combination with the similarity of the resonance positions would make selective, homonuclear decoupling experiments unproductive (even assuming a spectrometer with $^{31}P$ homonuclear decoupling hardware could be located).

Application of homonuclear chemical-shift correlation spectroscopy to a biological system is shown in Figs. 12 and 13, where spectra of the perchloric acid extract from unfertilized eggs of *Strongylocentrotus purpuratas* (sea urchin) are shown. The two-dimensional spectrum indicates the presence of ATP along with four diphosphate metabolites. The data set in Figs. 12 and 13 was symmetrized (see Section II,C), so the cross-peaks appear to be rectangular unless there is peak overlap. The expanded spectrum (Fig. 13) reveals the cross-peak labeled e in Fig. 12 actually consists of two overlapping peaks. The weak doublets centered at $-10.6$ and $-10.8$ ppm, respectively, are coupled to doublets hidden under the large doublet at $-9.25$ ppm. Detailed spectral analysis can be made with certainty from the two-dimensional spectrum.

Homonuclear correlated two-dimensional spectra may prove to be particularly useful for resonance assignment of *in vivo* systems. In order to resolve spin–spin couplings in one-dimensional spectra, the transverse magnetization must persist for at least $J^{-1}$. There are many examples in the literature where this condition is not met and homonuclear $^{31}P-^{31}P$ couplings are obscured. In contrast, modulation of the $^{31}P$ precessional frequencies during the detection period of the two-dimensional experiment can be adequately digitized during a time of only $(4J)^{-1}$ (Bax and Freeman, 1981). The two-dimensional data set will display and correlate spin-coupling information that is unresolved in the one-dimentional spectrum.

## B. Heteronuclear $^{31}P-^{1}H$ Chemical-Shift Correlated Spectra

Section II,A demonstrates how homonuclear chemical-shift correlated spectra provide additional information for NMR studies of phosphorus cell metabolites. Because the homonuclear method requires the presence of $^{31}P-^{31}P$ spin–spin coupling, no data are obtained for monophosphorylated

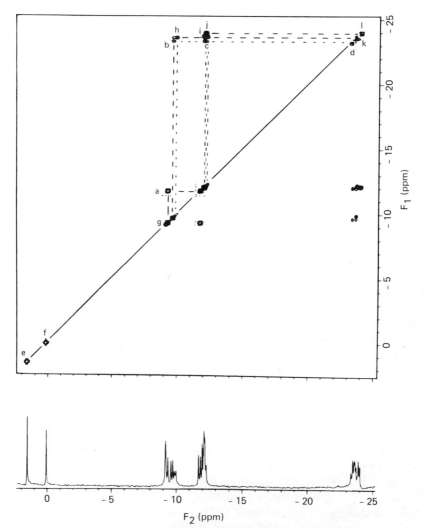

**Fig. 11.** Two-dimensional correlated spectrum of the following mixture: ATP, ADP, AMP, inorganic phosphate, pyrophosphate, adenosine tetraphosphate (A4P), diadenosine 5′,5″-tetraphosphate (AP4A), and GdCl$_3$. The labeling is consistent with Fig. 10 with the following additions: diagonal peaks **k** and **l** (−23.7 and −23.9 ppm) represent the $\beta$- and $\gamma$-phosphorus resonances of A4P and the $\beta$ resonances of AP4A, respectively; the off- diagonal peak **h** correlates the $\delta$-P of A4P (−9.9 ppm) and the $\gamma$-P of A4P (−23.7 ppm); **i** correlates the $\alpha$-P of A4P (−12.2 ppm) and $\beta$-P of A4P (−23.7 ppm); and **j** correlates the $\alpha$-P of AP4A (−12.1 ppm) and the $\beta$-P of AP4A (−23.9 ppm). From Van Divender and Hutton (1982).

16. Two-Dimensional ³¹P NMR

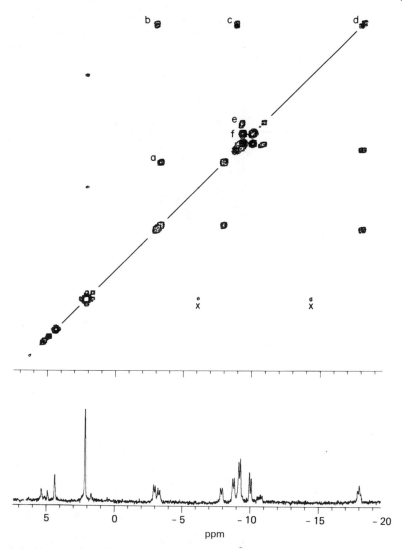

**Fig. 12.** Two-dimensional correlated spectrum of extracted, unfertilized sea urchin eggs. The pH of the mixture was 7.95. The labeling is the same as in Fig. 10 with the following exceptions: cross-peak **f** indicates that the doublets at $-9.25$ and $-10.0$ ppm are due to the diphosphate NAD; and peak **e** consists of two overlapping cross-peaks, indicating the weak doublets centered at $-10.6$ and $-10.8$ ppm are coupled to two doublets under the large doublet at $-9.25$ ppm. This spectrum was obtained in collaboration with Dr. D. K. Jemiolo and Professor R. B. Martin.

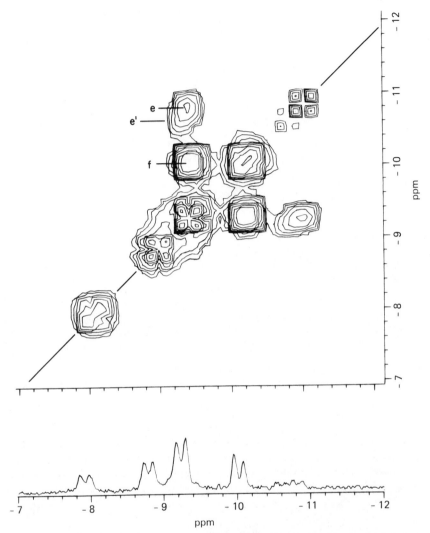

**Fig. 13.** Expansion of the two-dimensional spectrum in Fig. 12. The cross-peak labeled e in Fig. 12 overlaps with another cross-peak e'. The cross-peaks e and e' indicate two other diphosphates, most likely NADP and another dinucleotide or nucleoside diphospho sugar, are present in the extract.

molecules. A heteronuclear $^{31}$P–$^{1}$H chemical-shift correlated technique utilizing $^{3}J_{^{31}P-^{1}H}$ coupling pathways has been demonstrated for cellular phosphates (Bolton and Bodenhausen, 1979, 1982a; Bodenhausen and Bolton, 1980; Bolton, 1981a). The $^{31}$P–$^{1}$H heteronuclear method has been

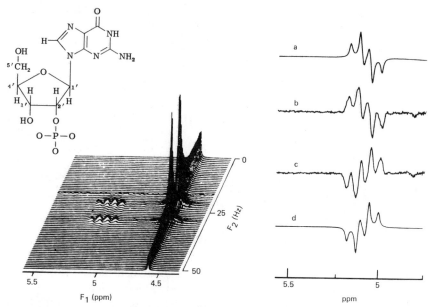

**Fig. 14.** (Left) Heteronuclear, correlated absolute-value spectrum of guanosine 2'-monophosphate. The $F_2$ dimension contains the $^1$H-coupled $^{31}$P spectrum and the $F_1$ dimension contains the $^{31}$P-coupled $^1$H spectrum. The large peaks at 4.6 ppm ($F_1$) are unsuppressed axial peaks. (Right) Phase-sensitive slices for each phosphorus transition **b** and **c** in the 2'-GMP two-dimensional spectrum. Spectra **a** and **d** are simulated for comparison with the experimental slices **b** and **c**. From Bolton and Bodenhausen (1979). Copyright 1979 American Chemical Society.

reviewed (Bolton, 1981b). This two-dimensional experiment produces a spectrum that correlates the $^{31}$P and $^1$H chemical shifts of nuclei with mutual heteronuclear scalar coupling networks. An example of a stacked-plot, heteronuclear, two-dimensional spectrum of a guanosine–2'-monophosphate may be seen in Fig. 14 (left). A projection onto the $F_2$ axis recreates the $^1$H-coupled $^{31}$P spectrum. Likewise, a projection onto the $F_1$ axis yields the $^{31}$P-coupled $^1$H spectrum.

Because the $^{31}$P and $^1$H chemical shifts are correlated, additional data are available for the assignment of monophosphorylated cell metabolites. The $\alpha$-$^{31}$P resonances of many molecules with phosphate chains will also appear in the heteronuclear chemical-shift correlation map. Cross-sections of the $^{31}$P dimension contain $^1$H–$^1$H coupled spectra. Obtaining $^1$H-NMR parameters from $^{31}$P cross-sections in the two-dimensional data set has several advantages for biological systems. Only protons that are spin–spin coupled to a phosphorus appear in the two-dimensional slices. Ribose protons in nucleotides, oligonucleotides, and other phosphorylated molecules resonate

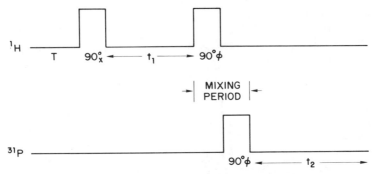

**Fig. 15.** Pulse sequence used to generate heteronuclear chemical-shift correlated two-dimensional spectra. The phase of the mixing pulses may be cycled to cancel axial peaks and allow quadrature detection in both dimensions.

near water in the $^1$H spectrum. The intense water peak makes detection of the ribose protons difficult by causing distortions from receiver overload, dynamic-range problems, and overlap in the region of interest. Similarly, for the case of enzyme–substrate studies and for *in vivo* systems, large proton backround resonances will obscure the proton peaks of interest. These problems are eliminated with the heteronuclear two-dimensional technique. Conformational analysis, utilizing the Karplus equations, is possible from the data contained in the $^1$H dimension and metabolite levels are available from intensities in the $^{31}$P dimension.

The pulse sequence used to obtain the heteronuclear two-dimensional data is shown in Fig. 15. The spin decoupler is used to generate the 90° proton preparation pulse. Consider the behavior of the proton magnetization of a heteronuclear AX spin system during the evolution period. Two proton vectors precess in the *XY* plane of the rotating frame of reference in opposite directions:

$$\begin{aligned} v_{13} &= (v + J/2)t_1 \\ v_{24} &= (v - J/2)t_1 \end{aligned} \quad (9)$$

where $v$ is the proton resonant frequency and $J$ the $^{31}$P–$^1$H scalar coupling constant. Thus the phase angle separating the vectors is a function of $t_1$. The second proton 90° pulse converts the precessing transverse magnetization back to longitudinal, or Z, magnetization. The proton spin populations now reflect the proton magnetization precession during the evolution period. The resulting proton spin populations are modulated as a function of $t_1$, that is,

$$\begin{aligned} \mathbf{M}_{13} &= -\mathbf{M}_0 \cos(v_{13}t_1) \\ \mathbf{M}_{24} &= -\mathbf{M}_0 \cos(v_{24}t_1) \end{aligned} \quad (10)$$

where $\mathbf{M}_{13}$ and $\mathbf{M}_{24}$ are the longitudinal magnetization of the protons and $\mathbf{M}_0$ is the equilibrium proton magnetization vector (Bax, 1982). The phosphorus and proton spins have common energy levels. Accordingly, the phosphorus spin populations are also altered as a function of the evolution period $t_1$. In this way, the $^{31}$P resonances become amplitude modulated as $t_1$ is incremented and the proton spin properties are encoded into the $^{31}$P free induction decays collected during the detection period. The amplitude modulation may be converted to phase modulation to affect quadrature detection in both dimensions (Bolton and Bodenhausen, 1979; Bolton, 1981b).

Before the $^{31}$P mixing pulse, the phosphorus magnetization is polarized into two components aligned along the $Z$ axis that are 180° apart, the $+\frac{1}{2}$ and $-\frac{1}{2}$ spin states. In the phase-corrected mode, cross-sections or subspectra corresponding to the $^{31}$P lines will be 180° out of phase (Fig. 14, right). Subtraction of the two proton cross-sections corresponding to the phosphorus $+\frac{1}{2}$ and $-\frac{1}{2}$ spin states has the following advantages. The signal-to-signal ratio is improved as both random noise and coherent $t_1$ noise are reduced. Computer simulation of the two proton subspectra corresponding to each $^{31}$P spin state allows the precise determination of $^1$H–$^1$H coupling constants from the two-dimensional cross-sections. The difference of the simulated subspectra will equal the difference of the experimental subspectra when the proper $^1$H spectral parameters are used in the subspectra calculations.

Phosphoserine provides a stringent test of the two-dimensional technique (Bolton, 1981a). Not only is the phosphorus resonance in phosphoserine coupled to three nonequivalent protons, but the two methylene protons are also strongly coupled to each other. Figure 16 contains simulated $^{31}$P spectra using the parameters in Table I. The agreement with the experimental two-dimensional difference spectra in Fig. 17 is impressive. The application of heteronuclear two-dimensional spectroscopy permits one to obtain accurate proton coupling constants that may be used to evaluate the conformation of phosphorylated cellular metabolites.

The heteronuclear two-dimensional experiment described here provides

**TABLE I**

$^1$H-NMR Parameters for Phosphoserine[a]

| Chemical shift (ppm) | | $J$ (Hz) | | | |
|---|---|---|---|---|---|
| α | 4.65 | $J\alpha\beta$ | 5.7 | $J\alpha P$ | 0.4 |
| β | 4.78 | $J\alpha\beta'$ | 3.3 | $J\beta P$ | 6.7 |
| β' | 4.87 | $J\beta\beta'$ | −11 | $J\beta' P$ | 6.7 |

[a] From Bolton (1981a).

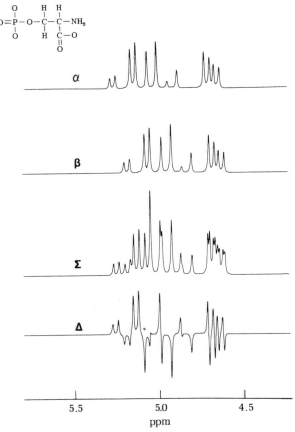

**Fig. 16.** Simulations of the two subspectra for phosphoserine using the parameters in Table I. Subspectra $\alpha$ and $\beta$ correspond to the phosphorus spin states. Their sum $\Sigma$ is the conventional $^1$H spectrum, whereas their difference $\Delta$ corresponds to experimental two-dimensional difference spectrum. From Bolton (1981a).

only information from protons near the phosphorus nucleus. This restriction is eliminated by the relayed coherence-transfer two-dimensional technique (Bolton, 1982; Bolton and Bodenhausen, 1982b). The relayed-coherence experiment combines the homonuclear correlated and heteronuclear correlated pulse sequences. For the case of an AMX spin system, where X is the phosphorus nucleus, A and M are protons, $J_{MX} >$ or $< 0$, and $J_{AX} = 0$, the proton pulses would transfer magnetization from the A spin to the M spin. The heteronuclear transfer would then be made from the M spin to the X spin. The two-dimensional data set obtained from this technique permits

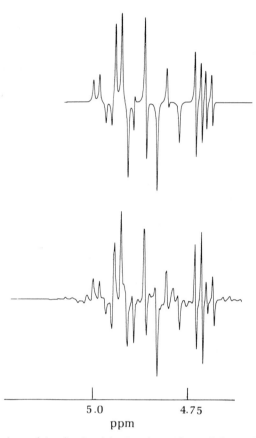

**Fig. 17.** Comparison of the simulated (top) and experimental (bottom) two-dimensional difference spectra of phosphoserine. From Bolton (1981a).

the evaluation of $^1$H-NMR parameters for any proton coupled to a proton that is directly coupled to the phosphorus.

Figure 18 shows the application of relayed coherence transfer to phosphothreonine, an $AMQ_3X$ spin system (Bolton and Bodenhausen, 1982b). The contour plot of the two-dimensional data set in Fig. 18 has correlated all the protons of phosphothreonine with the $^{31}$P resonance. Figure 19 compares the simulated $^1$H spectrum of phosphothreonine with a phase-sensitive cross-section, two-dimensional data set. Quantitative analysis of coupling constants can be obtained for protons not directly coupled to the $^{31}$P nucleus. The relayed coherence-transfer experiment makes possible analysis

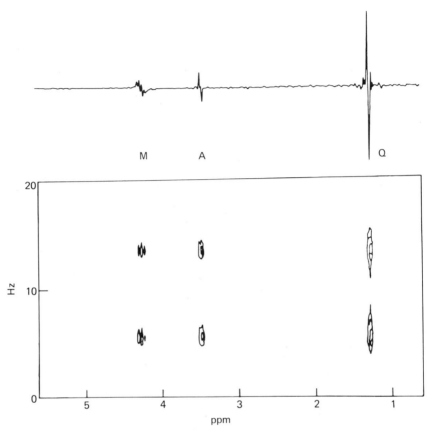

**Fig. 18.** Heteronuclear two-dimensional relayed coherence-transfer spectrum of phosphothreonine. The absolute-value contour plot (bottom) contains the phosphorus doublet on the vertical axis and the proton spectrum on the horizontal axis. The scalar coupling patterns for protons indirectly coupled to the phosphate group can be evaluated from the cross-section (top). From Bolton and Bodenhausen (1982b).

of molecular conformation in regions of a molecule well removed from the phosphorus group.

## C. Chemical-Exchange Correlated Spectra

A two-dimensional experiment has been described for correlating the chemical shifts of nuclei undergoing chemical exchange (Jeener *et al.*, 1979; Macura *et al.*, 1981). The pulse sequence for generating a homonuclear correlated exchange spectrum is shown in Fig. 20. At the end of the

16. Two-Dimensional $^{31}$P NMR

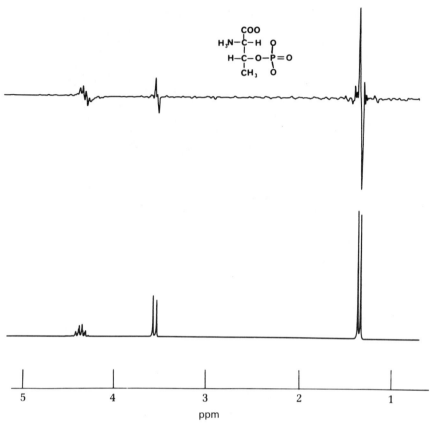

**Fig. 19.** Comparison of the cross-section in Fig. 20 with the computer-simulated proton spectrum. From Bolton and Bodenhausen (1982b).

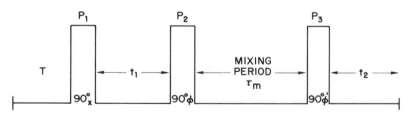

**Fig. 20.** Pulse sequence used to produce chemical-exchange correlated two-dimensional spectra. The phases of $P_2$, $P_3$, and the receiver may be cycled to suppress axial peaks, $J$ cross-peaks, and allow quadrature detection in both dimensions (Jeener *et al.*, 1979; Macura *et al.*, 1981).

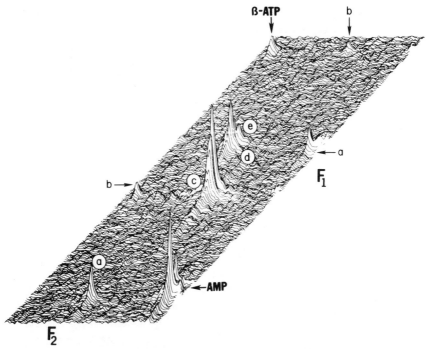

**Fig. 21.** Two-dimensional exchange correlated spectrum for an equilibrium solution of AMP, ADP, and ATP in the presence of 13,000 units of adenylate kinase (rabbit muscle). The mixing time for this experiment was 1 s and the temperature was 30°C. The diagonal runs from right to left, and peaks owing to AMP, the $\gamma$-phosphate of ATP, and the $\beta$-phosphate of ADP **c**; the $\alpha$-phosphates for ATP and ADP **d**; and the $\beta$-phosphate of ATP lie on the diagonal. Cross-peaks **a**, **b**, and **e** map out the chemical-exchange pathways. This spectrum was generously provided by Dr. R. S. Balaban and co-workers at the National Institutes of Health.

preparation period $T$, a 90° pulse generates transverse magnetization. Each spin precesses at its characteristic resonant frequency during the evolution period. $P_2$ begins the mixing period by converting the $X$ component of the $XY$ transverse magnetization back into the $Z$ axis. Throughout the course of the mixing period $\tau_m$, dynamic processes exchange the $Z$ magnetization between nonequivalent sites. The exchange may be inter- or intramolecular. The third 90° pulse ends the mixing period and produces transverse magnetization. The precessional frequency of the spins now reflects their new environment because of the exchange. The free induction decay is digitized in the detection period. Incrementing the evolution period $t_1$ creates a two-dimensional data set. The precession of the exchanged spins during $t_2$ is modulated by their original precessional frequencies during $t_1$. The two-dimensional spectrum maps out all the exchange processes simultaneously.

Cross-peaks, correlating exchanging nuclei A and X, will only be observed when the exchange rate constant $k$ is slow on the NMR time scale, that is, $k_{AX} \ll |v_A - v_X|$.

Figure 21 illustrates an application of two-dimensional exchange correlated spectroscopy. This stacked-plot presentation maps the phosphate chemical exchange catalyzed by adenylate kinase among ATP, ADP, and AMP. The adenylate interconversion proceeds as

$$2 \text{ ADP} \xrightleftharpoons{\text{adenylate kinase}} \text{AMP} + \text{ATP}$$

Inspection of the two-dimensional spectrum allows discrimination of the exchange pathways. The cross-peak marked **a** shows that the AMP phosphate is in exchange with the $\alpha$-phosphate of ADP (**d**). The cross-peak labeled **b** correlates the $\beta$-phosphorus of ATP with the $\beta$-phosphorus of ADP (**c**). Cross-peak **e** is barely resolved from the diagonal peaks **d** with which it correlates and indicates exchange of the $\alpha$-ATP and $\alpha$-ADP phosphates. The exchange correlated experiment simultaneously portrays the exchange network. This method will be particularly useful for defining exchange processes in systems with many sites. Peak overlap, which would be detrimental to one-dimensional techniques, does not hamper analysis of the two-dimensional data.

## IV. Concluding Remarks

Two-dimensional spectroscopy can be applied to $^{31}$P magnetic resonance to provide analysis of complex spectra that may be difficult or impossible to obtain with one-dimensional techniques. Because the instrument time required to generate two-dimensional spectra is significantly longer than the time used to obtain a conventional spectrum, two-dimensional experiments should be employed only when less time-consuming experiments are ambiguous. $^{31}$P-$^{1}$H heteronuclear correlated two-dimensional spectra are able to provide detailed $^{1}$H parameters in cases where conventional $^{1}$H spectra would be fruitless.

NMR instrumentation and software has progressed to the point that two-dimensional data acquisition, processing, and plotting can be completely automated. Advances in computer hardware and software development, along with improvements in sensitivity, enable two-dimensional experiments to be conducted on a routine basis. These instrumental gains show no sign of abating, making the application of two-dimensional spectroscopy even more attractive in the future.

Multiple quantum and zero-quantum, two-dimensional experiments can be applied to $^{31}$P magnetic resonance. These techniques have the capability to further simplify complicated single-quantum, two-dimensional spectra. It is clear that the continuing advances in two-dimensional spectroscopy will allow the experimentalist to unravel complex spin systems that would otherwise be undecipherable.

## References

Aue, W. P., Bartholdi, E., and Ernst, R. R. (1976). *J. Chem. Phys.* **64,** 2229–2247.
Bauman, R., Wider, G., Ernst, R. R., and Wüthrich, K. (1981). *J. Magn. Reson.* **44,** 402–406.
Bax, A. (1982). "Two-Dimensional Nuclear Magnetic Resonance in Liquids." Reidel Publ., Dordrecht, Netherlands.
Bax, A., and Freeman, R. (1981). *J. Magn. Reson.* **44,** 542–551.
Bax, A., Freeman, R., and Morris, G. (1981). *J. Magn. Reson.* **42,** 164–168.
Bodenhausen, G., and Bolton, P. H. (1980). *J. Magn. Reson.* **39,** 399–412.
Bodenhausen, G., Freeman, R., Niedermeyer, R., and Turner, P. L. (1977). *J. Magn. Reson.* **26,** 373–379.
Bolton, P. H. (1981a). *J. Magn. Reson.* **45,** 239–253.
Bolton, P. H. (1981b). *In* "Biomolecular Stereodynamics" (R. H. Sarma, ed.), Vol. 2, pp. 437–453. Adenine Press, New York.
Bolton, P. H. (1982). *J. Magn. Reson.* **48,** 336–340.
Bolton, P. H., and Bodenhausen, G. (1979). *J. Am. Chem. Soc.* **101,** 1080–1084.
Bolton, P. H., and Bodenhausen, G. (1982a). *J. Magn. Reson.* **46,** 306–318.
Bolton, P. H., and Bodenhausen, G. (1982b). *Chem. Phys. Lett.* **89,** 139–144.
Ciampolini, M., Dapporto, P., Dei, A., Nardi, N., and Zanobini, F. (1982). *Inorg. Chem.* **21,** 489–495.
DeMarco, A., and Wüthrich, K. (1976). *J. Magn. Reson.* **24,** 201–208.
Ernst, R. R. (1975). *Chimia* **29,** 179–183.
Freeman, R., and Morris, G. A. (1979). *Bull. Magn. Reson.* **1,** 5–26.
Hyde, E. M., Kennedy, J. D., Shaw, B. L., and McFarlane, W. (1977). *J. Chem. Soc., Dalton Trans.* **16,** 1571–1576.
Jeener, J., Meier, B. H., Bachmann, P., and Ernst, R. R. (1979). *J. Chem. Phys.* **71,** 4546–4553.
Macura, S., Huang, Y., Suter, D., and Ernst, R. R. (1981). *J. Magn. Reson.* **43,** 259–281.
Nagayama, K., Kumar, A., Wüthrich, K., and Ernst, R. (1980). *J. Magn. Reson.* **40,** 305–322.
Navon, G., Ogwa, S., Shulman, R. G., and Yamane, E. (1977). *Proc. Natl. Acad. Sci. U.S.A.* **74,** 371–377.
Oldfield, E., and Meadows, M. (1978). *J. Magn. Reson.* **31,** 327–335.
Van Divender, J., and Hutton, W. C. (1982). *J. Magn. Reson.* **48,** 272–279.
Wagner, G., Kumar, A., and Wüthrich, K. (1981). *Eur. J. Biochem.* **114,** 375–389.

CHAPTER 17

# Identification of Diseased States by Phosphorus-31 NMR

### Michael Bárány
Department of Biological Chemistry
Health Sciences Center
College of Medicine
University of Illinois at Chicago
Chicago, Illinois

### Thomas Glonek
Nuclear Magnetic Resonance Laboratory
Chicago College of Osteopathic Medicine
Chicago, Illinois

| | |
|---|---|
| I. Introduction | 512 |
| II. Skeletal Muscle | 513 |
|     A. Qualitative Analysis | 513 |
|     B. Quantitative Analysis | 517 |
|     C. Intact Human Subject | 524 |
| III. Heart | 530 |
| IV. Kidney | 533 |
| V. Brain | 534 |
| VI. Eye | 537 |
|     A. Lens | 537 |
|     B. Cornea | 539 |
|     C. Vitreous Humor | 540 |
| VII. Mammalian Fluids | 540 |
|     A. Amniotic Fluid | 540 |
|     B. Human Saliva | 540 |
|     C. Cerebrospinal Fluid | 542 |
| VIII. Perspectives | 543 |
|     References | 544 |

## I. Introduction

Identification of diseased states by $^{31}$P NMR originates from the observation of Burt et al. (1976) that intact dystrophic chicken muscle contains an extra resonance not present in normal chicken muscle. The compound giving rise to this signal has been isolated and identified as a phosphodiester, L-serine ethanolamine phosphate (Chalovich et al., 1977). At the same time, it was found that intact normal human leg muscle contains another phosphodiester, *sn*-glycerol 3-phosphorylcholine (GPC),[1] and that this compound is missing in diseased human muscles, notably in Duchenne muscular dystrophy (Bárány et al., 1977; Chalovich et al., 1979). This suggested that GPC may be used as a marker for human muscle diseases.

In addition to the qualitative differences between normal and diseased tissue detectable by $^{31}$P NMR, the technique readily quantitates the level of common phosphate metabolites (e.g., ATP, PCr) in tissues. Because these compounds play a key role in cellular metabolism, changes in their concentration may provide another index concerning the healthy state of the tissue.

By recording the $^{31}$P-NMR spectrum of intact tissue as a function of time, various enzyme activities may be measured (Burt et al., 1977; Gadian and Radda, 1981; Bárány et al., 1982). This may lead to the diagnosis of a disease correlated with a specific enzyme defect (Ross et al., 1981; Edwards et al., 1982).

With the availability of topical $^{31}$P-NMR instruments capable of obtaining highly resolved spectra of a human forearm or leg, the identification of diseased states by $^{31}$P NMR entered a new chapter (Cresshull et al., 1981; Leigh et al., 1981; Ross et al., 1981). That such a device is capable of accessing tissue biochemistry noninvasively, efficiently, and without risk to the subject (Oxford Research Systems, 1979) makes it a powerful tool indeed.

In this chapter we discuss several clinical applications of $^{31}$P NMR. The main emphasis is placed on advances in skeletal muscle (since the review of Glonek et al., 1981), but advances in heart, kidney, brain, eye, and mammalian fluids are also covered. The resonances we deal with are from the phosphorus compounds soluble in the cytoplasm. At the time of this writing, the phospholipids in the membranes of intact tissues are not amenable for $^{31}$P-NMR analysis under the high-resolution conditions cus-

---

[1] Abbreviations: GPC, *sn*-glycerol 3-phosphorylcholine; GPE, *sn*-glycerol 3-phosphorylethanolamine; PCr, phosphocreatine; SP, sugar phosphates; XMP, xanthine monophosphate; DN, dinucleotides; NS, nucleoside diphospho sugars; PCA, perchloric acid; NC, noncollagenous; CSF, cerebrospinal fluid; $T_1$, spin–lattice relaxation time; NOE, nuclear Overhauser enhancement.

tomarily employed. Furthermore, the phosphorus of nucleic acids and of phosphoproteins is present in concentrations below the sensitivity of the $^{31}$P method in the majority of cells.

## II. Skeletal Muscle

### A. Qualitative Analysis

#### 1. Phosphodiesters

GPC and GPE are the hydrolytic products of the major membrane phospholipids, phosphatidylcholine and phosphatidylethanolamine, respectively. The enzymes responsible for the degradation of these phospholipids, phospholipase $A_1$ and $A_2$, are present in human muscle. Hereditary muscular dystrophy, a lethal disease of young children, is generally assumed to be a membrane disease (Rowland, 1980). Therefore, $^{31}$P-NMR studies on

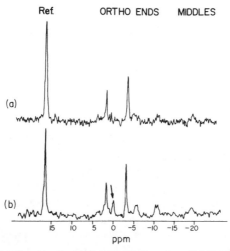

**Fig. 1.** Comparison of proton-decoupled $^{31}$P-NMR spectra (36.4 MHz) of normal (b) and nemaline rod-diseased (a) human quadriceps. The muscles were placed into 10-mm tubes that were spun at 45 Hz at 31°C. Spectrometer conditions: sweep width, 2500 Hz; data points, 4096; cycling time, 832 ms; pulse width, 4.5 $\mu$s; 1.2-Hz line broadening. The spectra show the signal average of 2160 scans. "ORTHO, ENDS, and MIDDLES" refer to the characteristic regions of the $^{31}$P spectrum. Peak assignments (ppm) in the muscle spectrum from left to right: the external methylene diphosphonate reference compound (16.3), SP (3.7), $P_i$ (1.7), PCr ($-3.2$); the phosphate groups of ATP: $\gamma$ ($-5.6$), $\alpha$ ($-10.7$), $\beta$ ($-19.1$). Chemical-shift data are relative to 85% $H_3PO_4$ and follow the IUPAC convention. The arrow indicates the presence of GPC in the normal but not the diseased muscle. From Bárány et al. (1977).

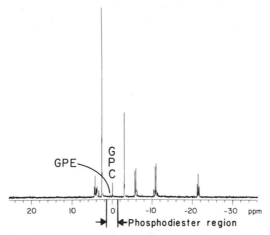

**Fig. 2.** Proton-decoupled $^{31}$P-NMR spectrum (81.0 MHz) from a healthy human gastrocnemius muscle PCA extract. The extract was placed into a 12-mm tube that was spun at 18 Hz at 24.5°C. Spectrometer conditions: sweep width, ±2500 Hz; data points, 16,384; cycling time, 1.64 s; pulse width, 9 μs (45° flip angle); 0.6-Hz line broadening; number of scans, 28,100. Peak assignments (ppm) in the spectrum from left to right: SP (4.8–3.3), $P_i$ (1.7), GPE (0.37), GPC (−0.13), PCr (−3.2); the phosphate groups of ATP: $\gamma$ (−5.4 to −6.1), $\alpha$ (−10.4 to −11.0), $\beta$ (−21.0 to −21.9). The chemical-shift scale follows the IUPAC convention and is given relative to the resonance position of 85% $H_3PO_4$.

phosphodiesters may contribute to a better understanding of the nature of human muscular disorders.

GPC is a major resonance in $^{31}$P spectra of intact normal human leg muscle (lower part of Fig. 1); it resonates at −0.1 ppm. In nemaline rod disease, GPC is not visible (Fig. 1a).

In PCA muscle extracts, the $^{31}$P-NMR spectrum is more highly resolved than that from the corresponding intact muscle (Bárány and Glonek, 1982). This allows for greater information content from the spectrum, resulting in the detection of relevant minor differences that are obscured in the spectrum of the intact muscle. GPE is detectable in PCA extracts of normal human muscle (Fig. 2). The peak height of GPE (0.37 ppm) is only a small fraction of that of GPC (−0.13 ppm). Figure 2 also illustrates the phosphodiester resonance band, which covers the region from approximately 1 to −1 ppm. In addition to GPC and GPE, this region contains other related phosphodiesters such as glycerol 3-phosphorylserine, glycerol 3-phosphorylinositol, and serine ethanolamine phosphate; however, their occurrence in human muscular diseases is rare.

GPC is a major resonance in muscle extracts of a Kugelberg–Welander patient (Fig. 3). Its peak height is equal to or higher than those of the $\gamma$-phosphate of ATP. Because the ATP concentration in muscle remains

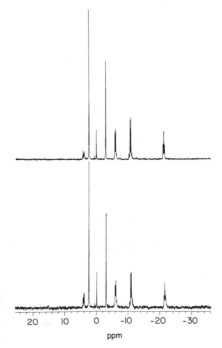

Fig. 3. Phosphorus-31 spectra (81.0 MHz) from PCA extracts of left (top) and right (bottom) gastrocnemius muscles of a patient with Kugelberg–Welander disease.

essentially constant so long as phosphocreatine is present (because the creatine phosphokinase catalyzed reaction is shifted far toward the synthesis of ATP), the $\gamma$-phosphate of ATP can be used as a reference for qualitative estimation of GPC levels in muscle biopsy samples that still exhibit a phosphocreatine resonance. For instance, in a healthy muscle extract (Fig. 2) the GPC concentration is less than half that in the extract of the Kugelberg–Welander patient. The precision of the $^{31}$P-NMR analysis is indicated by the close similarity between spectra from extracts of the left and right gastrocnemius muscles of the patient with Kugelberg–Welander disease.

By the same criteria, GPC is elevated in Charcot–Marie–Tooth disease; its peak height is 1.5-fold that of the $\gamma$-phosphate (Fig. 4). In contrast, the GPC concentration is reduced in several diseases. In a myopathy of unknown etiology, the GPC peak height is only one-third (Fig. 5), and in congenital myotonia it is only a small fraction of that of the $\gamma$-phosphate (Fig. 6).

## 2. Other Metabolites

High-resolution $^{31}$P spectroscopy readily separates the resonance signals of the phosphorylated sugars that appear in the downfield portion of the

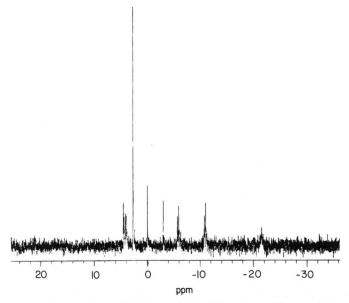

**Fig. 4.** Phosphorus-31 spectrum (81.0 MHz) from a PCA extract of the quadriceps muscle of a patient with Charcot–Marie–Tooth disease.

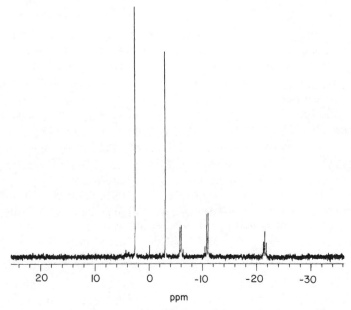

**Fig. 5.** Phosphorus-31 spectrum (81.0 MHz) from a PCA extract of the quadriceps muscle of a patient with myopathy of unknown etiology.

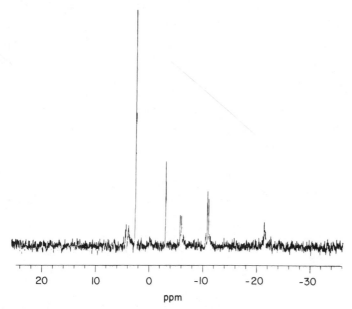

**Fig. 6.** Phosphorus-31 spectrum (81.0 MHz) from a PCA extract of the vastus muscle of a patient with congenital myotonia.

spectrum between 3 and 5 ppm. Several times we observed a double resonance in the sugar phosphate region that coresonated with added fructose 1,6-diphosphate. Occasionally, the double resonance was so strong that fructose 1,6-diphosphate could be identified without marker addition (Fig. 7); recording the expanded scale of the sugar phosphate region made the identification unequivocal (Fig. 8). We have found high concentrations of fructose 1,6-diphosphate in extracts of diseased muscles with diagnosis of "myotonic dystrophy," "polymyositis," and "collagen vascular disease."

Occasionally, we noticed the presence of inorganic pyrophosphate ($-7.0$ ppm), an indicator for protein-synthesis activity of muscle, in diseased human muscle extracts (Bárány and Glonek, 1982). The presence of phosphorylcholine (3.3 ppm) may be correlated with increased lecithin turnover.

## B. Quantitative Analysis

Validation of $^{31}$P NMR as a quantitative analytical procedure requires the determination of $T_1$ and NOE values of the muscle extract phosphates (Table I). The $T_1$ values of Table I can be compared with the data determined on pure aqueous solutions in the case of five of the compounds. The

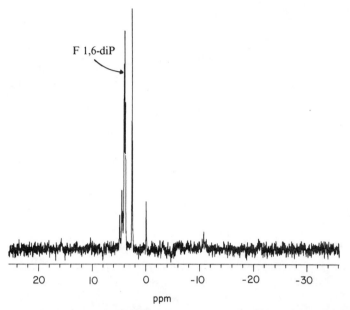

**Fig. 7.** Phosphorus-31 spectrum (81.0 MHz) from a PCA extract of a diseased human gastrocnemius muscle showing very elevated fructose 1,6-diphosphate (F 1,6-diP) signals.

published $T_1$ values of 10.4 s at pH 10.2 and 4.0 s at pH 7.2 for 0.1 $M$ tetramethylammonium orthophosphate (Morgan and Van Wazer, 1975) are close to the inorganic orthophosphate $T_1$ values of Table I. For adenosine monophosphate (0.1 $M$ in phosphate), the reported $T_1$ values of 7.4 s at

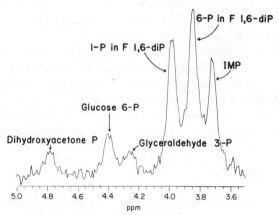

**Fig. 8.** Expanded $^{31}$P spectrum of the sugar phosphate resonance band in the region of the two fructose 1,6-diphosphate (F 1,6-diP) signals. Signals from other hexose, pentose, and triose phosphates are also indicated in the figure.

### TABLE I

31P-NMR Parameters of the Principal Muscle Phosphates in Degassed Perchloric Acid Extracts at 37.7°C with Potassium as the Countercation[a]

| Compound | pH | Concentration (mol %) | Chemical shift (ppm) | $T_1$ (s) | NOE (factor) |
|---|---|---|---|---|---|
| Triose phosphates | 10.2 | 4.7 | 4.65 | 12.0 ± 0.15 | 1.00 |
|  | 7.2 |  | 4.13 | 3.4 ± 0.08 | 1.00 |
| Hexose phosphates | 10.2 | 4.4 | 4.47 | 7.6 ± 0.17 | 1.00 |
|  | 7.2 |  | 3.61 | 1.4 ± 0.07 | 1.00 |
| Inosine 5′-monophosphate | 10.2 | 2.7 | 3.78 | 10.0 ± 0.09 | 1.00 |
|  | 7.2 |  | 3.11 | 2.1 ± 0.08 | 1.23 |
| Adenosine 5′-monophosphate | 10.2 | 1.8 | 3.75 | 6.7 ± 0.32 | 1.00 |
|  | 7.2 |  | 3.44 | 2.6 ± 0.01 | 1.22 |
| Phosphorylcholine | 10.2 | 0.2 | 3.33 | 8.4 ± 0.41 | 1.07 |
|  | 7.2 |  | 2.99 | 4.2 ± 0.38 | 1.24 |
| Inorganic orthophosphate | 10.2 | 30.2 | 2.63 | 9.3 ± 0.13 | 1.00 |
|  | 7.2 |  | 2.20 | 1.0 ± 0.01 | 1.00 |
| Glycerol 3-phosphorylethanolamine | 10.2 | 0.2 | 0.81 | 14.6 ± 0.29 | 1.21 |
|  | 7.2 |  | 0.31 | 13.9 ± 0.31 | 1.22 |
| Glycerol 3-phosphorylcholine | 10.2 | 2.9 | −0.13 | 14.2 ± 0.16 | 1.90 |
|  | 7.2 |  | −0.13 | 12.7 ± 0.10 | 1.82 |
| Phosphocreatine | 10.2 | 13.6 | −3.12 | 9.0 ± 0.06 | 1.00 |
|  | 7.2 |  | −2.89 | 6.8 ± 0.11 | 1.00 |
| ATP |  |  |  |  |  |
| α-phosphate | 10.2 | 31.1 | −10.92 | 6.4 ± 0.11 | 1.22 |
|  | 7.2 |  | −10.87 | 6.2 ± 0.07 | 1.24 |
| β-phosphate | 10.2 | — | −20.45 | 10.2 ± 0.16 | 1.00 |
|  | 7.2 |  | −20.50 | 7.4 ± 0.16 | 1.11 |
| γ-phosphate | 10.2 | — | −5.80 | 11.2 ± 0.21 | 1.00 |
|  | 7.2 |  | −6.19 | 2.6 ± 0.13 | 1.07 |
| ADP |  |  |  |  |  |
| α-phosphate | 10.2 | 5.1 | −10.61 | 7.2 ± 0.03 | 1.17 |
|  | 7.2 |  | −10.61 | 4.5 ± 0.04 | 1.20 |
| β-phosphate | 10.2 | — | −6.11 | 9.7 ± 0.60 | 1.15 |
|  | 7.2 |  | −6.33 | 2.5 ± 0.02 | 1.24 |
| Dinucleotides | 10.2 | 3.1 | −11.37 | 6.0 ± 0.53 | 2.20 |
|  | 7.2 |  | −11.36 | 5.9 ± 0.41 | 2.20 |

[a] The $T_1$ values were measured at 40.5 MHz for 31P by the inversion–recovery method using the procedure of Morgan and Van Wazer (1975), where the delay time employed was at least five times the longest $T_1$ in the spectrum. The inversion-recovery data were fitted by a least-squares analysis to the function $\ln(I_\infty - I_t) = K_\tau$ by a subroutine of the Transform Technology NTCFT software package for the determination of the $T_1$ values. NOE values were measured by comparing signal areas measured with and without broadband irradiation of the 1H-resonance spectrum (Noggle and Schirmer, 1971). Samples were $N_2$ purged and vacuum degassed for the $T_1$ and NOE determinations.

pH 11.2 and 1.6 s at pH 6.8, with potassium as the counterion, are in agreement with the data of Table I when the proper allowance is made for the oscillatory change of relaxation time with the pH of ionized compounds (Morgan and Van Wazer, 1975; Glonek and Van Wazer, 1976; McCain and Markley, 1980). Such allowance for the $T_1$ values of the various resonances of the potassium salts of ADP and ATP (0.1 $M$ in phosphate) also indicates an agreement with Table I. The reported $T_1$ values at pH 11.2 are for ADP-$\alpha$ 11.7 and -$\beta$ 11.3 s; ATP-$\alpha$ 9.1, -$\beta$ 13.3, and -$\gamma$ 13.5 s. Furthermore, a 0.1 $M$ aqueous solution of GPC exhibits a $T_1$ of 13–14 s, depending on the other solutes present (Glonek and Van Wazer, 1976). These comparisons seem to indicate that the $T_1$ values of Table I have broad significance at relatively low aqueous concentrations for the selected pH values and with potassium as the counterion for the ionic species.

In general, the NOE values of the phosphates measured in the muscle extracts indicated little or no increase in signal area on irradiation of the protons, a finding consistent with the previously reported data from these laboratories on intact muscles and their extracts (Burt et al., 1976). Exceptions in the series of Table I are GPC and the dinucleotides, both of which exhibit large NOE values, the dinucleotide enhancement being essentially at the theoretical limit of 2.235 (Yeagle et al., 1975). Presumably this enhancement is due to a tight molecular conformation, maintained in the extract medium, in which some hydrogen atoms are sufficiently close to the phosphorus to cause this effect. Other phosphates showing about 20% enhancement are the nucleotide monophosphates and phosphorylcholine at pH 7.2 and GPE, as well as the $\alpha$-phosphorus of ATP and the $\beta$-phosphorus of ADP at both pH values.

Because $T_1$ relaxation time values of phosphorus-31 in organophosphates are long in aqueous solutions (Morgan and Van Wazer, 1975; Glonek and Van Wazer, 1976; Glonek et al., 1976; McCain and Markley, 1980), it is not practical to gather $^{31}$P-NMR profile data from such samples under conditions where magnetic resonance saturation of the sample does not occur (i.e., conditions where the delay time between pulses greater than or equal to five times the longest $T_1$ of the sample). Instead, spectrometer conditions are chosen so as to optimize signal-to-noise ratios (S/N) per unit of instrument time, and these always involve delay times much shorter than that which is spectroscopically ideal. The resultant consequences manifested in the $^{31}$P extract spectra are that data are gathered far from the spin-resonance equilibrium condition, and saturation of the separate phosphorus resonance signals occurs. For accurate quantitation, therefore, it is necessary to calibrate the spectra against known quantities of phosphate standards. The prefered method of calibration involves addition of standards containing known concentrations of phosphate metabolites listed in Table II, in small

## TABLE II
### Phosphate Profiles of Diseased Human Muscles[a]

| Disease | NC protein (mg/g) | Total phosphate (μmol/g) | Total phosphate (μmol/174 mg NC protein) | GPC | GPE (μmol/174 mg NC protein) | ATP | ADP | NAD | PCr | P$_i$ | SP | n |
|---|---|---|---|---|---|---|---|---|---|---|---|---|
| Healthy | 174 ± 9 | 51.3 ± 7.0 | 51.3 ± 7.0 | 1.5 ± 0.2 | 0.1 ± 0.1 | 5.3 ± 0.5 | 1.3 ± 0.2 | 0.8 ± 0.3 | 10.5 ± 3.2 | 12.0 ± 4.5 | 7.1 ± 3.9 | 4 |
| Duchenne | 107 | 17.3 | 28.1 | <0.1 | <0.1 | 2.7 | 1.2 | 0.6 | 5.6 | 8.3 | 2.5 | 2 |
| Becker | 98 | 15.3 | 27.2 | 0.2 | <0.1 | 3.1 | 1.2 | 0.7 | 7.4 | 6.1 | 0.4 | 1 |
| Facioscapulohumeral dystrophy | 123 | 29.2 | 41.3 | 0.4 | 0.1 | 4.2 | 1.1 | 0.8 | 8.7 | 9.5 | 6.2 | 2 |
| Congenital myotonia | 136 | 29.1 | 37.2 | 0.1 | <0.1 | 3.9 | 1.3 | 0.6 | 7.4 | 7.8 | 6.4 | 2 |
| Myopathy of unknown etiology | 145 | 30.7 | 36.8 | 0.5 | 0.1 | 4.1 | 1.0 | 0.5 | 8.3 | 7.8 | 4.8 | 1 |
| Charcot–Marie–Tooth | 118 | 31.3 | 46.1 | 3.1 | <0.1 | 3.5 | 1.2 | 0.9 | 8.7 | 11.1 | 8.5 | 2 |
| Kugelberg–Welander | 105 | 33.0 | 54.7 | 3.0 | 0.3 | 5.6 | 1.1 | 0.7 | 10.6 | 14.7 | 5.7 | 2 |
| Meningomyelocele | 102 | 30.0 | 51.1 | 2.1 | 0.2 | 2.9 | 1.0 | 0.4 | 13.2 | 15.4 | 8.7 | 2 |
| Cerebral palsy | 121 | 27.1 | 39.0 | 0.1 | <0.1 | 4.4 | 1.3 | 0.8 | 5.3 | 8.4 | 7.8 | 2 |
| Amyotrophy After encephalitis | 127 | 21.1 | 28.9 | 0.2 | <0.1 | 3.1 | 0.9 | 0.6 | 5.9 | 5.2 | 5.2 | 1 |
| Of unknown etiology | 116 | 19.7 | 29.6 | 1.6 | 0.1 | 2.5 | 0.9 | 0.5 | 7.1 | 7.4 | 3.1 | 2 |

[a] Results for healthy muscles are given plus or minus standard deviation; for duplicate muscles the average values which were within 30% are given. Muscles used and age of patients: Healthy, quadriceps and other leg muscles (10, 16, 27, and 39 years old); Duchenne, quadriceps (4, 9); Becker, quadriceps (23); facioscapulohumeral dystrophy, shoulder muscles (22, 23); congenital myotonia, vastus (2, 4); myopathy of unknown etiology, quadriceps (18); Charcot–Marie–Tooth, quadriceps (37, 53); Kugelberg–Welander, gastrocnemius (39); meningomyelocele, soleus (0.2, 12); cerebral palsy, soleus (8, 10); amyotrophy after encephalitis, vastus (14); amyotrophy of unknown etiology, quadriceps (28, 54).

measured increments, to PCA-extract samples from which quantitative $^{31}$P profile data had previously been gathered and then redetermining the spectrum. Such a method insures that the microenvironment sensed by the added standard P atoms will be essentially the same as that sensed by the P atoms of the extracted organophosphates, and relative-area measurements ought to be comparable. When such determinations were performed on PCA extracts, the observed changes in integrated peak areas were proportional to the concentration of the phosphorus-containing component added to the sample.

An example of the effects of $T_1$ and NOE on relative integrated phosphorus-31 signal areas is given for the molecule GPC in PCA extracts as they are ordinarily prepared for analysis (i.e., without $N_2$ purging and vacuum degassing). Allowing 71-s delay between acquisitions (the maximum sample $T_1$ of 14.2 s is assumed), with proton coupling, an extract GPC signal yielded a relative signal intensity equivalent to 6.58% of the total phosphorus. With proton decoupling, which sharpens signals by eliminating $^1H-^{31}P$ spin-coupled multiplets and, consequently, significantly improves spectral resolution, the relative signal area was increased to 7.83% owing to the NOE (Noggle and Schirmer, 1971). With a period between pulses equal to the acquisition time (1.64 s), however, the relative signal was reduced to 6.56% of the total phosphorus. The rapid pulse rate plus the effects of proton decoupling cause the GPC signal to be underestimated by a factor of 6.56/6.58. The correct value for the GPC signal area then becomes the measured integral times 1.003. Such a correction factor is insignificant.

For quantitation of phosphate metabolites, (Table II), the human muscle biopsies were extracted with 30 volumes of 3% PCA at 0°C for 10 min to allow maximal extraction of the phosphodiesters. Carrier-free $^{32}$P was added to the extract, before centrifugation, and the counts in the subsequent workup steps (Bárány and Glonek, 1982) were used to determine the percent recovery of the muscle phosphates. The total phosphate content of the extract was determined after wet digestion and the NC protein content of the muscle was determined in the PCA-insoluble residue (Chalovich et al., 1979). The distribution of phosphates in the extract was determined by $^{31}$P-NMR spectroscopy at 81 MHz for $^{31}P_i$ under conditions as described in the legend of Fig. 2. The signal areas were analyzed as described (Bárány and Glonek, 1982).

Table II lists the GPC content of various diseased human muscles. Because the NC protein content of all diseased muscles was decreased considerably as compared to that of the value of 174 mg NC protein per gram for healthy muscle (column 2, Table II), the GPC content is expressed as micromoles per 174 mg NC protein. Such a normalization permits detection of specific differences in phosphate profiles (Glonek et al., 1981).

Muscle from healthy children and adults contains, on average, 1.5 μmol GPC per 174 mg NC protein. In agreement with our previous report (Chalovich et al., 1979), no GPC was detectable in muscles of Duchenne children. In Becker's dystrophy, 0.2 μmol GPC was detected, whereas in facioscapulohumeral dystrophy, 0.4 μmol GPC was found. Both values are considerably less than those of healthy muscle. GPC was also diminished to 0.1 μmol in muscles of two young children with congenital myotonia, and GPC was decreased to 0.5 μmol in the quadriceps of an 18-year-old patient with a myopathy of unknown etiology. These results indicate that several myogenic disorders may be characterized by reduced GPC content.

GPC exhibits a broad concentration range in muscles of patients with neurogenic disorders. The diester was increased to 3 μmol in Charcot–Marie–Tooth and Kugelberg–Welander diseases. Moreover, in muscles of children with meningomyelocele there was as much as 2.1 μmol of GPC. In contrast, two children with cerebral palsy and one child with amyotrophy after encephalitis possessed only 0.1 and 0.2 μmol GPC, respectively. Amyotrophy of unknown etiology, however, showed normal GPC content (1.6 μmol). This variation in GPC indicates that the nervous system may control lecithin metabolism in muscle, and that diseases of cerebral origin may exert an inhibitory effect on this metabolism.

The GPE content of healthy muscles is low, 0.1 μmol per 174 mg NC protein (Table II). In most of the diseased muscles, GPE is not detectable; in a few cases, its concentration is at the normal level. In meningomyelocele GPE is increased to 0.2 μmol and in Kugelberg–Welander disease to 0.3 μmol. The results of Table II clearly show that the GPE content of a muscle is only a small fraction of its GPC content. We determined the phosphatidylcholine and phosphatidylethanolamine content of normal human muscles and found an average of 3.4 and 1.8 μmol per 174 mg NC protein, respectively. This 2:1 ratio between the two phospholipids differs greatly from the approximately 10:1 ratio between GPC and GPE in muscle. From these data, it appears that the catabolism of phosphatidylcholine and phosphatidylethanolamine is not synchronized in muscle.

Table II also lists the common phosphate metabolites in muscle, in terms of micromoles per 174 mg NC protein. The ATP content is reduced to half of the normal value in only two cases, Duchenne dystrophy and the amyotrophy of unknown etiology. The ADP content remains essentially unchanged in all cases. Because in the PCA extract the major source of ADP is that bound to actin in muscle ($\sim 0.9$ μmol ADP per 174 mg NC protein), the constancy of ADP indicates that the protein-degradation process in muscle diseases does not affect actin levels relative to those of other NC proteins. Similarly, NAD remains unaffected by the various diseases. Evaluation of the phosphocreatine (PCr) content of the muscles is not possible because during the course of 10-min PCA extraction of the muscles, significant PCr breakdown occurs. From our experience with NMR, fresh, intact healthy muscle contains 1–2 μmol $P_i$ per 174 mg NC protein, and thus a higher concentration of $P_i$ in the perchloric extract can be interpreted to represent PCr in the original muscle. The concentration of SP in diseased muscles corresponds well to those in healthy muscle, indicating no selective impairment of glycolysis by most of the diseases examined; the exceptions are Becker and Duchenne dystrophies, where the sugar phosphates are reduced greatly.

The NC protein content of all diseased muscles is decreased compared to that of healthy muscle (column 2, Table II), reflecting the wasting of muscle in the diseases shown. Concomitant with the decrease in protein there is a decrease in the total phosphate content of the muscles (column 3, Table II). No good correlation between loss of total protein and loss of total phosphate can be seen; thus in most cases, when the total protein is normalized to 174 mg, the total phosphate remains well below the 51.3-μmol normal value. This suggests that the disease process perturbs phosphate metabolism to a greater extent than it does protein metabolism.

The data presented in Table II demonstrate the usefulness of the [31]P-NMR method for quantification of muscle phosphate metabolites. Dawson *et al.* (1977) showed that the phosphate concentrations determined by [31]P NMR are in good agreement with those measured by chemical methods. These *in vitro* data are significant in view of the current commercial availability of NMR instrumentation capable of measuring phosphorus-containing metabolites in human arms or legs. At the time of this writing, the degree of signal resolution obtained in *in vivo* human limb spectra is far below that obtained from muscle biopsy extract spectra.

## C. Intact Human Subject

Advances in [31]P-NMR technology allow one to study the muscles of healthy and diseased human subjects. The method, called topical magnetic resonance (Gordon *et al.*, 1980), consists of modifying the main magnetic field $B_0$ with static field gradients, so that the field is homogeneous only in a centered sensitive volume that is surrounded by inhomogeneous field gradients. High-resolution signals are detected only from the sensitive volume, whereas beyond these limits broad signals caused by the rapidly changing magnetic field are produced. When applied to the whole body, the total signal will contain both narrow and broadened lines that must be separated from each other in order to isolate the high-resolution spectrum of the sensitive volume. The broad components are eliminated by the convolution

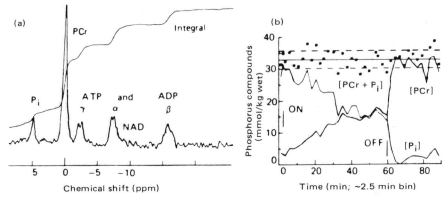

**Fig. 9.** (a) Phosphorus-31 spectrum of a resting human forearm taken at 32.5 MHz in the 20-cm TMR32 spectrometer at Oxford Research Systems. 800 scans were taken at 2-s intervals, and the spectrum was enhanced by convolution differencing. (b) Changes in concentration of PCr and $P_i$ during and following 58 min of complete arterial occlusion. Asterisks indicate the sum of [PCr] + [$P_i$] with mean plus or minus standard deviation (SD). From Cresshull et al. (1981).

difference technique of Campbell et al. (1973) with line broadenings from a few hertz to 200 Hz.

Current clinical instruments utilize the superconducting magnets of Oxford Research Systems (1.5–1.9 T)., at frequencies of 24.3–32.5 MHz, with sensitive volumes of 3–25 ml (Chance et al., 1980; Ross et al., 1981). It takes only 2–5 min to obtain reasonably good spectra. Ample evidence exists that the power applied is well below that of the recommended standards for biological studies (Oxford Research Systems, 1979; Chance et al., 1980).

The muscles usually examined belong to the flexor compartment of the forearm. They are surrounded only by the ulna and radius bones, the subcutaneous fat, and the skin. None of these tissues contains soluble phosphate metabolites to an extent that could interfere with the analysis of the muscle's $^{31}$P spectrum.

Figure 9a illustrates the $^{31}$P spectrum of a resting human forearm. Five resonances are clearly seen: three from the phosphate groups of ATP and one each from PCr and $P_i$. These resonance areas are large enough to allow integration. Figure 9b shows the effect of a 58-min anoxia on PCr and $P_i$ levels. Concomitant with the fall in PCr, the rise in $P_i$ takes place. On restoring the blood flow, PCr rapidly increases and $P_i$ decreases.

Ross et al. (1981) were the first to apply topical $^{31}$P NMR for the identification of human muscle diseases. A patient who had a lifelong history of rapid onset of fatigue and elevated levels of serum creatine phosphokinase and aldolase was subjected to $^{31}$P-NMR analysis. Figures 10 and 11 illustrate the changes occurring in the forearm muscles of the patient

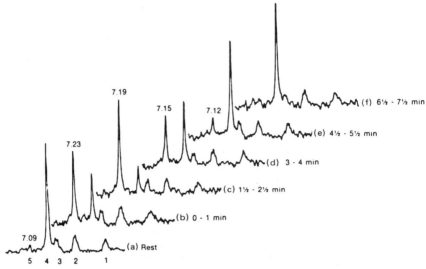

**Fig. 10.** $^{31}$P-NMR spectra of a patient with McArdle's syndrome, showing the effects of ischemic exercise. Peak assignments: 1, 2, and 3, the $\beta$-, $\alpha$-, and $\gamma$-phosphates of ATP; 4, PCr; 5, $P_i$. The pH values are given above each inorganic phosphate signal. The first spectrum (a) was recorded at rest before exercise; subsequent spectra (b–f) were recorded during the periods shown in which 0 min corresponds to the time at which exercise was started. Exercise was maintained during the period from 0 to $\frac{3}{4}$ min, but arterial occlusion was maintained for up to 3 min. Arterial flow was restored after this period. From Ross *et al.* (1981), by permission of *The New England Journal of Medicine.*

and the normal subject, respectively, during ischemic exercise. PCr decreases and $P_i$ increases much more rapidly in the patient than in the normal subjects. Nevertheless, the most striking abnormality in the patient, as compared with the control, is the relative constancy of intramuscular pH (from the chemical shift of the $P_i$ resonance the muscle pH can be estimated; Hoult *et al.*, 1974; Burt *et al.*, 1976). In the patient, the pH oscillates between values of 7.09 and 7.23 (Fig. 10). In contrast, in the normal subject the pH decreases from 7.04 to 6.43 (Fig. 11). That the pH did not fall below 7.0, despite fatigue during ischemic exercise, indicates the lack of lactic acid production in the patient's muscle. This should happen when glycolysis is stopped because of the absence of a key enzyme. Indeed, enzymatic analysis of the patient's muscle revealed the absence of glycogen phosphorylase. Thus the disease could be identified as McArdle's syndrome, an inborn error of metabolism caused by the lack of glycogen phosphorylase activity in skeletal muscle. (The rapid hydrolysis of PCr is understandable in McArdle's disease because in the absence of glycolysis, the resynthesis of ATP requires extensive use of PCr.)

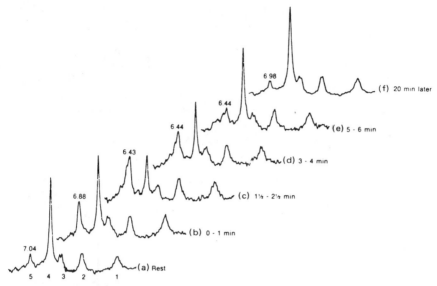

**Fig. 11.** ³¹P-NMR spectra of a control subject, showing the effect of ischemic exercise. Peak assignments are the same as in Fig. 10. The pH values are given above each inorganic phosphate signal. The first spectrum (a) was recorded at rest before exercise; subsequent spectra (b–f) were recorded during the periods shown in which 0 min corresponds to the time at which exercise was started. Exercise was maintained during the period from 0 to 1½ min, but arterial occlusion was maintained for up to 3 min. Arterial flow was restored after this period. From Ross *et al.* (1981), by permission of *The New England Journal of Medicine.*

Edwards *et al.* (1982) diagnosed phosphofructokinase deficiency by ³¹P NMR (Fig. 12). In the normal subject, exercise produces no increase in the level of sugar phosphates and decreases the pH slightly, from 7.18 to 7.05. In the phosphofructokinase-deficient patient, exercise increases the sugar phosphate concentration 50-fold without any change in pH. These data are consistent with the following biochemical mechanism: In the patient's muscle, exercise turns on glycogen breakdown to yield glucose 1-phosphate, glucose 6-phosphate, and fructose 6-phosphate. These three sugar phosphates accumulate because in the absence of phosphofructokinase, the subsequent step of glycolysis, the conversion of fructose 6-phosphate to fructose 1,6-diphosphate, does not take place. Hence, ³¹P NMR is able to differentiate phosphofructokinase deficiency from the McArdle syndrome through the difference in the level of sugar phosphates.

The magnetic saturation characteristics of PCr and $P_i$ in human muscle are very similar (Gadian *et al.,* 1981); therefore, changes in the ratio of PCr/$P_i$ peak areas provide another parameter for the characterization of diseased states. In normal resting muscles the PCr/$P_i$ ratio varies from 10 to

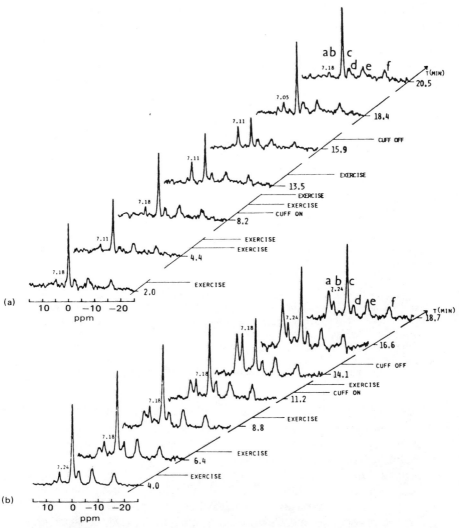

**Fig. 12.** Phosphorus-31 spectra and muscle pH values obtained from forearm muscles of a patient with phosphofructokinase deficiency. Peak assignments: a, SP; b, $P_i$; c, PCr; d, e, f, $\gamma$-, $\alpha$-, and $\beta$-phosphates of ATP, respectively. From Edwards et al. (1982).

20 (Chance et al., 1980, 1981; Radda et al., 1982), whereas in myopathic muscles it varies from 2 to 4 (Gadian et al., 1981; Radda et al., 1982). The decreased ratio is caused by the elevated $P_i$ concentration in the resting state, and it may be taken to be a direct consequence of the lesion in oxidative phosphorylation in the patient. This is confirmed by electron micrographs of the muscle that show gross abnormality of the mitochondrial structure.

# 17. Identification of Disease by $^{31}$P NMR

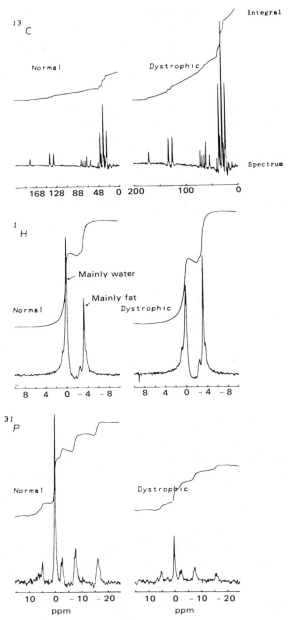

**Fig. 13.** $^{13}$C, $^{1}$H, and $^{31}$P spectra obtained from forearm muscles of a patient with Duchenne muscular dystrophy and a normal boy of similar age. From Edwards *et al.* (1982).

The increased phosphate peak is a very sensitive indicator of tissue hypoxia, and as such it may be used to assess the extent of peripheral vascular diseases of leg muscles (Chance et al., 1980). In this disease, insufficient blood flow, and therefore insufficient tissue oxygenation, causes muscle damage that may eventually lead to amputation of the affected limb. This type of $^{31}$P-NMR diagnosis was further improved by determining the relationship between work output in the exercising human limb and the steady-state capability of oxidative phosphorylation as measured by the PCr/$P_i$ ratio (Chance et al., 1981). A decrease in the slope of plots is indicative for the peripheral vascular flow decrease, causing diminished oxygen delivery and oxidative phosphorylation.

A comparison of human leg spectrum with that of the forearm clearly shows a GPC peak in the leg and no peak in the arm (Chance et al., 1981). This is in agreement with the observations on PCA-extracted human muscle biopsy samples (Chalovich et al., 1979).

The forearm muscle of a patient with Duchenne muscular dystrophy was also investigated by topical NMR (Edwards et al., 1982). Figure 13 shows that the common phosphate metabolites are reduced in the dystrophic patient as compared with those in the normal subject. A detailed analysis of the data (Table III in Edwards et al., 1982) shows that the PCr/$P_i$ ratio in the dystrophic muscle has a value of only 2.8 as compared with the value of 6.8 in normal muscle. However, as pointed out by the authors, the distal forearm muscle may not be ideal for studying Duchenne dystrophy because the muscles usually affected by this disease are located in the proximal part of the leg.

In addition to the $^{31}$P spectra, Fig. 13 also illustrates the $^{13}$C and $^1$H spectra of the dystrophic and normal muscles. The multiple resonances in the $^{13}$C spectra represent only one class of compounds—neutral fats. The $^1$H spectra show two peaks, one from the repeating methylene group protons of fatty acids in neutral fats and another peak from the water protons. Dystrophic muscle contains more fat and less water than normal muscle.

## III. Heart

Phosphorus-31 NMR of heart has been applied extensively to clinical medicine. During ischemia of the perfused mammalian heart, a rapid decrease in PCr, a slower decrease in ATP, and a marked increase in $P_i$ were noted (Garlick et al., 1977; Hollis et al., 1978; Salhany et al., 1979). Even more important was the finding that the increase in $P_i$ was correlated with its chemical shift, indicating acidosis in the heart; during a 25-min global

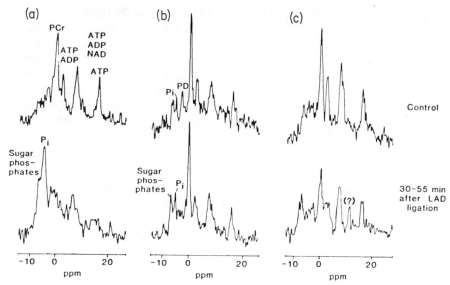

**Fig. 14.** (a) Spectra of a normally oxygenated and a regionally ischemic area of the left ventricle of a perfused heart. (b) Same sequence for a heart treated with verapamil and then subjected to LAD ligation. (c) Comparative sequence for chlorpromazine treatment. All spectra were obtained with a surface coil. Control spectra (at the top in each panel) were obtained before LAD ligation; spectra from the ischemic region (below) were obtained for the time interval indicated. The surface coil was placed in approximately the same location for control spectra and spectra from the ischemic region. From Nunnally and Bottomley (1981). Copyright 1981 by the American Association for the Advancement of Science.

ischemia the pH dropped 1.7 units, from pH 7.4 to 5.7; on reperfusion some recovery of pH took place (Jacobus *et al.*, 1977). It was suggested that the change in the chemical shift of $P_i$ resonance may be used to detect the extent of heart infarction (Gadian *et al.*, 1976). Hollis *et al.* (1977) ligated the left anterior descending coronary artery and thereby simulated infarction in perfused rabbit heart that could be detected by $^{31}P$ NMR. For the extensive literature of pH measurements in perfused heart by $^{31}P$ NMR, we refer to the articles by Jacobus *et al.* (1982) and Gadian *et al.* (1982). As pointed out by Hollis (1980), there are difficulties in locating and sizing an ischemic region of the heart from the area of $P_i$ resonance peaks. However, these studies opened new avenues for investigating the effect of agents to reduce ischemic necrosis. Thus Hollis (1979) reported that ATP level and pH are maintained at the control value in the KCl-arrested heart after 40 min of global ischemia. Bailey *et al.* (1982) showed that in the perfused rat heart, infusion of insulin prior to ischemia results in increased accessibility of glycogen to phosphorylase during the subsequent ischemia and thereby decreases ATP depletion.

The $^{31}$P NMR of the perfused heart was subsequently replaced by $^{31}$P NMR of the *in vivo* heart (Grove *et al.*, 1980). The abdomen of an anesthetized rat was opened, the radiofrequency coil was passed through the diaphragm and positioned around the heart ventricles. This technique was further improved by the *surface-coil* method (Ackerman *et al.*, 1980), whereby the coil can be placed on different regions of an organ (e.g., ischemic regions can be distinguished from healthy regions). Surface coils were used by Nunnally and Bottomley (1981) to assess the effect of pharmacological treatment on myocardial infarction by $^{31}$P NMR (Fig. 14). Ligation of the left anterior descending coronary artery hydrolyzes the ATP and PCr of rabbit heart to yield $P_i$ (Fig. 14a). Pretreatment of the heart with verapamil, a coronary vasodilating agent, results in the maintenance of nearly normal ATP and PCr levels in the ligated ischemic region (Fig. 14b). Chlorpromazine, which prevents irreversible mitochondrial damage, has much less but an observable effect (Fig. 14c). This experiment illustrates the value of $^{31}$P NMR in selecting potential drugs for myocardial therapy.

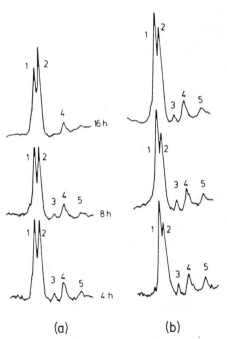

**Fig. 15.** $^{31}$P-NMR spectra of single human kidneys flushed with (a) saline or (b) hypertonic citrate solutions after 4, 8, and 16 h of cold ischemia. Peak assignments are as follows: 1, SP and AMP; 2, $P_i$; 3, $\gamma$-phosphate of ATP and $\beta$-phosphate of ADP; 4, $\alpha$-phosphate of ATP and ADP + NAD/NADH; 5, $\beta$-phosphate of ATP. From Bore *et al.* (1982).

## IV. Kidney

Sehr *et al.* (1977) and Ackerman *et al.* (1981) used $^{31}P$ NMR to measure ATP content and intrarenal pH in isolated perfused rat kidney in order to assess tissue viability and to find ways of prolonging its preservation for application to kidney transplantation. With the availability of the 20-cm-bore magnet, the English workers extended their work to the study of whole human kidneys *in vitro*.

$^{31}P$-NMR examinations were carried out keeping the kidneys under sterile conditions and in ice (Bore *et al.*, 1982). Figure 15 shows the effect of storage in two different media on the phosphate profile of kidney. Citrate flushing preserves the ATP and reduces the $P_i$ content as compared with saline flushing. The decrease in pH is also considerably less in the presence of citrate than with saline (Fig. 16). These experiments form the basis of $^{31}P$-NMR monitoring of kidneys before transplantation.

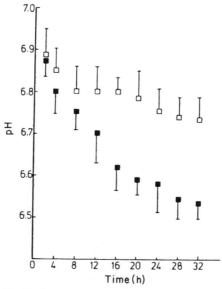

**Fig. 16.** Change of pH with time of saline- (■, $n = 4$) or citrate- (□, $n = 4$) flushed human kidneys during cold (4°C) ischemia. The pH values are expressed as mean and 1 SD. (vertical bars). From 12 h the differences were statistically significant ($p = .05$). From Bore *et al.* (1982).

## V. Brain

The phosphate profile of brain is similar to that of muscle (i.e., it contains high concentrations of PCr and ATP). Because brain is in a constant state of activity, its high-energy phosphate content depends on its blood supply, and, consequently, restrictions in the blood supply of the brain may lead to symptoms ranging from dizziness to coma and death.

Chance et al. (1978) pioneered $^{31}$P NMR of mouse brain *in vivo* and of freeze-dried rat brain. In addition to the presence of common phosphate metabolites, they also noted the presence of phosphodiesters in brain. Our studies reveal three major phosphodiester peaks in guinea pig brain slices (Fig. 17); only GPC has been identified. Major unknown phosphates are detected at 0.85 and −0.73 ppm. The unknown at 0.85 ppm exhibits a rate of hydrolysis at pH 7.2 at 31°C of 3% per minute, and one of the products was identified as GPE. This unknown is acid-labile. The unknown at −0.73 ppm is a saccharide phosphodiester.

Table III lists the phosphate compounds we have found in PCA extracts of rat cerebrum and cerebellum. A total of 32 phosphate resonances were separated; two of these (10.69 and 9.98 ppm) have not been described as constituents of brain tissue prior to this $^{31}$P-NMR analysis.

Figure 18 presents two $^{31}$P spectral profiles of brain tissue. The top spectrum was obtained from a rat brain PCA extract adjusted to pH 7.2. A

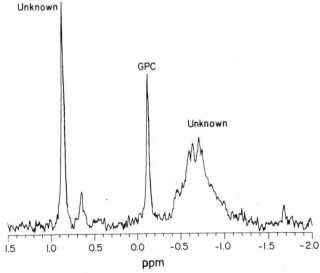

**Fig. 17.** $^{31}$P-NMR spectrum (81.0 MHz) of PCA-extract guinea pig brain slice phosphodiesters (N$_2$ incubated) at pH 10.2.

**TABLE III**

$^{31}$P-NMR Comparison of Phosphate Metabolite Levels in Normoxic Cerebral and Cerebellar Rat Tissues

| Phosphate compound | Chemical shift in PCA extract (ppm) | Amount (as percent of total P detected) | |
|---|---|---|---|
| | | Cerebrum | Cerebellum |
| Unknown | 10.69 | 0.15 | 0.13 |
| Unknown | 9.98 | 0.06 | 0.09 |
| Triose-P | 4.78 | ND[a] | 0.19 |
| Triose-P | 4.62 | 0.28 | ND |
| Glu 6-P + Gal 6-P | 4.46 | 0.81 | 0.73 |
| α-Glycero-P | 4.29 | 0.38 | 0.90 |
| Fruc 1,6-diP | 4.10–4.04 | 0.26 | 0.23 |
| 2,3-DPG | 3.93 | 0.50 | 0.19 |
| Ribose 5-P | 3.84 | 4.84 | 1.62 |
| IMP | 3.77 | 0.79 | 0.40 |
| AMP | 3.73 | 0.63 | 0.38 |
| XMP | 3.68 | 0.29 | ND |
| 2′P NADP | 3.55 | 0.20 | 0.60 |
| Unknown | 3.43 | 0.37 | 0.51 |
| Phosphorylcholine | 3.31 | 1.29 | 1.13 |
| Unknown | 3.13 | 0.32 | ND |
| $P_i$ | 2.60 | 5.19 | 10.63 |
| Glu 1-P | 2.28 | 0.26 | 0.28 |
| Gal 1-P | 2.13 | 0.17 | 0.25 |
| Anomeric sugar P | 1.84 | 0.07 | ND |
| | 1.39 | 0.06 | 0.26 |
| Unknown phosphodiester | 0.92 | 0.07 | ND |
| Acid-labile phosphodiester | 0.85 | 0.38 | 0.36 |
| GPE | 0.78 | 0.93 | 1.62 |
| GPC | −0.13 | 2.78 | 2.94 |
| Saccharide phosphodiesters | −0.77 | 0.87 | 1.45 |
| PCr | −3.12 | 19.36 | 24.05 |
| Unknown | −4.24 | 0.27 | 0.19 |
| ADP | α-10.61 | 3.14 | 2.75 |
| | β-6.11 | | |
| NAD + NADP | −11.37 | 6.36 | 6.09 |
| UDP-Glu | −12.81 | 0.40 | 0.56 |
| UDP-Gal | −13.10 | 0.38 | ND |
| ATP | α-10.92 | 45.30 | 39.67 |
| | β-21.45 | | |
| | γ-5.80 | | |

[a] ND, Not detected.

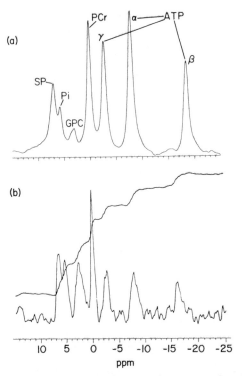

**Fig. 18.** ³¹P-NMR spectra of brain phosphates. (a) PCA-extract data (rat normoxic brain suitably filtered so as to simulate linewidths obtainable from *in vivo* analysis. (b) *In vivo* rabbit brain data, topical analysis (obtained through the courtesy of Oxford Research Systems, Ltd). In this figure, the $\delta$ scales follow the IUPAC convention with phosphocreatine being selected as the zero-shift reference substance, a convention currently followed by English researchers.

filter time constant introducing 70-Hz line broadening was used to simulate the natural linewidths obtained from *in vivo* analysis. The bottom spectrum, obtained through the courtesy of Oxford Research Systems, Ltd., was obtained from a living rabbit brain by topical NMR analysis using the Oxford TMR032 instrument. Note that the relative amplitudes of SP, $P_i$, and PCr signals are about the same in both spectra. The GPC signal, which is composed primarily of the resonance signals of GPC and GPE, is larger in the rabbit than in the rat; this is a species-specific difference. The high-resolution signals of ATP are greater by a factor of 3.0 in the rat PCA-extract spectrum than in the rabbit *in vivo* spectrum. Because there is no essential difference in the ATP content of rabbit and rat brain, the spectral difference is interpreted to indicate that in living brain tissue, two-thirds of the tissue's ATP is restricted in motion such that the signals become too broad to be detected under high-resolution conditions. One-third of the tissue's ATP has sufficient freedom within the cellular matrix to yield high-resolution NMR signals. Note that the β-phosphate group of ATP is shifted to higher parts-per-million values in the *in vivo* spectrum relative to that of the extract. This shift difference is primarily the result of magnesium binding to ATP in the intact tissue.

## VI. Eye

Phosphorus-31 NMR has been applied to the eye (Kopp *et al.*, 1982), and the clinical relevance of such a study is apparent. We investigated three parts of the eye: lens, cornea, and vitreous humor.

### A. Lens

A highly resolved $^{31}$P spectrum of a single bovine lens, incubated *in vitro*, is shown in Fig. 19. The prominent signals arise from the phosphates of ATP and $P_i$. In contrast, the PCr resonance is small. The sizeable SP resonance is split into two groupings. The signal from dinucleotides is well separated from that of the α-phosphate group of ATP. The uridine diphosphoryl–sugar resonance band, consisting of UDP-galactose, UDP-glucose, and UDP-mannose, appears between $-12.3$ and $-13.7$ ppm. GPC is a major

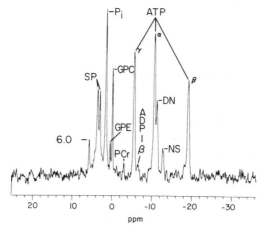

**Fig. 19.** $^{31}$P-NMR spectrum of a single bovine lens accumulated over a period of 15 min. The lens was maintained during the experiment by Earl's buffer (with glucose) at pH 7.4 at 37°C. The signal at 6 ppm is a new phosphorus metabolite.

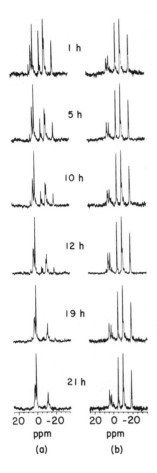

**Fig. 20.** 31P-NMR spectra of the intact rabbit lens from six time points during incubation in (b) control Earle's buffer and in (a) Earle's buffer with $1 \times 10^{-3}$ M dexamethasone, pH 7.4, 37°C.

and GPE a minor lens resonance. The phosphate that gives rise to the signal at 6.0 ppm is not yet characterized. Interestingly, the same signal appears in the lens of rat. Chronic feeding with environmental levels of Cd and Pb enhanced this resonance (S. J. Kopp and T. Glonek, unpublished). In the spectrum of intact dog lens, we observed resonances as far downfield as 10 and 18 ppm.

Figure 20 illustrates a time course for the incubation of rabbit lens in $1 \times 10^{-3}$ M dexamethasone; corresponding spectra taken during a control time course are also presented for comparison. With dexamethasone drug treatment, the lens hydrolyzes its ATP to AMP and $P_i$. Similar hydrolysis of ATP took place when the mammalian lens was exposed to 30 mM galactose, an agent which induces cataract formation within one day. Furthermore, galactose treatment of lens resulted in accumulation of α-glycerophosphate to levels approaching 40% of the total lens phosphorus.

## B. Cornea

$^{31}$P-NMR data of rabbit cornea and its PCA extract are summarized in Table IV. As in case of the lens, ATP and $P_i$ are the major phosphate components of this tissue and the PCr content is low. A novel finding is the high ribose 5-phosphate level that may indicate energy production in the cornea through the phosphogluconate oxidative pathway. The tissue also contains both GPC and GPE. Some minor resonances are observable only in the PCA extract and not in the intact cornea. Removal of the thin endothelial layer does little to alter the corneal phosphate profile; thus the intact tissue spectrum reflects the bulk constitution of the sample. In most instances the intact cornea and PCA extract data compare well. The chemical shifts of all the strong acid resonances are the same in the tissue and the extract; the exceptions are GPE and the $\beta$-phosphate of ATP. The latter difference is due to the MgATP complex in the tissue versus free ATP in the

TABLE IV

Rabbit Cornea $^{31}$P-NMR Data

| Phosphate compound | Chemical shift (ppm) | | Amount (as percent of the total P detected) | | |
|---|---|---|---|---|---|
| | | | | In perchloric acid extract | |
| | In intact cornea | In perchloric acid extract | In intact cornea | Intact cornea | Cornea minus endothelium |
| Unknown | | 4.85 | | 1.2 | 1.3 |
| Unknown | | 4.64 | | 0.4 | 0.4 |
| Glu 6-P | | 4.47 | | 1.0 | 0.7 |
| α-Glycero-P | | 4.28 | | 4.4 | 4.5 |
| Fruc 1,6-diP | | 4.08 | | 2.2 | 2.3 |
| Ribose 5-P | 3.50 | 3.84 | 23.1 | 5.9 | 4.5 |
| IMP | | 3.77 | | 0.3 | ND$^a$ |
| AMP | | 3.71 | | 2.7 | 2.3 |
| 2′-P NADP | | 3.54 | | 1.0 | 1.0 |
| Phosphorylcholine | | 3.31 | | 3.1 | 3.6 |
| $P_i$ | 1.43 | 2.61 | 14.3 | 13.9 | 13.5 |
| Unknown | | 1.85 | | 0.1 | 0.1 |
| GPE | 0.42 | 0.75 | 2.7 | 2.5 | 2.2 |
| GPC | −0.13 | −0.13 | 4.1 | 3.8 | 3.3 |
| PCr | −3.11 | −3.13 | 1.6 | 1.0 | ND |
| ATP | α, −10.65 | α, −10.8 | 41.9 | 43.0 | 48.4 |
| | β, −19.24 | β, −21.3 | | | |
| | γ, −5.62 | γ, −5.7 | | | |

$^a$ ND, Not detected.

extract. The shift change of $P_i$ results from dissociation of the weak acid proton in the alkaline extract medium (pH 10.2).

### C. Vitreous Humor

The phosphorus profile from the vitreous humor of the *Rhesus* monkey eye showed two major resonances, $P_i$ at 2.08 ppm and a downfield component at 8.16 ppm, in addition to minor components. The pH at 37°C measured by the shift position of the humor inorganic orthophosphate resonance was 7.21, which is in reasonably close agreement with the pH value of 7.26 measured by the glass electrode using the same sample.

## VII. Mammalian Fluids

Fluids play a major role in medical diagnosis because they reflect the metabolism of the surrounding tissue and they may be obtained from the patient without risk. We analyzed three human fluids: amniotic fluid, saliva, and cerebrospinal fluid.

### A. Amniotic Fluid

Figure 21 shows comparative phosphate profiles of human amniotic fluid taken during the third trimester. The control spectrum shows little more than the inorganic orthophosphate signal. (A profile of characteristic orthophosphate mono- and diesters is detectable in native amniotic fluid after 12-hr signal averaging.) The profile from the diabetic mother, however, shows additional resonances at 1.3 and 8.9 ppm. The 1.3-ppm resonance is characteristic of an anomeric sugar phosphate. The 8.9-ppm signal, the source of which is unidentified, probably arises from monoesterified sugar phosphates of polymeric sugar molecules.

### B. Human Saliva

Figure 22 shows the phosphate profiles of freeze-dried saliva resuspended in a buffer solution. The top spectrum was obtained from a patient with a history of dental caries and relatively good oral hygenic habits. The principal signal arises from salivary inorganic orthophosphate. The other resonances arise from various phosphomono- and diesters. The monoester resonance band appears at 3.95 ppm and corresponds to approximately 43% of the

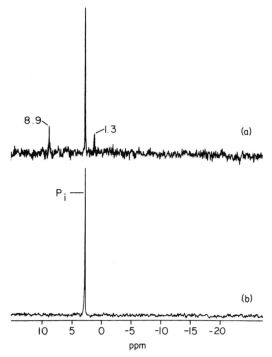

**Fig. 21.** (a) Phosphorus-31 NMR spectral profile of whole human amniotic fluid obtained during the third trimester. Amniotic fluid obtained from the diabetic mother gives rise to additional phosphorus resonance at 8.9 and 1.3 ppm. (b) Inorganic orthophosphate is the prominant resonance detected in the amniotic fluid from the healthy mother.

esterified phosphate component. The bands at 1.1 and 0.6 ppm correspond to two different sets of orthodiester functional groups. These spectral patterns are similar to those observed from yeast glycans, where the phosphate serves as a linkage between chains of polymerized sugar residues. The quantity present in the sample, however, precludes any possibility that these signals arise from sloughed cell-wall polysaccharides of the oral flora.

The bottom spectrum was obtained from the saliva of a patient exhibiting a caries-free oral cavity but who also had a history of poor oral hygiene as well as a well-documented diet rich in carbohydrates. In this case, the phosphorus profile shows all of the resonances observed in the profile from the control patient and, in addition, resonances to quite low field at 9.44 and 8.20 ppm and an additional resonance in the diester region at 0.67 ppm. At this stage of the investigation, it cannot be said what the nature of the molecules are that give rise to these additional resonances, except to say that

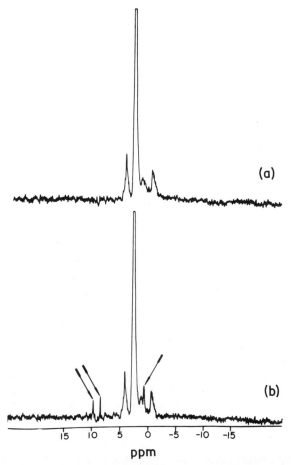

**Fig. 22.** Phosphorus-31 NMR spectral profiles of whole human saliva after it has been concentrated by lyophilization. The control spectrum (a) was obtained from a patient with a common history of dental caries. The other spectrum (b) was obtained from a patient with a caries-free oral cavity. Several phosphates, indicated by the arrows, are present in this profile that are not detectable in the control profile. The major resonance, which is off-scale in both profiles, arises from inorganic orthophosphate, the principal salivary phosphate.

because they appear in pelleted material, they most probably arise from relatively high-molecular-weight components.

## C. Cerebrospinal Fluid

The CSF from a healthy subject shows only the resonance of inorganic orthophosphate. As of this writing, we have not come upon a pathology involving the CSF that alters this simple phosphate profile. The resonance position of $P_i$, however, yields a pH value the same as that obtained *in vitro* with the glass electrode. Thus the $^{31}P$ spectrum of CSF will be useful in accessing the pH of CSF noninvasively when human-subject $^{31}P$-NMR devices become available. It is generally agreed among clinicians that to perform even this simple measurement without risk would be a significant advance in diagnosing disorders of the central nervous system.

## VIII. Perspectives

The data presented herein demonstrate the usefulness of the NMR method for the identification of tissue pathologies. Disease causes readily measurable alterations in quantitative tissue profiles that can be directly related to altered tissue biochemistry, specifically, that segment of tissue biochemistry involving the low-molecular-weight phosphorus-containing components of intermediate metabolism. In considering muscle tissue, a particularly useful portion of the $^{31}P$ profile is the phosphodiester resonance band, which is isolated from the signals of other tissue phosphates and, moreover, which is a sensitive indicator of neuromuscular diseases.

The tissue NMR results of the past decade are of special significance in view of the current commercial availability of NMR instrumentation capable of measuring phosphorus-containing metabolites *in vivo* in the human subject. Currently, the degree of signal resolution obtained from human biopsy extract spectra is well above that obtained from *in vivo* human-limb spectra. Furthermore, the magnets used thus far have a bore with a maximum of 20 cm, which limits clinical investigations to only a few parts of the human body. Naturally, with the expected technological advances, topical $^{31}P$ NMR will become a standard equipment in the hospital laboratory, and by the turn of the century the $P_i$, ATP, and PCr content of tissues will assist the physician in the diagnosis of diseases, assessment of drug effects, and selection of organs for transplantation. In addition to topical $^{31}P$ NMR, it appears that clinical phosphorus NMR will be incorporated into the currently existing NMR imaging instruments. The three-dimensional resolution of phosphorus in the body will open new avenues for the diagnosis of human diseases.

## Acknowledgments

We thank Dr. Stephen J. Kopp for his contribution to this work; Drs. Edward Abraham, Moris J. Danon, Irvin M. Siegel, and David L. Spencer for the human muscle specimens; and Donna M. Lattyak for typing the manuscript. A special thanks is given to the people of the Chicago College of Osteopathic Medicine for their generous funding of the NMR Laboratory and the support of its scientific program. This work was also supported by the Muscular Dystrophy Association, Inc., Shriners Hospitals for Crippled Children, grants ES-02397-02 and EY-03988-01 from the United States National Institutes of Health, and grant number 821070 from the American Heart Association.

## References

Ackerman, J. J. H., Grove, T. H., Wong, G. G., Gadian, D. G., and Radda, G. K. (1980). *Nature (London)* **283**, 167-170.
Ackerman, J. J. H., Lowry, M., Radda, G. K., Ross, B. D., and Wong, G. G. (1981). *J. Physiol. (London)* **319**, 65-79.
Bailey, I. A., Radda, G. K., Seymour, A. L., and Williams, S. R. (1982). *Biochim. Biophys. Acta* **720**, 17-27.
Bárány, M., and Glonek, T. (1982). In "Methods in Enzymology" (D. W. Frederikson and L. W. Cunningham, eds.), Vol. 85, pp. 624-676. Academic Press, New York.
Bárány, M., Burt, C. T., Labotka, R. J., Danon, M. J., Glonek, T., and Huncke, B. H. (1977). In "Pathogenesis of Human Muscular Dystrophies" (L. P. Rowland, ed.), pp. 337-340. Excerpta Medica, Amsterdam.
Bárány, M., Chalovich, J. M., Burt, C. T., and Glonek, T. (1982). In "Disorders of the Motor Unit" (D. L. Schotland, ed.), pp. 697-713. Wiley, New York.
Bore, P. J., Chan, L., Gadian, D. G., Radda, G. K., Ross, B. D., Styles, P., and Taylor, D. J. (1982). In "Intracellular pH: Its Measurement, Regulation, and Utilization in Cellular Functions" (R. Nuccitelli and D. W. Deamer, eds.), pp. 527-535. Alan R. Liss, Inc., New York.
Burt, C. T., Glonek, T., and Bárány, M. (1976). *J. Biol. Chem.* **251**, 2584-2591.
Burt, C. T., Glonek, T., and Bárány, M. (1977). *Science* **195**, 145-149.
Campbell, I. D., Dobson, C. M., Williams, R. J. P., and Xavier, A. V. (1973). *J. Magn. Reson.* **11**, 172-181.
Chalovich, J. M., Burt, C. T., Cohen, S. M., Glonek, T., and Bárány, M. (1977). *Arch. Biochem. Biophys.* **182**, 683-689.
Chalovich, J. M., Burt, C. T., Dannon, M. J., Glonek, T., and Bárány, M. (1979). *Ann. N.Y. Acad. Sci.* **317**, 649-669.
Chance, B., Nakase, Y., Bond, M., Leigh, J. S., Jr., and McDonald, G. (1978). *Proc. Natl. Acad. Sci. U.S.A.* **75**, 4925-4929.
Chance, B., Eleff, S., and Leigh, J. S., Jr. (1980). *Proc. Natl. Acad. Sci. U.S.A.* **77**, 7430-7434.
Chance, B., Eleff, S., Leigh, J. S., Jr., Sokolow, D., and Sapega, A. (1981). *Proc. Natl. Acad. Sci. U.S.A.* **78**, 6714-6718.
Cresshull, I., Dawson, M. J., Edwards, R. H. T., Gadian, D. G., Gordon, R. E., Radda, G. K., Shaw, D., and Wilkie, D. R. (1981). *J. Physiol. (London)* **317**, 18P.
Dawson, M. J., Gadian, D. G., and Wilkie, D. R. (1977). *J. Physiol. (London)* **267**, 703-735.
Edwards, R. H. T., Wilkie, D. R., Dawson, M. J., Gordon, R. E., and Shaw, D. (1982). *Lancet* **i**, 725-731.

## 17. Identification of Disease by $^{31}$P NMR

Gadian, D. G., and Radda, G. K. (1981). *Annu. Rev. Biochem.* **50,** 69–83.
Gadian, D. G., Hoult, D. I., Radda, G. K., Seeley, P. J., Chance, B., and Barlow, C. (1976). *Proc. Natl. Acad. Sci. U.S.A.* **73,** 4446–4448.
Gadian, D. G., Radda, G. K., Ross, B. D., Hockaday, J., Bore, P. J., Taylor, D. J., and Styles, P. (1981). *Lancet* **ii,** 774–775.
Gadian, D. G., Radda, G. K., Dawson, M. J., and Wilkie, D. R. (1982). *In* "Intracellular pH; Its Measurements, Regulation, and Utilization in Cellular Functions" (R. Nuccitelli and D. W. Deamer, eds.), pp. 61–77. Alan R. Liss, Inc., New York.
Garlick, P. B., Radda, G. K., Seeley, P. J., and Chance, B. (1977) *Biochem. Biophys. Res. Commun.* **74,** 1256–1262.
Glonek, T., and Van Wazer, J. R. (1976). *J. Phys. Chem.* **80,** 639–643.
Glonek, T., Wang, P. J., and Van Wazer, J. R. (1976). *J. Am. Chem. Soc.* **98,** 7968–7973.
Glonek, T., Burt, C. T., and Bárány, M. (1981). *NMR: Basic Princ. Prog.* **19,** 121–159.
Gordon, R. E., Hanley, P. E., Shaw, D., Gadian, D. G., Radda, G. K., Styles, P., Bore, P. J., and Chan, L. (1980). *Nature (London)* **287,** 736–738.
Grove, T. H., Ackerman, J. J. H., Radda, G. K., and Bore, P. J. (1980). *Proc. Natl. Acad. Sci. U.S.A.* **77,** 299–302.
Hollis, D. P. (1979). *Bull. Magn. Reson.* **1,** 27–37.
Hollis, D. P. (1980). *In* "Biological Magnetic Resonance" (L. J. Berliner and J. Reuben, eds.), Vol. 2, pp. 1–44. Plenum, New York.
Hollis, D. P., Nunnally, R. L., Jacobus, W. E., and Taylor, G. J., IV (1977). *Biochem. Biophys. Res. Commun.* **75,** 1086–1091.
Hollis, D. P., Nunnally, R. L., Taylor, G. J., IV, Weisfeldt, M. L., and Jacobus, W. E. (1978). *J. Magn. Reson.* **29,** 319–330.
Hoult, D. I., Busby, S. J. W., Gadian, D. G., Radda, G. K., Richards, R. E., Seeley, P. J. (1974). *Nature (London)* **252,** 285–287.
Jacobus, W. E., Taylor, G. J., IV, Hollis, D. P., and Nunnally, R. L. (1977). *Nature (London)* **265,** 756–758.
Jacobus, W. E., Pores, I. H., Lucas, S. K., Kallman, C. H., Weisfeldt, M. L., and Flaherty, J. T. (1982). *In* "Intracellular pH; Its Measurement, Regulation, and Utilization in Cellular Functions" (R. Nuccitelli and D. W. Deamer, eds.), pp. 537–565. Alan R. Liss, Inc., New York.
Kopp, S. J., Glonek, T., and Greiner, J. V. (1982). *Science* **215,** 1622–1625.
Leigh, J. S., Jr., Chance, B., and Eleff, S. (1981). *Fed. Proc., Fed. Am. Soc. Exp. Biol.* **40,** 586.
McCain, D. C., and Markley, J. M. (1980). *J. Am. Chem. Soc.* **102,** 5559–5565.
Morgan, W. E., and Van Wazer, J. R. (1975). *J. Am. Chem. Soc.* **97,** 6347–6352.
Noggle, J. H., and Schirmer, R. E. (1971). "The Nuclear Overhauser Effect." Academic Press, New York.
Nunnally, R. L., and Bottomley, P. A. (1981). *Science* **211,** 177–180.
Oxford Research Systems (1979). "Topical Magnetic Resonance Spectroscopy, Safety Aspects." ORS, Oxford, England.
Radda, G. K., Bore, P. J., Gadian, D. G., Ross, B. D., Styles, P., Taylor, D. J., and Morgan-Hughes, J. (1982). *Nature (London)* **295,** 608–609.
Ross, B. D., Radda, G. K., Gadian, D. G., Rocker, G., Esiri, M., and Falconer-Smith, J. (1981). *N. Engl. J. Med.* **304,** 1338–1342.
Rowland, L. P. (1980). *Muscle Nerve* **3,** 3–20.
Salhany, J. M., Pieper, G. M., Wu, S., Todd, G. L., Clayton, F. C., and Eliot, R. S. (1979). *J. Mol. Cell. Cardiol.* **11,** 601–610.
Sehr, P. A., Radda, G. K., Bore, P. J., and Sells, R. A. (1977). *Biochem. Biophys. Res. Commun.* **77,** 195–202.
Yeagle, P. L., Hutton, W. C., and Martin, R. B. (1975). *J. Am. Chem. Soc.* **97,** 7175–7177.

# PART 4

*Selected Data*

CHAPTER 18

# Appendixes: Selective Compilation of $^{31}$P-NMR Data

## David G. Gorenstein
## Dinesh O. Shah

Department of Chemistry
University of Illinois at Chicago
Chicago, Illinois

| | | |
|---|---|---|
| Appendix I. | Doubly Connected Phosphorus (P$<$) | 550 |
| Appendix II. | Triply Connected Phosphorus (P$\leq$) | 551 |
| Appendix III. | Quadruply Connected Phosphorus ($>$P$<$) | 563 |
| Appendix IV. | Quintuply Connected Phosphorus ($>$P$<$) | 586 |
| Appendix V. | Sextuply Connected Phosphorus ($\geq$P$\leq$) | 588 |
| | References | 589 |

The following information is a representative sampling of data from the $^{31}$P-NMR literature from the period 1960 to June 1982. The format and classification scheme of the compilation has been kept as close to Van Wazer's (in Crutchfield *et al.*, 1967) extensive compilation of $^{31}$P data covering the period from 1960 to August 1966. An additional compilation of $^{31}$P data covering the period from December 1960 to June 1982 may be found in Gorenstein (1983).

Chemical shifts are now reported with the opposite sign convention from earlier literature in keeping with the new, generally accepted convention. Thus positive chemical shifts are now reported *downfield* from 85% $H_3PO_4$. Tables of additional $^{31}$P data may be found in the preceding chapters.

## APPENDIX I

### Doubly Connected Phosphorus (P⟨)

| Formula | Chemical shift $\delta^{31}P$ (ppm from $H_3PO_4$) | Coupling constant $J$ (Hz) | | | Reference[a] |
|---|---|---|---|---|---|
| $P(C_6H_5)_2^- K^+$ | −17.4 | | | | 1 |
| $P(i\text{-}C_3H_5)_2^- K^+$ | 23.2 | | | | 1 |
| [benzannulated cyclopentene with :P:] | 81.7 | $^1J_{PH} = 166.6$ | | | 2 |
| $F_2C=PH$ | −201.4 | $^2J_{PF} = 213.6, 823$ | | | 3 |
| | | $^2J_{PH}$ | $^4J_{PNNCH}$ | $^3J_{PCCCH}$ | |
| [H₃C-N=N-P cyclopentane-fused] | 195.2 | | 1.4 | | 4 |
| + HCl | 204.1 | | 1.6 | | 4 |
| [H₃C, CH₃ substituted N=N-P ring] | 223.0 | 32.0 | 1.1 | 1.8 | 4 |

[a] Key to references is at the end of Appendix V.

## Appendix II

### Triply Connected Phosphorus (P≡)

| Formula | Chemical shift $\delta^{31}P$ (ppm from $H_3PO_4$) | Coupling constant $J$ (Hz) | Reference |
|---|---|---|---|
| **1.** $PX_3$ | | | |
| $PF_3$ | 97.0 | 1400, 1420 | 5 |
| $PCl_3$ | 215–220 | | 5 |
| $PBr_3$ | 222–227.4 | | 5 |
| $PI_3$ | 178.0 | | 5 |
| $PH_3$ | −238 to −241 | 180–182 | 5 |
|  | −235.1 (−90°C) | 186.4 | 6 |
| **2.** $PRX_2$ (R = organic substituent, X = halogen) | | | |
| [structure: P bonded to phenyl ring with $H_\alpha$, $H_\beta$ substituents and $F_2$] | 208.3 | $J_{PF} = 1174$ $J_{PH_\alpha} = 12.5$ $J_{PH_\beta} = 1.5$ | 5 |
| $P(CH_3)Cl_2$ | 191–193, 202.2 | 16–16.9 | 5, 7 |
| [Newman projection: P with $CH_3$, $CH_3$, Cl, Cl, H substituents] | 202.2 (−140°C) | $^2J_{PC} = 13.6$ (RT) | 7 |
| [Newman projection: P with $CH_3$, $H_3C$, Cl, Cl, H substituents] | 192.0 (−140°C) | $^2J_{PC} = 13.6$ (RT) | 7 |

(*Continued*)

## Appendix II (Continued)

| Formula | Chemical shift $\delta^{31}P$ (ppm from $H_3PO_4$) | Coupling constant $J$ (Hz) | Reference |
|---|---|---|---|
| **3.** $PR_2X$ | | | |
| $P(CH_3)_2Cl$ | 92 to 96.0 | | 5 |
| $P(C_2H_5)_2Cl$ | 119.0 | | 5 |
| $P(CH_3)_2Br$ | 87.9 to 92 | | 5 |
| **4.** $PRZ_2$ ($Z$ = miscellaneous substituents) | | | |
| $PH_2(CH_3)$ | −163.5 | $^1J_{PH} = 210$ <br> $^2J_{PCH} = 4.1$ | 5 |

$$P\!\left(\!\!\begin{array}{c}\phantom{Z}\\ Z\end{array}\!\!\right)\!H_2$$

| | | $J_{PH}$ (aromatic) | | Coupling constants $J$ (Hz) | |
|---|---|---|---|---|---|
| Z | Chemical shift $\delta^{31}P$ (ppm from $H_3PO_4$) | Ortho | Meta | $^1J_{PH}$ | Reference |
| H | −122.3, −122.0 | 6.72 | 1.56 (para) | 199.5, 196–207 | 8, 5 |
| $CH_3O$ | −125.79 | 6.81 | 1.1 | 197.3 | 8 |
| $CH_3$ | −124.48 | 6.96 | 1.65 | 197.7 | 8 |
| F | −125.44 | 6.39 | 1.05 | 198.7; $J_{PF} = 3.14$ | 8 |
| Cl | −124.13 | 5.96 | 2.31 | 200.3 | 8 |
| Br | −124.08 | | | 201 | 8 |
| $(CH_3)_2N$ | −126.12 | 7.14 | 1.085 | 199.2 | 8 |
| **5.** $PR_2Z$ | | | | | |
| $PH(CH_3)_2$ | −98.5 to −99.5 | | | $J_{PH} = 188$–$210$ <br> $J_{PCH} = \pm 3.6$ | 5 |
| $PH(C_2H_5)_2$ | −55 | | | $J_{PH} = 190$ | 5 |
| $PH(C_3H_7)_2$ | −73 | | | $J_{PH} = 178$ | 5 |
| $PH(C_6H_5)_2$ | −41.1 | | | $J_{PH} = 214$ to $239$ <br> $J_{PCH} = 7.9 \pm 0.2$ | 5 |
| $P_2H_4$ | −204.0 (−80°C) | | | $J_{PP} = 108.2$ <br> $^1J_{PH} = 186.5$ <br> $^2J_{PH} = 11.9$ | 6 |

| | | | |
|---|---|---|---|
| 6. $PR_3$ | | | |
| $P(CH_3)_3$ | $-61$ to $-62$, $-63.3$ | | 5, 9 |
| $P(CH_3)_2 C_6H_5$ | $-46$ to $-47.0$, $-47.55$ | $J_{PCH} = 2.66$ | 5, 10 |
| $P(CH_3)_2[C(CH_3)_3]$ | $-28.66$ | | 10 |
| $P(CH_3)(C_6H_5)_2$ | $-28.0$, $-27.8$ | | 5, 10 |
| $P(CH_2CH_3)_3$ | $-19$ to $-20.4$, $-20.1$ | $J_{PCH} = \pm 13.7$<br>$J_{PCCH} = \pm 0.5$ | 5, 10 |
| $P(C_2H_5)(C_6H_5)_2$ | $-12$ to $-13.5$, $-12.3$, $-12.5$ | | 5, 10, 11 |
| $P(n\text{-}C_3H_7)_3$ | $-33$, $-33.0$ | | 5, 12 |
| $P(n\text{-}C_4H_9)_3$ | $-32.3$ to $-33.4$, $-32$ | | 5, 10 |
| $P(C_6H_5)_2 [C(CH_3)_3]$ | $17.32$ | | 10 |
| $P(C_6H_5) [C(CH_3)_3]_2$ | $37.99$ | | 5 |
| $P(n\text{-}C_4H_9)(C_6H_5)_2$ | $-17.1$, $-17.1$ | | 12, 13 |
| $P(C_6H_5)_3$ | $-5.6$ to $-8$ | $J_{CP}$<br>$-12.51$ (ipso)<br>$+19.65$ (ortho)<br>$+6.80$ (meta)<br>$0.33$ (para) | 5, 14 |
| $(4\text{-}CH_3C_6H_4)_3 P$ | $-8.0$, $-6.8$ | | 15, 16 |
| $(4\text{-}CH_3OC_6H_4)_3 P$ | $-10.2$, $-9.8$ | | 15, 17 |
| $(4\text{-}FC_6H_4)_3 P$ | $-9.0$, $-8.8$, $-9.4$ | | 15, 17, 18 |
| $(4\text{-}ClC_6H_4)_3 P$ | $-8.5$, $-9.2$ | | 15, 19 |
| $(4\text{-}BrC_6H_4)_3 P$ | $-8.2$ | | 15, 19 |
| $(3\text{-}CH_3C_6H_4)_3 P$ | $-5.3$ | | 15 |
| $(2\text{-}CH_3C_6H_4)_3 P$ | $-30.0$, $-30.2$ | | 15, 18 |
| $(C_6H_5)_2 PCH_2P (C_6H_5)_2$ | $-23.6$ | | 20 |
| $(C_6H_5)_2 P CH_2CH_2 P (C_6H_5)_2$ | $-12.5$ | | 20 |
| $(C_6H_5)_2 P (CH_2)_3 P(C_6H_5)_2$ | $-17.3$ | | 20 |
| $(CH_3)_2 PCH_2CH_2 P (CH_3)_2$ | $-49.4$ | | 9 |
| $(C_6H_5)_2 PHC{=}CH_2$ | $-11.7$, $-13.8$ | | 21, 22 |

*(Continued)*

## Appendix II (Continued)

| Formula | Chemical shift $\delta^{31}P$ (ppm from $H_3PO_4$) | Coupling constant $J$ (Hz) | | | | Reference |
|---|---|---|---|---|---|---|
| | | $J_{PA}$ | $J_{PB}$ | $J_{PM}$ | | |
| (M)H$_3$C\C=C/H(A) \ (C$_6$H$_5$)$_2$P/  \H(B) | −0.9 | 25.3<br>26.5 | 10.7<br>10.5 | 9.2 | | 23 |
| (M)H\C=C/CH$_3$(A) \ (C$_6$H$_5$)$_2$P/  \CH$_3$(B) | −28.3 | 0.3 | 0.5 | 4.0 | | 23,24 |

| R$_2$PHC=HR′ | | | | | | | | |
|---|---|---|---|---|---|---|---|---|
| R | R′ | Geometry | | $^1J_{PC\text{-}1}$ | $^2J_{PC\text{-}2}$ | $^3J_{PC\text{-}3}$ | $^1J_{PC_{ipso}}$ | $^2J_{PC_{ortho}}$ |
| C$_6$H$_5$ | CH$_3$ | Z | −32.1 | 6.7 | 34.8 | 22.9 | 9.7 | 19.2 | 25 |
| C$_6$H$_5$ | CH$_3$ | E | −13.8 | 6.5 | 33.9 | 13.8 | 11.0 | 18.3 | 25 |

| Formula | Chemical shift $\delta^{31}P$ | $^1J_{PC\text{-}1}$ | $^2J_{PC\text{-}2}$ | $^3J_{PC\text{-}3}$ | $J_{CP}$ (aromatic) | Reference |
|---|---|---|---|---|---|---|
| (C$_6$H$_5$)$_2$P—H | −40.6 | | | | | 26 |
| (C$_6$H$_5$)$_2$P$\overset{1}{C}$H$_2\overset{2}{C}$H$_2\overset{3}{C}$H$_2\overset{4}{C}$H$_3$ | | −13.7 | +16.1 | 12.4 | −14.8 (ipso)<br>+18.1 (ortho)<br>+5.4 (meta)<br>0.0 (para) | 27 |
| (C$_6$H$_5$)$_2$PCH$_2$CH$_2$CH$_3$ | −17.6 | | | | | 23 |
| H$_5$C$_6$P (cyclohexyl-P ring) | | −14.8 | +2.4 | 2.5 | −19.1 (ipso)<br>+14.5 (ortho)<br>+4.7 (meta) | 27 |

| Structure | | | | Ref |
|---|---|---|---|---|
| H₅C₆P (cyclopentyl) | | −14.0 | −4.7 | −25.0 (ipso)<br>+15.6 (ortho)<br>+4.8 (meta) | 27 |
| H₅C₆P (cyclobutyl, dimethyl) | | +0.6 | | −35.4 (ipso)<br>+15.6 (ortho) | 27 |
| H₅C₆—P (4-tBu-cyclohexyl) | 32.92 | 11.6 | 0 | 19.4 (ipso)<br>11.9 (ortho) | 28 |
| H₅C₆—P (4-tBu-cyclohexyl, axial) | 38.62 | 8.9 | 5.1 | 15.6 (ipso)<br>15.6 (ortho)<br>7.6 (meta) | 28 |
| RP (phosphole), R = $CH_2C_6H_5$ | 8.0 | | | | 29 |
| RP (phosphole), R = $CH_2CH_2C_6H_5$ | 5.8 | | | | 29 |
| 3,4-dimethyl-1-($CH_2C_6H_5$)-phosphole | −3.0 | | | $^4J_{PCH_3} = 3.0$<br>$^2J_{PCH} = 37.2$ | 29 |

*(Continued)*

## Appendix II (Continued)

| $R^A$ | $R^B$ | $R^C$ | $R^D$ | Chemical shift $\delta^{31}P$ (ppm from $H_3PO_4$) | Coupling constants $J_{PH}$ (Hz) | | | | | Reference |
|---|---|---|---|---|---|---|---|---|---|---|
| | | | | | $CH_3$ | H-2 | H-3 | H-4 | H-5 | |
| $CH_2C_6H_5$ | H | $CH_3$ | H | 11.5 | 3 | | | | | 29 |
| $CH_3$ | $CH_3$ | H | H | −7.3 | 11 | | | | | 29 |
| $CH_3$ | H | $CH_3$ | H | −6.9 | 3.3 | 41 | 12.5 | | 40.0 | 29 |
| $CH_3$ | H | $CH_3$ | $CH_3$ | −20.2 | 3 | 41 | | | 41 | 29 |

| Formula | | | | | | | | | | |
|---|---|---|---|---|---|---|---|---|---|---|
| $(C_6H_5CH_2)P$ (cyclopentane) | | | | −14.4 | | | | | | 29 |
| $(C_6H_5CH_2)P$ (cyclopentene) | | | | 0 | | | | | | 29 |
| $H_5C_6P$ (cyclopentene) | | | | −25.3 | | | | | | 30 |
| $H_5C_6$–P with $H_3C$ substituents (2,5-dimethyl) | | | | 11.4 | | $^1J_{PC\text{-}2} = 6$; $^2J_{PC\text{-}3} = 6$ | | | | 30 |

| Compound | δ | $^1J_{PCH_3}$ | Ring CH$_3$ | Ring carbon | Ref. |
|---|---|---|---|---|---|
| (2,5-dimethyl-phospholene, H$_3$C/H$_5$C$_6$) | 24.2 | | $^1J_{PC-2} = 4$; $^2J_{PC-2-CH_3} = 25$ | | 30 |
| H$_3$CP (2,5-dimethylphosphole) | | 20 | $^3J_{PC} = 4$ | $^2J_{PC} = 7.5$ | 31 |
| H$_3$CP | −33.5 | | | | 32 |
| H$_3$CP | −50.7 | | | | 32 |

| | δ | Coupling constant $J$ (Hz) | Ref. |
|---|---|---|---|
| **8. PX(OZ)$_2$** | | | |
| ClP(O-4,5,6-O) | 153–153.9 | $^2J_{PC-4} = 2.4$; $^2J_{PC-6} = 2.4$; $^3J_{PC-5} = 5.2$ | 5, 33 |
| P(OC$_2$H$_5$)$_2$(CN) | 117.5 | | 34 |
| **9. PR$_2$(OR) and PR(OR)$_2$** | | | |
| P(C$_6$H$_5$)$_2$(OCH$_3$) | 115.6 | $J_{PCH} = 8.3$ | 5 |
| P(CH$_3$)(OCH$_3$)$_2$ | 200.8 | $J_{POCH} = 10.9$ | 5 |
| P(C$_6$H$_5$)(OCH$_3$)$_2$ | 159 | $J_{POCH} = 5.3$ | 5 |

*(Continued)*

## Appendix II (Continued)

$$R \rightarrow P \begin{array}{c} O-4-R^2/R^4 \\ | \\ O-5-R^1/R^3 \end{array}$$

| R | R$^1$ | R$^2$ | R$^3$ | R$^4$ | Chemical shift $\delta^{31}$P (ppm from H$_3$PO$_4$) | $J_{PCH_3}$ (exocyclic) | Coupling constants $J$ (Hz) | | | | Reference |
|---|---|---|---|---|---|---|---|---|---|---|---|
| | | | | | | | $^2J_{PC-4}$ | $^2J_{PC-5}$ | $^2J_{PC}$ (exocyclic) | $^3J_{POCC}$ | |
| (CH$_3$)$_3$C | CH$_3$ | H | H | H | 207.4 | 11.6 | 6.4 | 6.6 | 43.7 | 3.4 | 35 |
| (CH$_3$)$_3$C | H | H | CH$_3$ | H | 201.5 | 11.6 | 8.8 | 7.6 | 44.9 | 4.5 | 35 |
| (C$_6$H$_5$)CH$_2$ | CH$_3$ | H | H | H | 190.4 | 6.9 | 9.7 | 8.4 | 49.2 | <3 | 35 |
| (C$_6$H$_5$)CH$_2$ | H | H | CH$_3$ | H | 185.4 | 6.2 | 9.8 | 8.0 | 47.6 | 3.7 | 35 |
| (C$_6$H$_5$) | CH$_3$ | H | H | H | 169.0 | | 9.5 | 8.6 | | <3 | 35 |
| (C$_6$H$_5$) | H | H | CH$_3$ | H | 165.2 | | 8.9 | 7.9 | | 3.7 | 35 |

(C$_2$H$_5$O)$_2$PHC=CHR
   1    2   3

| R | Geometry | Chemical shift $\delta^{31}$P (ppm from H$_3$PO$_4$) | Coupling constants $J$ (Hz) | | | Reference |
|---|---|---|---|---|---|---|
| | | | $^1J_{PC-1}$ | $^2J_{PC-2}$ | $^3J_{PC-3}$ | |
| (CH$_3$)$_3$C | Z | | 29.2 | 4.7 | 1.4 | 25 |
| C$_6$H$_6$ | E | | 24.3 | 33 | 10.5 | 25 |

**10.** P(OR)$_3$

| | | | | $^3J_{POCH} = 9.7$ | | |
|---|---|---|---|---|---|---|
| P(OCH$_3$)$_3$ | | 140–141 | | | | 5 |
| P(OCH$_3$)$_2$(OC$_6$H$_5$) | | 135.2 | | | | 5 |
| P(OC$_2$H$_5$)$_3$ | | 137–140 | | | | 5 |
| P(OC$_3$H$_7$)$_3$ | | 137.9 | | | | 5 |
| P(OC$_6$H$_5$)$_3$ | | 125–129 | | | | 5 |

H₃CO—P structure with positions 4, 5 bearing R¹, R², R³, R⁴ (five-membered ring)

| R¹ | R² | R³ | R⁴ | Chemical shift δ³¹P (ppm from H₃PO₄) | $^2J_{POCH_3}$ | $^3J_{POCH\text{-}1}$ | $^3J_{POCH\text{-}2}$ | $^3J_{POCH\text{-}3}$ | $^3J_{POCH\text{-}4}$ | Reference |
|---|---|---|---|---|---|---|---|---|---|---|
| H | H | H | H | 131.2 | 10.7 | 2.0 | 8.8 | 8.8 | 2.0 | |
| H | CH₃ | H | CH₃ | 136.7 | 10.4 | 0.6 | 0.6 | 2.9 | 0.5 | |
| H | CH₃ | CH₃ | H | 132.5 | 10.9 | 2.2 | 0.5 | 0.5 | 2.2 | |
| CH₃ | H | H | CH₃ | 146.4 | 12.3 | 0.4 | 8.2 | 8.2 | 0.4 | |
| CH₃ | CH₃ | CH₃ | CH₃ | 148.0 | 12.5 | <0.3 | <0.3 | <0.3 | <0.3 | |

| R¹ | R² | R³ | R⁴ | | $^2J_{PC\text{-}4}$ | $^2J_{PC\text{-}5}$ | $^2J_{PCH\text{-}3}$ | $^3J_{PCR_1R_2}$ | $^3J_{PCR_2R_3}$ | Reference |
|---|---|---|---|---|---|---|---|---|---|---|
| H | H | H | H | | 9.2 | 8.1 | — | — | — | 36 |
| H | CH₃ | H | CH₃ | | 7.6 | 9.2 | 8.6 | <1 | 5.3 | 36 |
| H | CH₃ | CH₃ | H | | 7.9 | 7.9 | 10.5 | — | 3.9 | 36 |
| CH₃ | H | H | CH₃ | | 8.4 | 8.4 | 20.0 | <1 | — | 36 |
| CH₃ | CH₃ | CH₃ | CH₃ | | 7.5 | 7.5 | 14.9 | <1 | 3.6 | 36 |

MeO—P six-membered ring with positions 4, 5, 6 bearing R¹, R²

| R¹ | R² | Chemical shift δ³¹P (ppm from H₃PO₄) | $^2J_{PC\text{-}4}$ | $^2J_{PC\text{-}6}$ | $^2J_{POCH_3}$ | $^3J_{PC\text{-}5}$ | $^4J_{PCH_3}$ | Reference |
|---|---|---|---|---|---|---|---|---|
| H | H | 130.1 | 1.3 | 1.3 | 18.0 | 5.2 | | 33, 37 |
| CH₃ | H | 123.5 | 1.5 | 1.5 | 18.3 | 5.2 | 1.5 | 33, 37 |
| H | CH₃ | 129.8 | | | | | | 33 |
| CH₃ | CH₃ | 122.6, 123, 122.7 | 1.3 | 1.3 | 18.7 | 4.7 | 1.2 (equatorial) (axial <1) | 33, 37 |

(*Continued*)

## Appendix II (Continued)

| Formula | Chemical shift $\delta^{31}P$ (ppm from $H_3PO_4$) | Coupling constant $J$ (Hz) | Reference |
|---|---|---|---|
| **11.** $PX_2(SR)$ and $PX(SR)_2$ | | | |
| $P(SCH_3)Cl_2$ | 206.0 | | 5 |
| $P(SC_6H_5)Cl_2$ | 204.2 | | 5 |
| **12.** $P(OR)(SR)_2$, $P(SR)_3$, $PZ(SZ)_2$, $PZ_2(SZ)$, and $P(SZ)_3$ | | | |
| $P(OCH_3)(SC_4H_9)_2$ | 162.1 | | 5 |
| $P(SCH_3)_3$ | 124.1–125.6 | | 5 |
| [cyclic structure with $CH_3S-P$ and $CH_3$] | 179.5 | $^2J_{PC-4} = 4.4$<br>$^2J_{PC-5} = 3.7$<br>$^3J_{PCH_3} = 11.7$ | 39 |
| **13.** $PR(N{<})_2$ and $PR_2(N{<})$ | | | |
| $P(CH_3)[N(CH_3)_2]_2$ | 86–86.4 | 10–12 | 5 |
| $P(C_6H_5)[N(CH_3)_2]_2$ | 100.3 | | 5 |
| [cyclic NEt–PPh–NEt structure] | 86.8 | | 40 |

### Coupling constants $J_{PC}$ (Hz)

| | Phenyl | | | | Benzyl | | | | Amino | |
|---|---|---|---|---|---|---|---|---|---|---|
| | C-1 | Ortho | Meta | Para | C-1 | Ortho | Meta | Para | $CH_2$ | $CH_3$ |
| $Ph_2PN(CH_2Ph)_2$ | 14.4 | 20.4 | 3.9 | 0 | 1.7 | 0.5 | 0 | 0 | 14.4 | |
| $PhP[N(CH_2Ph)_2]_2$ | | | 5.7 | 1.1 | 1.9 | 1.2 | 0 | 0 | 15.0 | |
| $PhP[N(CH_2CH_3)_2]_2$ | | | 3.1 | 2.0 | | | | | 16.9 | 3.2 |

|  | 42.1 | 19.1 | 5.1 | 0 | 8.4 | 23.5 | 41 |

**14.** $P(N{<}){_2}X$

R—N(4)—PCl—N(6)—C⁷(C⁸H₃)₃, with tBu on N and 5-position

| R | Chemical shift $\delta^{31}P$ (ppm from $H_3PO_4$) | Coupling constants $J$ (Hz) | | | | | Reference |
|---|---|---|---|---|---|---|---|
|  |  | $^2J_{PC\text{-}4}$ | $^3J_{PC\text{-}5}$ | $^2J_{PC\text{-}6}$ | $^2J_{PC\text{-}7}$ | $^3J_{PC\text{-}8}$ |  |
| H | 152.0 | 5.3 | 1.0 | 5.3 | 20.5 | 7.2 | 42 |
| CH₃ | 153 | 8.9 | 2.0 | 5.6 | 23.0 | 16.7 | 42 |

**15.** $P(N{<}){_2}(OR)$ and $P(N{<})(OR){_2}$

$(CH_3)_2N-P$ with O(4)-C-R² R⁴ / O(6)-C-R¹ R³, positions 4,5,6

| R¹ | R² | R³ | R⁴ | Chemical shift $\delta^{31}P$ (ppm from $H_3PO_4$) | Coupling constants $J$ (Hz) | | | | | Reference |
|---|---|---|---|---|---|---|---|---|---|---|
|  |  |  |  |  | $^3J_{PNCH_3}$ | $^2J_{PC\text{-}4}$ | $^2J_{PC\text{-}5}$ | $^2J_{PNC}$ | $^3J_{POCC}$ |  |
| CH₃ | H | CH₃ | H | 147.3 | 8.2 | 7.5 | 7.5 | 19.8 | 3.0 | 35 |
| CH₃ | H | H | H | 142.5 | 8.2 | 8.1 | 9.5 | 19.8 | 4.1 | 35 |
| H | H | CH₃ | H | 83.7 | 8.2 | 3.5 | 6.3 | 19.7 | 3.5 | 35 |
| (CH₃)₃ | H | H | H | 84.0 | 8.1 | 4.0 | 8.7 | 19.7 | 4.0 | 35 |
| H | H | (CH₃)₃C | H |  | 8.1 |  |  |  |  | 35 |

(*Continued*)

## Appendix II (Continued)

Structure:

tBu–N(6)–XP–N(4)(tBu)–R (six-membered ring with positions 4,5,6)

| X | R | Chemical shift $\delta^{31}P$ (ppm from $H_3PO_4$) | Coupling constants $J$ (Hz) | | | | |
|---|---|---|---|---|---|---|---|
| | | | $^2J_{PC-4}$ | $^3J_{PC-5}$ | $^2J_{PC-6}$ | $^2J_{PC-7-(tBu)}$ | $^3J_{PC-8}$ | Reference |
| $OCH_2CH_3$ | H | 117.0 | 3.8 | 0.6 | 3.8 | 20.8 | 15.6 | 42 |
| $OCH_3$ | $CH_3$ (cis) | 120.4 | 4.8 | <0.6 | 4.5 | 18.9 | 15.1 | 4 |
| $OCH_3$ | $CH_3$ (trans) | 118.0 | 7.8 | 5.1 | 4.7 | ($J_{POCH_3}$ = 17.7) 27.8 ($^2J_{POCH_3}$ = 23.6) | 14.4 | 42 |
| $OC_6H_5$ | $CH_3$ (cis) | 120.5 | 4.4 | <1.5 | 4.4 | 19.3 | 15.0 | 42 |
| $OC_6H_5$ | $CH_3$ (trans) | 117.5 | 8.4 | 5.6 | 4.4 | 29.3 | 15.0 | 42 |

### 16. $P(N{<}\!)_3$

| Formula | Coupling constant $J$ (Hz) | Reference |
|---|---|---|
| $P[N(CH_3)_3]_3$ | $J_{PNCH_3}$ = 9 | 5 |
| $(CH_3)_2N-P$ (with imidazolidine ring: $H_3C-N$—$N-CH_3$) | $J_{PNCN(CH_3)_2}$ = 17.4 at ~30°C; $J_{PNCH_3}$ = 9.1; $J_{PNCH_3}$ = 23.8 | 4 |

## Appendix III

### Quadruply Connected Phosphorus (>P<)

#### A. Phosphoryl Compounds (OP≡)

| Formula | Chemical shift $\delta^{31}P$ (ppm from $H_3PO_4$) | Coupling constant $J$ (Hz) | | Reference |
|---|---|---|---|---|
| **1.** $OPX_3$ | | | | |
| $OPF_3$ | −35.5 | | | 5 |
| $OPCl_3$ | 1.9–5.4 | 1080–1055 | | 5 |
| $OPBr_3$ | −102 to −104.3 | | | 5 |
| **2.** $OPRX_2$ | | | | |
| $OP(CH_3)F_2$ | 26.8–27.4 | 1270, 1090 | | 5 |
| $OP(CH_3)Cl_2$ | 43.5–44.5 | | | 5 |
| $OP(CH_3)Br_2$ | 8.5 | 14.7 | | 5 |
| $OP(C_2H_5)Cl_2$ | 52–54.8 | | | 5 |
| $OP(C_6H_5)Cl_2$ | 33.7–34.5 | | | 5 |
| $O(X)_2PR$    R    X | | $^1J_{PC}$ | $^2J_{PC}$ | |
| $C_2H_5$    F | 29.2 | | | 5 |
| $HC=CH_2$    F | 10.6 | | | 43 |
| $C≡CH$    F | −21.9 | 196.7 | 2.8 | 43 |
| **3.** $OPR_2X$ | | | | |
| $OP(CH_3)_2F$ | 66.3 | $J_{PF} = 990$ | | 5 |
| $OP(CH_3)_2Cl$ | 62.8–68.3 | $J_{PH} = 19$ | | 5 |
| $OP(CH_3)_2Br$ | 50.7 | | | |
| $OP(C_6H_5)_3Cl$ | 42.7–44.5 | | | 5 |
| **4.** $OPRZ_2$ and $OPR_2Z$ | | | | |
| $OPH(CH_3)_2$ | 63.2 | $J_{PH} = 704$ | | 5 |
| $OPH(C_6H_5)_2$ | 22.9 | $J_{PH} = 490$ | | 5 |

(*Continued*)

## Appendix III (Continued)

| Formula | Chemical shift $\delta^{31}P$ (ppm from $H_3PO_4$) | Coupling constant $J$ (Hz) | Reference |
|---|---|---|---|
| **5. $OPR_3$** | | | |
| $OP(CH_3)_3$ | 36.2 | $J_{PC} = 68.3$ | 5, 44 |

$R_2P(O)HC=CHR'$ with positions 1, 2, 3

| | | | Chemical shift $\delta^{31}P$ (ppm from $H_3PO_4$) | Coupling constants $J$ (Hz) | | | | | Reference |
|---|---|---|---|---|---|---|---|---|---|
| R | R' | Geometry | | $^1J_{PC-1}$ | $^2J_{PC-2}$ | $^3J_{PC-3}$ | $^1J_{PC_{ipso}}$ | $^2J_{PC_{ortho}}$ | |
| $C_6H_5$ | $CH_3$ | Z | 21.3 | 101.0 | 0 | 7.3 | 103.7 | 9.2 | 25 |
| | | E | 23.8 | 102.8 | 1.9 | 18.3 | 110.0 | 10.1 | 25 |

Formula:

[structure: P with two $C(CH_3)_3$ groups and $O$, bracketed with $[C(CH_3)_3]$ counterion]

Chemical shift $\delta^{31}P$ (ppm from $H_3PO_4$): $-1.3$

$^3J_{PCH_3} = 17.4$

Reference: 45

[structure with numbered positions: Y—P—X with substituents at positions 2, 3, 4, 5, 6, 9, 10, and $C^7H_3$, H]

| | | Chemical shift $\delta^{31}P$ (ppm from $H_3PO_4$) | Coupling constants $J_{P-C}$ (Hz) | | | | | | | Reference |
|---|---|---|---|---|---|---|---|---|---|---|
| X | Y | | C-2,4 | C-3 | C-5,9 | C-6,10 | C-7 | X | Y | |
| O | $CH_3$ | | 59.4 | 6.3 | 4.6 | 3.6 | 23.0 | | 40.9 | 44, 46 |
| $CH_3$ | O | | 59.4 | 10.0 | 2.2 | 4.4 | 12.6 | 36.9 | | 44, 46 |

564

## Formula

[Structure: H with C-3, C-4, C-2, C-5, C-6, 7CH₃, Y-P-X]

| X | Y |
|---|---|
| O | Ph |
| Ph | O |

| | C-2 | C-4 | C-3 | C-5 | C-6 | C-7 | |
|---|---|---|---|---|---|---|---|
| | 63.3 | 52.5 | 11.9 | 4.4 | 3.2 | 28.3 | 44 |
| | 63.1 | 52.3 | 16.0 | 2.6 | 4.5 | 16.5 | 44 |

## Formula

| | | | | | |
|---|---|---|---|---|---|
| H₃C(O)P (cyclopentane) | 65.8 | | | | 47 |
| H₃C(O)P (cyclopentene) | 66.5 | | | | 47 |

| | C-1 | C-2 | C-3 | C-4 | |
|---|---|---|---|---|---|
| Ph(O)P (cyclopentane, 1,2) | 66.8 | 7.9 | | | 44 |
| Ph(O)P (cyclopentene, 1,2,3) | 71.0 | 10.4 | 24.0 | 92.0 | 44 |
| Ph(O)P (cyclohexane, 1,2,3) | 65.2 | 6.0 | 6.8 | | 44 |

*(Continued)*

**Appendix III (Continued)**

| Formula | Chemical shift $\delta^{31}P$ (ppm from $H_3PO_4$) | Coupling constant $J$(Hz) | | | | Reference |
|---|---|---|---|---|---|---|
| | | $J_{PC-2,6}$ | $J_{PC-4}$ | $J_{PC-7}$ | $J_{PCC_4H}$ | |
| [structure with positions 7,4,5,6,3,2, P=O] | | 63 | 35 | 4 | 28 | 48 |
| [structure with positions 7,8,4,5,6,3,2, P=O] | | 63 | 47 | | | 48 |
| 6. $OPX_2(OZ)$ | | | | | | |
| $OP(OC_2H_5)F_2$ | 20.9–21.2 | | $J_{PF}$ = 1010, 1015 | | | 5 |
| $OP(OC_2H_5)Cl_2$ | 3.4–6.4 | | | | | 5 |
| $OP(OC_2H_5)Cl_2$ | 1.5–1.8 | | | | | 5 |
| 7. $OPX(OZ)_2$ | | | | | | |
| $OP(OCH_3)_2Cl$ | 6.4 | | $J_{PH}$ = 13.65 | | | 5 |
| $O(Cl)P$ [cyclic structure with two $CH_3$ groups] | −4.85 | | | | | 49 |
| 8. $OPZ_2(OZ)$ | | | | | | |
| $OPH_2(OCH_3)$ | 19 | | $J_{PH}$ = 575, 13 | | | 5 |
| $OPH_2(OC_2H_5)$ | 15 | | $J_{PH}$ = 567 | | | 5 |
| 9. $OPZ(OZ)_2$ | | | | | | |
| $OPH(OCH_3)_2$ | 9.8–12.8 | | $J_{PH}$ = 696 to 710 $^3J_{POCH}$ = 14, 12.2 | | | 5 |
| $OPH(OC_2H_5)_2$ | 6.2–8.0 | | $J_{PH}$ = 670, 690 $^3J_{POCH}$ = 9.9 | | | 5 |

| | | | |
|---|---|---|---|
| **10.** OPR$_2$(OZ) | | | |
| OP(CH$_3$)$_2$(OC$_2$H$_5$) | 50.3 | | 5 |
| R(O)P structure with CH$_3$, CH$_3$, H$_3$C, O | | | |
| R | | | |
| CH$_3$ | 75.1 | | 50 |
| C$_6$H$_5$ | 59.4 | | 50 |
| O(R$_2$)—OC$_6$H$_4$–p-NO$_2$ | | | |
| R | | | |
| ClCH$_2$ | 36 | | 51 |
| n-C$_4$H$_9$ | 60 | | 51 |
| CH$_3$ | 54 | | 51 |
| **11.** OPR(OZ)$_2$ | | | |
| OP(H)(OCH$_3$)$_2$ | | $J_{P^{17}O} = +220$ | 52 |
| | | $J_{P^{17}OCH_3} = +88$ | |
| OP(CH$_3$)(OCH$_3$)$_2$ | 32.3–32.6 | | 5 |
| OP(CH$_3$)(OC$_6$H$_5$)$_2$ | 22.0–23.5 | | 5 |
| O(C$_2$H$_5$O)$_2$PR | | | |
| R | | | |
| C$_2$H$_5$ | 32.8 | | 5 |
| HC=CH | 17.3 | $^1J_{PC} = 182.1$ | 43 |
| | | $^2J_{CP} = 1.9$ | |
| | | $^2J_{POC} = 5.9; \ ^3J_{POCC} = 7.1$ | |

*(Continued)*

## Appendix III (Continued)

| Formula | Chemical shift $\delta^{31}P$ (ppm from $H_3PO_4$) | Reference |
|---|---|---|
| $O(R)P(OC_2H_5)_2$ | | |
| R | | |
| Me | 29.6 | 5 |
| Bu | 31.7 | 5 |
| Et | 32.6 | 5 |
| tBu | 39.4 | 5 |
| C(Me)$_2$OH | 27.6 | 5 |
| CCl$_3$ | 6.5 | 5 |

| X | Y | R$^1$ | R$^2$ | R$^3$ | R$^4$ | Chemical shift $\delta^{31}P$ (ppm from $H_3PO_4$) | $^1J_{PC}$ | $^3J_{PC}$ | $^2J_{PC}$ | Reference |
|---|---|---|---|---|---|---|---|---|---|---|
| O | H | CH$_3$ | H | CH$_3$ | H | 2.9 | | | | 53 |
| H | O | CH$_3$ | H | CH$_3$ | H | −1.3 | | | | 53 |
| O | H | CH$_3$ | H | H | H | 2.8 | | | | 53 |
| H | O | CH$_3$ | H | H | H | −1.7 | | | | 53 |
| O | CH$_3$ | CH$_3$ | H | CH$_3$ | H | 19.4 | | | | 53 |
| CH$_3$ | O | CH$_3$ | H | CH$_3$ | H | 28.0 | | | | 53 |
| O | CH$_3$ | CH$_3$ | H | H | H | 23, 20.4 | 32.8 | 6.8 | 6.5 | 5, 54 |
| | | | | | | | | | 6.8 | |
| CH$_3$ | O | CH$_3$ | H | H | H | 28.0, 27.7 | 145.0 | 5.4 | 5.9 | 5, 53 |
| | | | | | | | | | 6.3 | |

$O(C_2H_5O)_2PR$

| R | Coupling constants $J$ (Hz) | | | | | | | Reference |
|---|---|---|---|---|---|---|---|---|
| | $^1J_{P,C-1}$ | $^2J_{P,C-2}$ | $^3J_{P,C-3}$ | $^4J_{P,C-4}$ | $^2J_{P,C-\alpha}$ | $^3J_{P,C-\beta}$ | | |
| $^2CH_3{}^1CH_2$ | 142.6 | 6.7 | | | 6.4 | 5.6 | | 55 |
| $^3CH_3{}^2CH_2{}^1CH_2$ | 140.4 | 5.2 | 16.2 | | 6.2 | 5.9 | | 55 |
| $^4CH_3{}^3CH_2{}^2CH_2{}^1CH_2$ | 140.9 | 5.1 | 16.3 | 1.2 | 6.3 | 5.6 | | 55 |
| $^5CH_3{}^4CH_2{}^3CH_3{}^2CH_2{}^1CH_2$ | 140.7 | 4.7 | 15.8 | 0.9 | 6.4 | 5.6 | | 55 |
| 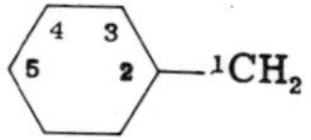 ![cyclohexyl-CH2] (4,3,2,5 numbered cyclohexyl)—$^1CH_2$ | 138.7 | 4.0 | 10.7 | 0.8 | 6.4 | 5.8 | | 55 |

$O(C_2H_5O)_2PCH_2C_6H_4X$

| X | Coupling constants $J$ (Hz) | | | | | | | | Reference |
|---|---|---|---|---|---|---|---|---|---|
| | $^1J_{PCH_2}$ | $^2J_{PC-1}$ | $^3J_{PC-2}$ | $^4J_{PC-3}$ | $^5J_{PC-4}$ | $^4J_{PC-5}$ | $^3J_{PC-6}$ | $^2J_{PC-\alpha}$ | $^3J_{PC-\beta}$ | |
| H | 137.0 | 8.7 | 6.8 | 2.9 | 3.6 | 2.9 | 6.8 | 6.6 | 5.6 | 55 |
| 2-CH$_3$ | 137.4 | 9.3 | 6.7 | 3.4 | 4.0 | 3.3 | 5.6 | 6.5 | 5.6 | 55 |
| 3-CH$_3$ | 137.1 | 9.0 | 6.7 | 3.0 | 3.4 | 2.7 | 6.7 | 6.5 | 5.7 | 55 |
| 4-CH$_3$ | 138.2 | 9.1 | 6.8 | 2.3 | 3.9 | 2.3 | 6.8 | 6.4 | 5.8 | 55 |
| 2,5-(CH$_3$)$_2$ | 137.2 | 9.3 | 6.8 | 3.4 | 3.9 | 3.7 | 5.6 | 6.5 | 5.7 | 55 |
| 3-F | 137.0 | 8.8 | 6.6 | 3.2 | 3.5 | 2.9 | 7.0 | 6.6 | 5.7 | 55 |
| 4-F | 137.6 | 8.8 | 6.8 | 2.9 | 3.8 | 2.9 | 6.8 | 6.6 | 5.6 | 55 |

(*Continued*)

**Appendix III** (*Continued*)

| Formula | Chemical shift $\delta^{31}P$ (ppm from $H_3PO_4$) | Coupling constants $J$ (Hz) | | | Reference |
|---|---|---|---|---|---|
| | | $^1J_{PC\text{-}1}$ | $^2J_{PCC\text{-}2,6}$ | $^3J_{PCCC}$ | |
| $O(CH_3O)_2PR$ | | | | | |
| R | | | | | |
| HO–⬜ (cyclobutyl) | | 164.3 | 0.6 | 6.5 | 56 |
| HO–⬠ (cyclopentyl) | | 169.0 | 8.4 | 13.0 | 56 |
| HO–⬡ (cyclohexyl) | | 165.5 | 2.8 | 13.0 | 56 |
| HO–C(CH$_2$)$_n$ (cycloalkyl) | | 159.6 | 4.4 | 10.9 | 56 |
| $n$ = 7 | | | | | |
| 8 | | 161.2 | 4.6 | 8.8 | 56 |
| 9 | | 156.8 | 4.5 | 6.7 | 56 |
| 11 | | 157.1 | 5.2 | 6.7 | 56 |
| 12 | | 157.9 | 4.9 | 5.7 | 56 |

$^3J_{POCH} = 7.2 \pm 0.4$ (all Rs)

| | | | |
|---|---|---|---|
| $(O_3PCH_3)^{2-} + 2H^+$ | 29.8–32.1 | $J_{PH} = 15, 17$ | 5 |

[Structure: adenosine triphosphate showing ribose with HO, CH$_2$O–P$_\alpha$(O$^-$)=O–O–P$_\beta$(O$^-$)=O–CH$_2$P$_\gamma$(O$^-$)=O–O(H), with adenine base (NH$_2$-substituted purine)]

| | | | |
|---|---|---|---|
| | $\alpha = -11.2$<br>$\beta = 8.2$ } pH 6.5<br>$\gamma = 13.8$ | | |
| | $\alpha = -11.4$<br>$\beta = 10.9$ } pH 10<br>$\gamma = 12.4$ | | 5 |

| | | Coupling constants $J$ (Hz) | |
|---|---|---|---|
| **12.** OPXR(OZ) | | | |
| OP(CH$_3$)(OC$_2$H$_5$)Cl | 39.5 | | 5 |
| OP(C$_6$H$_5$)(OC$_6$H$_5$)Cl | 25.1 | | 5 |
| **13.** OPHR(OZ) | | | |
| [O$_2$PH(CH$_3$)]$^-$ + H$^+$ | 35 | $J_{PH} = 557$<br>$J_{PCH} = 15.5$ | 5 |
| **14.** OP(OR)$_3$ and P(OR)$_4^+$ | | | |
| OP(OCH$_3$)$_3$ | $-2.4$ to $+2.5$ | $J_{POCH} = \pm 8.7$ | 5 |
| OP(OC$_2$H$_5$)$_3$ | $-0.5$ to $-1.2$ | $J_{POCCH} = \pm 0.9$<br>$^2J_{PC} = 5.9$<br>$^3J_{PC} = 6.5$ | 5 |
| OP(OC$_4$H$_9$)$_3$ | $-0.6$ to $-1.0$ | | 5, 57 |
| OP(OC$_6$H$_4$–p–CH$_3$)$_3$ | $-16.0$ | $J_{PH} = +51.5 \pm 1.0$ | 58 |
| [P(OCH$_3$)$_4$]$^+$ + SbCl$_6^-$ | | $J_{PH} = -18$ | 5 |
| [P(OC$_6$H$_5$)$_4$]$^+$ + (OC$_6$H$_5$)$^-$ | | | 5 |
| OP(OP)$_3$ | | | |

*(Continued)*

**Appendix III (Continued)**

| X | Y | R¹ | R² | R³ | R⁴ | Chemical shift $\delta^{31}P$ (ppm from $H_3PO_4$) | $J_{PCH_3}$ | $J_{PH-4a}$ | $J_{PH-4e}$ | $J_{PH-5e}$ | $J_{PH-5a}$ | Reference |
|---|---|---|---|---|---|---|---|---|---|---|---|---|
| O | OCH₃ | CH₃ | H | H | CH₃ | −7.1 | | | | | | 53 |
| OCH₃ | O | CH₃ | H | H | CH₃ | −5.0 | | | | | | 53 |
| O | OC₆H₅ | CH₃ | H | CH₃ | CH₃ | −15.1 | 2.7 | | | | | 59 |
| OC₆H₅ | O | CH₃ | H | CH₃ | CH₃ | −12.4 | 2.7 | | | | | 59 |
| O | OC₆H₅ | H | H | H | H | 126.0 (from $P_4O_6$) | | 2.9 | 21.9 | 2.7 | | 60 |
| O | OC₆H₅ | H | CH₃ | CH₃ | H | 126.5 (from $P_4O_6$) | | 1.7 | 22.5 | | 0.9 | 60 |

| Formula | | | | | | | | | | | | |
|---|---|---|---|---|---|---|---|---|---|---|---|---|
| $(O_4P)^{3-} + 3H^+$ | | | | | | 0.0 | | | | | | 5 |
| $(O_4P)^{3-} + Na^+ + 2H^+$ | | | | | | 0.5 to −1.3 | | | | | | 5 |
| $(O_4P)^{3-} + 2Na^+ + H^+$ | | | | | | 3.1−3.5 | | | | | | 5 |
| $(O_4P)^{3-} + 3Na^+$ | | | | | | 5.4−6.0 | | | | | | 5 |
| $H_n^+(PO_4)^{3-}$ (pH 7.8, 5 m$M$ MgCl₂) | | | | | | 2.32 | | | | | | 61 |
| $(O_3POPO_3)^{4-}$ | | | | | | −10.6 to −11.8 | | | | | | 5 |
| $(O_3POPO_3)^{4-} + 2Na^+ + 2H^+$ | | | | | | −9.5 to −10.3 | | | | | | 5 |
| $(O_3POPO_3)^{4-} + 4Na^+$ | | | | | | −5.5 to −7 | | | | | | 5 |
| $(O_3POPO_3)^{4-}$ (pH 7.8, 5 m$M$ MgCl₂) | | | | | | −4.95 | | | | | | 61 |
| $CH_3C(O)OPO_3H^-$ (pH 7.4) | | | | | | −2.0 | | | | | | 61 |
| $^-O_2CC(H)(NH_3^+)CH_2C(O)OPO_3H^-$ (pH 7.4) | | | | | | −11.6 | | | | | | 62 |

| | | |
|---|---|---|
| (dAMP structure) | 2.84 ± 0.07 | 5 |
| (dIMP structure) | 3.13 | 5 |
| (dCMP structure) | 3.65 | 5 |

*(Continued)*

## Appendix III (*Continued*)

| Formula | Chemical shift $\delta^{31}P$ (ppm from $H_3PO_4$) | Coupling constant $J$ (Hz) | Reference |
|---|---|---|---|
| ATP structure (adenosine 5′-triphosphate): $CH_2O-P_\alpha(=O)(O^-)-O-P_\beta(=O)(O^-)-O-P_\gamma(=O)(O^-)-O^-$, ribose with OH, HO, and adenine ($NH_2$) base (pH 7.8, 5 m$M$ $MgCl_2$) | $\alpha = -9.66$<br>$\beta = -5.63$ | $J_{P_\alpha P_\beta} = 18.0$ | 61 |
| $[O_3P_\alpha OP_\beta(O_2)OP_\alpha O_3]^{5-} + 5H^+$ | $\alpha = -11.5$<br>$\beta = -23.9$ | $J_{P_\alpha P_\beta} = 16.7$ | 5 |
| $[O_3P_\alpha OP_\beta(O_2)OP_\alpha O_3]^{5-} + 3Na^+ + 2H^+$ | $\alpha = -10.3$<br>$\beta = -22.5$ | $J_{P_\alpha P_\beta} = 18.2$ | 5 |
| $[O_3P_\alpha OP_\beta(O_2)OP_\alpha O_3]^{5-} + 5Na^+$ | $\alpha = -4.7$<br>$\beta = -19.0, -18$ | $J_{P_\alpha P_\beta} = 19.4, 17$ | 5 |
| $[O_3P_\alpha OP_\beta(O_2)OP_\gamma O_2(NH_2)]^{4-} + 3Na^+ + NH_4^+$ | $\alpha = -5.4$<br>$\beta = -20.7,$<br>$\gamma = -0.2$ | $J_{P_\alpha P_\beta} = 18$ | 5 |
| $[O_3P_\alpha OP_\beta(O_2)OP_\beta(O_2)OP_\alpha O_3]^{6-} + 6H^+$ | $\alpha = -11.5$<br>$\beta = -23.9$ | $J_{P_\alpha P_\beta} = 16.7,$<br>$J_{P_\beta P_\beta} = 15.9$ | 5 |
| $[O_3P_\alpha OP_\beta(O_2)OP_\beta(O_2)OP_\alpha O_3]^{6-} + 4Na^+ + 2H^+$ | $\alpha = -10.4$<br>$\beta = -22.5$ | $J_{P_\alpha P_\beta} = 18.7,$<br>$J_{P_\beta P_\beta} = 16.5$ | 5 |
| $[O_3P_\alpha OP_\beta(O_2)OP_\beta(O_2)OP_\alpha O_3]^{6-} + 6Na^+$ | $\alpha = -6.0, -5$<br>$\beta = -21.0, -13$ | $J_{P_\alpha P_\beta} = 19.9,$<br>$J_{P_\beta P_\beta} = 16.7$ | 5 |

[ATP structure diagram showing adenosine triphosphate with labeled $P_\alpha$, $P_\beta$, $P_\gamma$ phosphate groups]

(ATP)

| Compound | Chemical shift | Coupling | Ref. |
|---|---|---|---|
| ATP (pH 7.8, 5 mM MgCl$_2$) | $\alpha = -10.31$ | $J_{P_\alpha P_\beta} =$ | 61 |
|  | $\beta = -18.57$ | $J_{P_\beta P_\gamma} = 15.4$ |  |
|  | $\gamma = -5.07$ |  |  |
| ApppA (pH 7.8, 5 mM MgCl$_2$) | $\alpha = -10.98$ | $J_{P_\alpha P_\beta} = 16.7$ | 61 |
|  | $\beta = -22.09$ |  |  |
| AppppA (pH 7.8, 5 mM MgCl$_2$) | $\alpha = -10.90$ | $J_{P_\alpha P_\beta} = 16.3$ | 61 |
|  | $\beta = -21.49$ | $J_{P_\beta P_\beta} = 13.5$ |  |
| 15. OP[R or X]$_2$(SZ), OP[R or X](SZ)$_2$, and OP[R or X](OZ)(SZ) |  |  |  |
| OP(CH$_3$)$_2$(SC$_3$H$_7$) | 52.7 |  | 5 |
| OP(OCH$_3$)(SCH$_3$)Cl | 37.1 |  | 5 |
| 16. OP(OZ)$_2$(SZ) and OP(OZ)(SZ)$_2$ |  |  |  |
| [O$_3$PS]$^{3-}$ + 3Na$^+$ | 61.0–61.9 |  | 5 |
| OP(OC$_2$H$_3$)$_2$(SCH$_3$) | 28.6 |  | 5 |
| OP(OC$_2$H$_5$)(SC$_2$H$_5$)$_2$ | 53.5 |  | 5 |
| 17. OP(SZ)$_3$ |  |  |  |
| OP(SC$_2$H$_5$)$_3$ | 61.3–61.4 |  | 5 |
| OP(SC$_6$H$_5$)$_3$ | 55.2 |  | 5 |
| 18. OPX$_2$(NZ$_2$) and OPX(NZ$_2$)$_2$ |  |  |  |
| OP[N(CH$_3$)$_2$]Cl$_2$ | 16.1–18.1 | $J_{PH} = 15.$ | 5 |
| OP[N(CH$_3$)$_2$]F | 40. | $J_{PF} = 975$ | 5 |

(*Continued*)

## Appendix III (Continued)

| X | Y | Chemical shift $\delta^{31}P$ (ppm from $H_3PO_4$) | Coupling constant $J_{PH}$ (Hz) | $^3J_{PNCH_3}$ | Reference |
|---|---|---|---|---|---|
| O | $OCH_3$ | | 12 ($OCH_3$) | 10 | 63 |
| $OCH_3$ | O | | 12 ($OCH_3$) | 11 | 63 |
| O | $OC_2H_5$ | 5.9 | | 11 | 63 |
| $OC_2H_5$ | O | 5.9 | | 11.0 | 63 |
| O[(CH$_3$)$_2$N]P | | 7.1 | | | 49 |

| Y | X | R | Chemical shift $\delta^{31}P$ (ppm from $H_3PO_4$) | Reference |
|---|---|---|---|---|
| O= | $N(CH_3)_2$ | $CH_3$ | 3.5 | 53 |
| $N(CH_3)_2$ | O | $CH_3$ | 6.6 | 53 |
| O | $N(CH_3)_2$ | H | 3.5 | 53 |
| $N(CH_3)_2$ | O= | H | 6.6 | 53 |

| R | R' | Coupling constants $J$ (Hz) | | | Reference |
|---|---|---|---|---|---|
| | | $J_{PH_A}$ | $J_{PH_B}$ | $J_{PH_Y}$ | |
| $C_6H_5O$ | $C_6H_5$ | 2.3 | 23.5 | 21.35 | 64 |
| $C_2H_5O$ | $i$-$C_3H_5$ | 2.51 | 21.76 | 23.68 | 64 |

| $R^1$ | $R^2$ | X | Y | Chemical shift $\delta^{31}P$ (ppm from $H_3PO_4$) | Reference |
|---|---|---|---|---|---|
| $CH_3$ | H | O | $NHC_6H_5$ | −2.6 | 65 |
| H | $CH_3$ | $NHC_6H_5$ | O | −1.0, 1.1 | 65, 66 |
| H | $CH_3$ | O | $NHC_6H_5$ | −4.5, 3.5 | 65, 66 |

**B.** Thiophosphoryl compounds (SP⟨)

| Formula | Chemical shift $\delta^{31}P$ (ppm from $H_3PO_4$) | Reference |
|---|---|---|
| **1.** $SPX_3$ | | |
| $SPCl_3$ | | |
| $SPRX_2$ and $SPZX_2$ | 28.8–30.8 | 5 |
| $SP(CH_3)Cl_2$ | 79.4–80.9 | 5 |

(Continued)

**Appendix III (Continued)**

| Formula | | Chemical shift $\delta^{31}P$ (ppm from $H_3PO_4$) | Coupling constants $J$ (Hz) | | Reference |
|---|---|---|---|---|---|
| **2. $SP(X)_2R$** | | | | | |
| R | X | | $^1J_{PC}$ | $^2J_{PC}$ | |
| $C_2H_5$ | Cl | 95.4 | | | 5 |
| $HC=CH_2$ | Cl | 69.5 | 109.5 | 3.2 | 43 |
| $C_2H_5$ | F | 110.7 | | | 67 |
| $HC=CH_2$ | F | 90.9 | 151.3 | 7.9 | 43 |

Formula

**3. $SPR_2X$**    $SP(R)(-\overset{\overset{2}{CH_3}}{\underset{\underset{3}{CH_3}}{C}}-H)Br$

| | Chemical shift $\delta^{31}P$ (ppm from $H_3PO_4$) | Coupling constants $J$ (Hz) | | | Reference |
|---|---|---|---|---|---|
| R | | $^2J_{PH-1}$ | $^3J_{PH-2}$ | $^3J_{PH-3}$ | |
| $CH_3$ | 93.2 | −6.18 | 23.04 | 23.17 | 68 |
| $C_2H_5$ | 110.2 | −5.62 | 22.00 | 22.24 | 68 |
| $i$-$C_3H_7$ | 124.1 | −3.71 | 21.21 | 21.25 | 68 |
| $t$-$C_4H_9$ | 135.6 | −4.6 | 20.35 | 20.60 | 68 |

| Formula | Chemical shift $\delta^{31}P$ (ppm from $H_3PO_4$) | Reference |
|---|---|---|
| **5. $SPR_3$** | | |
| $SP(CH_3)_3$ | 59.1 | 5 |
| $SP(C_2H_5)_3$ | 51.9–54.5 | 5 |
| $SP(C_6H_5)_3$ | 39.9–43.5 | 5 |

$R_2P(S)HC=CHR'$

| R | R' | Geometry | Chemical shift $\delta^{31}P$ (ppm from $H_3PO_4$) | Coupling constants $J$ (Hz) | | | | | | Reference |
|---|---|---|---|---|---|---|---|---|---|---|
| | | | | $^1J_{PC-1}$ | $^2J_{PC-2}$ | $^3J_{PC-3}$ | $^1J_{PC_{ipso}}$ | $^2J_{PC_{ortho}}$ | | |
| $C_6H_5$ | $CH_3$ | Z | 28.0 | 84.5 | 0 | 11.0 | 80.4 | 7.3 | | 25 |
| | | E | 35.4 | 88.2 | 0 | 25.7 | 85.8 | 8.2 | | 25 |
| $C_6H_5$ | $C_6H_5$ | Z | 30.6 | 81.7 | 2.7 | | 85.4 | 11.0 | | 25 |
| | | E | 37.1 | 86.3 | 5.4 | | 87.2 | 10.0 | | 25 |

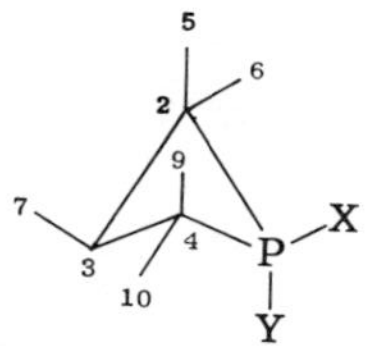

| X | Y | Coupling constants $J_{PC}$ (Hz) | | | | | | Reference |
|---|---|---|---|---|---|---|---|---|
| | | C-2,4 | C-3 | C-5,9 | C-6,10 | C-7 | | |
| S | Ph | 47.9 | 5.4 | 1.7 | 4.2 | 20.9 | | 44 |
| Ph | S | 47.3 | 6.9 | 2.5 | 2.2 | 21.5 | | 44 |

(Continued)

**Appendix III (Continued)**

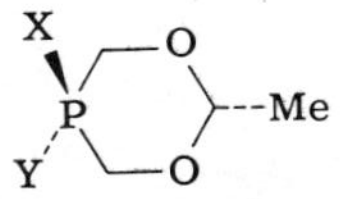

[Structure: 1,3-dioxa-2-phosphorinane with X, Y substituents at P and Me at C-2]

| X | Y | Chemical shift $\delta^{31}P$ (ppm from $H_3PO_4$) | Coupling constants $J$ (Hz) | | | Reference |
|---|---|---|---|---|---|---|
| | | | $^2J_{PCH}$ | $^4J_{POCCH}$ | $^3J_{PCCH_3}$ | |
| S | Ph | 24 | 6.8 | 0.8 | 16 | 69 |
| Ph | S | 27 | 13.4 | 1.1 | 18 | 69 |

Formula

8. $SPR(OR)_2$, $SPR(OR)X$
$S(C_2H_5O)_2PHC=CH_2$ | 83.7 | $^1J_{PC} = 148.3$ $^2J_{PC} = 7.3$ | 43 |

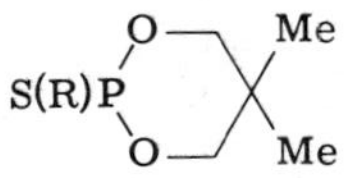

[Structure: $S(R)P$ 1,3-dioxa-2-phosphorinane with two Me groups at C-5]

| R | Chemical shift $\delta^{31}P$ (ppm from $H_3PO_4$) | Coupling constants $J$ (Hz) | | Reference |
|---|---|---|---|---|
| | | $^3J_{POCH_{eq}}$ | $^3J_{POCH_{ax}}$ | |
| $CH_3$ | 95.3 | 23.2 | 5.3 | 70 |
| $C_6H_5$ | 86.0 | 22.0 | 5.2 | 70 |

**10.** SP(OZ)$_3$

$$S(C_6H_5O)P\diagup_O^O\diagdown$$

| Formula | R | Chemical shift $\delta^{31}$P (ppm from H$_3$PO$_4$) | Coupling constants $J$ (Hz) | Reference |
|---|---|---|---|---|
| OP(R)(SCH$_2$CH$_3$)$_2$ | C$_6$H$_5$ | 53.7 | $^3J_{POCH_{eq}}$ = 19.8;  $^3J_{POCH_{ax}}$ = 6.8 | 70 |
| | C$_6$H$_{11}$ | 52.5 | | 71 |
| **11.** SPR$_2$(SZ) | C$_6$H$_5$ | 100.0 | $J_{PSCH}$ = 13.8 | 71 |
| SP(R)(SCH$_2$CH$_3$)$_3$ | C$_6$H$_5$ | 80.5 | $J_{PSCH}$ = 15.7 | 71 |
| | C$_6$H$_{11}$ | 102.0 | | 71 |

| | Chemical shift $\delta^{31}$P | | | $^1J_{PC}$ | | $^3J_{PSCH_2}$ | |
|---|---|---|---|---|---|---|---|
| | R = CH$_3$ | R = (CH$_3$)$_3$C— | R = CH$_3$ | R = (CH$_3$)$_3$C— | | | |
| ![ring1] (5-membered dithiolane P=S) | 97.6 | 129.0 | 56.7 | 46.3 | | | 72 |
| ![ring2] (6-membered dithiane P=S) | 59.1 | 100.8 | 58.5 | 50.1 | 16.8, 24.6 (R = CH$_3$) | | 72 |
| PrS—P(=S)—SPr—R | 77.6 | — | 61.3 | — | | | 72 |
| C$_6$H$_5$(S)P(O-N(CH$_3$)) oxazolidine | 94.3 | | | | $^2J_{PNCH_3}$ = 12.6; $^2J_{PC-4}$ = 6.3; $^2J_{PC-5}$ = 9.5; $^2J_{PCH_3}$ = 3.8; $^3J_{PH-1}$ = 10.0; $^3J_{PH-2}$ = 14.3; $^3J_{PH-3}$ = 6.0; $^3J_{PH-4}$ = 16.8 | | 74 |

(*Continued*)

## Appendix III (Continued)

| $R^1$ | $R^2$ | A | B | Chemical shift $\delta^{31}P$ (ppm from $H_3PO_4$) | Coupling constants $J$ (Hz) | | | | Reference |
|---|---|---|---|---|---|---|---|---|---|
| $CH_3$ | H | S | $NHC_6H_5$ | 62.5 | | | | | 65 |
| $CH_3$ | H | Se | $NHC_6H_5$ | 62.0 | | | | | 65 |
| H | $CH_3$ | $NHC_6H_5$ | S | 63.0 | | | | | 65, 73 |
| H | $CH_3$ | S | $NHC_6H_5$ | 59.5 | | | | | 65, 73 |
| H | $CH_3$ | $NHC_6H_5$ | Se | 62.5 | | | | | 65, 73 |
| H | $CH_3$ | Se | $NHC_6H_5$ | 60.0 | | | | | 65, 73 |
| H | $CH_3$ | S | $SCH_3$ | 88.5 | $^2J_{PCH_3} = 2.9$; | $^2J_{PC-4} = 8.8$ | | | 39, 65 |
| H | $CH_3$ | $SCH_3$ | S | 95.5 | $^2J_{PCH_3} = 4.4$; | $^2J_{PC-4} = 7.3$ | | | 39, 65 |
| H | $CH_3$ | S | $SeCH_3$ | 79.0 | $J_{PSe} = 437$ | | | | 65 |
| H | $CH_3$ | $SeCH_3$ | S | 88.0 | $J_{PSe} = 510$ | | | | 65 |
| | | | | | $J_{PSCH_3}$ | $^2J_{PC-4}$ | $^2J_{PC-6}$ | $^3J_{PC-5}$ | |
| H | $CH_3$ | Se | $SCH_3$ | 85.5 | 7.3 | 10.3 | 10.3 | 5.9 | 39 |
| H | $CH_3$ | $SCH_3$ | Se | 94.0 | 8.8 | 7.3 | 7.3 | 5.9 | 39 |
| H | $CH_3$ | S | Cl | 59.0 | 5.9 | 10.3 | 7.3 | 11.8 | 39 |
| H | $CH_3$ | Cl | S | 60.0 | 8.8 | 10.3 | 10.3 | 5.9 | 39 |

### C. Carbophosphoryl compounds ($\equiv$CP$<$ and $>$CP$=$)

| Formula | Chemical shift $\delta^{31}P$ (ppm from $H_3PO_4$) | Reference |
|---|---|---|
| $PR_4^+$ | | |
| [MeP(cyclopentadienyl)]$_3$ I$^-$ | 2.25 | 75 |

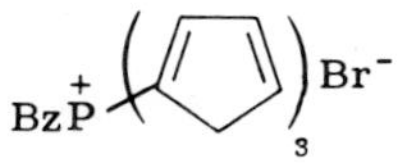

$BzP(\phantom{})_3 Br^-$

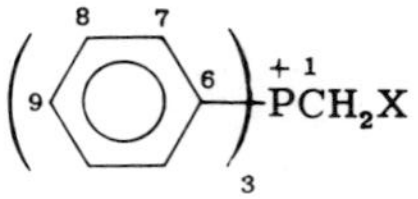

3.9

$(\phantom{})_6^{7}{}^{8}_{9}\!-\!\overset{+1}{P}CH_2X_3$

## Coupling constants $J_{PC}$ (Hz)

| X | C-6 | C-7 | C-8 | C-9 | C-1 | Reference |
|---|---|---|---|---|---|---|
| H | 88.6 | 10.7 | 12.8 | 3.0 | 56.8 | 76 |
| CH$_3$ | 86.0 | 9.7 | 12.5 | 3.0 | 52.6 | 76 |
| CH$_2$CH$_3$ | 85.6 | 10.0 | 12.4 | 2.9 | 49.6 | 76 |
| CH$_2$CH$_2$CH$_3$ | 85.8 | 9.9 | 12.5 | 3.0 | 50.3 | 76 |
| HC=CH$_2$ | 85.9 | 9.9 | 12.5 | 2.8 | 50.2 | 76 |
| C$_6$H$_5$ | 86.0 | 9.7 | 12.5 | 3.0 | 47.7 | 76 |

$CH_3(C_6H_5)_2\overset{+}{P}HC\!=\!CHR$

## Coupling constants $J$ (Hz)

| R | Geometry | Chemical shift $\delta^{31}P$ (ppm from H$_3$PO$_4$) | $^1J_{PC\text{-}1}$ | $^2J_{PC\text{-}2}$ | $^3J_{PC\text{-}3}$ | $^1J_{PC_{ipso}}$ | $^2J_{PC_{ortho}}$ | Reference |
|---|---|---|---|---|---|---|---|---|
| CH$_3$ | Z | 9.9 | 82.6 | 0 | 25.7 | 85.8 | 8.2 | 25 |
|  | E | 16.2 | 87.5 | 0 | 10.1 | 88.1 | 11.0 | 25 |
| C$_6$H$_5$ | Z | 11.6 | 82.6 | 4.5 |  | 88.6 | 11.0 | 25 |
|  | E | 18.4 | 90.0 | 0 |  | 89.5 | 11.0 | 25 |

(*Continued*)

**Appendix III** (*Continued*)

### Coupling constants $J_{PC}$ (Hz)

| X | Y | C-2,4 | C-3 | C-5,9 | C-6,10 | C-7 | X | Y | Reference |
|---|---|---|---|---|---|---|---|---|---|
| CH$_3$ | CH$_3$ | 45.2 | 11.2 | 2.5 | 3.7 | 18.1 | 29.1 | 34.7 | 44 |
| CH$_3$ | Ph | 45.2 | 10.2 | 2.1 | 3.7 | 17.6 | 31.1 | | 44 |
| Ph | CH$_3$ | 45.3 | 10.5 | 3.1 | 3.4 | 22.2 | | 35.9 | 44 |

### Chemical shift $\delta^{31}P$ (ppm from H$_3$PO$_4$)

### Coupling constant $J$ (Hz)

| Formula | | |
|---|---|---|
| | 21 | 77 |

$^1J_{PC} = 47.6; {}^2J_{PC} = 4.3; {}^3J_{PC} = 15.4$

| R | X | | $^1J_{PC-2}$ | $^3J_{PC-4}$ | $^1J_{PR}$ | Reference |
|---|---|---|---|---|---|---|
| CH$_3$ | I | 35.27 | 40.43 | 7.35 | 74.25 | 78 |
| C$_2$H$_5$ | I | 37.39 | 37.63 | 7.35 | 71.37 | 78 |
| C$_6$H$_5$CH$_2$ | Br | 35.21 | 36.80 | 6.44 | | 78 |
| P(X)$_4^-$ | | | | | | |
| P(CN)$_2$X$_2$ | | | | | | |
| $\frac{X}{Br}$ | | 165 | | | | 79 |
| I | | 172 | | | | 79 |

### D. Imidophosphoryl compounds (—NP≡)

| X | Chemical shift $\delta^{31}P$ (ppm from H$_3$PO$_4$) | Reference |
|---|---|---|
| | (C$_6$H$_5$)$_3$P=N—C$_6$H$_4$X | |
| 4-CH$_3$ | 2.60 | 80 |
| 4-C$_6$H$_5$ | 3.19 | 80 |
| 4-F | 3.05 | 80 |
| 4-Cl | 4.00 | 80 |
| 4-Br | 3.79 | 80 |
| 4-I | 3.90 | 80 |
| 4-COOC$_2$H$_5$ | 5.33 | 80 |
| 4-OCH$_3$ | 2.06 | 80 |
| 4-N(CH$_3$)$_3^+$ (anion: I$^-$) | 5.70 | 80 |
| 4-NO$_2$ | 7.45 | 80 |

## APPENDIX IV

Quintuply Connected Phosphorus (>P<)

| Formula | R | Chemical shift $\delta^{31}P$ (ppm from $H_3PO_4$) | Coupling constant $J$ (Hz) | Reference |
|---|---|---|---|---|
| $PF_5$ | | −35.1 | $J_{PF} = 916-1010$ | 5 |
| $PBr_5$ | | −101 | | 5 |
| $P(CH_3)F_4$ | | −29.9, −29.6 | $J_{PF} = 975-967$ | 5 |
| $P(CH_3)_2F_3$ | | 8.0, 8.95 | $J_{PF_{ax}} = 787, 772$; $J_{PF_{eq}} = 975, 960$ | 5 |
| $P(CH_3)_3F_2$ | | −15.8 | $J_{PF} = 541$ | 5 |
| $P(C_6H_5)_5$ | | −89 | | 5 |
| $(C_2H_5O)_3P$ (furan-like) | | −30.1 | | 5 |
| $(C_6H_5)(C_2H_5O)_2P$ (phenanthrene dioxy) | | −16.3 to −18.0 | | 5 |
| $H_5C_6-P$ (spirobifluorene-like) | | −85 | | 5 |
| (catechol-N-R) | $CH_3$ | −26.0 | $J_{PCH_3} = 18.25$ | 81 |
| | Ph | −34.5 | $J_{PNH} = 18.00$ | 81 |

| Structure | R / Ar | $\delta$ | $J$ | Ref. |
|---|---|---|---|---|
| Me(NMe$_2$)P(O-C(CF$_3$)$_2$-C(CF$_3$)$_2$-O) | | −1.5 | | 82 |
| Ph(EtO)P-C(R)(Me)$_2$-C(Me)$_2$-OEt (phosphetane) | H | −36.0 | | 83 |
| | CH$_3$ | −36.0 | | 83 |
| Me$_2$(Me)P-OMe | | −88.0 (−90°C) | $J_{PC_{eq}} = 113.0$ | 84 |
| | | −89.0 (80°C) | $J_{PC_{eq/ax}} = 88.0$ | 84 |
| Ph-P(OCH$_2$CH$_2$N)H (bicyclic) | | −45.7 | $J_{PH} = 694$ | 85 |
| Ar-P(OCH$_2$CH$_2$N)H | Ar = C$_{10}$H$_{11}$ | −46.3 | $J_{PH} = 701$ | 85 |
| Ar-P(OCH$_2$CH$_2$N)H | Ar = p-MeOC$_6$H$_4$ | −49.6 | $J_{PH} = 706$ | 85 |

## APPENDIX V

### Sextuply Connected Phosphorus ($\geqslant P \leqslant$)

| Formula | Chemical shift $\delta^{31}P$ (ppm from $H_3PO_4$) | Coupling constant $J$ (Hz) | Reference[a] |
|---|---|---|---|
| $KPF_6$ | −143.7 | $J_{PF} = 710$ | 67 |
| $NH_4PF_6$ | −145.0 | $J_{PF} = 708$ | 67 |
| $[N(CH_3)_4]^+[PF_6]^-$ | | $J_{PF} = 711$ | 67 |
| $[(C_6H_5O)_4P]^+[PF_6]^-$ | −28.0 −148.0 | $J_{PF} = 720$ | 86 |
| $[(C_2H_5)_2NH_2]^+[C_2H_5PF_5]^-$ | | $J_{PF_{eq}} = 835$ $J_{PF_{ax}} = 710$ | 67 |

benzodioxaphosphole with substituents $X^1, X^2, X^3, X^4$:

| | | | |
|---|---|---|---|
| $X^1 = N_3$; $X^2 = X^3 = X^4 = Cl$ | −157.0 | | 87 |
| $X^1 = X^2 = N_3$; $X^3 = X^4 = Cl$ | −132.0 | | 87 |
| $X^1 = X^2 = X^3 = N_3$ | −143.0 | | 87 |
| $X^1 = X^2 = X^3 = N_3$; $X^4 = Cl$ | −122.0 | | 87 |

[a] Key to references: 1, Issleib and Kummel (1965); 2, Quin and Orton (1979); 3, Eshtiagh-Hosseini et al. (1979); 4, von Weinmaier et al. (1979); 5, see primary references in Crutchfield et al. (1967); 6, Junkes et al. (1972); 7, Dutasta and Robert (1975); 8, Maier (1974); 9, Tolman (1970); 10, Mann et al. (1971); 11, Grim et al. (1967a); 12, Grim and Keiter (1970); 13, Grim and Wheatland (1969); 14, Quin and Somers (1972); 15, Grim and Yankowsky (1977); 16, Tolman et al. (1974); 17, McFarlane (1969); 18, De Ketelaere et al. (1969); 19, Maier (1964); 20, Grim et al. (1974); 21, Grim et al. (1967b); 22, Maier (1972); 23, Grim et al. (1980); 24, Kosovtsev et al. (1971); 25, Duncan and Gallagher (1981); 26, Evangelidou-Tsolis et al. (1974); 27, Gray and Cremer (1974); 28, MacDonell et al. (1978); 29, Quin et al. (1973a); 30, Quin and Stocks (1977); 31, Quin et al. (1973b); 32, Breen et al. (1972); 33, Haemers et al. (1973a); 34, Jones and Coskran (1971); 35, Bentrude and Tan (1976); 36, Pouchoulin et al. (1976); 37, Haemers et al. (1973b); 38, White et al. (1970); 39, Okruszek and Stec (1975); 40, Hutchins et al. (1972); 41, Gray and Nelson (1980); 42, Michalski et al. (1978); 43, Althoff et al. (1978); 44, Gray and Cremer (1972); 45, Quast and Heuschmann (1978); 46, Gray and Cremer (1971) 47, Orton et al. (1979); 48, Wetzel and Kenyon (1974); 49, Majoral et al. (1972); 50, Moedritzer and Miller (1982); 51, Bel'skii et al. (1972); 52, McFarlane and McFarlane (1978); 53, Mosbo and Verkade (1977); 54, Adamcik et al. (1974); 55, Ernst (1977); 56, Buchanan and Morin (1977); 57, Weigert and Roberts (1973); 58, Ramirez et al. (1979); 59, Majoral and Navech (1971); 60, Hall and Malcolm (1972); 61, Plateau et al. (1981); 62, Fossel et al. (1981); 63, Harrison et al. (1975); 64, Roca et al. (1976); 65, Stec et al. (1976); 66, Stec and Lopusinski (1973); 67, Reddy and Schmutzler (1970); 68, Peters and Haegele (1978); 69, Arbuzov et al. (1978); 70, Dutasta et al. (1974); 71, Yoshifuji et al. (1973); 72, Martin and Robert (1981); 73, Stec and Okruszek, (1976); 74, Robert and Weichmann (1978); 75, Allen et al. (1972); 76, Gray (1973a,b); 77, Jenkins et al. (1973); 78, Rampal et al. (1981); 79, Schmidpeter and Zwascha (1979); 80, Bödeker et al. (1980); 81, Malavaud and Barrans (1975); 82, Volkholz et al. (1978); 83, Denney et al. (1972); 84, Schmidbaur et al. (1974); 85, Bonningue et al. (1979); 86, Peake et al. (1971); 87, Skowronska et al. (1979).

## References

Adamcik, R. D., Chang, L. L., and Denney, D. B. (1974). *J. Chem. Soc., Chem. Commun.*, p. 986.
Allen, D. W., Hutley, B. G., and Mellor, T. J. (1972). *J. Chem. Soc., Perkin Trans. 2* pp. 63–67.
Althoff, W., Fild, M., Rieck, H.-P., and Schmutzler, R. (1978). *Chem. Ber.,* **111,** 1845–1856.
Arbuzov, B. A., Erastov, O. A., Khetagurova, S. S., Zyablikova, T. A., and Arshinova, R. P. (1978). *Bull. Acad. Sci. USSR, Div. Chem. Sci., (Engl. Transl.)* **27,** 1682–1685.
Bel'skii, V. E., Kudryautseva, L. A., and Ivanov, B. E. (1972). *J. Gen. Chem. USSR (Engl. Transl.)* **42,** 2421–2427.
Bentrude, W. G., and Tan, H.-W. (1976). *J. Am. Chem. Soc.,* **98,** 1850–1859.
Bödeker, J., Köckritz, P., Köppel, H., and Radeglia, R. (1980). *J. Prakt. Chem.,* **322,** 735–741.
Bonningue, C., Brazier, J. F., Houalla, D., and Osman, F. H. (1979). *Phosphorus Sulfur,* **5,** 291–298.
Breen, J. J., Engel, J. F., Meyers, D. K., and Quin, L. D. (1972). *Phosphorus Relat. Group V Elem.* **2,** 55–59.
Buchanan, G. W., and Morin, F. G. (1977). *Can. J. Chem.* **55,** 2885–2892.
Cox, R. H., and Newton, M. G. (1972). *J. Am. Chem. Soc.,* **94,** 4212–4217.
Crutchfield, M. M., Dungan, C. H., Letcher, L. H., Mark, V., and Van Wazer, J. R. (1967). *Top. Phosphor. Chem.* **5,** 1–457.
De Ketelaere, R. F., Muylle, E., Vanermen, W., Claeys, E., and van der Kalen, G. P. (1969). *Bull. Soc. Chim. Belg.* **78,** 219–227.
Denney, D. B., Denney, D. Z., Hall, C. D., and Marsi, K. L. (1972). *J. Am. Chem. Soc.* **94,** 245–249.
Duncan, M., and Gallagher, M. J. (1981). *Org. Magn. Reson.* **15,** 37–42.
Dutasta, J. P., and Robert, J. B. (1975). *J. Chem. Soc., Chem. Commun.* pp. 747–748.
Dutasta, J. P., Grand, A., Robert, J. B., and Taieb, M. (1974). *Tetrahedron Lett.* pp. 2659–2662.
Ernst, L. (1977). *Org. Magn. Reson.* **9,** 35–43.
Eshtiagh-Hosseini, H., Kroto, H. W., and Nixon, J. F. (1979). *J. Chem. Soc., Chem. Commun.* pp. 653–654.
Evangelidou-Tsolis, E., Ramirez, F., Pilot, J. F., and Smith, C. P. (1974). *Phosphorus Relat. Group V Elem.* **4,** 109–119.
Gorenstein, D. G. (1983). *Progr. NMR Spectrosc.* **16,** 1–98.
Gray, G. A. (1973a). *J. Am. Chem. Soc.* **95,** 5092–5094.
Gray, G. A. (1973b). *J. Am. Chem. Soc.* **95,** 7736–7742.
Gray, G. A., and Cremer, S. E. (1971). *Tetrahedron Lett.* 3061–3064.
Gray, G. A., and Cremer, S. E. (1972). *J. Org. Chem.* **37,** 3458–3469.
Gray, G. A., and Cremer, S. E. (1974). *J. Chem. Soc., Chem. Commun.* pp. 451–452.
Gray, G. A., and Nelson, J. H. (1980). *Org. Magn. Reson.* **14,** 8–13.
Grim, S. O., and Keiter, R. L. (1970). *Inorg. Chim. Acta* **4,** 56–60.
Grim, S. O., and Wheatland, D. A. (1969). *Inorg. Chem.* **8,** 1716–1719.
Grim, S. O., and Yankowsky, A. W. (1977). *Phosphorus Sulfur* **3,** 191–195.
Grim, S. O., Keiter, R. L., and McFarlane, W. (1967a). *Inorg. Chem.* **6,** 1133–1137.
Grim, S. O., McFarlane, W., and Davidoff, E. F. (1967b). *J. Org. Chem.* **32,** 781–784.
Grim, S. O., Briggs, W. L., Barth, R. C., Tolman, C. A., and Jesson, J. P. (1974). *Inorg. Chem.* **13,** 1095–1100.
Grim, S. O., Molenda, R. P., and Mitchell, J. D. (1980). *J. Org. Chem.* **45,** 250–252.

Haemers, M., Ottinger, R., Zimmermann, D., and Reisse, J. (1973a). *Tetrahedron Lett.* pp. 2241–2244.
Haemers, M., Ottinger, R., Zimmermann, D., and Reisse, J. (1973b). *Tetrahedron* **29,** 3539–3545.
Hall, L. D., and Malcolm, R. B. (1972). *Can. J. Chem.* **50,** 2092–2101.
Harrison, J. M., Inch, T. D., and Lewis, G. J. (1975). *J. Chem. Soc., Trans. Perkin 1* pp. 1892–1902.
Hutchins, R. O., Maryanoff, B. E., Albrand, J. P., Cogne, A., Gagnaire, D., and Robert, J. B. (1972). *J. Am. Chem. Soc.* **94,** 9151–9158.
Issleib, K., and Kümmel, R. (1965). *J. Organomet. Chem.* **3,** 84–91.
Jenkins, R. N., Freedman, L. D., and Bordner, J. (1973). *J. Cryst. Mol. Struct.* **3,** 103–114.
Jones, C. E., and Coskran, K. J. (1971). *Inorg. Chem.* **10,** 1536–1537.
Junkes, P., Baudler, M., Dobbers, J., and Rackwitz, D. (1972). *Z. Naturforsch., B: Anorg. Chem., Org. Chem., Biochem., Biophys., Biol.* **27B,** 1451–1456.
Kosovtsev, V. V., Timofeeva, T. N., Ionin, B. I., and Chistokletov, V. N. (1971). *Zh. Obshch. Khim.* **41,** 2638–2642; *J. Gen. Chem. USSR (Engl. Transl.)* **41,** 2671–2675.
Macdonell, G. D., Berlin, K. D., Baker, J. R., Ealick, S. E., van der Helm, D., and Marsi, K. L. (1978). *J. Am. Chem. Soc.* **100,** 4535–4540.
McFarlane, H. C. E., and McFarlane, W. (1978). *J. Chem. Soc., Chem. Commun.* pp. 531–532.
McFarlane, W. (1969). *Org. Magn. Reson.* **1,** 3–9.
Maier, L. (1964). *Helv. Chim. Acta.* **47,** 2137–2140.
Maier, L. (1972). *In Organic Phosphorus Compounds* (G. M. Kosolapoff and L. Maier, eds.), Vol. **1,** p. 156. Wiley (Interscience), New York.
Maier, L. (1974). *Phosphorus* **4,** 41–47.
Majoral, J.-P., and Navech, J. (1971). *Bull. Soc. Chim. Fr.* pp. 2609–2612.
Majoral, J.-P., Pujol, R., and Navech, J. (1972). *Bull. Soc. Chim. Fr.* pp. 606–610.
Malavaud, C., and Barrans, J. (1975). *Tetrahedron Lett.* **35,** 3077–3080.
Mann, B. E., Masters, C., Shaw, B. L., Slade, R. M., and Stainbank, R. E. (1971). *Inorg. Nucl. Chem. Lett.* **7,** 881–885.
Martin, J., and Robert J. B. (1981). *Org. Magn. Reson.* **15,** 17–93.
Michalski, J., Mikolajczak, J., Pakulski, M., and Skowronska, A. (1978). *Phosphorus Sulfur* **4,** 233–234.
Moedritzer, K., and Miller, R. (1982). *J. Org. Chem.* **47,** 1530–1534.
Mosbo, J. A., and Verkade, J. G. (1977). *J. Org. Chem.* **42,** 1549–1555.
Okruszek, A., and Stec, W. J. (1975). *Z. Naturforsch., B: Anorg. Chem., Org. Chem.* **30B,** 430–436.
Orton, W. L., Mesch, K. A., and Quin, L. D. (1979). *Phosphorus Sulfur* **5,** 349–357.
Peake, S. C., Fild, M., Hewson, M. J. C., and Schmutzler, R. (1971). *Inorg. Chem.* **10,** 2723–2727.
Peters, W. and Hagele, G. (1978). *Phosphorus Sulfur* **4,** 149–153.
Plateau, P., Mayaux. J.-F., and Blanquet, S. (1981). *Biochemistry* **20,** 4654–4662.
Pouchoulin, G., Llinas, J. R., Buono, G., and Vincent, E. J. (1976). *Org. Magn. Reson.* **8,** 518–521.
Quast, H., and Heuschmann, M. (1978). *Angew. Chem., Int. Ed. Engl.* **17,** 867–868.
Quin, L. D., and Orton, W. L. (1979). *J. Chem. Soc., Chem. Commun.* pp. 401–402.
Quin, L. D., and Somers, J. H. (1972). *J. Org. Chem.* **37,** 1217–1222.
Quin, L. D., and Stocks, R. C. (1977). *Phosphorus Sulfur* **3,** 151–156.
Quin, L. D., Borleske, S. G., and Engel, J. F. (1973a). *J. Org. Chem.* **38,** 1858–1866.
Quin, L. D., Borleske, S. G., and Stocks, R. C. (1973b). *Org. Magn. Reson.* **5,** 161–162.

Ramirez, F., Prasad, V. A. V., and Maracek, J. F. (1979). *J. Am. Chem. Soc.* **96**, 7269–7275.
Rampal, J. B., Macdonell, G. D., Edasery, J. P., Berlin, K. D., Rahman, A., van der Helm, D., Pietrusiewicz, K. M. (1981). *J. Org. Chem.* **46**, 1156–1165.
Reddy, C. S., and Schmutzler, R. (1970). *Z. Naturforsch., B: Anorg. Chem., Org. Chem., Biochem., Biophys., Biol.* **25B**, 1199–1214.
Robert, J. B., and Weichmann, H. (1978). *J. Org. Chem.* **43**, 3031–3035.
Roca, C., Kraemer, R., Majoral, J.-P., Navech, J., Brault, J. P., and Savignac, P. (1976). *Org. Magn. Reson.* **8**, 407–412.
Schmidbaur, H., Buchner, W., and Kohler, F. H. (1974). *J. Am. Chem. Soc.* **96**, 6208–6210.
Schmidpeter, A., and Zwaschka, F. (1979). *Angew, Chem., Int. Ed. Engl.* **18**, 411–412.
Skowronska, A., Pakulski, M., and Michalski, J. (1979). *J. Am. Chem. Soc.* **101**, 7412–7413.
Stec, W. J., and Lopusinski, A. (1973). *Tetrahedron* **29**, 547–551.
Stec, W. J., and Okruszek, A. (1975). *J. Chem. Soc., Perkin Trans. 1* pp. 1828–1832.
Stec, W. J., Okruszek, A., Lesiak, K., Uznanski, B., and Michalski, J. (1976). *J. Org. Chem.* **41**, 227–233.
Tolman, C. A. (1970). *J. Am. Chem. Soc.* **92**, 2956–2965.
Tolman, C. A., Meakin, P. Z., Linder, D. L., and Jesson, J. P. (1974). *J. Am. Chem. Soc.* **96**, 2762–2774.
Volkholz, M., Stelter, O., and Schmultzler, R. (1978). *Chem. Ber.* **111**, 890–900.
Weigert, F. J., and Roberts, J. D. (1973). *Inorg. Chem.* **12**, 313–316.
von Weinmaier, J. H., Luber, J., Schmidpeter A., and Pohl, S. (1979). *Agnew. Chem.* **91**, 442–443.
Wetzel, R. B., and Kenyon, G. L. (1974). *J. Am. Chem. Soc.* **96**, 5189–5198.
White, D. W., Bertrand, R. D., McEwen, G. K., and Verkade, J. G. (1970). *J. Am. Chem. Soc.* **92**, 7125–7135.
Yoshifuji, M., Nakayama, S., Okazaki, R., and Inamoto, N. (1973). *J. Chem. Soc., Perkin Trans. 1* **19**, 2065–2068.

# Index

## A

Acceptor tRNAs, spectral comparisons of, 268–275, *see also* tRNA
Acetyl-CoA synthetase, $^{31}$P-NMR spectra of, 189
G-Actin, high-affinity binding sites of, 83–84
Actinomycin D, intercalation model for, 302
Active-site
  labeling, 130
  serine residue, 130
Acyclic phosphites, $^{31}$P chemical shifrs of, 18
Acyl migration
  lysophospholipids and, 438–440
  pH dependence of, 441
Additive constants, examples of, 24
Adenosine diphosphate, $^{31}$P-NMR shifts of, 114, *see also* ADP
Adenosine triphosphate, *see also* ATP; ATPase
  glycolytic enzymes in synthesis of, 92
  $^{31}$P-NMR spectra of, 114
Adenylate cyclase, stereochemical consequences of, 220–224
Adenylate kinase, 74
  computer calculation of chemical-exchange effects of, 99–100
  $^{31}$P-NMR measurements of, 92
  reaction, 90–92
Adenylyl-3′,5′-adenosine, $J_{3'P}$ coupling constants for, 51
β-P(ADP) signals, merging to single peak, 92
β-P(ADP)⇌β-P(ATP) exchange, 89
α-P(MgADP) resonance, 81, 91–92
  for E·MgADP·NO$_3^-$·creatine complexes, 71
ADPβS, chiral thiophosphates and, 176

Alkaline phosphatase, $^{31}$P-NMR spectra of, 124
Alternating B-DNA, sequence dependence of, 245
Amniotic fluid, $^{31}$P-NMR diagnosis of, 540
Anistropic Brownian diffusion, simulated $^{31}$P NMR of, 453
Arginine kinase, 71–73
  computer simulation of $^{31}$P-NMR spectra of, 97
  $^{31}$P-NMR spectra of, 87–90
Aryl esters, three-bond coupling constants of, 48
Aspartate aminotransferase, 135
*Aspergillus niger*, 137
β-P(ATP) chemical shift, 87–88
β-P(ATP) resonance, 95, 98
β-P(ATP) signals, linewidth of, 89
($S_p$)-ATPαS, $^{17}$O-labeled AMPS conversion to, 190
($R_p$)ATPβS, P$_\beta$ signals of, 193–194
β-P(MgATP) resonances, 71
ATPases, membrane embedded, 126
Autocorrelation function, for phosphorus–proton dipolar interactions, 320
*Azotobacter vinelandii*, 137

## B

β,γ-bridging ($^{18}$O) oxygen, 108
Biomembranes, $^{31}$P-NMR spectra for, 461–472
Biophosphate molecules, chiral carbon center in, 184

Bloch equation, spin–spin interactions and, 63–64
Bond-angle effects, in $^{31}$P chemical shifts in nontetracoordinated phosphorus compounds, 11–20, 23
Bovine liver tRNA, $^{31}$P-NMR spectra of, 267
Brain, $^{31}$P-NMR studies of, 534–537
Breathing, of double-stranded nucleic acids, 370–371
*Brevibacterium liquefaciens* enzyme, adenylate cyclase from, 222–225
Brownian diffusion, simulated $^{31}$P-NMR spectra for, 453

## C

Carbamoyl-phosphate synthetase, relaxation data for, 165
Carbon-13–phosphorus-31 coupling constants, 50
Carbonyl compound, reaction with *N*-phenyl phosphoramidates, 205–210
Casein, 106, 140–141
Casein spin–spin lattice relaxation times, 343
Cerebrospinal fluid, $^{31}$P-NMR analysis of, 542–543
Charcot–Marie–Tooth disease, 516
Chemical-exchange effects, line shapes and, 62
Chemical shielding tensor, in solid-state $^{31}$P NMR, 403–408
Chemical-shift anisotropy mechanism
  for nonbridging substitutions, 115–117
  nuclear spin and, 403–404
  for nucleic acids, 373
  of ordered systems, 449–452
  for $^{31}$P relaxation, 352–353, 356–357, 362
  in spin–lattice and spin–spin relaxation, 66
Chemical shifts
  doubly connected phosphorus, 550
  quadruply connected phosphorus, 563–585
  quintuply connected phosphorus, 586–587
  for representative model compounds, 115
  sextuply connected phosphorus, 588
  and shielding by electron cloud, 113–116
  triply connected phosphorus, 561–562
Chicken erythrocyte DNA, $^{31}$P-NMR spectra of, 240–241
Chiral carbon center, in biophosphate molecules, 185

Chiral *meso*-[$^{17}$O]hydrobenzoin, 205
Chiral [$^{16}$O,$^{17}$O,$^{18}$O]phosphate esters
  in enzymatic phosphoryl transfer reactions, 199–230
  phosphate diesters and, 205–210
Chiral phosphate monoesters, configurational analysis of, 191
Chiral [$^{16}$O,$^{18}$O]phosphodiesters, first synthesis of, 205
Chiral thiophosphates, 175–195
  biochemical applications of, 177
  configurational analysis of, 191–192
  $^{31}$P chemical shifts of diasteromeric pairs of, 186
  $^{31}$P-NMR analysis of ($R_p$)-ATPβS derived from, 195
  stereochemical problems with, 175–177
Chromatin, DNA relaxation in, 390–391
Chromium(III), as paramagnetic probe, 158
Chromium(III)–nucleotide complexes, studies with, 166–169
Chromium(III)– and cobalt(III)–nucleotide complexes, 156–159
  properties of, 157–160
  studies with, 169–172
Cr(H$_2$O)$_4$ ATP complexes, 157
CMC, *see* Critical micelle concentrations
Cobalt(III)–nucleotide complexes, studies with, 160–166
Co(NH$_3$)$_4$ ATP complexes, 157
  relaxation data for, 165
β,γ-Co(NH$_3$)$_4$ ATP complex, $^{31}$P-NMR spectrum of, 157
Cobra venom phospholipase, A$_2$, 428, 448
Coenzyme analogs, $^{31}$P-NMR spectra of, 78
Collagen vascular disease, 517
Configurational analysis, stereochemical studies using, 221
Conformation problems, solution of, 317–318
Correlated spectroscopy experiment, 484–485
COSY experiment, 485
Coupling constants
  doubly connected phosphorus, 550
  phosphoproteins and, 116
  quadruply connected phosphorus, 563–585
  quintuply connected phosphorus, 586–587
  sextuply connected phosphorus, 588
  triply connected phosphorus, 551–562
Covalently attached coenzymes, 109–111
Cow's milk caseins, phosphoproteins in, 342

Creatin kinase, 71–73
  computer simulation of $^{31}$P-NMR spectra of, 97
  $^{31}$P-NMR spectra of, 87–90
Critical micelle concentrations, determination of, 435–538, *see also* Phospholipids in micelles
Cross-polarization–dipole decoupling, 402–403
CSA mechanism, see Chemical-shift anisotropy mechanism
Cyclic [$^{16}$O,$^{18}$O]AMP diastereomers, synthesis of, 207
Cyclic dAMP, enzymatic and nonenzymatic hydrolyses of, 227
Cyclic hydrobenzoin triesters, hydrogenolysis of, 204–205
Cyclic nucleotide phosphodiesterase, oxygen chiral study of, 224–227
Cyclic phosphoramidates, hydrolysis and hydrogenolysis of, 203–204, 208

# D

Deoxyribonucleic acid, *see* DNA
Diadenosine pentaphosphate, 74
Diastereomeric phosphoramidates, 22
DIFP, *see* Diisopropyl fluorophosphate
Dihydrofolate, reduction to tetrahydrofolate, 77
Dihydrofolate reductase, 77–81, 148
Diisopropyl fluorophosphate, 108, 130
Diisopropylphosphoryl derivatives of serine proteases, $^{31}$P-NMR titration data for, 130
Dioleoylphosphatidylcholine, simulated $^{31}$P-NMR spectra for, 456
Dipolar relaxation, nucleic acids and, 351–366
Dipole–dipole interaction, and correlated motion of magnetic dipoles, 319–320
Dipole–dipole interaction-mediated populations, nuclear Overhauser effect and, 322–323
Diseased human muscles
  GPC and, 512–514, 520–523
  phosphate profiles of, 521
Diseased human subjects, topical magnetic resonance method of study of, 524–525
Diseased-state identifications, by $^{31}$P-NMR, 511–543

DNA
  A form of, see A-DNA *below*
  B form of, see B-DNA *below*
  binding of, 379
  calf thymus, 380–381
  in chromatin and nucleosomes, 390–391
  daunorubicin and, 397
  double-stranded, *see* Double-stranded DNA
  dynamic behavior of in solution, 255–256
  ethidium and, 397
  nucleosomal, 256–258
  orientation of phosphodiester groups in, 416
  quinacrine and, 397
  relaxation studies of, 377–386
  sodium DNA, 410–414
    B form in, 414
  tetralysine and, 397
  torsional deformation models of, 382
  two-state geometric parameters for, 337
  X-ray fiber diffractions and, 401
DNA backbone
  plausible motional models for $^{13}$C and $^{31}$P relaxation in, 383
  torsion angles, 372, 384
DNA complexes
  filamentous virus as, 419
  studies of, 418–421
DNA conformations, 233–260
  biological and genetic significance of, 259–260
  hydration effects on, 409–410
DNA helix axis, D coordinate frame and, 366
DNA restriction fragments, $^{31}$P spin–lattice relaxation times for, 336
DNA structure, X-ray fiber diffraction patterns and, 401, 408–409
DNA synthesis, inhibition of, 299
A-DNA, 410–414
  orientation of shielding tensor of, 412–413
  phosphodiester backbone of, 413–414
  sequence dependence in, 249
B-DNA, 414–418
  base-sequence dependence of backbone configuration in, 417–418
  sequence dependence of, 240–245
C-DNA, sequence dependence of, 246–249
Z-DNA
  in gene regulation, 260
  sequence dependence of, 250–253

Double helix, stabilization of by ethidium ion, 306–310
Double-stranded DNA, *see also* Double-stranded nucleic acids
 conformational transitions of, 253–255
 molecular motions in, 377
 phosphorus relaxation times for, 334–341
 salt-induced transitions in, 253–255
 sequence dependence in, 239–253
Double-stranded nucleic acids
 anisotropic diffusion model of, 372
 breathing of, 370–371
 probable motions in, 369–370
Double-stranded polynucleotides, relaxation studies of, 386–387
$d_\pi$–$p_\pi$ bond-order increase, $d$-orbital occupation increase and, 25
*Drosophila melanogaster*, polytene chromosomes of, 259
Drug–double helical nucleic acid complexes, 300–310
Drug–nucleic acid complexes, $^{31}$P-NMR spectra of, 299–315
Duchenne muscular dystrophy, 512, 529

## E

Electron closed shielding, chemical shifts and, 113–116
Environmental effects, on $^{31}$P chemical shifts, 27–31
Enzymatic catalytic intermediates, in phosphoryl-transferring reactions, 107–108
Enzymatic catalysis, structure and mechanism in, 155
Enzymatic phosphoryl-transfer reactions, chiral [$^{16}$O,$^{17}$O,$^{18}$O]phosphate esters in stereochemical course of, 199–230
Enzyme·AP$_5$A complex, $^{31}$P-NMR spectra of, 74–75
Enzyme·MgADP·NO$_3^-$·arginine complex, 71
Enzyme·MgADP·creatine complex, $^{31}$P-NMR spectra of, 71–72
Enzyme·MgATP·arginine complex, 89
Enzyme·NADP$^+$·methotrexate, 80
Enzyme-bound equilibrium mixtures, $^{31}$P-NMR spectra of, 87–96
Enzyme-bound substrates, equilibrium constant for interconversion of, 95
Enzyme-complex $^{31}$P-NMR studies, theoretical considerations in, 59–67

Enzyme complexes, 57–101, *see also* Enzyme systems
 experimental results in, 67–69
 line shapes and chemical exchange effects in, 61–64
 paramagnetic probes of with phosphorus-containing compounds, 155–172
 $^{31}$P-NMR spectral parameter changes in, 59–61
 relaxation times for, 64–67
 spin–spin couplings in, 60
Enzyme–phosphorus complex, linewidth of, 124–125
Enzyme·trimethylprim, 80
Enzymes, active sites of, 144–145
Enzyme systems, *see also* Enzyme complexes
 metabolites, coenzymes, and inhibitors in, 69–86
Equilibrium mixture, $^{31}$P-NMR spectra of, 91–94
*Escherichia coli*
 glutamine synthase, 140
 glutamine synthetase, active site of, 168
 succinyl CoA synthetase, $^{31}$P-NMR spectra of, 119, 128
 tRNA$^{Tyr}$
  chemical-exchange line-broadening effects in, 286
  cloverleaf model of, 274
  $^{31}$P-NMR spectra of, 270
Ethidium, DNA complexing with, 397
Ethidium ion
 double-helix stabilization by, 306–310
 in poly(A)·oligo(U) complex, 301
 structural interpretation of effects of on tRNA, 311–315
Eukaryotic DNA sequences, in gene expression, 259–260
Eye, $^{31}$P-NMR study of, 537–540

## F

FAD, *see* Flavin adenine dinucleotide
fd, *see* Filamentous virus
Fermi-contact contribution, in $^1J_{PC}$ coupling constants, 39
Filamentous virus, as DNA complex, 419
Flavin adenine dinucleotide, 109–110
Flavin mononucleotide, 109, 137
Flavodoxins, 136–137
Flavoproteins, 136–137

Index 597

FMN, see Flavin mononucleotide
Fourier-transform NMR technology, 58, 106
Free induction decay, cross-polarization and, 403

**G**

Gene expression, eukaryotic DNA sequences in, 259–260
Gene regulation, Z-DNA in, 260
Glucose 1-phosphate, glycogen breakdown to, 131
Glutamate-43, nucleophilic attack of, on 5'-phosphorus atom of substrate, 229
Glutamine synthetase, 140
Glycerate kinase, 73–74
rac-Glycerol 3-phosphate, $^{31}$P-NMR spectra of, 86
Glycogen, breakdown to glucose 1-phosphate, 131
Glycogen phosphorylase, 131–135, 139
Glycolytic enzymes, in ATP manufacture, 93
GPC (sn-Glycerol-3-phosphorylcholine), in human muscle, 512–514, 520–523
Guanosine 3'-monophosphate, 85

**H**

Halobacterium cutirubrum, purple membrane of, 466–467
Hamiltonian, for nucleic acid interactions with fluctuating fields, 350
Hartman–Hahn condition, 402
Heart, $^{31}$P NMR of, 530–532
Helix–coil transitions
 for DNA, 237–239
 for RNA, 235–237
Helix twisting, dynamics of, 369
Heteronuclear dipolar interaction, in spin–lattice and spin–spin relaxation, 66
Heteronuclear $^{31}$P–$^{1}$H chemical-shift correlated spectra, 497–506
High-resolution $^{31}$P NMR, of tRNAs, 265–294
Histidine residues, 85
Histones, 139–140
Homonuclear chemical-shift correlated spectra, 492–497
Homonuclear dipole interactions, in spin–lattice and spin–spin relaxation, 66

Hybridization, in variation of $^{1}J_{PH}$ coupling constants, 38–39
Hydrogen bonding, chemical shifts and, 113

**I**

Imidazole ring, electrophilic phosphorus atom and, 108
Intercalating drug–double-helical polynucleic acid complexes, 301–310
International Union of Pure and Applied Chemistry convention, 7
Isoalloxazine ring, 109

**J**

Jeener pulse sequence, 485–488
Jump models, nucleic acids and, 363–366

**K**

Karplus-like relationship, between HCOP and CCOP dihedral angles and three-bond coupling constants, 50
Kidney, $^{31}$P NMR in evaluation of, 533
Kinases
 in catalysis of reversible transfer of δ-P of ATP, 69–70
 chemical shifts and coupling constants of nucleotides bound to, 70–71
 magnesium ion and, 70

**L**

Lactobacillus casei, 77
 dihydrofolate reductase, 148
Laocoön-type spectral-simulation program, 48
Lipid-containing viruses, in DNA complex, 420
Lipid polymorphism, 461–464
Lysophospholipids
 acyl and phosphoryl migration in, 438–442
 in monomeric and micellar states, 424

**M**

MgAMP·PNP complex, temperature-dependent equilibrium in, 81
Magnesium dependence of tRNA$^{Phe}$ $^{31}$P spectra, 280
Mammalian fluids, $^{31}$P-NMR studies of, 540–543
Manganese ions, and tRNA$^{Phe}$ $^{31}$P spectra, 280

Membrane-active agents, effects of, 468–470
Membrane lipid, motional states of phosphodiesterase moiety of, 450–452
Membranes
  phospholipids in, see Phospholipids in membranes
  properties of, 447–448
Messenger ribonucleoprotein, RNA relaxation in, 391
Metal-ion binding
  phosphoamino acids in, 140–144
  proteins, 112
  sites, 145–148
Metal–metal distances, for various complexes with pyrophosphatase, 171–172
Met–AMP complex
  chemical-shift changes for, 77
  resonance of, 95
Methionyl-tRNA synthetase, 76–77, 94, 96
Methylene groups, $^{31}$P chemical shifts as function of bridging in, 18
Methyl ester mixture, $^{31}$P-NMR spectrum of, 229
Methyl esters, $^{31}$P-NMR spectra of, 219
Methyl phosphonium ion, $^1J_{PC}$ coupling constants for, 40
Micelles, phospholipids in, see Phospholipids in micelles
Molecular motions, of nucleic acids, 353–368
Monomeric phospholipids, micellization of by detergents, 435–438
Motional rates, for phospholipids in membranes, 450–452
Muscle
  contractile state, regulation of, 138
  diseased, see Diseased human muscles
  phosphates, $^{31}$P-NMR parameters of, 519
  protein-synthesis activity in, 517
  proteins, 138
  skeletal, see Skeletal muscle
Myocardial infarction, $^{31}$P-NMR assessment of, 532
Myosin, 138
  subfragment-1, ATPase reaction catalyzed by, 80–81
Myotonic dystrophy, 517

## N

NADP$^+$ (Nicotinamide adenine dinucleotide phosphate), $^{31}$P-NMR spectra of, 78

NHDP$^+$ (Nicotinamide hypoxanthine dinucleotide phosphate), 79
NMR, see Nuclear magnetic resonance
NOE, see Nuclear Overhauser effect
Nonenzymatic $S_N2(P)$ reactions, complexity of, 199–200
Nuclear magnetic resonance
  first detection of, 1
  Fourier-transform, see Fourier-transform NMR
  phosphorus-31, see Phosphorus-31 NMR
Nuclear magnetic resonance line shapes, computer calculation of, 96–100
Nuclear magnetic resonance spectrometers
  development of, 1–2
  high-field superconducting, 2
  signal-averaging Fourier-transform, 2
Nuclear magnetic resonance spectroscopic information, area embraced by, 2–3
Nuclear Overhauser effect, 319, 322–328, 353, 355, 357–360, 373–377, 385, 388–389
  in diseased skeletal muscle, 520–522
  for nucleic acids, 351, 359
  in phospholipids in micelles, 431–433
  single-frequency, 324–325
  for skeletal muscle extract phosphates, 517
Nucleic acid–drug complexes, relaxation studies in, 395–397
Nucleic acid phosphorus, DD vector orientation of, 372
Nucleic acid–protein systems, relaxation studies for, 390–396
Nucleic acids, see also DNA; RNA; tRNA
  conformation and dynamics of, 317–344
  distribution of correlation times for, 359–360
  double-stranded, see Double-stranded nucleic acids
  dynamics of, 373–376
  electrostatic mode of action in, 310
  free internal diffusion with overall isotropic orientation, 354–359
  jump models and, 363–366
  lattice models of, 366
  molecular dynamics–derived correlation function models and, 367–368
  non-model, 367
  nuclear Overhauser effect for, 351, 359
  pharmacological activity of, 299
  phosphodiester groups in, 421

Index 599

phosphorus atoms in, 421
$^{31}$P-NMR chemical shifts in, 299
random isotope motion of, 354
relaxation behavior of, 349–397
relaxation parameters of, 350–353
restricted internal diffusion of, 360–363
and rotational tumbling of a cylinder, 359
spectral densities and molecular motions of, 353–368
structure and molecular motions in, 368–372
Nucleoside [α-$^{16}$O,$^{18}$O]diphosphates, 215–217
Nucleoside triphosphate–nucleoside diphosphate exchange, 98–99
Nucleosomal DNA, 256–258, see also DNA
Nucleosomes, DNA relaxation in, 390–391
Nucleotide coupling constant, 50
Nucleotidyl cyclases, stereochemistry of, 222
Nucleotidyl polymerases, stereochemistry of, 220–222

## O

Oligodeoxynucleotide duplexes, temperature dependence for, 237
Oligonucleotides
 phosphodiester as fundamental unit of, 317
 phosphorus relaxation times for, 328–334
One-bond P–X coupling constants, 38
One-bond spin–spin coupling constants, 39
One-displacement reaction, covalent intermediate and, 230
Ordered systems
 $^{31}$P NMR of, 449–457
 types of phases in, 454–457
Oriented DNA fibers, studies of, 408–418
Ovalbumin
 discovery of, 105
 in metal-ion binding, 143–144
 $^{31}$P-NMR spectra of, 143
Overhauser effect, see Nuclear Overhauser effect
Oxygen chiral ester approach, to stereochemical studies, 199–230
Oxygen chiral nucleotide ester, nuclease-catalyzed hydrolysis of, 228
Oxygen chiral phosphate esters
 configurational analysis of, 210–220
 phosphate monoesters and, 202–205
 stereochemical studies using, 221
 synthesis of, 210

Oxygen-17 quadrupolar effects, 181–185
Oxygen-18, shielding effect of, 211
Oxygen-18 isotope shifts, 178–181

## P

$p$-orbital unbalancing term, $^{31}$P chemical shifts and, 26
$^{31}$P, see Phosphorus-31
PADPR-OMe (2′-Phosphoadenosine-5′-diphosphoribose), 79
Paramagnetic probes, of enzyme complexes, 155–172
PAS, see Principal-axis system
$N$-Phenyl phosphoramidates, reaction with carbonyl compounds, 205–207
Phosphate diesters, 205–210
[$^{16}$O,$^{18}$O]Phosphate diesters, 212–215
Phosphate ester POCH dihedral angles, 45
Phosphate esters, configurational analysis of, 47–50
Phosphate monoesters, 202–205
[$^{16}$O,$^{17}$O,$^{18}$O]Phosphate monoesters, 217–220
Phosphates
 oxygen isotope enrichment of, 106
 PCCH dihedral angles in, 46
 in skeletal muscle, 517–521
Phosphatidic acid, 442
Phosphatidylglycerophosphate, $^{31}$P-NMR spectra of, 466
Phosphines
 correlations for, 24
 $^{1}J_{PC}$ coupling constants for, 40
 $^{31}$P chemical shifts of, 16
Phosphites
 acyclic, 18
 POCH dihedral angles in, 45
Phosphoamino acids
 enzymes carrying, 108
 in metal-ion binding, 140–144
Phosphodentine protein, 141–142
Phosphodiester backbone, of A-DNA, 413–414
Phosphodiester group, conformation of, 317
Phosphodiesters
 as fundamental unit of oligonucleotides, 317
 $^{31}$P resonances in, 237
 in skeletal muscle, 513–515
Phosphoenolpyruvate-dependent phosphotransferase system, 126–127
Phosphofructokinase deficiency, 527–528
Phosphoglucomutase, 125–126

3-Phosphoglycerate kinase, 73–74, 93–94
Phosphohistidine, 108
$N^3$-Phosphohistidine HPr, 127
Phospholipase $A_2$, 442–443
Phospholipases, specificity and kinetics of, 442–444
Phospholipid chemical shift, pH dependence of, 430
Phospholipids
  in aqueous solution with Triton X-100, 429
  in chloroform–methanol and analogs, 427
  icellization of by detergents, 435–438
  in potassium cholate, 426
Phospholipids in membranes
  biomembranes and, 461–472
  membrane-active agents and, 468–470
  obtaining spectra for, 457–460
  $^{31}$P NMR of, 447–452
  relaxation times for, 470–472
Phospholipids in micelles
  nuclear Overhauser effect and, 431–433
  $^{31}$P NMR of, 423–444
  quantitative analysis of, 431–433
  solubilization by detergents, 433–435
  spectral characteristics of, 425–431
  spin–lattice relaxation times and, 431–433
Phosphomonoesters
  conformational parameters of, 318
  hydrolysis of, 190
Phosphonium ions, $^1J_{PC}$ coupling constants for, 40–42
Phosphoproteins
  classification of, 106–112
  conformation and dynamics of, 317–344
  as control mechanisms, 341
  coupling constants for, 116
  phosphorus relaxation methods for, 341–344
  pH titration of, 118–120
  $^{31}$P-NMR parameters measured for, 146–147
  $^{31}$P-NMR studies of, 105–149
  relaxation behavior of, 116–118
Phosphorinane esters, bond angles in, 23
Phosphorothioates, hydrolysis of, 208–210
Phosphorus, first $^{31}$P-NMR signals of, 1
Phosphorus atom, deshielding of, 15
Phosphorus bond, electronegativity of, 114
Phosphorus–carbon bonding orbital, 39
Phosphorus-containing compounds, paramagnetic probes of enzyme complexes with, 155–172

Phosphorus–hydrogen coupling constants, 41–42
Phosphorus magnetic relaxation measurements, 317–344
Phosphorus nucleus, spin ½ of, 113
Phosphorus–oxygen–carbon–hydrogen bonds, pyrophosphate $^{31}$P nuclei and, 78
Phosphorus–proton dipolar interactions, autocorrelation function for, 320
Phosphorus–proton nuclear Overhauser effect, 322–328
Phosphorus relaxation methods, 317–344
  experimental observations made by, 319
  for phosphoproteins, 341–344
Phosphorus relaxation times, 328–344
  for double-stranded DNA, 334–341
  joint conformation model, 330–334
  for oligonucleotides, 328–334
Phosphorus signal, shielding of, 25
Phosphorus-31, deshielding of, 26
Phosphorus-31 chemical shifts, 7–33
  of acyclic phosphites, 18
  and asymmetry in charge distribution, 9
  bond-angle effects in, 13–20, 23
  of diastereomeric pairs in chiral thiophosphates, 186
  in diastereomeric phosphoramidates, 22
  $d_\pi$–$p_\pi$ bond-order increase and, 25
  extrinsic effects in, 27–31
  as a function of bridging methylene groups, 18
  hydrogen-bonding donors and, 31–32
  intrinsic effects in, 31–32
  for model phosphate diester, 12
  in nucleic acids, 299–300
  $p$-orbital unbalancing term and, 26
  of phosphines, 16
  in phosphoryl compounds, 10–14
  polar and resonance effects in, 23–27
  salt effect on, 28
  stereoelectronic effects in, 19–23
  stereoelectronic origin of, 281–283
  of substituted phenylphosphoric dichlorides, 24
  temperature dependence of, 237, 271–273
  temperature effect on, 29–30
  torsion-angle effect in, 19–20, 23
Phosphorus-31 chemical-shift parameters, 17
Phosphorus-31 chemical-shift theory, applications of, 9–32

# Index

Phosphorus-31 chemical-shift titration curve, 32
Phosphorus-31 NMR
  of brain, 534–537
  configurational information from, 211
  in diseased state identifications, 511–543
  of eye, 537–540
  of heart, 530–532
  of kidney, 533
  of mammalian fluids, 540–543
  of phospholipids in membranes, 447–452
  of phospholipids in micelles, 423–444
  solid-state, see Solid-state $^{31}$P NMR
  two-dimensional, 479–510
Phosphorus-31 NMR spectra
  chemical shifts measured in, 113–116
  of ordered systems, 449–457
Phosphorus-31 NMR spectroscopic probes, theory of, 7
Phosphorus-31 NMR spectroscopy
  of DNA and RNA secondary structures, 233–260
  expansion of, 2
Phosphorus-31 NMR studies
  analog use in, 121–122
  of enzyme complexes, 57–101
  of phosphoproteins, 105–149
Phosphorus-31 shielding, for gauche conformation, 22
Phosphorus-31 spectral differences
  of acceptor tRNAs, 268–275
  structural bases for, in tRNA species, 273–275
Phosphorus-31 spectral perturbations, on drug binding, 300
Phosphorus-31 spin–spin coupling constants, 37–51
  directly bonded, 38–42
  three-bond, 43–51
  two-bond, 42–43
  utility of, 47–48
Phosphoryl compounds, $^{31}$P chemical shifts in, 10–14
Phosphoryl group, 106
Phosphoryl migration, lysophospholipids and, 438–440
Phosphoryl moiety, of prosthetic groups, 131–137
Phosphoryl transfer enzymes (kinases), 69–76

Phosphoryl-transferring reactions, enzyme–phosphoryl intermediate in, 107
Phosphoserine subspectra, simulations of, 504–505
Phosvitin, 141–142
POCH, see Phosphorus–oxygen–carbon–hydrogen bonds
Poly(A)·oligo(U) (1:1), $^{31}$P-NMR spectra for, 303
Poly(A)·oligo(U)·ethidium complex, 301–308
Poly(dAdT)·poly(dAdT)
  B-to-C transition in, 253
  ethanol or B-to-A transition in, 255
  peaks for fibers of, 418
  $^{31}$P-NMR spectra of, 237–238, 240
Polydeoxynucleotides, temperature dependence of, 237
Poly(dG-m$^5$dC)·poly(dG-m$^5$dC)
  B-to-Z transition in, 252, 254
  $^{31}$P-NMR spectra of, 240
Polymyositis, 517
Positional isotope exchange, 107–108
Potassium cholate, $^{31}$P-NMR chemical shifts of phospholipids in, 426
Principal-axis system
  chemical-shielding tensor and, 404–405
  orientation of, 406
Prochiral phosphorus centers, 185–190
Pro-prochiral phosphorus centers, 187–190
Pro-pro-prochiral phosphorus centers, 190–195
Prosthetic groups, phosphoryl moieties of, 131–137
Proton–phosphorus spin–spin coupling, 116
Purine–pyrimidine deoxynucleotide duplexes, $^{31}$P-NMR spectra of, 243
Purine–pyrimidine polydeoxynucleotides, $^{31}$P-NMR signals in, 238
Purple membrane, of Halobacterium cutirubrum, 466–467
Pyridoxal phosphate, 110
  as electrophile, 133–135
  resonances of, 132
Pyridoxal phosphate–containing enzymes, 135–136
Pyrophosphatase complexes, metal–metal distances for, 171–172
Pyruvate kinase, 73, 93–97

## R

Rabbit cornea, $^{31}$P-NMR data for, 539
Regulatory phosphorylates, 111–112
Regulatory phosphorylation sites, characterization of, 138–140
Relaxation studies
  of DNA, 377–386
  in nucleic acid–drug complexes, 396–397
  of RNA, 386–388
  of tRNA, 389
  in viruses, 395–396
Relaxation times, *see also* Spin–lattice relaxation times
  for enzyme complexes, 64–67
  for phospholipids in membranes, 470–472
  for phospholipids in micelles, 431–433
Ribonuclease T$_1$, 84–86
Ribonucleic acid, *see* RNA
Ribosomes, RNA relaxation in, 391–395
Rigid phosphodiester, $^{31}$P chemical shift and, 449
RNA
  A form of, 371
  dynamic behavior of in solution, 255–256
  helix–coil transitions in, 235–237
  relaxation studies in, 386–387
  in ribosomes and messenger ribonucleoprotein, 391–395
RNA conformations, 233–260
  biological significance of, 259–260
RNase, enzyme-inhibitor complexes of, 85

## S

$S_p$ diastereomer of methyl phosphate, synthesis of, 204
Saliva
  $^{31}$P-NMR analysis of, 540–542
  proteins, 142
Salt linkages, downfield shifts and, 113
Serine
  dehydratase, 136
  proteases, diisopropylphosphoryl derivatives and, 130
  residues, 111
Serine-14 residue, phosphorylation of, 139
Single-stranded polynucleotides, relaxation studies of, 387–388
Skeletal muscle
  phosphates in, 517–519
  phosphodiesters in, 513–515
  phosphorylated sugars in, 515–517
  qualititative analysis of, 513–517
  quantitative analysis of, 517–524
Sodium DNA, A form of, 410–414, *see also* DNA
Solid-state $^{31}$P NMR, 401–421, *see also* Phosphorus-31 NMR
  basic concepts of, 402–408
  chemical-shielding tensor in, 403–408
  cross-polarization–dipole decoupling in, 402–403
Sonicated poly(dAdT)·poly(dAdT), ethanol dependence of $^{31}$P-NMR spectra of, 250
Sonicated polydeoxynucleotides, $^{31}$P-NMR parameters of, 244
*Specialist Periodical Reports,* 3
Spectral densities, of nucleic acids, 353–368
Spectral parameter changes, in enzyme complexes, 59–61
Spermine effects, in tRNA$^{Phe}$ spectrum, 291–292
Sphingomyelin, phospholipases and, 444
Spin–lattice relaxation, mechanism in, 64–65
Spin–lattice relaxation times
  for casein, 342–343
  for phospholipids in micelles, 431–433
Spin–lattice relaxation-time simulations, 338
Spin–spin coupling
  chemical-exchange effects and, 62
  proton–phosphorus, 116
Spin–spin coupling constants, 37–51, *see also* Phosphorus-31 spin–spin coupling constants
Spin–spin relaxation, mechanism in, 64–65
Spin–spin relaxation times, versus temperature, for DNA restriction fragments, 336
Spin–spin relaxation-time simulations, 339–341
Staphylococcal nuclease, oxygen chiral study of, 227–230
Stereochemical problems, with chiral thiophosphates, 175–177
Stereochemical studies, oxygen chiral approach to, 199–230
Stereoelectronic effects, in $^{31}$P chemical shifts, 19–23
Stereospecific phosphorylation, 190
Succinyl-CoA synthase, 108, 127–129
Succinyl phosphate resonance, 129

Index 603

**T**

TATA box, 260
Tetrahydrofolate, dihydrofolate reduction to, 77–78
*Tetrahymena* phosphonolipids, 464–465
Thionicotinamide adenosine dinucleotide phosphate, 79
Thiophosphorylase, $^{31}$P-NMR spectra of, 134
Three-bond spin–spin coupling constants, 43–51
Three-state jump model, 330–334
  relative potential energy profile for, 333
Threonine
  phosphate, 116
  residues, 111
Thymidine 5′-(4-nitrophenyl[$^{17}$O,$^{18}$O]-phosphate), hydrolysis of, 228
TNADP$^+$, 79
Topical magnetic resonance method, for diseased human muscle evaluation, 524–525
Torsion-angle contour map, 19–20
Transfer RNA, *see* tRNA
Triosephosphate isomerase, interconversion activity of, 86
Triton X-100, 425, 429, 431, 442
  micelle concentration of, 438
tRNA [Transfer ribonucleic acid(s)]
  conformational transitions in, 287–291
  ethidium ion effects on, 311–315
  high-resolution $^{31}$P-NMR spectroscopy of, 265–294
  magnesium-dependent AC-loop conformational change in, 290
  peak shifts in, 288
  relaxation studies for, 389
tRNA·ethidium spectra, 311–315
tRNA peaks, temperature sensitivity of, 289
tRNA$^{Met}$, aminoacylation of, 76
tRNA$^{Phe}$ linewidths, 289
tRNA$^{Phe}$ spectra, *see also Escherichia coli* tRNA; Yeast tRNA$^{Phe}$ spectra
  assignment of $^{31}$P signals in, 281–286
  manganese dependence of, 276–280
  manganese effects in, 280
  spermine effects in, 291–292
tRNA$^{Tyr}$ chemical shifts, temperature dependence of, 271–273
Tropomyosin, 138

Tryptophanase, 136
Twist-boat conformation, coupling constants for, 49
Two-bond spin–spin coupling constants, 42–43
Two-dimensional data set
  graphic display of, 491–492
  processing of, 489–490
Two-dimensional $^{31}$P NMR, 479–510, *see also* Phosphorus-31 NMR
  application of, 492–509
  chemical-exchange correlated spectra of, 506–509
  collection of data set in, 489
  defined, 480–483
  general principles of, 484–492
  heteronuclear $^{31}$P–$^1$H chemical-shift correlated spectra and, 497–506
  historical summary of, 480
  homonuclear chemical-shift correlated spectra and, 492–497
  instrument requirement for, 483–484
  Jeener pulse sequence in, 484–488
Two-site exchange, computer-simulated spectra for, 63–65
Two-state jump model, 329–330

**V**

Varian Associates, 1
Vicinal $^3J_{HCCH}$ coupling constants, 43
Viruses, DNA and RNA relaxation in, 395–396, *see also* Filamentous virus
Vitamin B$_2$, 109–110
Vitamin B$_6$, 110
Vitreous humor, $^{31}$P-NMR profile of, 540

**W**

Wobble-in-a-cone model, nucleic acids and, 362

**X**

*Xenopus laevis*, 141–142

**Y**

Yeast phenylalanine tRNA, $^{31}$P-NMR spectra of, 269

Yeast tRNA, $^{31}$P-NMR spectra of, 266
Yeast tRNA$^{Phe}$
  cloverleaf model of, 274
  three-dimensional structure of, 275

Yeast tRNA$^{Phe}$·*E. coli*·tRNA$_2^{Glu}$ complex,
  spectral changes in, 292–294
Yeast tRNA$^{Phe}$ spectra, effect of perturbations
  and modifications on, 284–285